THE DNA MOLECULE: STRUCTURE AND PROPERTIES

A SERIES OF BOOKS IN BIOLOGY
Cedric I. Davern, EDITOR

THE DNA MOLECULE

STRUCTURE AND PROPERTIES

ORIGINAL PAPERS, ANALYSES, AND PROBLEMS

David Freifelder

BRANDEIS UNIVERSITY

W. H. FREEMAN AND COMPANY
San Francisco

Library of Congress Cataloging in Publication Data

The DNA molecule.

Includes bibliographies and index.
1. Deoxyribonucleic acid—Addresses, essays, lectures.
I. Freifelder, David Michael, 1935–
QP624.D16   547′.596   77-2768
ISBN 0-7167-0287-8
ISBN 0-7167-0286-X pbk.

Printed in the United States of America

9 8 7 6 5 4 3 2 1

CONTENTS

PREFACE

This collection of important original papers on the elucidation of the DNA structure is for undergraduate courses in molecular biology. I believe that students should be exposed to the original work at this level for two reasons. First, students get much more out of the information in the text if they understand the experimental basis for the facts they are reading about. Second, I think that a future scientist should acquire a feeling for the ways in which ideas evolve and experiments are conceived. In textbooks, everything is synthesized in a straightforward manner, without mention of the false leads, the experiments that yielded the unexpected results, the mistakes that, when recognized, led one's thinking in new directions, and the controversies in which differing viewpoints were resolved. It seems to me that the best way to appreciate this is by reading the original papers in which researchers have communicated their results and ideas to the scientific community and to read these papers in chronological order. In addition—the pedagogical reasons aside—the history of the development of our understanding of DNA structure makes a good story.

In this book I hope to convey to students not only the important facts that were uncovered but also some of the excitement of the growth of molecular biology and the role of emotions,

hunches, and controversy in scientific discovery. To this end I have assembled a number of the most important papers on DNA structure, omitting others that require detailed understanding of complicated physical techniques. The papers selected are complemented by my own narrative, which I have written with the following objectives in mind: to summarize certain papers not included in this volume; to describe some of the beliefs held before the articles that are reprinted in the following pages first appeared; to relate how the scientific community reacted to these papers; to discuss how the papers have added to our understanding of DNA structure.

I view this book as a supplement to a course in molecular biology, molecular genetics, or nucleic acids. The biochemistry student, geneticist, or student of molecular biology should be able to understand the content without difficulty. For those who are unfamiliar with molecular biology or for the beginner, I have added several appendixes that describe briefly the biological phenomena. At the end of each section are question sets for the various papers. For some of the questions the student is required to recall facts or points of view presented in the paper itself; in others, he is obliged to consider additional problems that might be raised by the information contained in the paper. The more difficult questions are indicated by a bullet (•). Also I have incorporated into my own narrative occasional questions for further thought and discussion.

I am indebted to the authors of the papers who supplied me with reprints for reproduction and original electron micrographs. Cedric Davern read the first draft of the book and encouraged me to continue in the endeavor. Several undergraduates at Brandeis University, including Mike Cohen, Jeff Danzig, Steve Golombek, Alberto Kriger, and Gregg Silverman read the third draft for clarity. Dorothy Freifelder read the entire manuscript for style, consistency, and logic.

November 1977

David Freifelder

THE DNA MOLECULE: STRUCTURE AND PROPERTIES

DNA: THE GENETIC MATERIAL

The History of the Identification of DNA as the Genetic Substance

The discovery of the structure of DNA by James Watson and Francis Crick was certainly one of the most influential events in biology in the past century. To many people it marks the birth of modern molecular biology. At the time that Watson and Crick were pursuing the structure of the DNA molecule, DNA was already known to be the material of which genes consist, and it was understood that its structure would have deep significance. It is worth reviewing the events leading up to this important discovery, since in this account one can see the insights as well as the accumulation and destruction of misconceptions that so often mark the development of our scientific understanding.

Early History

A little more than a century ago a German histologist Wilhelm His told his nephew, Friedrich Miescher, that the final solution of the problems of tissue development could be solved only by chemistry. So began the long and fruitful line of investigation that produced the discovery and identification of nucleic acids. Although Miescher had a strong interest in medicine, his uncle's influence led him to study chemistry. After obtaining his doctorate, Miescher joined Hoppe-Seyler's laboratory, which was the first laboratory in Europe to be concerned exclusively with biochemistry. Miescher was interested in the chemistry of the cell nucleus and therefore chose to study human pus cells (obtained from surgical bandages), since the lymphoid cells in pus have a large nucleus-to-cytoplasm ratio and thus provide a good source of nuclear material. In his studies of the chemical composition of nuclei, Miescher extracted purified nuclei with either concentrated salt solutions or dilute alkali and isolated a gelatinous material that had not been previously observed. Chemical analysis of this substance showed that it contained organic phosphorus, which at this time was only known to occur in the phospholipid lecithin. Miescher named this new substance *nuclein*.

Nuclein was approximately 70% protein and was probably what we now call *nucleoprotein* or *chromatin*. Four years later Miescher isolated a purer sample of nuclein from sperm heads, and also a new nuclear protein later called *protamine*.

Although credited with the discovery of nucleic acid, Miescher never actually isolated it in pure form. He did, however, recognize that either nuclein *was* a substance of very high molecular weight, or contained such a substance. This conclusion may have produced the long-lasting but incorrect view that nuclein provides the framework for the chromosome structure.

Ten years later the chemistry of nuclein was taken up by Albrecht Kossel who discovered that nuclein contained the organic bases adenine, guanine, thymine, and cytosine (see Appendix A). For this work he was awarded the Nobel Prize. In 1889 Altman deproteinized the nuclein obtained from yeast and called the deproteinized substance *nucleic acid.* In 1901 the base, uracil, which we now know to be in RNA but not in DNA* was found by A. Ascoli in yeast nucleic acid (which has a very high ratio of RNA to DNA). In 1891 Kossel found a sugar in nucleic acid that was identified as D-ribose in 1908 by P. A. Levene and W. A. Jacobs. Then in 1929 Levene found and identified 2'-deoxy-D-ribose, the sugar in DNA.

Not until about 1910 was it realized that there were two classes of nucleic acids: DNA, the thymus type, which contains deoxyribose; and RNA, the yeast type, which contains ribose.† It was not recognized, however, that both types could be found in the same cells; rather, it was thought that DNA was the animal nucleic acid, and RNA the plant nucleic acid.

In 1924, R. Feulgen developed a chemical indicator for DNA within cells, i.e., a reaction with deoxyribose that yields a bright red compound. With it he demonstrated by light microscopy that DNA is in the nuclei and, in fact,

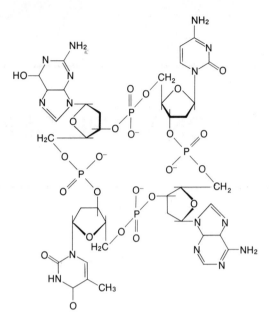

Figure I-1

The hypothetical tetranucleotide structure of DNA. [From *Molecular Genetics* by G. S. Stent and R. Calendar, 2nd ed. W. H. Freeman and Company. Copyright © 1978.]

in the chromosomes of all species. However, except for a few laboratories in Germany, nucleic acids still attracted little attention. The possibility that nucleic acid might carry genetic information seems not to have entered anyone's mind. In fact it was later proposed that DNA is a simple repeating tetranucleotide (Levine and Bass, 1931) a structure quite incapable of containing genetic information (Fig. I-1). Therefore, the idea evolved that DNA was either responsible for maintaining chromosome structure or somehow involved in chromosome physiology.

The Transforming Principle Is DNA

The development of the current idea that DNA is the genetic material began with an observation in 1928 by Fred Griffith, who was studying the bacterium responsible for human pneumonia—i.e., *Streptococcus pneumoniae* (or *Pneumococcus*). The virulence of this bacterium was known to be dependent on a surrounding polysaccharide capsule that protects it from the defense systems of the body. This

---

*RNA is an abbreviation for ribonucleic acid; DNA is deoxyribonucleic acid.

†The thymus of mammals is very rich in DNA but contains very little RNA, whereas yeast has a great excess of RNA. Since the early isolation procedures did not distinguish DNA from RNA and since purification was difficult, it was thought that thymus contained exclusively DNA, and yeast exclusively RNA.

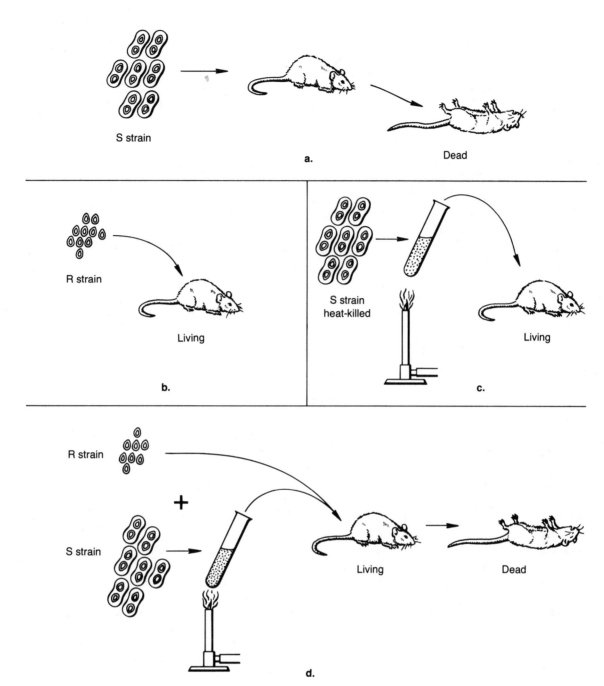

Figure I-2
The Griffith experiment. [From *Molecular Genetics* by G. S. Stent and R. Calendar, 2nd ed. W. H. Freeman and Company. Copyright © 1978.]

capsule also causes the bacterium to produce smooth-edged (S) colonies on an agar surface. It was known that mice were normally killed by the S-type (Fig. I-2a). Griffith isolated a rough-edged (R) colony mutant, which proved to be both nonencapsulated and nonlethal (Fig. I-2b). Heat-killed S was nonlethal, but a mixture of live R and heat-killed S was lethal (Fig. I-2c, d). Furthermore, the bacteria isolated from a mouse that had died from such a mixed infection were only S-type, i.e., the live R had somehow been replaced by or *transformed* to the S-type. Several years later it was shown that the mouse itself was not needed to mediate this transformation because if a mixture of R and heat-killed S were grown in a

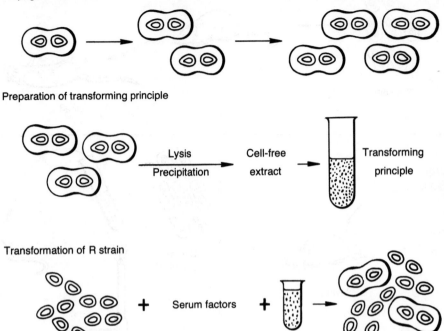

Figure I-3
The transformation experiment. [From *Molecular Genetics* by G. S. Stent and R. Calendar, 2nd ed. W. H. Freeman and Company. Copyright © 1978.]

culture fluid, living S cells were produced. A possible explanation for this surprising phenomenon was that the R cells restored the viability of the dead S cells; but this idea was eliminated by the observation that the heat-killed S culture could be replaced by an *extract* that was prepared from broken S cells and was free both of intact cells and of the capsular polysaccharide (Fig. I-3). Hence it was concluded that the cell extract contained a *transforming principle,* the nature of which was unknown.

The next development occurred 15 years later when Oswald Avery, Colin MacLeod, and Maclyn McCarty (1944) partially purified the transforming principle from the cell extract and demonstrated that it was DNA. These workers modified known schemes for isolating DNA and prepared samples of DNA from S bacteria. They added this DNA to a live R bacterial culture, and after a period of time placed a sample of the R bacterial culture on an agar surface and allowed it to grow to form

colonies. Some of the colonies (about 1 in 10⁴) that grew were S type (Fig. I-3). To show that this was a permanent genetic change, they dispersed many S colonies and placed them on a second agar surface. The resulting colonies were again S type. If an R colony was dispersed, only R-type were found. Hence the R colonies had retained the R character, whereas the transformed S colonies bred true as S-type. Because S and R colonies differed by a polysaccharide coat on the S bacteria, the ability of purified polysaccharide to transform was also tested, but no transformation was observed. Since the procedures for isolating DNA that were in use at the time produced DNA containing many impurities, it was necessary to provide evidence that the transformation was actually caused by the DNA.

This evidence was provided by the following four procedures. (1) Chemical analysis showed that the major component was a deoxyribose-containing nucleic acid. (2) Physical measurements showed that the sample

TABLE I-1
The inactivation of transforming principle by crude enzyme preparations

Crude enzyme preparations	Enzymatic activity			
	Phosphatase	Tributyrin esterase	Depolymerase for deoxyribonucleate	Inactivation of transforming principle
Dog intestinal mucosa	+	+	+	+
Rabbit bone phosphatase	+	+	−	−
Swine kidney phosphatase	+	−	−	−
Pneumococcus autolysates	−	+	+	+
Normal dog and rabbit serum	+	+	+	+

contained a highly viscous substance having the properties of DNA. (3) Experiments demonstrated that transforming activity is not lost by reaction with either (a) purified proteolytic (protein-hydrolyzing) enzymes—trypsin, chymotrypsin, or a mixture of both—or (b) with ribonuclease (an enzyme that depolymerizes RNA). (4) It was demonstrated that treatment with materials known to contain DNA depolymerizing activity (DNase) inactivated the transforming principle. The fourth point, which is the most critical, is worth examining further since Avery, MacLeod, and McCarty, unable to purchase purified DNase as we now can, had to prove it was the DNase activity in the various fluids tested that destroyed the transforming activity.

Some of the data they used to make this argument are shown in Table I-1 and Fig. I-4. In Table I-1 the results of incubating transforming principle with five crude enzyme preparations are shown. For each preparation the activity of three enzymes—phosphatase, tributyrin esterase, and DNA depolymerase—was measured and correlated with the ability to inactivate transforming principle. Since rabbit bone phosphatase contained phosphatase and esterase activity but failed to prevent transformation, these enzymes could not be responsible for the inactivation. Instead, loss of transformation correlated with the presence of DNA depolymerase, but since the preparations were impure and probably contained many other enzyme activities, these results could not be taken as proof. Figure I-4 shows the loss of viscosity of the transforming principle produced by dog and rabbit sera that were

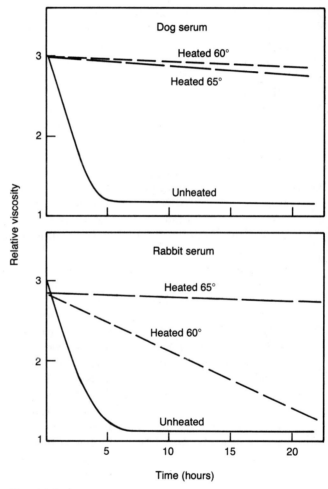

Figure I-4
Relative viscosity of transforming principle as a function of time of incubation in three samples of dog and rat sera: (1) unheated, (2) heated for 30 minutes at 60°C, (3) heated for 30 minutes at 65°C. The decrease in viscosity is due to depolymerization of DNA. When the viscosity is constant, the DNase in the particular serum must have been inactivated. [From O. Avery, C. MacLeod, and M. McCarty, *J. Expt. Med.* **79** (1944), 137–158.]

pretreated by heating to different temperatures. Since the viscosity was presumably due to DNA, this viscosity decrease was taken as a measure of the DNase activity and the loss (upon heating) of the ability of the sera to reduce viscosity was presumably a result of loss of the DNase activity by heat inactivation of the enzyme. These heated sera were also tested for the ability to inactivate transforming principle, and it was found that the temperature dependence of depolymerizing activity followed that of inactivation of transformation. The final point was that the addition of sodium fluoride to any preparation that had the ability both to inactivate transformation and to depolymerize DNA always simultaneously removed both activities. Thus, Avery, MacLeod, and McCarty stated, "the fact that transforming activity is destroyed only by preparations containing depolymerase for desoxyribonucleic acid and the further fact that in both instances the enzymes concerned are inactivated at the same temperature and inhibited by fluoride provide additional evidence for the belief that the active principle is a nucleic acid of the desoxyribose type." The caution with which they stated their inference should be noted.

Avery, MacLeod and McCarty avoided stating explicitly that DNA was *the* genetic material and concluded at the end of their work only that nucleic acids have "biological specificity, the chemical basis of which is as yet undetermined." The problem they faced in persuading the scientific community to accept their conclusion was that, whatever the genetic material was, it had to be a substance capable of enormous variation in order to contain the information carried by the huge number of genes, but DNA was still considered to be a tetranucleotide. This classification certainly seemed incompatible with the strong contention that DNA was the sole component of the genetic material. Furthermore, the general consensus was that genes were made of chromosomal protein—an idea that, in the course of a 40-year period, had logically evolved from the recognition that protein composition and structure varied greatly between organisms. For this reason the transformation work had little impact and those people who supported the protein theory posed the following two alternative explanations for the results: (1) the transforming prin-

ciple might not be DNA but rather one of the proteins invariably contaminating DNA samples; (2) DNA somehow affected capsule formation directly by acting in the metabolic pathway for biosynthesis of the polysaccharide and permanently altering this pathway.

Point (1) had already been discounted in the original work because the experiments showed insensitivity to proteolytic enzymes and sensitivity to DNase. However, since the DNase was not a pure enzyme and its activity against the transforming principle was only inferred by correlating it to other activities against DNA in the same preparation, the possibility was not eliminated conclusively. One year later Rollin Hotchkiss repeated the transformation experiment, using a DNA sample whose protein content was only 0.02% and found that this extensive purification did not reduce the transforming activity. This result supported the view of Avery, MacLeod, and McCarty but of course still did not prove it. The second alternative, however, was cleanly eliminated, also by Hotchkiss, by an experiment in which he transformed a penicillin-sensitive strain of *Pneumococcus* to penicillin-resistance, using the DNA from a penicillin-resistant strain. Since penicillin-resistance is totally distinct from the rough-smooth character, this showed that the transforming ability of DNA was not limited to capsule synthesis. Interestingly enough, most biologists still remained unconvinced that DNA is the genetic material and it was not until five years later—when Erwin Chargaff (1950) showed that a wide variety of chemical structures in DNA were possible, thus allowing biological specificity—that this idea became accepted (see below).

The tetranucleotide hypothesis had arisen from the belief that DNA contained equimolar quantities of adenine, thymine, guanine, and cytosine. This incorrect conclusion was formulated for two reasons. First, in the chemical analysis of DNA, the technique used to separate the bases before identification did not resolve them very well, so that the quantitative analysis was poor. Second, the DNA analyzed was usually isolated from eukaryotes (and in eukaryotes the four bases are nearly equimolar) or from bacteria whose DNA happened to have nearly equal amounts. Using the DNA of a wide variety of organisms, Chargaff applied the newly developed technique of paper chromatography to separate the bases and ultraviolet absorbance measurements of the separated base to determine the concentration of each. He thereby showed that the molar concentrations of the bases could vary widely.* Thus it was demonstrated that DNA could have variable composition, a primary requirement for a genetic substance. Upon publication of Chargaff's results, the tetranucleotide hypothesis quietly died and the DNA-gene idea began to catch on. Shortly afterward, Alfred Mirsky and Hans Ris in one laboratory, and R. Vendrely (1955) and A. Boivin in another found that, for a wide variety of organisms, the somatic cells have twice the DNA content of the germ cells, a characteristic to be expected of the genetic material, given the tenets of classical chromosome genetics. Although it could apply just as well to any component of chromosomes, once this result was revealed, the hereditary nature of DNA rapidly became the fashionable idea. Although no additional data had been released by Avery, MacLeod, and McCarty, objections to their work were no longer heard.

Bacteriophage Genes Are Carried on DNA

While the transformation experiments were being done, a group of physicists and biologists who were strongly influenced by Max Delbrück (a former nuclear physicist who had turned to biology) were actively studying

---

*It is interesting to note that many conceptual changes in science follow technological advances. In this series of experiments Chargaff was able to obtain high precision in base analysis because the new technique of paper chromatography allowed almost perfect separation of the four bases.

bacteriophages (*phages* for short), viruses that grow exclusively in bacteria (see Appendix B). Delbrück had not only organized the Phage Group, a small collection of phage workers who frequently met at the Cold Spring Harbor Laboratory on Long Island, New York, but also had convinced these people to join forces and work on a single phage type so that experimental results from different laboratories could be compared. The bacterium, *Escherichia coli*, was chosen because it was nonpathogenic and easy to grow. The phages were taken from Milislav Demerec's collection of *E. coli* phages, called T-phages, numbered T1, T2 ··· T7; of these, the T-even phages and especially T2 and T4 became the preferred objects of study.

An early stage in the infection of *E. coli* by a phage—namely, adsorption of the phage to the bacterial cell wall—was being studied by Thomas Anderson with the use of the new technique of electron microscopy. He observed that phage T2 consisted of a head and a long tail and that the tip of the tail was the point of attachment to the bacterial cell wall. This fact led Alfred Hershey and Martha Chase in 1952 to perform an important experiment called the *blendor experiment* in which phage T2 was first allowed to adsorb to *E. coli* and was then stripped from the bacteria wall by violent agitation in a Waring blendor. With this experiment Hershey and Chase showed that when the bacterium *Escherichia coli* is infected with phage T2, the following occurs: (1) the major fraction of the phage DNA enters the cell; (2) a significant fraction of the injected DNA appears in progeny phage; (3) most of the phage protein does not enter the cell; (4) little or none of the original phage protein appears in progeny phage. Thus this experiment provided final proof for the genetic nature of DNA. These findings are presented by Hershey and Chase in Article 1.

From *The Journal of General Physiology*, Vol. 36, No. 1, 39–56, 1952. Reproduced with permission of the authors and publisher.

INDEPENDENT FUNCTIONS OF VIRAL PROTEIN AND NUCLEIC ACID IN GROWTH OF BACTERIOPHAGE*

By A. D. HERSHEY AND MARTHA CHASE

(*From the Department of Genetics, Carnegie Institution of Washington, Cold Spring Harbor, Long Island*)

(Received for publication, April 9, 1952)

The work of Doermann (1948), Doermann and Dissosway (1949), and Anderson and Doermann (1952) has shown that bacteriophages T2, T3, and T4 multiply in the bacterial cell in a non-infective form. The same is true of the phage carried by certain lysogenic bacteria (Lwoff and Gutmann, 1950). Little else is known about the vegetative phase of these viruses. The experiments reported in this paper show that one of the first steps in the growth of T2 is the release from its protein coat of the nucleic acid of the virus particle, after which the bulk of the sulfur-containing protein has no further function.

Materials and Methods.—Phage T2 means in this paper the variety called T2H (Hershey, 1946); T2*h* means one of the host range mutants of T2; UV-phage means phage irradiated with ultraviolet light from a germicidal lamp (General Electric Co.) to a fractional survival of 10^{-5}.

Sensitive bacteria means a strain (H) of *Escherichia coli* sensitive to T2 and its *h* mutant; resistant bacteria B/2 means a strain resistant to T2 but sensitive to its *h* mutant; resistant bacteria B/2*h* means a strain resistant to both. These bacteria do not adsorb the phages to which they are resistant.

"Salt-poor" broth contains per liter 10 gm. bacto-peptone, 1 gm. glucose, and 1 gm. NaCl. "Broth" contains, in addition, 3 gm. bacto-beef extract and 4 gm. NaCl.

Glycerol-lactate medium contains per liter 70 mM sodium lactate, 4 gm. glycerol, 5 gm. NaCl, 2 gm. KCl, 1 gm. NH_4Cl, 1 mM $MgCl_2$, 0.1 mM $CaCl_2$, 0.01 gm. gelatin, 10 mg. P (as orthophosphate), and 10 mg. S (as $MgSO_4$), at pH 7.0.

Adsorption medium contains per liter 4 gm. NaCl, 5 gm. K_2SO_4, 1.5 gm. KH_2PO_4, 3.0 gm. Na_2HPO_4, 1 mM $MgSO_4$, 0.1 mM $CaCl_2$, and 0.01 gm. gelatin, at pH 7.0.

Veronal buffer contains per liter 1 gm. sodium diethylbarbiturate, 3 mM $MgSO_4$, and 1 gm. gelatin, at pH 8.0.

The HCN referred to in this paper consists of molar sodium cyanide solution neutralized when needed with phosphoric acid.

---

* This investigation was supported in part by a research grant from the National Microbiological Institute of the National Institutes of Health, Public Health Service. Radioactive isotopes were supplied by the Oak Ridge National Laboratory on allocation from the Isotopes Division, United States Atomic Energy Commission.

39

Adsorption of isotope to bacteria was usually measured by mixing the sample in adsorption medium with bacteria from 18 hour broth cultures previously heated to 70°C. for 10 minutes and washed with adsorption medium. The mixtures were warmed for 5 minutes at 37°C., diluted with water, and centrifuged. Assays were made of both sediment and supernatant fractions.

Precipitation of isotope with antiserum was measured by mixing the sample in 0.5 per cent saline with about 10^{11} per ml. of non-radioactive phage and slightly more than the least quantity of antiphage serum (final dilution 1:160) that would cause visible precipitation. The mixture was centrifuged after 2 hours at 37°C.

Tests with DNase (desoxyribonuclease) were performed by warming samples diluted in veronal buffer for 15 minutes at 37°C. with 0.1 mg. per ml. of crystalline enzyme (Worthington Biochemical Laboratory).

Acid-soluble isotope was measured after the chilled sample had been precipitated with 5 per cent trichloroacetic acid in the presence of 1 mg./ml. of serum albumin, and centrifuged.

In all fractionations involving centrifugation, the sediments were not washed, and contained about 5 per cent of the supernatant. Both fractions were assayed.

Radioactivity was measured by means of an end-window Geiger counter, using dried samples sufficiently small to avoid losses by self-absorption. For absolute measurements, reference solutions of P^{32} obtained from the National Bureau of Standards, as well as a permanent simulated standard, were used. For absolute measurements of S^{35} we relied on the assays (±20 per cent) furnished by the supplier of the isotope (Oak Ridge National Laboratory).

Glycerol-lactate medium was chosen to permit growth of bacteria without undesirable pH changes at low concentrations of phosphorus and sulfur, and proved useful also for certain experiments described in this paper. 18-hour cultures of sensitive bacteria grown in this medium contain about 2×10^9 cells per ml., which grow exponentially without lag or change in light-scattering per cell when subcultured in the same medium from either large or small seedings. The generation time is 1.5 hours at 37°C. The cells are smaller than those grown in broth. T2 shows a latent period of 22 to 25 minutes in this medium. The phage yield obtained by lysis with cyanide and UV-phage (described in context) is one per bacterium at 15 minutes and 16 per bacterium at 25 minutes. The final burst size in diluted cultures is 30 to 40 per bacterium, reached at 50 minutes. At 2×10^8 cells per ml., the culture lyses slowly, and yields 140 phage per bacterium. The growth of both bacteria and phage in this medium is as reproducible as that in broth.

For the preparation of radioactive phage, P^{32} of specific activity 0.5 mc./mg. or S^{35} of specific activity 8.0 mc./mg. was incorporated into glycerol-lactate medium, in which bacteria were allowed to grow at least 4 hours before seeding with phage. After infection with phage, the culture was aerated overnight, and the radioactive phage was isolated by three cycles of alternate slow (2000 G) and fast (12,000 G) centrifugation in adsorption medium. The suspensions were stored at a concentration not exceeding 4 μc./ml.

Preparations of this kind contain 1.0 to 3.0×10^{-12} μg. S and 2.5 to 3.5×10^{-11} μg. P per viable phage particle. Occasional preparations containing excessive amounts of sulfur can be improved by absorption with heat-killed bacteria that do not adsorb the phage. The radiochemical purity of the preparations is somewhat uncertain, owing to the possible presence of inactive phage particles and empty phage membranes. The presence in our preparations of sulfur (about 20 per cent) that is precipitated by antiphage serum (Table I) and either adsorbed by bacteria resistant to phage, or not adsorbed by bacteria sensitive to phage (Table VII), indicates contamination by membrane material. Contaminants of bacterial origin are probably negligible for present purposes as indicated by the data given in Table I. For proof that our principal findings reflect genuine properties of viable phage particles, we rely on some experiments with inactivated phage cited at the conclusion of this paper.

The Chemical Morphology of Resting Phage Particles.—Anderson (1949) found that bacteriophage T2 could be inactivated by suspending the particles in high concentrations of sodium chloride, and rapidly diluting the suspension with water. The inactivated phage was visible in electron micrographs as tadpole-shaped "ghosts." Since no inactivation occurred if the dilution was slow

TABLE I

Composition of Ghosts and Solution of Plasmolyzed Phage

Per cent of isotope]	Whole phage labeled with		Plasmolyzed phage labeled with	
	P^{32}	S^{35}	P^{32}	S^{35}
Acid-soluble	—	—	1	—
Acid-soluble after treatment with DNase	1	1	80	1
Adsorbed to sensitive bacteria	85	90	2	90
Precipitated by antiphage	90	99	5	97

he attributed the inactivation to osmotic shock, and inferred that the particles possessed an osmotic membrane. Herriott (1951) found that osmotic shock released into solution the DNA (desoxypentose nucleic acid) of the phage particle, and that the ghosts could adsorb to bacteria and lyse them. He pointed out that this was a beginning toward the identification of viral functions with viral substances.

We have plasmolyzed isotopically labeled T2 by suspending the phage (10^{11} per ml.) in 3 M sodium chloride for 5 minutes at room temperature, and rapidly pouring into the suspension 40 volumes of distilled water. The plasmolyzed phage, containing not more than 2 per cent survivors, was then analyzed for phosphorus and sulfur in the several ways shown in Table I. The results confirm and extend previous findings as follows:—

1. Plasmolysis separates phage T2 into ghosts containing nearly all the sulfur and a solution containing nearly all the DNA of the intact particles.

2. The ghosts contain the principal antigens of the phage particle detectable by our antiserum. The DNA is released as the free acid, or possibly linked to sulfur-free, apparently non-antigenic substances.

3. The ghosts are specifically adsorbed to phage-susceptible bacteria; the DNA is not.

4. The ghosts represent protein coats that surround the DNA of the intact particles, react with antiserum, protect the DNA from DNase (desoxyribonuclease), and carry the organ of attachment to bacteria.

5. The effects noted are due to osmotic shock, because phage suspended in salt and diluted slowly is not inactivated, and its DNA is not exposed to DNase.

TABLE II

Sensitization of Phage DNA to DNase by Adsorption to Bacteria

Phage adsorbed to		Phage labeled with	Non-sedimentable isotope, *per cent*	
			After DNase	No DNase
Live bacteria.............................		S^{35}	2	1
" "		P^{32}	8	7
Bacteria heated before infection..............		S^{35}	15	11
" " " "		P^{32}	76	13
Bacteria heated after infection................		S^{35}	12	14
" " " "		P^{32}	66	23
Heated unadsorbed phage: acid-soluble P^{32}	70°.......	P^{32}	5	
	80°.......	P^{32}	13	
	90°.......	P^{32}	81	
	100°.......	P^{32}	88	

Phage adsorbed to bacteria for 5 minutes at 37°C. in adsorption medium, followed by washing.

Bacteria heated for 10 minutes at 80°C. in adsorption medium (before infection) or in veronal buffer (after infection).

Unadsorbed phage heated in veronal buffer, treated with DNase, and precipitated with trichloroacetic acid.

All samples fractionated by centrifuging 10 minutes at 1300 *G*.

Sensitization of Phage DNA to DNase by Adsorption to Bacteria.—The structure of the resting phage particle described above suggests at once the possibility that multiplication of virus is preceded by the alteration or removal of the protective coats of the particles. This change might be expected to show itself as a sensitization of the phage DNA to DNase. The experiments described in Table II show that this happens. The results may be summarized as follows:—

1. Phage DNA becomes largely sensitive to DNase after adsorption to heat-killed bacteria.

2. The same is true of the DNA of phage adsorbed to live bacteria, and then

heated to 80°C. for 10 minutes, at which temperature unadsorbed phage is not sensitized to DNase.

3. The DNA of phage adsorbed to unheated bacteria is resistant to DNase, presumably because it is protected by cell structures impervious to the enzyme.

Graham and collaborators (personal communication) were the first to discover the sensitization of phage DNA to DNase by adsorption to heat-killed bacteria.

The DNA in infected cells is also made accessible to DNase by alternate freezing and thawing (followed by formaldehyde fixation to inactivate cellular enzymes), and to some extent by formaldehyde fixation alone, as illustrated by the following experiment.

Bacteria were grown in broth to 5×10^7 cells per ml., centrifuged, resuspended in adsorption medium, and infected with about two P^{32}-labeled phage per bacterium. After 5 minutes for adsorption, the suspension was diluted with water containing per liter 1.0 mM $MgSO_4$, 0.1 mM $CaCl_2$, and 10 mg. gelatin, and recentrifuged. The cells were resuspended in the fluid last mentioned at a concentration of 5×10^8 per ml. This suspension was frozen at -15°C. and thawed with a minimum of warming, three times in succession. Immediately after the third thawing, the cells were fixed by the addition of 0.5 per cent (v/v) of formalin (35 per cent HCHO). After 30 minutes at room temperature, the suspension was dialyzed free from formaldehyde and centrifuged at 2200 G for 15 minutes. Samples of P^{32}-labeled phage, frozen-thawed, fixed, and dialyzed, and of infected cells fixed only and dialyzed, were carried along as controls.

The analysis of these materials, given in Table III, shows that the effect of freezing and thawing is to make the intracellular DNA labile to DNase, without, however, causing much of it to leach out of the cells. Freezing and thawing and formaldehyde fixation have a negligible effect on unadsorbed phage, and formaldehyde fixation alone has only a mild effect on infected cells.

Both sensitization of the intracellular P^{32} to DNase, and its failure to leach out of the cells, are constant features of experiments of this type, independently of visible lysis. In the experiment just described, the frozen suspension cleared during the period of dialysis. Phase-contrast microscopy showed that the cells consisted largely of empty membranes, many apparently broken. In another experiment, samples of infected bacteria from a culture in salt-poor broth were repeatedly frozen and thawed at various times during the latent period of phage growth, fixed with formaldehyde, and then washed in the centrifuge. Clearing and microscopic lysis occurred only in suspensions frozen during the second half of the latent period, and occurred during the first or second thawing. In this case the lysed cells consisted wholly of intact cell membranes, appearing empty except for a few small, rather characteristic refractile bodies apparently attached to the cell walls. The behavior of intracellular P^{32} toward DNase, in either the lysed or unlysed cells, was not significantly different from that shown in Table III, and the content of P^{32} was only slightly less after lysis. The phage liberated during freezing and thawing was also titrated in this experiment. The lysis occurred without appreciable liberation of phage in suspensions frozen up to and including the 16th minute, and the 20 minute sample yielded only five per bacterium. Another sample of the culture formalinized at 30 minutes, and centrifuged without freezing, contained 66 per cent of the P^{32} in non-sedimentable form. The yield of extracellular phage at 30 minutes was 108 per bacterium, and the sedimented material consisted largely of formless debris but contained also many apparently intact cell membranes.

TABLE III
Sensitization of Intracellular Phage to DNase by Freezing, Thawing, and Fixation with Formaldehyde

	Unadsorbed phage frozen, thawed, fixed	Infected cells frozen, thawed, fixed	Infected cells fixed only
Low speed sediment fraction			
Total P^{32}............................	—	71	86
Acid-soluble..............................	—	0	0.5
Acid-soluble after DNase....................	—	59	28
Low speed supernatant fraction			
Total P^{32}............................	—	29	14
Acid-soluble..............................	1	0.8	0.4
Acid-soluble after DNase....................	11	21	5.5

The figures express per cent of total P^{32} in the original phage, or its adsorbed fraction.

We draw the following conclusions from the experiments in which cells infected with P^{32}-labeled phage are subjected to freezing and thawing.

1. Phage DNA becomes sensitive to DNase after adsorption to bacteria in buffer under conditions in which no known growth process occurs (Benzer, 1952; Dulbecco, 1952).

2. The cell membrane can be made permeable to DNase under conditions that do not permit the escape of either the intracellular P^{32} or the bulk of the cell contents.

3. Even if the cells lyse as a result of freezing and thawing, permitting escape of other cell constituents, most of the P^{32} derived from phage remains inside the cell membranes, as do the mature phage progeny.

4. The intracellular P^{32} derived from phage is largely freed during spontaneous lysis accompanied by phage liberation.

We interpret these facts to mean that intracellular DNA derived from phage is not merely DNA in solution, but is part of an organized structure at all times during the latent period.

Liberation of DNA from Phage Particles by Adsorption to Bacterial Fragments.—The sensitization of phage DNA to specific depolymerase by adsorption to bacteria might mean that adsorption is followed by the ejection of the phage DNA from its protective coat. The following experiment shows that this is in fact what happens when phage attaches to fragmented bacterial cells.

TABLE IV

Release of DNA from Phage Adsorbed to Bacterial Debris

	Phage labeled with	
	S³⁵	P³²
Sediment fraction		
Surviving phage	16	22
Total isotope	87	55
Acid-soluble isotope	0	2
Acid-soluble after DNase	2	29
Supernatant fraction		
Surviving phage	5	5
Total isotope	13	45
Acid-soluble isotope	0.8	0.5
Acid-soluble after DNase	0.8	39

S^{35}- and P^{32}-labeled T2 were mixed with identical samples of bacterial debris in adsorption medium and warmed for 30 minutes at 37°C. The mixtures were then centrifuged for 15 minutes at 2200 G, and the sediment and supernatant fractions were analyzed separately. The results are expressed as per cent of input phage or isotope.

Bacterial debris was prepared by infecting cells in adsorption medium with four particles of T2 per bacterium, and transferring the cells to salt-poor broth at 37°C. The culture was aerated for 60 minutes, M/50 HCN was added, and incubation continued for 30 minutes longer. At this time the yield of extracellular phage was 400 particles per bacterium, which remained unadsorbed because of the low concentration of electrolytes. The debris from the lysed cells was washed by centrifugation at 1700 G, and resuspended in adsorption medium at a concentration equivalent to 3×10^9 lysed cells per ml. It consisted largely of collapsed and fragmented cell membranes. The adsorption of radioactive phage to this material is described in Table IV. The following facts should be noted.

1. The unadsorbed fraction contained only 5 per cent of the original phage particles in infective form, and only 13 per cent of the total sulfur. (Much of this sulfur must be the material that is not adsorbable to whole bacteria.)

2. About 80 per cent of the phage was inactivated. Most of the sulfur of this phage, as well as most of the surviving phage, was found in the sediment fraction.

3. The supernatant fraction contained 40 per cent of the total phage DNA (in a form labile to DNase) in addition to the DNA of the unadsorbed surviving phage. The labile DNA amounted to about half of the DNA of the inactivated phage particles, whose sulfur sedimented with the bacterial debris.

4. Most of the sedimentable DNA could be accounted for either as surviving phage, or as DNA labile to DNase, the latter amounting to about half the DNA of the inactivated particles.

Experiments of this kind are unsatisfactory in one respect: one cannot tell whether the liberated DNA represents all the DNA of some of the inactivated particles, or only part of it.

Similar results were obtained when bacteria (strain B) were lysed by large amounts of UV-killed phage T2 or T4 and then tested with P^{32}-labeled T2 and T4. The chief point of interest in this experiment is that bacterial debris saturated with UV-killed T2 adsorbs T4 better than T2, and debris saturated with T4 adsorbs T2 better than T4. As in the preceding experiment, some of the adsorbed phage was not inactivated and some of the DNA of the inactivated phage was not released from the debris.

These experiments show that some of the cell receptors for T2 are different from some of the cell receptors for T4, and that phage attaching to these specific receptors is inactivated by the same mechanism as phage attaching to unselected receptors. This mechanism is evidently an active one, and not merely the blocking of sites of attachment to bacteria.

Removal of Phage Coats from Infected Bacteria.—Anderson (1951) has obtained electron micrographs indicating that phage T2 attaches to bacteria by its tail. If this precarious attachment is preserved during the progress of the infection, and if the conclusions reached above are correct, it ought to be a simple matter to break the empty phage membranes off the infected bacteria, leaving the phage DNA inside the cells.

The following experiments show that this is readily accomplished by strong shearing forces applied to suspensions of infected cells, and further that infected cells from which 80 per cent of the sulfur of the parent virus has been removed remain capable of yielding phage progeny.

Broth-grown bacteria were infected with S^{35}- or P^{32}-labeled phage in adsorption medium, the unadsorbed material was removed by centrifugation, and the cells were resuspended in water containing per liter 1 mM $MgSO_4$, 0.1 mM $CaCl_2$, and 0.1 gm. gelatin. This suspension was spun in a Waring blendor (semimicro size) at 10,000 R.P.M. The suspension was cooled briefly in ice water at the end of each 60 second running period. Samples were removed at intervals, titrated (through antiphage serum) to measure the number of bacteria capable of yielding phage, and centrifuged to measure the proportion of isotope released from the cells.

The results of one experiment with each isotope are shown in Fig. 1. The data for S^{35} and survival of infected bacteria come from the same experiment, in which the ratio of added phage to bacteria was 0.28, and the concentrations

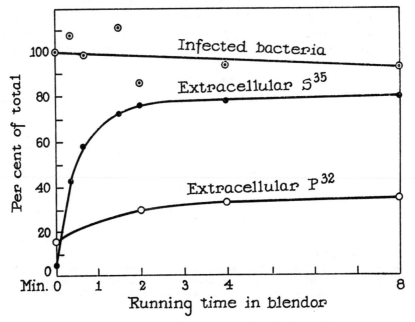

FIG. 1. Removal of S^{35} and P^{32} from bacteria infected with radioactive phage, and survival of the infected bacteria, during agitation in a Waring blendor.

of bacteria were 2.5×10^8 per ml. infected, and 9.7×10^8 per ml. total, by direct titration. The experiment with P^{32}-labeled phage was very similar. In connection with these results, it should be recalled that Anderson (1949) found that adsorption of phage to bacteria could be prevented by rapid stirring of the suspension.

At higher ratios of infection, considerable amounts of phage sulfur elute from the cells spontaneously under the conditions of these experiments, though the elution of P^{32} and the survival of infected cells are not affected by multiplicity of infection (Table V). This shows that there is a cooperative action among phage particles in producing alterations of the bacterial membrane which weaken the attachment of the phage. The cellular changes detected in this way may be related to those responsible for the release of bacterial components from infected bacteria (Prater, 1951; Price, 1952).

A variant of the preceding experiments was designed to test bacteria at a later stage in the growth of phage. For this purpose infected cells were aerated in broth for 5 or 15 minutes, fixed by the addition of 0.5 per cent (v/v) commercial formalin, centrifuged, resuspended in 0.1 per cent formalin in water, and subsequently handled as described above. The results were very similar to those already presented, except that the release of P^{32} from the cells was slightly less, and titrations of infected cells could not be made.

The S^{35}-labeled material detached from infected cells in the manner described possesses the following properties. It is sedimented at 12,000 G, though less completely than intact phage particles. It is completely precipitated by

TABLE V

Effect of Multiplicity of Infection on Elution of Phage Membranes from Infected Bacteria

Running time in blendor	Multiplicity of infection	P^{32}-labeled phage		S^{35}-labeled phage	
		Isotope eluted	Infected bacteria surviving	Isotope eluted	Infected bacteria surviving
min.		*per cent*	*per cent*	*per cent*	*per cent*
0	0.6	10	120	16	101
2.5	0.6	21	82	81	78
0	6.0	13	89	46	90
2.5	6.0	24	86	82	85

The infected bacteria were suspended at 10^9 cells per ml. in water containing per liter 1 mM $MgSO_4$, 0.1 mM $CaCl_2$, and 0.1 gm. gelatin. Samples were withdrawn for assay of extracellular isotope and infected bacteria before and after agitating the suspension. In either case the cells spent about 15 minutes at room temperature in the eluting fluid.

antiphage serum in the presence of whole phage carrier. 40 to 50 per cent of it readsorbs to sensitive bacteria, almost independently of bacterial concentration between 2×10^8 and 10^9 cells per ml., in 5 minutes at 37°C. The adsorption is not very specific: 10 to 25 per cent adsorbs to phage-resistant bacteria under the same conditions. The adsorption requires salt, and for this reason the efficient removal of S^{35} from infected bacteria can be accomplished only in a fluid poor in electrolytes.

The results of these experiments may be summarized as follows:—

1. 75 to 80 per cent of the phage sulfur can be stripped from infected cells by violent agitation of the suspension. At high multiplicity of infection, nearly 50 per cent elutes spontaneously. The properties of the S^{35}-labeled material show that it consists of more or less intact phage membranes, most of which have lost the ability to attach specifically to bacteria.

2. The release of sulfur is accompanied by the release of only 21 to 35 per

cent of the phage phosphorus, half of which is given up without any mechanical agitation.

3. The treatment does not cause any appreciable inactivation of intracellular phage.

4. These facts show that the bulk of the phage sulfur remains at the cell surface during infection, and takes no part in the multiplication of intracellular phage. The bulk of the phage DNA, on the other hand, enters the cell soon after adsorption of phage to bacteria.

Transfer of Sulfur and Phosphorus from Parental Phage to Progeny.—We have concluded above that the bulk of the sulfur-containing protein of the resting phage particle takes no part in the multiplication of phage, and in fact does not enter the cell. It follows that little or no sulfur should be transferred from parental phage to progeny. The experiments described below show that this expectation is correct, and that the maximal transfer is of the order 1 per cent

Bacteria were grown in glycerol-lactate medium overnight and subcultured in the same medium for 2 hours at 37°C. with aeration, the size of seeding being adjusted nephelometrically to yield 2×10^8 cells per ml. in the subculture. These bacteria were sedimented, resuspended in adsorption medium at a concentration of 10^9 cells per ml., and infected with S^{35}-labeled phage T2. After 5 minutes at 37°C., the suspension was diluted with 2 volumes of water and resedimented to remove unadsorbed phage (5 to 10 per cent by titer) and S^{35} (about 15 per cent). The cells were next suspended in glycerol-lactate medium at a concentration of 2×10^8 per ml. and aerated at 37°C. Growth of phage was terminated at the desired time by adding in rapid succession 0.02 mM HCN and 2×10^{11} UV-killed phage per ml. of culture. The cyanide stops the maturation of intracellular phage (Doermann, 1948), and the UV-killed phage minimizes losses of phage progeny by adsorption to bacterial debris, and promotes the lysis of bacteria (Maaløe and Watson, 1951). As mentioned in another connection, and also noted in these experiments, the lysing phage must be closely related to the phage undergoing multiplication (*e.g.*, T2H, its *h* mutant, or T2L, but not T4 or T6, in this instance) in order to prevent inactivation of progeny by adsorption to bacterial debris.

To obtain what we shall call the maximal yield of phage, the lysing phage was added 25 minutes after placing the infected cells in the culture medium, and the cyanide was added at the end of the 2nd hour. Under these conditions, lysis of infected cells occurs rather slowly.

Aeration was interrupted when the cyanide was added, and the cultures were left overnight at 37°C. The lysates were then fractionated by centrifugation into an initial low speed sediment (2500 *G* for 20 minutes), a high speed supernatant (12,000 *G* for 30 minutes), a second low speed sediment obtained by recentrifuging in adsorption medium the resuspended high speed sediment, and the clarified high speed sediment.

The distribution of S^{35} and phage among fractions obtained from three cultures of this kind is shown in Table VI. The results are typical (except for the excessively good recoveries of phage and S^{35}) of lysates in broth as well as lysates in glycerol-lactate medium.

The striking result of this experiment is that the distribution of S^{35} among the fractions is the same for early lysates that do not contain phage progeny, and later ones that do. This suggests that little or no S^{35} is contained in the mature phage progeny. Further fractionation by adsorption to bacteria confirms this suggestion.

Adsorption mixtures prepared for this purpose contained about 5×10^9 heat-killed bacteria (70°C. for 10 minutes) from 18 hour broth cultures, and

TABLE VI

Per Cent Distributions of Phage and S^{35} among Centrifugally Separated Fractions of Lysates after Infection with S^{35}-Labeled T2

Fraction	Lysis at $t = 0$ S^{35}	Lysis at $t = 10$ S^{35}	Maximal yield	
			S^{35}	Phage
1st low speed sediment	79	81	82	19
2nd " " "	2.4	2.1	2.8	14
High speed "	8.6	6.9	7.1	61
" " supernatant	10	10	7.5	7.0
Recovery	100	100	96	100

Infection with S^{35}-labeled T2, 0.8 particles per bacterium. Lysing phage UV-killed *h* mutant of T2. Phage yields per infected bacterium: <0.1 after lysis at $t = 0$; 0.12 at $t = 10$; maximal yield 29. Recovery of S^{35} means per cent of adsorbed input recovered in the four fractions; recovery of phage means per cent of total phage yield (by plaque count before fractionation) recovered by titration of fractions.

about 10^{11} phage (UV-killed lysing phage plus test phage), per ml. of adsorption medium. After warming to 37°C. for 5 minutes, the mixtures were diluted with 2 volumes of water, and centrifuged. Assays were made from supernatants and from unwashed resuspended sediments.

The results of tests of adsorption of S^{35} and phage to bacteria (H) adsorbing both T2 progeny and *h*-mutant lysing phage, to bacteria (B/2) adsorbing lysing phage only, and to bacteria (B/2*h*) adsorbing neither, are shown in Table VII, together with parallel tests of authentic S^{35}-labeled phage.

The adsorption tests show that the S^{35} present in the seed phage is adsorbed with the specificity of the phage, but that S^{35} present in lysates of bacteria infected with this phage shows a more complicated behavior. It is strongly adsorbed to bacteria adsorbing both progeny and lysing phage. It is weakly adsorbed to bacteria adsorbing neither. It is moderately well adsorbed to bacteria adsorbing lysing phage but not phage progeny. The latter test shows that the S^{35} is not contained in the phage progeny, and explains the fact that the S^{35} in early lysates not containing progeny behaves in the same way.

The specificity of the adsorption of S^{35}-labeled material contaminating the phage progeny is evidently due to the lysing phage, which is also adsorbed much more strongly to strain H than to B/2, as shown both by the visible reduction in Tyndall scattering (due to the lysing phage) in the supernatants of the test mixtures, and by independent measurements. This conclusion is further confirmed by the following facts.

TABLE VII

Adsorption Tests with Uniformly S^{35}-Labeled Phage and with Products of Their Growth in Non-Radioactive Medium

Adsorbing bacteria	Per cent adsorbed				
	Uniformly labeled S^{35} phage		Products of lysis at $t = 10$	Phage progeny (Maximal yield)	
	+ UV-h	No UV-h			
	S^{35}	S^{35}	S^{35}	S^{35}	Phage
Sensitive (H).....................	84	86	79	78	96
Resistant (B/2)...................	15	11	46	49	10
Resistant (B/2h).................	13	12	29	28	8

The uniformly labeled phage and the products of their growth are respectively the seed phage and the high speed sediment fractions from the experiment shown in Table VI.

The uniformly labeled phage is tested at a low ratio of phage to bacteria: +UV-h means with added UV-killed h mutant in equal concentration to that present in the other test materials.

The adsorption of phage is measured by plaque counts of supernatants, and also sediments in the case of the resistant bacteria, in the usual way.

1. If bacteria are infected with S^{35} phage, and then lysed near the midpoint of the latent period with cyanide alone (in salt-poor broth, to prevent readsorption of S^{35} to bacterial debris), the high speed sediment fraction contains S^{35} that is adsorbed weakly and non-specifically to bacteria.

2. If the lysing phage and the S^{35}-labeled infecting phage are the same (T2), or if the culture in salt-poor broth is allowed to lyse spontaneously (so that the yield of progeny is large), the S^{35} in the high speed sediment fraction is adsorbed with the specificity of the phage progeny (except for a weak nonspecific adsorption). This is illustrated in Table VII by the adsorption to H and B/2h.

It should be noted that a phage progeny grown from S^{35}-labeled phage and containing a larger or smaller amount of contaminating radioactivity could not be distinguished by any known method from authentic S^{35}-labeled phage, except that a small amount of the contaminant could be removed by adsorption to bacteria resistant to the phage. In addition to the properties already mentioned, the contaminating S^{35} is completely precipitated with the phage by antiserum, and cannot be appreciably separated from the phage by further fractional sedimentation, at either high or low concentrations of electrolyte. On the other hand, the chemical contamination from this source would be very small in favorable circumstances, because the progeny of a single phage particle are numerous and the contaminant is evidently derived from the parents.

The properties of the S^{35}-labeled contaminant show that it consists of the remains of the coats of the parental phage particles, presumably identical with the material that can be removed from unlysed cells in the Waring blendor. The fact that it undergoes little chemical change is not surprising since it probably never enters the infected cell.

The properties described explain a mistaken preliminary report (Hershey *et al.*, 1951) of the transfer of S^{35} from parental to progeny phage.

It should be added that experiments identical to those shown in Tables VI and VII, but starting from phage labeled with P^{32}, show that phosphorus is transferred from parental to progeny phage to the extent of 30 per cent at yields of about 30 phage per infected bacterium, and that the P^{32} in prematurely lysed cultures is almost entirely non-sedimentable, becoming, in fact, acid-soluble on aging.

Similar measures of the transfer of P^{32} have been published by Putnam and Kozloff (1950) and others. Watson and Maaløe (1952) summarize this work, and report equal transfer (nearly 50 per cent) of phosphorus and adenine.

A Progeny of S^{35}-Labeled Phage Nearly Free from the Parental Label.—The following experiment shows clearly that the obligatory transfer of parental sulfur to offspring phage is less than 1 per cent, and probably considerably less. In this experiment, the phage yield from infected bacteria from which the S^{35}-labeled phage coats had been stripped in the Waring blendor was assayed directly for S^{35}.

Sensitive bacteria grown in broth were infected with five particles of S^{35}-labeled phage per bacterium, the high ratio of infection being necessary for purposes of assay. The infected bacteria were freed from unadsorbed phage and suspended in water containing per liter 1 mM $MgSO_4$, 0.1 mM $CaCl_2$, and 0.1 gm. gelatin. A sample of this suspension was agitated for 2.5 minutes in the Waring blendor, and centrifuged to remove the extracellular S^{35}. A second sample not run in the blendor was centrifuged at the same time. The cells from both samples were resuspended in warm salt-poor broth at a concentration of 10^8 bacteria per ml., and aerated for 80 minutes. The cultures were then lysed by the addition of 0.02 mM HCN, 2×10^{11} UV-killed T2, and 6 mg. NaCl per ml. of culture. The addition of salt at this point causes S^{35} that would otherwise be eluted (Hershey *et al.*, 1951) to remain attached to the bacterial debris. The lysates were fractionated and assayed as described previously, with the results shown in Table VIII.

The data show that stripping reduces more or less proportionately the S^{35}-content of all fractions. In particular, the S^{35}-content of the fraction containing most of the phage progeny is reduced from nearly 10 per cent to less than 1 per cent of the initially adsorbed isotope. This experiment shows that the bulk of the S^{35} appearing in all lysate fractions is derived from the remains of the coats of the parental phage particles.

Properties of Phage Inactivated by Formaldehyde.—Phage T2 warmed for 1 hour at 37°C. in adsorption medium containing 0.1 per cent (v/v) commercial formalin (35 per cent HCHO), and then dialyzed free from formalde-

TABLE VIII

Lysates of Bacteria Infected with S^{35}-Labeled T2 and Stripped in the Waring Blendor

Per cent of adsorbed S^{35} or of phage yield:	Cells stripped		Cells not stripped	
	S^{35}	Phage	S^{35}	Phage
Eluted in blendor fluid..............	86	—	39	—
1st low-speed sediment..............	3.8	9.3	31	13
2nd " " "	(0.2)	11	2.7	11
High-speed "	(0.7)	58	9.4	89
" " supernatant..............	(2.0)	1.1	(1.7)	1.6
Recovery............................	93	79	84	115

All the input bacteria were recovered in assays of infected cells made during the latent period of both cultures. The phage yields were 270 (stripped cells) and 200 per bacterium, assayed before fractionation. Figures in parentheses were obtained from counting rates close to background.

hyde, shows a reduction in plaque titer by a factor 1000 or more. Inactivated phage of this kind possesses the following properties.

1. It is adsorbed to sensitive bacteria (as measured by either S^{35} or P^{32} labels), to the extent of about 70 per cent.

2. The adsorbed phage kills bacteria with an efficiency of about 35 per cent compared with the original phage stock.

3. The DNA of the inactive particles is resistant to DNase, but is made sensitive by osmotic shock.

4. The DNA of the inactive particles is not sensitized to DNase by adsorption to heat-killed bacteria, nor is it released into solution by adsorption to bacterial debris.

5. 70 per cent of the adsorbed phage DNA can be detached from infected cells spun in the Waring blendor. The detached DNA is almost entirely resistant to DNase.

These properties show that T2 inactivated by formaldehyde is largely incapable of injecting its DNA into the cells to which it attaches. Its behavior in the experiments outlined gives strong support to our interpretation of the corresponding experiments with active phage.

DISCUSSION

We have shown that when a particle of bacteriophage T2 attaches to a bacterial cell, most of the phage DNA enters the cell, and a residue containing at least 80 per cent of the sulfur-containing protein of the phage remains at the cell surface. This residue consists of the material forming the protective membrane of the resting phage particle, and it plays no further role in infection after the attachment of phage to bacterium.

These facts leave in question the possible function of the 20 per cent of sulfur-containing protein that may or may not enter the cell. We find that little or none of it is incorporated into the progeny of the infecting particle, and that at least part of it consists of additional material resembling the residue that can be shown to remain extracellular. Phosphorus and adenine (Watson and Maaløe, 1952) derived from the DNA of the infecting particle, on the other hand, are transferred to the phage progeny to a considerable and equal extent. We infer that sulfur-containing protein has no function in phage multiplication, and that DNA has some function.

It must be recalled that the following questions remain unanswered. (1) Does any sulfur-free phage material other than DNA enter the cell? (2) If so, is it transferred to the phage progeny? (3) Is the transfer of phosphorus (or hypothetical other substance) to progeny direct—that is, does it remain at all times in a form specifically identifiable as phage substance—or indirect?

Our experiments show clearly that a physical separation of the phage T2 into genetic and non-genetic parts is possible. A corresponding functional separation is seen in the partial independence of phenotype and genotype in the same phage (Novick and Szilard, 1951; Hershey et al., 1951). The chemical identification of the genetic part must wait, however, until some of the questions asked above have been answered.

Two facts of significance for the immunologic method of attack on problems of viral growth should be emphasized here. First, the principal antigen of the infecting particles of phage T2 persists unchanged in infected cells. Second, it remains attached to the bacterial debris resulting from lysis of the cells. These possibilities seem to have been overlooked in a study by Rountree (1951) of viral antigens during the growth of phage T5.

SUMMARY

1. Osmotic shock disrupts particles of phage T2 into material containing nearly all the phage sulfur in a form precipitable by antiphage serum, and capable of specific adsorption to bacteria. It releases into solution nearly all

the phage DNA in a form not precipitable by antiserum and not adsorbable to bacteria. The sulfur-containing protein of the phage particle evidently makes up a membrane that protects the phage DNA from DNase, comprises the sole or principal antigenic material, and is responsible for attachment of the virus to bacteria.

2. Adsorption of T2 to heat-killed bacteria, and heating or alternate freezing and thawing of infected cells, sensitize the DNA of the adsorbed phage to DNase. These treatments have little or no sensitizing effect on unadsorbed phage. Neither heating nor freezing and thawing releases the phage DNA from infected cells, although other cell constituents can be extracted by these methods. These facts suggest that the phage DNA forms part of an organized intracellular structure throughout the period of phage growth.

3. Adsorption of phage T2 to bacterial debris causes part of the phage DNA to appear in solution, leaving the phage sulfur attached to the debris. Another part of the phage DNA, corresponding roughly to the remaining half of the DNA of the inactivated phage, remains attached to the debris but can be separated from it by DNase. Phage T4 behaves similarly, although the two phages can be shown to attach to different combining sites. The inactivation of phage by bacterial debris is evidently accompanied by the rupture of the viral membrane.

4. Suspensions of infected cells agitated in a Waring blendor release 75 per cent of the phage sulfur and only 15 per cent of the phage phosphorus to the solution as a result of the applied shearing force. The cells remain capable of yielding phage progeny.

5. The facts stated show that most of the phage sulfur remains at the cell surface and most of the phage DNA enters the cell on infection. Whether sulfur-free material other than DNA enters the cell has not been determined. The properties of the sulfur-containing residue identify it as essentially unchanged membranes of the phage particles. All types of evidence show that the passage of phage DNA into the cell occurs in non-nutrient medium under conditions in which other known steps in viral growth do not occur.

6. The phage progeny yielded by bacteria infected with phage labeled with radioactive sulfur contain less than 1 per cent of the parental radioactivity. The progeny of phage particles labeled with radioactive phosphorus contain 30 per cent or more of the parental phosphorus.

7. Phage inactivated by dilute formaldehyde is capable of adsorbing to bacteria, but does not release its DNA to the cell. This shows that the interaction between phage and bacterium resulting in release of the phage DNA from its protective membrane depends on labile components of the phage particle. By contrast, the components of the bacterium essential to this interaction are remarkably stable. The nature of the interaction is otherwise unknown.

8. The sulfur-containing protein of resting phage particles is confined to a

protective coat that is responsible for the adsorption to bacteria, and functions as an instrument for the injection of the phage DNA into the cell. This protein probably has no function in the growth of intracellular phage. The DNA has some function. Further chemical inferences should not be drawn from the experiments presented.

REFERENCES

Anderson, T. F., 1949, The reactions of bacterial viruses with their host cells, *Bot. Rev.*, **15,** 464.

Anderson, T. F., 1951, *Tr. New York Acad. Sc.*, **13,** 130.

Anderson, T. F., and Doermann, A. H., 1952, *J. Gen. Physiol.*, **35,** 657.

Benzer, S., 1952, *J. Bact.*, **63,** 59.

Doermann, A. H., 1948, *Carnegie Institution of Washington Yearbook, No. 47*, 176.

Doermann, A. H., and Dissosway, C., 1949, *Carnegie Institution of Washington Yearbook, No. 48*, 170.

Dulbecco, R., 1952, *J. Bact.*, **63,** 209.

Herriott, R. M., 1951, *J. Bact.*, **61,** 752.

Hershey, A. D., 1946, *Genetics*, **31,** 620.

Hershey, A. D., Roesel, C., Chase, M., and Forman, S., 1951, *Carnegie Institution of Washington Yearbook, No. 50*, 195.

Lwoff, A., and Gutmann, A., 1950, *Ann. Inst. Pasteur*, **78,** 711.

Maaløe, O., and Watson, J. D., 1951, *Proc. Nat. Acad. Sc.*, **37,** 507.

Novick, A., and Szilard, L., 1951, *Science*, **113,** 34.

Prater, C. D., 1951, Thesis, University of Pennsylvania.

Price, W. H., 1952, *J. Gen. Physiol.*, **35,** 409.

Putnam, F. W., and Kozloff, L.,1950, *J. Biol. Chem.*, **182,** 243.

Rountree, P. M., 1951, *Brit. J. Exp. Path.*, **32,** 341.

Watson, J. D., and Maaløe, O., 1952, *Acta path. et microbiol. scand.*, in press.

Questions

ARTICLE 1 Hershey and Chase

1. How was it shown that ^{32}P represents DNA and ^{35}S represents protein?

2. What is the argument that DNA enters the cell?

3. Why do you think that 100% of the ^{35}S is not stripped off by blending?

4. Why do you think that 30% of the ^{32}P is removed by blending?

5. What is the significance of the fact that the bacteria still yield phage after blending?

6. What explanation can be given for the fact that only a fraction of the injected ^{32}P is transferred to progeny DNA?

7. Suppose that neither ^{32}P nor ^{35}S could be removed by blending. Could any conclusion have been drawn? How does the result of the transfer experiment affect your answer?

•8. Suppose that when the Hershey-Chase experiment was finished, the following results had been obtained. After blending and centrifugation the ^{35}S was in the pellet and the ^{32}P was in the supernatant. Microscopic observation showed that, after blending, all of the bacteria had been broken into small fragments. Examination of the pellet showed that there were no whole cells but just phage shells attached to small fragments of something (you don't know what). What else would you have to know, measure, or determine in order to conclude that DNA is the genetic material of the phage? What reservations, if any, do you have about your conclusion?

References

Avery, O., C. MacLeod, and M. McCarty. 1944. "Studies on the Chemical Nature of the Substance Inducing Transformation of Pneumococcal Types." *J. Expt. Med.* **79**, 137–157.

Chargaff, E. 1950. "Chemical Specificity of Nucleic Acids and Mechanism of Their Enzymatic Degradation," *Experientia* **6**, 201–209. Base composition analysis is described.

Levene, P. A., and L. W. Bass. 1931. *Nucleic Acid.* Chemical Catalog Company. The tetranucleotide hypothesis is clearly stated.

Vendrely, R. 1955. "The Deoxyribonucleic Acid of the Nucleus." In E. Chargaff and J. N. Davidson, Eds., *The Nucleic Acids*, Vol. 2, pp. 155–180. Academic Press. The DNA content of haploid and diploid cells is reviewed.

Additional Readings

Chargaff, E. 1963. *Essays on Nucleic Acids*, Elsevier.

Davidson, J. N., and E. Chargaff, Eds. 1955. *The Nucleic Acids*, "Introduction," Vol. 1, pp. 1–8. Academic Press.

Stent, G., and R. Calendar. 1978. *Molecular Genetics*, 2nd ed. W. H. Freeman and Company. The history of the transformation problem is described in great detail.

Ts'o, P. O. P. 1974. *Basic Principles in Nucleic Acid Chemistry*. Academic Press. An excellent historical review appears at the beginning of this book.

THE DOUBLE HELIX

The Elucidation of the
Three-Dimensional Structure of DNA

As we saw in the preceding section, the chemical structure of DNA remained the subject of scientific debate for decades. Then in 1950 when Erwin Chargaff made his careful analysis of the base composition of a number of DNA samples isolated from different organisms, he demonstrated that the composition of DNA varied widely between organisms. His work not only disproved the tetranucleotide model but also yielded the significant observation that the molar ratios—adenine to thymine, and guanine to cytosine—are 1.00. This was called the *equivalence rule* or the *base pairing rule* and was an essential fact that James Watson and Francis Crick used later to elucidate the structure of DNA by X-ray diffraction analysis.

The formulation of the structure of DNA unfolded gradually, largely at the Cavendish laboratory of Cambridge University. It followed the development, also at the Cavendish, of X-ray diffraction analysis as a tool in biology. Since 1921 various workers had used X-ray diffraction to determine the structure of the crystals of simple salts. In the early 1940's Dorothy Crowfoot first applied this technique to biological molecules and worked out the structure

of penicillin, a relatively simple molecule. William T. Astbury had been interested in DNA primarily as a curiosity, and its very high density led him to propose that the nucleotide bases were very close to one another, and were probably stacked in a single chain one above the other. By X-ray diffraction he confirmed the stacked structure and determined the internucleotide spacing of 3.4 angstroms (Å). His photographs, however, were not clear enough to show any greater structural detail. Maurice Wilkins and Rosalind Franklin, also at Cambridge, then developed a method for preparing highly oriented DNA fibers that yielded X-ray photographs of high quality, thus allowing a considerable amount of information about the structure of DNA to be obtained. Watson, a visitor in the Cavendish laboratory, had also been working with Crick on the structure of DNA but had obtained only photos of low quality. At one point they saw one of Franklin's X-ray photographs that contained several essential features lacking in their own pictures and in a few exciting weeks they had worked out the structure of DNA. They describe this analysis in Article 2.

ARTICLE 2

From *Nature*, Vol. 171, No. 4356, 737–738, 1953. Reproduced with permission of the authors and publisher.

MOLECULAR STRUCTURE OF NUCLEIC ACIDS

A Structure for Deoxyribose Nucleic Acid

WE wish to suggest a structure for the salt of deoxyribose nucleic acid (D.N.A.). This structure has novel features which are of considerable biological interest.

A structure for nucleic acid has already been proposed by Pauling and Corey[1]. They kindly made their manuscript available to us in advance of publication. Their model consists of three intertwined chains, with the phosphates near the fibre axis, and the bases on the outside. In our opinion, this structure is unsatisfactory for two reasons: (1) We believe that the material which gives the X-ray diagrams is the salt, not the free acid. Without the acidic hydrogen atoms it is not clear what forces would hold the structure together, especially as the negatively charged phosphates near the axis will repel each other. (2) Some of the van der Waals distances appear to be too small.

Another three-chain structure has also been suggested by Fraser (in the press). In his model the phosphates are on the outside and the bases on the inside, linked together by hydrogen bonds. This structure as described is rather ill-defined, and for this reason we shall not comment on it.

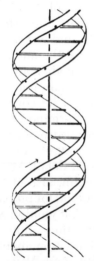

We wish to put forward a radically different structure for the salt of deoxyribose nucleic acid. This structure has two helical chains each coiled round the same axis (see diagram). We have made the usual chemical assumptions, namely, that each chain consists of phosphate diester groups joining β-D-deoxyribofuranose residues with 3′,5′ linkages. The two chains (but not their bases) are related by a dyad perpendicular to the fibre axis. Both chains follow right-handed helices, but owing to the dyad the sequences of the atoms in the two chains run in opposite directions. Each chain loosely resembles Furberg's[2] model No. 1; that is, the bases are on the inside of the helix and the phosphates on the outside. The configuration of the sugar and the atoms near it is close to Furberg's 'standard configuration', the sugar being roughly perpendicular to the attached base. There

This figure is purely diagrammatic. The two ribbons symbolize the two phosphate—sugar chains, and the horizontal rods the pairs of bases holding the chains together. The vertical line marks the fibre axis

is a residue on each chain every 3·4 A. in the z-direction. We have assumed an angle of 36° between adjacent residues in the same chain, so that the structure repeats after 10 residues on each chain, that is, after 34 A. The distance of a phosphorus atom from the fibre axis is 10 A. As the phosphates are on the outside, cations have easy access to them.

The structure is an open one, and its water content is rather high. At lower water contents we would expect the bases to tilt so that the structure could become more compact.

The novel feature of the structure is the manner in which the two chains are held together by the purine and pyrimidine bases. The planes of the bases are perpendicular to the fibre axis. They are joined together in pairs, a single base from one chain being hydrogen-bonded to a single base from the other chain, so that the two lie side by side with identical z-co-ordinates. One of the pair must be a purine and the other a pyrimidine for bonding to occur. The hydrogen bonds are made as follows: purine position 1 to pyrimidine position 1; purine position 6 to pyrimidine position 6.

If it is assumed that the bases only occur in the structure in the most plausible tautomeric forms (that is, with the keto rather than the enol configurations) it is found that only specific pairs of bases can bond together. These pairs are: adenine (purine) with thymine (pyrimidine), and guanine (purine) with cytosine (pyrimidine).

In other words, if an adenine forms one member of a pair, on either chain, then on these assumptions the other member must be thymine; similarly for guanine and cytosine. The sequence of bases on a single chain does not appear to be restricted in any way. However, if only specific pairs of bases can be formed, it follows that if the sequence of bases on one chain is given, then the sequence on the other chain is automatically determined.

It has been found experimentally[3,4] that the ratio of the amounts of adenine to thymine, and the ratio of guanine to cytosine, are always very close to unity for deoxyribose nucleic acid.

It is probably impossible to build this structure with a ribose sugar in place of the deoxyribose, as the extra oxygen atom would make too close a van der Waals contact.

The previously published X-ray data[5,6] on deoxyribose nucleic acid are insufficient for a rigorous test of our structure. So far as we can tell, it is roughly compatible with the experimental data, but it must be regarded as unproved until it has been checked against more exact results. Some of these are given in the following communications. We were not aware of the details of the results presented there when we devised our structure, which rests mainly though not entirely on published experimental data and stereochemical arguments.

It has not escaped our notice that the specific pairing we have postulated immediately suggests a possible copying mechanism for the genetic material.

Full details of the structure, including the conditions assumed in building it, together with a set of co-ordinates for the atoms, will be published elsewhere.

We are much indebted to Dr. Jerry Donohue for constant advice and criticism, especially on interatomic distances. We have also been stimulated by a knowledge of the general nature of the unpublished experimental results and ideas of Dr. M. H. F. Wilkins, Dr. R. E. Franklin and their co-workers at King's College, London. One of us (J. D. W.) has been aided by a fellowship from the National Foundation for Infantile Paralysis.

J. D. WATSON
F. H. C. CRICK
Medical Research Council Unit for the
Study of the Molecular Structure of
Biological Systems,
Cavendish Laboratory, Cambridge.
April 2.

[1] Pauling, L., and Corey, R. B., *Nature*, 171, 346 (1953); *Proc. U.S. Nat. Acad. Sci.*, 39, 84 (1953).
[2] Furberg, S., *Acta Chem. Scand.*, 6, 634 (1952).
[3] Chargaff, E., for references see Zamenhof, S., Brawerman, G., and Chargaff, E., *Biochim. et Biophys. Acta*, 9, 402 (1952).
[4] Wyatt, G. R., *J. Gen. Physiol.*, 36, 201 (1952).
[5] Astbury, W. T., Symp. Soc. Exp. Biol. 1, Nucleic Acid, 66 (Camb. Univ. Press, 1947).
[6] Wilkins, M. H. F., and Randall, J. T., *Biochim. et Biophys. Acta*, 10, 192 (1953).

Questions

ARTICLE 2 Watson and Crick

1. The X-ray data showed that the double helix has a constant diameter. How would Watson and Crick reconcile this finding with the fact that purines and pyrimidines have different sizes? How were they helped in this view by Chargaff's work?

2. How many hydrogen bonds are there in an adenine-thymine pair? A guanine-cytosine pair? Why are adenine-cytosine and guanine-thymine pairs improbable? Why are purine-purine and pyrimidine-pyrimidine pairs improbable? (Disregard Chargaff's work in answering this).

3. Why did Watson and Crick conclude that the strands are antiparallel?

4. How did they calculate the number of bases per turn?

5. The X-ray picture indicated helicity. What fact suggested a *double* helix?

6. Why did Watson and Crick conclude that the whole DNA molecule is self-complementary?

References

Chargaff, E. 1950. "Chemical Specificity of Nucleic Acids and Mechanisms of their Enzymatic Degradation," *Experientia* **6**, 201–209.

Franklin, R. E., and R. Gosling. 1953. "Molecular Configuration in Sodium Thymonucleate." *Nature* **171**, 740–741.

Wilkins, M. H. F., A. R. Stocker, and H. R. Wilson. 1953. "Molecular Structure of Desoxypentose Nucleic Acids." Nature **171**, 738–740.

Note: The papers by Franklin and Gosling, and by Wilkins, Stocker, and Wilson were published together with Article 2, and also report information about the structure of DNA derived from X-ray diffraction photographs. They have not been reprinted because they are more difficult to read, requiring a more detailed understanding of the X-ray technique. Maurice Wilkins shared the Nobel prize with Watson and Crick for this work.

Additional Readings

Davidson, J. N. 1972. *The Biochemistry of the Nucleic Acids*, Academic Press.

Watson, J. D. 1968. *The Double Helix*. Atheneum.

THE DNA OF AN INDIVIDUAL PHAGE OR BACTERIUM AS A SINGLE MOLECULE

Methods of Isolating DNA

For many years it was believed that the chromosomes of all organisms (both eukaryotes and microorganisms such as phages and bacteria) consisted of many small DNA molecules held together by some unknown linker. To understand the reason for this incorrect idea one must examine the methods for isolating DNA.

When eukaryotic cells and bacteria are broken open (usually by the addition of detergents such as sodium dodecyl sulfate), it is reasonably simple to separate DNA from most of the cellular material because a mixture of 70% ethanol and 0.2 M NaCl causes DNA to precipitate as sticky fibers that adhere both to one another and to a glass rod. Thus, by stirring a cell extract to which ethanolic NaCl has been added, a "spool" of DNA fibers can be collected on the glass rod. RNA and protein form a nonfibrous, flocculent precipitate that does not spool, and most other cell constituents remain in solution. Since many impurities are often trapped in the spooled fibers, the precipitated DNA is usually redissolved and then respooled. Nonetheless, since many proteins bind tightly to DNA, a considerable amount of protein always remains.

Two procedures have mainly been used for deproteinizing DNA. In the first the DNA solution is shaken vigorously with a chloroform-n-octanol (C-O) mixture. Because this mixture is immiscible with water, shaking produces a suspension of tiny droplets of aqueous and organic phases. Proteins tend to precipitate at aqueous-organic interfaces so that if the two phases are separated (e.g., by centrifugation), a heavy protein precipitate remains at the interface while the aqueous layer contains almost pure DNA. Some years ago, C-O was replaced by water-saturated phenol because C-O treatment does not cause most bacteriophages to release their DNA but phenol treatment does. Since phenol and water are also immiscible, shaking a DNA solution with phenol deproteinizes it in much the same way as the C-O procedure does. The added advantage of the phenol treatment is that some proteins, most lipoproteins, and some RNA pass into the phenol phase, thus producing purer DNA. Unfortunately, in both procedures, the vigorous shaking usually fragments the DNA although this drawback was not recognized at first. The result was that the molecular weight M of isolated DNA,

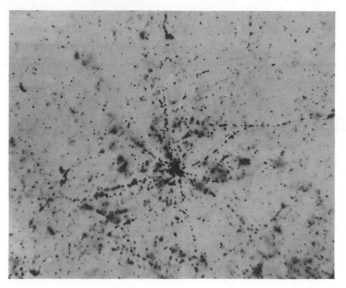

Figure III-1

A star. A bacteriophage labeled with ^{32}P is embedded in a nuclear emulsion. All ^{32}P β tracks originate from a single point, which allows the number of tracks (rays) per star to be counted. To count all rays it is necessary to focus the microscope up and down in order to recognize those that are not in the plane of the source. [Courtesy of Charles A. Thomas.]

though the highest observed for any known molecule, was far less than the DNA content of the cell or phage from which it was isolated. Early values of M for phage and bacterial DNA ranged from 0.5–2.0×10^6; by the late 1950's values as high as 7×10^6 were reported. However, the DNA content of the phages studied at that time was 50–300×10^6, that of bacteria was 10^{10}, and of eukaryotic chromosomes 10^{10}–10^{11} so that it was reasonable to conclude that DNA consisted of *subunits* of reasonably low M.* Furthermore, since extensive deproteinization decreased M, it was thought that these units might be held together by *protein linkers*. Considerable effort was extended to identify or purify these linkers—but totally without success (since they do not exist).

The Star Experiment

An important study of the size of DNA molecules known as the star experiment was

*The DNA content was determined, first by taking a sample of phages, bacteria, or cells whose number was approximately known, and then by measuring, with chemical tests, the weight of DNA in the sample. This procedure is not of course a molecular weight measurement.

carried out by Cyrus Levinthal and Charles Thomas, using a novel autoradiographic technique. *E. coli* phage T2 (see Appendix B) was grown in the presence of the radioisotope ^{32}P (as phosphate) so that each phage contained about 50 ^{32}P atoms. The phages were then embedded in nuclear emulsion and stored for about a week. Since the half life of ^{32}P is about two weeks, several decays occurred in each phage during the storage time. As each ^{32}P atom decayed, a β particle was emitted that produced a long straight track of exposed grains in the film. Because the phage has very small dimensions, all tracks from a single phage appear to originate from a single point. The resulting pattern of tracks was called a *star* and the individual tracks were called *rays* (Fig. III-1).

Levinthal and Thomas compared the patterns resulting from phages and from DNA isolated from the phages. They found that the DNA sample had more stars than the phage sample, indicating that each phage, on the average, had more than one DNA molecule. In confirmation of this conclusion it was shown that these stars had on the average a smaller number of rays; it was clear that the isolated DNA molecules were smaller than the DNA content of a single phage. Analysis of the data led to the conclusion that phage T2 contains one large subunit of DNA called the *big piece* and several small subunits termed the *little pieces*. As we shall see shortly, this conclusion was incorrect, although it was a logical one to draw from the data; since breakage of DNA molecules by hydrodynamic shear forces had not yet been discovered (see the following paragraph), the fact that phage T2 actually contains a single DNA molecule was not suspected. However, the experiment was extremely important, since it suggested that many DNA molecules might be very large (some having a molecular weight as great as 80 million).

The Breakage of DNA by Hydrodynamic Shear Forces

The results of the Levinthal-Thomas work were surprising to most geneticists, because genetic experiments with phage T2 had indicated that the T2 genes were contained in two or three linkage groups, a fact that had been interpreted to mean that the phage DNA consisted of two or three subunits. Since the star experiment showed that more subunits were present, it was hypothesized either that some of the DNA pieces contained no genetic information or that most of the genes had not yet been found by the geneticists. Later, adding to the confusion, George Streisinger and his colleagues used more genetic markers and found that all T2 genes were in a single linkage group. One possible solution to this new dilemma was to assume that all subunits but one were genetically inactive, an explanation that was generally felt to be unsatisfactory. Finally everyone realized that some element of the available data had to be incorrect when Julian Fleischman, a graduate student in Paul Doty's laboratory at Harvard, added to the complexity by carrying out careful viscometric and sedimentation studies of T2 DNA, which showed that the molecular weight of T2 DNA *as isolated* was approximately 10×10^6—*but* the DNA content of the phage was approximately 130×10^6! Here were the little pieces, but now where was the big piece?

Because sedimentation analysis was poorly understood and still considered to be suspect by many people, workers in various labora-

tories then set out to determine whether hydrodynamic analysis could actually detect a single large molecule in a mixture of large and small molecules. Few people suspected what the solution to the paradox would be and hence most were surprised when Peter Davison showed that DNA could be fragmented by the hydrodynamic shear forces* encountered in simple laboratory operations such as pipetting. Davison had noticed that the viscosity of DNA solutions often decreased after they had been poured or shaken. He correctly guessed that high-molecular-weight DNA could be broken by hydrodynamic shear forces generated during passage of the solution through a small orifice (such as the capillary of a viscometer or the hypodermic needle used to fill the analytic ultracentrifuge cells that were used in determining M by sedimentation rate). Davison describes this important discovery in Article 3.

The Star Experiment Redone

As we see in Article 3, Davison's work indicated that the DNA used in the star experiment had probably been partially degraded by pipetting and that Fleischman's DNA, if intact after isolation, had been broken during subsequent handling or filling of the centrifuge cell. The star experiment definitely needed to be done again. Also a second experiment would be aided by (1) the timely discovery by Alfred Hershey that DNA could be fractionated according to molecular weight by chromatography on methylated albumin, and (2) the development, by Roy Britten and Richard Roberts, of zonal centrifugation in a concentration gradient (see Appendixes C and D for a description of these methods). Thus, isolated DNA could be checked for homogeneity and accidentally broken molecules could be removed. With these procedures and the information provided by Davison, the star experiment was repeated, and it was demonstrated that phage T2 contains a single DNA molecule. The experiment is described in Ar-

---

*Hydrodynamic shear force is the force on a molecule (or any particle) that is in a liquid, different parts of which are moving in the same direction but at different velocities. The movement of each part of a molecular chain in relation to other parts causes the chain to extend into a nearly linear configuration and, if the force is great enough, causes it to break roughly in half.

ticle 4 by P. F. Davison, R. Hede, C. Levinthal, and me.

A Eukaryotic Chromosome Might Contain One DNA Molecule

The new star experiment showed that a T2 phage and a T2 DNA molecule contained the same number of phosphorus atoms. Therefore, the DNA of the phage is a single molecule. This experiment then led everyone to expect that all phages would have a single DNA molecule, although this supposition was never systematically explored; however, accumulated data have continued to uphold it. Even more important, though, it suggested to a few people that larger organisms such as bacteria and perhaps even mammalian chromosomes might contain a single DNA molecule. The first serious proposal that eukaryote DNA might be very large was made by Davison. He noticed that when trout sperm are lysed* by the addition of sodium dodecyl sulfate (a detergent), the released DNA yields a solution having enormous viscosity and that this viscosity decreases significantly each time the solution passes through a viscometer. From the initial viscosity measurement he estimated that the DNA might have a molecular weight of at least 10^9, whereas DNA purified from the trout sperm by standard techniques usually had a molecular weight of about 5×10^6. From the DNA content per sperm cell and the known number of trout chromosomes, Davison suggested that each chromosome might contain a single DNA molecule.

Isolation of Unbroken Bacterial DNA

Following Davison's work, researchers in various laboratories tried to isolate a single DNA molecule from a bacterium by minimizing hydrodynamic shear forces. The first intact bacterial DNA molecule was isolated by John Cairns, who used a special technique that eliminated virtually all shear forces. He thereby obtained the striking autoradiograms shown in Articles 5 and 6.

———————————

*The term *lysis* refers to breaking open a cell and releasing its contents.

For several years all attempts in other laboratories to isolate and describe intact DNA molecules by other methods failed. Cedric Davern, however, finally showed that if shearing is rigorously avoided, a fraction of the *E. coli* DNA molecules can be isolated as intact molecules. He describes his results in Article 7. (See also Article 10 in the next section, by M. Meselson, F. W. Stahl, and J. Vinograd for useful background to Davern's paper.)

The next objective was to obtain an electron micrograph of the entire intact bacterial chromosome. This was widely felt to be important because it would confirm the work by Cairns, whose difficult autoradiographic procedure had failed to work successfully in any laboratory but his own. It was believed that such an achievement would be possible because Albrecht Kleinschmidt had devised a procedure enabling him to visualize DNA by electron microscopy (see Section IV, page 81 for further details).

Kleinschmidt made several attempts to obtain a micrograph of a bacterial chromosome, but his results yielded only tangled molecules whose length could not be measured. Loren MacHattie, Ken Berns, and Charles Thomas also obtained electron micrographs of the DNA isolated from the bacterium *Hemophilus influenzae*. These molecules were highly compact, appeared to have no more than two free ends and usually none, and had a total length roughly equivalent to that expected for a single molecule. However, most of the strands appeared as multiple loops coming from a single point, so that one could not be sure that the DNA did not consist of units all linked to some yet unidentified structure. A modification of the Kleinschmidt technique was then employed by Hans Bode and Harold Morowitz, who obtained micrographs of untangled DNA of the bacterium *Mycoplasma hominis* (Fig. III-2).

By 1967, even prior to the work by Bode and Morowitz, most workers in the DNA field had begun to accept the idea that the bacterial genome (i.e., the set of genes) resides in a single DNA molecule. Furthermore, it was thought that all bacteria probably contained circular DNA (see Section VII for a detailed discussion of circularity). However, the chromosomes of animal cells seemed to be

Figure III-2

A drawing from an electron micrograph of a DNA molecule isolated from *Mycoplasma hominis*. The contrast in the original micrograph is inadequate for reproduction. [Reprinted from H. R. Bode and H. J. Morowitz, *J. Mol. Biol.* **23** (1967), 191–199.]

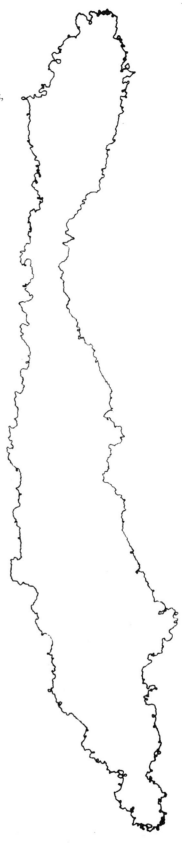

more complicated. For instance, if we calculate the replication rate (i.e., the number of bases incorporated into new DNA per minute) from the DNA content per bacterial or mammalian chromosome, this rate has to be approximately ten times greater in mammalian cells than in bacteria. Since it had been assumed (incorrectly) that the rate of movement of a replication fork* is nearly the same in all organisms, the logical conclusion was that a chromosome would contain either several DNA molecules, each with a single replication fork, or one DNA molecule *with several forks*. In Section IV (page 83), in which we discuss molecular weight determination, we present the evidence from the laboratory of Bruno Zimm that eukaryote chromosomes probably contain a single DNA molecule, demonstrating that the first alternative is not possible. We will not discuss this question further in this volume, but note only that we now know that animal DNA polymerases carry out the DNA synthesis much more slowly than the bacterial polymerase and that it is the presence of about 1000 replication forks per DNA molecule that allows a high net replication rate in animal cells.

---

*A replication fork is a site at which the new DNA strands are being synthesized.

ARTICLE 3

From the *Proceedings of the National Academy of Sciences*, Vol. 45, No. 11, 1560–1568, 1959. Reproduced with permission of the author.

THE EFFECT OF HYDRODYNAMIC SHEAR ON THE DEOXYRIBONUCLEIC ACID FROM T₂ AND T₄ BACTERIOPHAGES*

By Peter F. Davison

Department of Biology, Massachusetts Institute of Technology

Communicated by Francis O. Schmitt, September 28, 1959

In 1956, Levinthal[1] reported autoradiographic studies of T_2 and T_4 bacteriophage deoxyribonucleic acid (DNA). He found that the DNA in each phage particle was present in the form of several chains, one of molecular weight about 45×10^6 and six or more of molecular weight about 12×10^6. Attempts to differentiate the large and the small pieces of phage DNA by ultracentrifugation have been unsuccessful. For example, in sedimentation velocity experiments Fleischman[2] found a single peak with a sedimentation coefficient of 30 to 35 S(vedbergs) (the lower value was found on deproteinized preparations); Meselson, Stahl, and Vinograd[3] reported the DNA to band with a Gaussian distribution, implying a uniform molecular weight, in equilibrium experiments in a cesium chloride gradient.

With the ultraviolet absorption system in the Spinco Model E ultracentrifuge, DNA solutions below 0.001 per cent concentration can be studied, and sedimentation coefficients ranging from 8–50 S have been reported.[4, 5] Calculating from the formula which Doty, McGill, and Rice[6] derived from studies on calf thymus DNA, the large and the small pieces of phage DNA could have sedimentation coefficients about 42 and 26 S. These values may not be accurate since they are deduced from an unwarranted extrapolation of the formula, and, moreover, recent studies have shown that molecular weights and physical constants for DNA from different sources cannot be correlated;[7, 8] nevertheless, the failure to observe any complexity in the sedimentation diagrams has raised doubts about the *in vivo* existence of the "large piece," which could be an experimental artifact.

Thomas and Knight[9] have measured the sedimentation coefficient of the "large piece" using a partition centrifuge cell. They obtained a value of 40–58 S. The disparity between these and Fleischman's results is sufficient to provoke inquiry. The experiments reported below show that the investigations of Fleischman were probably made on DNA which had been inadvertently degraded. They also show that any physical experiment on phage DNA (and possibly DNA from other sources) must be attended by much greater manipulative precautions than have usually been employed.

Experimental.—Four bacteriophage samples were studied, of which two were given by Dr. H. Van Vunakis (T_2 and T_4), one by Dr. J. Tomizawa (T_2), and of which the fourth (T_2) was prepared by the author. Each preparation was finally purified by differential centrifugation until a clean translucent pellet was obtained. The phage were resuspended in neutral phosphate buffer at a suitable concentration (about 2×10^{12} phage per ml). Results from all these preparations were essentially identical.

Experiments were performed in a Spinco Model E ultracentrifuge using a cell with a 30 mm centerpiece. The optical density of the photographs was measured on a Joyce-Loebl microdensitometer. The density tracings were evaluated by the method of Schumaker and Schachman[5] to give the differential distribution curve. It should be emphasized that no attempt was made to bring the area under the different parts of the curves accurately into accord with the heights in the integral curves. The differential curve was drawn merely to give the best fit through the points derived from at least two of the optical density tracings.

Shear degradation:† Thomas[10] showed that the large piece of T_4 DNA was degraded during deproteinization, presumably by the shaking involved. Flow birefringence experiments on a shockate of T_2 phage demonstrated that some irreversible changes could be effected in the elongated DNA molecules by the hydrodynamic shear imposed by the Couette apparatus.[11] Both of these results suggested that hydrodynamic shear gradients of the same magnitude as those commonly imparted to solutions by such simple laboratory manipulation as stirring or pipetting could cause a reduction in the molecular weight of long DNA molecules.

To confirm these observations, T_2 or T_4 phage in 3 M cesium or sodium chloride were osmotically shocked (Anderson)[12] by the rapid addition of 20 volumes of 0.005 M versene pH 8.5, or cold water. Versene was employed in the case of phage preparations which had been treated with pancreatic deoxyribonuclease, to ensure that no enzymatic degradation occurred. Cesium or sodium chloride was then added to the shockate to molar concentration to minimize any association of protein and DNA through salt bonds; later it was found that these salts could be substituted by 0.2 M phosphate buffer. For many studies the DNA was banded in a cesium chloride gradient in a preparative centrifuge to free it from protein.[13] The DNA recovered from the band was diluted with water to lower the salt concentration and directly examined in the ultracentrifuge. No significant differences were detected between solutions treated in this way and the untreated shockates.

The solutions were sucked slowly into a syringe, a hypodermic needle was attached to the syringe, and the solutions were injected into the centrifuge cell at a controlled rate to subject them to a measurable average shear gradient.   Figure 1 shows the effects of different shears.   The mean shear rate for the passage through the hypodermic needle was calculated, assuming that the flow was streamline, from Kroepelin's formula[14]

$$\bar{G} = \frac{8V}{3\pi r^3 t}$$

where V is the volume passing in time t, and r is the radius of the capillary.   (For

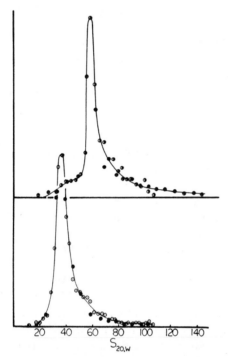

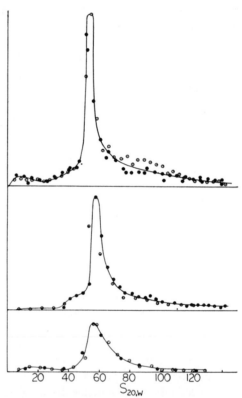

Fig. 1.—Distribution of sedimentation co-efficients in a 0.002% solution of DNA (recovered from banding a T_4 shockate in a cesium chloride gradient) in molar cesium chloride: (top) solution subjected to a shear gradient of 300 sec.$^{-1}$ during the filling of the cell; (bottom) solution subjected to 35,000 sec.$^{-1}$

Fig. 2.—Distribution of sedimentation co-efficients in a T_2 shockate in molar sodium chloride at the following DNA concentrations: (top) 0.0021%; (middle) 0.0014%; (bottom) 0.0007%.

example, forcing 1 ml through a No. 27 needle in 15 sec. imposes a mean shear gradient of 55,000 sec.$^{-1}$)

The sedimentation diagram of the slowly loaded sample is typical of a large number of shockates examined.   The addition of detergent or chymotrypsin to the solutions effected no obvious changes.   The sedimentation coefficient of the main peak varied from 50–56 S in sodium chloride solutions, and from 60–70 S in cesium chloride solutions (the upper values in each case corresponded to the most dilute solutions examined).   The main peak broadened noticeably as the concentration was lowered (Fig. 2).

The small peak or hump before the main peak varied in size and was assumed to indicate a small proportion of molecules ruptured in the turbulent solution at the moment of shocking. It rarely accounted for more than 10 per cent of the DNA and was frequently absent altogether. The long high molecular weight "tail" to the curve occasionally accounted for 25 per cent of the DNA, and, since its sedimentation coefficient was in some cases higher than that of the phage ghosts,[9] it was assumed that this DNA was associated and possibly crosslinked with protein. Support for this explanation was obtained from flow birefringence experiments indicating the presence of bonds which persisted despite the presence of strong salt in the solution, and which broke and reformed reversibly when the solution was briefly sheared in the apparatus. Further confirmation of the association of DNA and protein was afforded by the observation that from 25 per cent to 40 per cent of the DNA floated above the cesium chloride in the preparative banding experiments. Some of this DNA could be released by treatment with sodium dodecyl sulfate. The cause for the protein association is undetermined, but it is unlikely to be due to incomplete shocking since a plating assay revealed only 0.6 per cent survivors. Because only part of the DNA was free of protein the shockates could not be used to characterize the DNA of the phage, but they served to delimit the conditions under which the molecules were stable.

A series of experiments showed that when the titration of the solution was performed by a machine, a mean shear rate up to 25,000 sec.$^{-1}$ could be applied without any change in the sedimentation diagram. However, when the solution was injected by hand, even a shear rate as low as 16,000 sec.$^{-1}$ caused extensive breakdown. The integrity of the molecules could be safely preserved by injecting the solution slowly (1 ml/minute) by hand using a No. 22 needle (the largest the cell orifice will accommodate).

The difference between machine- and hand-pipetting might be explained by the plunger moving in a series of small jerks in the hand; however, a further feature of the sedimentation diagrams required explanation. If the flow in the needle is streamline, then the liquid through the center of the needle should suffer no shear gradient, and hence the sheared solution should show a mixture of degraded and undegraded molecules. In none of the five solutions examined was a bimodal curve detected. This could be understood if the rupture of the molecules occurred also in the turbulence at either end of the needle, where efficient mixing takes place and where high local shears could be present. Confirmation of this idea was obtained from another experiment in which half the contents of a cell was squirted through a fine needle; when this half of the solution was injected rapidly into the remainder of the solution in the cell, no undegraded molecules were detected, but when the sheared solution was added slowly to the unsheared half, a bimodal distribution with maxima at 29 and 53 S (in sodium chloride) was clearly resolved.

With the application of shears of the order of 20,000–40,000 sec.$^{-1}$ the DNA sedimented as a fairly symmetrical peak with a maximum about 30 S. However, by shearing more rapidly through a syringe needle the maximum could be dropped to 21 S; by 15 seconds' treatment in an Osterizer the maximum was dropped to 17 S.

No increase in the absorption of the DNA solutions at 260 mμ was detected accompanying the shear degradation.

Since the destructive shear in the case of turbulence cannot be evaluated, it is difficult accurately to define the conditions which must be avoided to maintain the DNA intact. However, it is likely that any violent agitation of the solution, filtration, pipetting, even chromatography through a fine adsorbent could cause a rupture of the DNA chain.

In an attempt to demonstrate the DNA degradation which Thomas observed in deproteinization experiments,[10] a T_2 shockate was examined before and after gently shaking thirty times in a half-filled test-tube (Fig. 3). The presence of an immis-

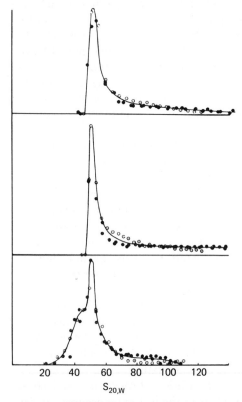

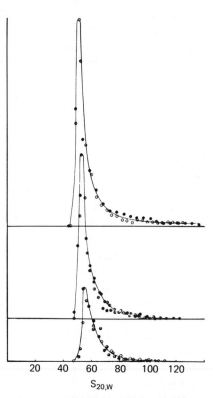

$S_{20,W}$ $S_{20,W}$

Fig. 3.—Effect of shaking T_2 shockate in 0.2 M pH 8 phosphate buffer (DNA concentration 0.001%): (top) untreated; (middle) shaken 30 times in test-tube; (bottom) shaken 30 times in test-tube in the presence of an equal volume of chloroform.

Fig. 4.—Distribution of sedimentation co-efficients in T_2 DNA released by treatment with pH 11 buffer. DNA concentrations: (top) 0.001%; (middle) 0.0007%; (bottom) 0.0004%.

cible liquid appears to be necessary for chain rupture to occur under these mild conditions of shaking.

Characterization of phage DNA: Since some of the DNA in the shockate appeared to be interacting with protein, methods were sought to prevent this interaction. Two methods are obvious: (*a*) the addition of excess competing molecules, or (*b*) raising the pH of the solution so that protein and DNA are similarly charged and mutually repel. Both methods were used and gave essentially similar diagrams, but the latter technique was the simpler, particularly when, on the advice of Dr. E. Freese, the osmotic shock treatment was omitted. The centrifuge cell was filled with borate or phosphate buffers of pH 11.0, and the requisite quantity of intact phage was added from an Agla micrometer syringe. The DNA concentration was kept below 0.002 per cent to lessen the likelihood of concentration-dependent interactions. The solution was mixed by gentle inversion and allowed to stand for 10 minutes at room temperature to allow the phage to burst. The characteristic sedimentation patterns obtained in this manner are shown in Figure 4. The absence of the hump on the low S_{20} side confirmed that the presence of low S_{20} material in the shockates was adventitious.

The high S "tail" on the diagrams, which was much smaller than in the case of the shockates, demonstrated a lower tendency to associate with protein. Sedimentation equilibrium experiments showed that more than 80 per cent of the DNA could be banded at a specific gravity of 1.7; so it may be concluded that these sedimentation diagrams are representative of the DNA in phage.

In a few examples there has been a suggestion of a hump on the high S_{20} side at about 64 S (in sodium chloride). However, since minor irregularities are constantly present in the densitometer traces, it cannot be claimed that there is any reliable indication of heterogeneity in the DNA beyond an undoubted skewness in the peak—and that could be accounted for by a hypersharpening of the boundary.

The DNA released by high pH appeared to be at least as sensitive to shear as in the shockates.

Discussion.—Throughout this paper it has been assumed that the lowering of the sedimentation coefficient corresponds to a lowering of the molecular weight of the DNA. This assumption is reasonable since: (*a*) it has been shown that for homologous DNA samples the value of $S_{20,w}$ decreases with molecular weight,[6, 7] and (*b*) from Thomas's experiments, the treatment which destroys the large piece of DNA[10] also produces material with a lowered sedimentation coefficient. Moreover, it is difficult to imagine that the decrease in $S_{20,w}$ could come about as a result of a change of shape since the shearing of the molecules is accompanied by a decrease in the specific viscosity.

A possibility to be considered is that the changes brought about by shear do not represent a scission of the DNA double helix, but a dissociation of aggregates. If this is the case, the aggregates are probably specific, since they are present in DNA liberated by high pH or shocking, and they do not reform in the solutions after shearing. However, the possibility seems unlikely, because the sedimentation coefficients of the products drop as the violence of the shearing is increased, and there is no evidence for a stable sub-unit. It must be concluded that DNA can be degraded by hydrodynamic shear.

The relative fragility of the DNA demonstrated in these experiments is quite understandable on theoretical grounds. Frenkel[15] in 1944 deduced that long polymers might be degraded by shears of the order of a few thousand reciprocal seconds, and Bestul and Belcher,[16] for example, demonstrated the rupture of synthetic polymers of 500,000 molecular weight at shears of 60,000 sec.$^{-1}$. Since it is only too easy to apply shears (calculated for streamline flow) of 20–30,000 sec.$^{-1}$ in volumetric pipettes or hypodermic syringes, rupture of long-chain molecules is to be expected. Although DNA is a double-strand molecule, it is not easy to predict whether or not the molecule could resist a lengthwise stress twice as great as a single polynucleotide strand. Also it is not possible to calculate accurately the stress applied by the viscous forces to a molecule traversing the streamlines. For example, it is dubious if

Stokes's equation can be applied since the solvent molecules and hydrated ions are by no means small compared with the diameter of the DNA helix. However, Levinthal,[17] employing reasonable values for the bond strengths, calculated the critical shear rates to be about 10,000 sec.$^{-1}$. The observed threshold of about 25,000 sec.$^{-1}$ is in reasonable accord, if it can be assumed that the rupture occurs in streamline flow. The degradation of DNA by shear has been briefly mentioned previously by Goldstein and Reichmann.[18]

The theoretical calculations show that the rupturing stress applied by the viscous forces is maximal in the center of the chain and increases as the square of the length of the molecules. The molecules should thus break roughly in half when the critical shear is applied, and a very long molecule should be halved repeatedly until a fairly uniform population of chains stable to the shear gradient is obtained.

No demonstration of heterogeneity in the undegraded phage DNA has been achieved, but the sharp skew peak may indicate a complexity which could be resolved at lower concentrations. Unfortunately the optical limitations at present preclude the employment of solutions much below 0.0005 per cent concentration for accurate analyses. It is known that self-sharpening effects persist in normal DNA preparations down to 0.01 per cent, and it may not be unexpected that in the presence of much longer chains the critical concentration should be much lower. However, it is also possible that the DNA population is essentially homogeneous (apart from aggregates). The Doty formula,[6] if it were applicable, would indicate a molecular weight about 1×10^8 (roughly the DNA content of a phage particle) for a 56 S molecule. If the DNA is normally present as one macromolecule, the autoradiographic experiments must be explained by degradation of this molecule, presumably during the manipulation of the emulsion. This appears unlikely since the application of controlled shear gradients did not suffice to produce a bimodal distribution. However, it is obvious that for an answer to the present difficulties some correlation of molecular weight and sedimentation coefficient is needed for the phage DNA.

Nothing in the experiments reported in this paper has shown that the DNA sedimenting in these sharp peaks is protein-free, although the fact that the material in some experiments has been banded at a specific gravity of 1.7 before study implies that any protein present must be in small proportion. Thus it could be postulated that protein bridges exist along the DNA macromolecule uniting DNA sub-units. It can also be postulated that the shear scission occurs at these or other weak points. If they are present, such bridges must be resistant to chymotrypsin. However, until there is some definite evidence, it seems most reasonable to assume that the hydrodynamic shear is transecting the DNA double helix.

It should perhaps be pointed out that the phage particle itself would suffer no embarrassment from the shear sensitivity of its DNA during the process of injecting these molecules into the bacterium, since the diameter of its tail is too small for the DNA to lie across the streamlines, even if there were a solvent present to impart a viscous drag.

The observed lability of the phage DNA poses the problem of how much the DNAs from other tissues might be degraded in the course of preparation. The relative uniformity of observed molecular weights in sperm and thymus DNA preparations, for example, is no proof that these molecules have not been degraded, since a fairly uniform distribution would be expected from the shearing action. The vigorous stirring, blending, or filtration to which most preparations of DNA are subjected would certainly degrade phage DNA. If DNA *in vivo* is much longer than is usually recognized, some of the present problems of correlating DNA structure and function would be simplified. Beiser, Pahl, Rosenkranz, and Bendich[19] found that specific pneumococcal transforming activities were chromatographically complex. Assume, for simplicity, that each activity relates to the presence of one enzyme. If the DNA molecule is very long and is ruptured during isolation or deproteinization to give chains of 8×10^6, average molecular weight (of which 1×10^6 would be needed to code for a protein of 50,000 molecular weight[20]), the activity might be found on molecules of variable length and composition. Little loss of activity need be envisaged since only one break in eight would interrupt the critical sequence; if the sequence lies close to the end of the gross macromolecule, the sequence would rarely be ruptured.

Furthermore, at present most speculations on chromosome structure assume that the DNA is of relatively uniform molecular weight (about $4-8 \times 10^6$). The evidence presented in this paper shows that the experimental grounds for this assumption are questionable. It is obvious that some of the earlier physical work on DNA must be repeated with suitable precautions to avoid the scission of any very long molecule which may be present.

Summary.—(1) Turbulence and high shear gradients have been shown to cause a decrease in the sedimentation coefficients of T_2 and T_4 bacteriophage deoxyribonucleic acid molecules in solution. (2) In sedimentation velocity experiments, carefully treated DNA solutions have shown a sharp peak, skew on the high $S_{20,w}$ side, with a maximum about 60 S.

The author thanks Dr. H. Van Vunakis and Dr. J. Tomizawa, who provided phage samples, and Dr. I. Tessman, for the plating assay of the phage. The author is also indebted to Dr. C. Levinthal for his interest and advice in the course of this work, and Mrs. K. Y. Ho for technical assistance.

* This research was aided by a grant, E-1469, from the National Institute of Allergy and Infectious Diseases, National Institutes of Health, U. S. Public Health Service.

† Throughout this paper the word "degradation" is used to denote a reduction in molecular weight—presumably by transverse scission of the DNA double helix.

[1] Levinthal, C., these Proceedings, **42**, 394 (1956).

[2] Fleischman, J. B., Ph.D. thesis, Harvard University (1959).

[3] Meselson, M., F. W. Stahl, and J. Vinograd, these Proceedings, **43**, 581 (1957).

[4] Shooter, K. V., and J. A. V. Butler, *Trans. Farad. Soc.*, **52**, 734 (1956).

[5] Schumaker, V. N., and H. K. Schachman, *Biochim. Biophys. Acta*, **23**, 628 (1957).

[6] Doty, P., B. B. McGill, and S. A. Rice, these Proceedings, **44**, 432 (1958).

[7] Butler, J. A. V., D. J. R. Laurence, A. B. Robins, and K. V. Shooter, *Proc. Roy. Soc.* (London), (A) **250**, 1 (1959).

[8] Hermans, J., and J. J. Hermans, *J. Phys. Chem.*, **63**, 170 (1959).

[9] Thomas, C. A., and J. H. Knight, these Proceedings, **45**, 332 (1959).

[10] Thomas, C. A., *J. Gen. Physiol.*, **42**, 503 (1959).

[11] Davison, P. F., unpublished experiments.

[12] Anderson, T. F., *J. Appl. Phys.*, **21**, 70 (1950).

[13] Meselson, M., and F. W. Stahl, these Proceedings, **44**, 671 (1958).

[14] Kroepelin, M., *Koll. Z.*, **47**, 294 (1929).

[15] Frenkel, J., *Acta Physicochim. U.R.S.S.*, **14**, 51 (1944).

[16] Bestul, A. B., and H. V. Belcher, *J. Appl. Phys.*, **24,** 1011 (1953).

[17] Levinthal, C., private communication.

[18] Goldstein, M., and M. E. Reichmann, *J. Am. Chem. Soc.*, **76,** 3337 (1954).

[19] Beiser, S. M., M. B. Pahl, H. S. Rosenkranz, and A. Bendich, *Biochim. Biophys. Acta*, **34,** 497 (1959).

[20] Morowitz, H. J., and R. C. Cleverdon, *Biochim. Biophys. Acta*, **34,** 578 (1959).

From the *Proceedings of the National Academy of Sciences*, Vol. 47, No. 8, 1123–1129, 1961. Reproduced with permission of the authors.

THE STRUCTURAL UNITY OF THE DNA OF T2 BACTERIOPHAGE*

By P. F. Davison, D. Freifelder,† R. Hede, and C. Levinthal

DEPARTMENT OF BIOLOGY, MASSACHUSETTS INSTITUTE OF TECHNOLOGY

Communicated by Francis O. Schmitt, June 15, 1961

The molecular weight of the DNA extracted from the phages T2 and T4 has been the subject of several investigations in recent years. Autoradiographic studies (Levinthal[1] and Levinthal and Thomas[2]) of P[32]-labeled DNA showed that the material after embedding in photographic emulsions consisted on the average of one large piece of DNA per phage particle plus smaller material. The large piece had a molecular weight of about 45 million and in these experiments comprised about 40 per cent of the total DNA of the cells. The smaller pieces were later shown by the same method (Thomas[3]) to have a molecular weight of about 12 million.

On the other hand, the molecular weight as measured by equilibrium sedimentation in cesium chloride (Meselson, Stahl, and Vinograd[4]) showed no evidence for any large piece of DNA, but rather a homogeneous distribution of pieces of size about 14 million. Sedimentation velocity, viscosity, and light-scattering measurements (Fleischman[5]) failed to detect any evidence of a large piece of DNA but showed that the material had a molecular weight of about 12 million and was homogeneous in size. The experimental demonstration (Davison[6]) that the DNA after its extraction from the phage is extremely susceptible to shear degradation, in addition to the theoretical analysis of the shear degradation process (Levinthal,[7] Levinthal and Davison[8]), resolved a part of the apparent difference. The mean molecular weight of a DNA preparation can be reduced by the shear forces resulting from velocity gradients in a solution and can, if the shearing is done sufficiently vigorously and in a uniform manner, result in an apparently homogeneous distribution. The fact that sufficiently large shear forces were produced in the hypodermic needle used to load the ultracentrifuge cell probably accounted for the reduced molecular weight observed in these experiments. Thus the uniformly small material observed in the centrifuge, viscosity and light-scattering experiments, could be explained as an artifact of the preparation of the material and the loading of the centrifuge cells. However, the original problem was now inverted, since no evidence of smaller pieces could be found in centrifuge runs on the undegraded material. Davison (unpublished results) showed by centrifuging sheared and unsheared DNA together that a small percentage of material of lower molecular weight could readily have been detected (assuming it had an unfolded structure). Furthermore, chromatographic analysis of T2 DNA extracted without shear degradation showed that the material was homogeneous with respect to its chromatographic behavior (Hershey and Burgi[9]). It seemed possible that the large piece originally observed in the autoradiographic studies was itself a degradation product of an even larger unit, which could be the complete DNA content of the virus. On the other hand, it was also possible that the DNA of the virus consisted of a small number, perhaps two or three, of large pieces, some of which were degraded by putting the material in the photographic emulsion.

Unfortunately the physical-chemical methods such as light-scattering or centrifugation are inadequate for measuring the molecular weights of such very large pieces of DNA, since there is no satisfactory theory to describe their behavior in solution. We have repeated the autoradiographic measurements under conditions which reduce the probability of shear degradation of the material and have also altered the method of purifying the phage so as to reduce possible damage to the DNA by irradiation from the P^{32} β-particles. The sedimentation properties of the material introduced onto the photographic emulsion were tested by zone sedimentation, using a D_2O-H_2O gradient to stabilize the boundaries. In addition, the same material was run on a chromatographic column prepared after the method of Mandell and Hershey.[10] As will be seen below, the results indicate that the phage DNA, extracted either by osmotic shock or by the phenol method, appears to consist of one DNA molecule per phage.

Materials and Methods.—(a) *Growth and purification of the phage:* *Escherichia coli* strain B was grown in a Tris-glucose synthetic medium, containing 4 μg of phosphorus as inorganic phosphate per ml, with sufficient P^{32} to bring the specific activity to approximately 45 microcuries per microgram phosphorus. Bacteria were allowed to multiply by a factor of 8 to a density of 2×10^8 cells per ml, prior to infection with phage T2r+ at a multiplicity of about one phage per bacterium. After 2 hrs the cells were lysed by chloroform and the lysate was filtered through Celite filter aid (Johns-Manville Products). Five milliliters of lysate was diluted fourfold with 0.01 M phosphate buffer (pH 7.4) and put onto a DEAE cellulose column (Peterson and Sober[11]), 1 cm in diameter and 2 cm high. The column was washed with 20 ml 0.01 M phosphate and then successively by the same buffer containing 0.1 M NaCl and 0.2 NaCl. Under these conditions, all of the viable phage remained attached to the column. The phage was eluted from the column with the same buffer containing 0.3 M NaCl and the bulk of the active phage was collected in about 2 ml of eluate. Labeled phage as prepared by this method was demonstrated to be free from P^{32} contaminants by the fact that over 95% of the radioactivity absorbed in 3 minutes to several volumes 5×10^8 per ml sensitive bacteria and less than 3% absorbed to T2 resistant cells. In addition, when zone sedimentation was performed with the intact phage, using the methods described in Levinthal and Davison,[8] essentially all of the radioactivity appeared in a peak sedimenting with the virus (Fig. 1).

The specific activity of the phage was measured from the death rate due to the decay of incorporated P^{32}, assuming 10 disintegrations produced one lethal hit (Stent and Fuerst[12]). This measurement indicated that each phage contained approximately 60 P^{32} atoms.

(b) *Extraction of DNA:* The DNA was liberated from the virus either by osmotic shock, in which 0.5 ml of phage in 4 M NaCl solution was rapidly diluted with 9.5 ml of distilled water, or by shaking a suspension of the phage, with added nonlabeled carrier phage, with water-saturated phenol (Mandell and Hershey[10]). The high concentration of DNA contributed by the carrier phage prevented shear degradation during phenol treatment.

(c) *Chromatography of the DNA:* The properties of DNA obtained by the phenol method were checked using a chromatographic column, according to the methods of Mandell and Hershey.[10]

(d) *Zone sedimentation:* The zone sedimentation of the DNA and the phage was carried out as described by Levinthal and Davison.[8]

(e) *Autoradiography:* In order to reduce the possible shear effect when mixing the DNA solution with the photographic emulsion, the liquid emulsion mixture previously used was avoided in these experiments. The DNA solutions were diluted with precautions to avoid shear degradation, and microdrops of a phage or DNA suspension were placed on an Ilford G-5 prepared photographic plate of thickness 100 microns, using an Agla microsyringe which could deliver accurately volumes of 1, 2, or 3 microliters. The drops were allowed to dry on the photographic plate and this was then covered by a layer of liquid G-5 emulsion, resulting in the material being embedded in a sandwich between two layers of emulsion. The photographic slides were then dried, stored, and developed after various lengths of time, according to the methods discussed by Levinthal and Thomas.[2] In all cases the intact phage, osmotically shocked phage, DNA extracted by the phenol method, and various fractions isolated from the main peak eluted from the Mandell-Hershey column were placed on the same photographic plate. In this way any variation in latent image

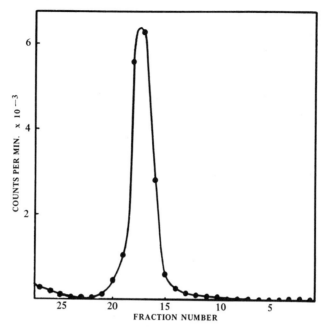

FIG. 1.—Distribution of P[32]-labeled phage in centrifuge tube after zone centrifugation 30 minutes at 12,000 rpm in Spinco model L.   Direction of sedimentation: left to right.

fading or in the development process would affect equally the intact phage and the various DNA suspensions. The plates were all counted by four observers, who for the most part were not aware of either the exposure time or the nature of the sample they were counting.

Results.—(a) *Chromatography:* When the DNA extracted from labeled phage mixed with excess unlabeled phage was chromatographed, the radioactivity was found to follow exactly the distribution in the eluate of the material absorbing at 260 mµ (Fig. 2).

(b) *Zone centrifugation:* When the P[32]-DNA released by shocking the freshly prepared heavily labeled phage was centrifuged, the distribution of radioactivity

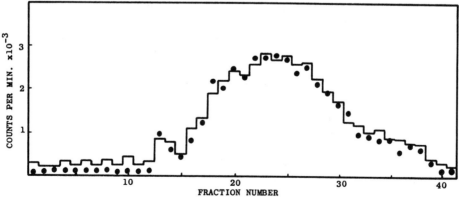

FIG. 2.—Optical density at 260 mµ (solid line) and radioactivity (indicated by dots) in fractions of a mixture of labeled and carrier phage DNA prepared by phenol method and eluted from Celite-methylated albumen column.

ARTICLE 4 / Davison, Freifelder, Hede, and Levinthal   49 was essentially the same as that of lightly labeled DNA (about one P^{32} per phage particle). This result indicates that the heavily labeled DNA shows no obvious evidence of damage or degradation resulting from growth in the highly radioactive medium.

When more phage was shocked 14 days after preparation and the DNA was centrifuged, the change of distribution of DNA expected from the chain scission by P^{32} decay was observed (Fig. 3).

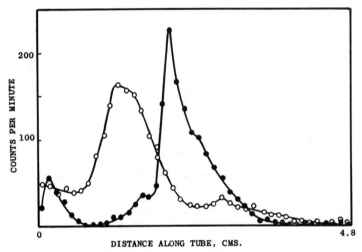

FIG. 3.—Distribution of P^{32}-labeled DNA in centrifuge tube after zone centrifugation 1 hour at 35,000 rpm on Spinco model L: (*a*) solid dots indicate DNA released from phage and centrifuged within 6 hours of P^{32} incorporation: (*b*) hollow circles refer to DNA released 17 days later from same phage as in (*a*).

(c) *Autoradiography results:* When the emulsion is examined in a microscope, stars can be seen resulting from several P^{32} tracks emanating from a point in the emulsion. The number of rays in such stars gives a measure of the amount of P^{32} at the point of origin and the number of stars per drop gives a measure of the number of pieces of P^{32}-containing material per volume of solution which was put on the emulsion.

(i) *Star number:* The intact phage and the osmotically shocked material were put on the photographic plate at a concentration which was the same for both and designed to give about 40 viable phage units per drop. For the count of the total number of star-forming units per drop the exposure was allowed to proceed to the point where the mean star size was 20 to 30. These stars were too large for the count of the number of rays per star, but they eliminated any ambiguity in the detection of the star-forming units. The number of star-forming units (that is, the number of large pieces of DNA) in the osmotically shocked material was approximately equal to the number of phage particles from which the DNA was extracted (Table 1). For the material which had been eluted from the column the star number could only be compared with the total radioactivity. The results indicated that there was the same number of star-forming units per unit of radioactivity put on the slide for the material which had been deproteinized by phenol and eluted from the column as there was for the intact phage, again indicating that for this material there was one large piece of DNA per phage particle.

TABLE 1

Counts of Autoradiographs of P^{32}-labeled Phage and DNA

Sample	A	B	C	D	E	F
(a) $f = 0.24$ No. of stars counted	44				14	
Mean rays per star	10.0				10.1	
(b) $f = 0.34$ No. of stars counted	61	70	75	29	71	31
Mean rays per star	13.3	12.2	13.9	12.6	12.7	13.0
(c) $f = 0.78$ No. of stars per drop*	39	43	29	162	204	96
Each count refers to a single drop	40	51	41	153	189	108
	44	49	37			106
(d) Radioactive counts per unit vol. of material plated	4.5	5.9	3.4	23.9	30.0	11.6
(e) Ratio (c)/(d)	9.2	8.1	10.6	6.6	6.6	8.9

Samples: A, intact phage.
      B, phage shocked with added carrier.
      C, phage shocked without cold carrier.
      D, E, F, samples from the leading, middle, and trailing fractions of the DNA eluted from the chromatographic column.
f indicates the fraction of the P^{32} disintegrations completed at the time of development of the plates.
  * In the course of counting the stars per drop, a rough count of the rays per star was also made. No evidence of a bimodal size distribution of stars was found.

(ii) *Star size:* The number of rays per star in these emulsions is considerably more difficult to determine with precision than the number of stars per drop. Use of the sandwich procedure consisting of a prepared slide plus a layer of solidified liquid emulsion results in a somewhat higher background than a complete layer of liquid emulsion. When the exposure was allowed to proceed to the point where the star size was approximately 10 to 15 per star, the counts of the same star made by various observers agreed to within one or two rays. This variation was due to occasional uncertainties as to whether a given track was produced by one or two β-particles, and there were also a few cases of uncertainty due to straight tracks passing near the center of the stars. These difficulties were approximately the same for all the preparations examined and contributed an inherent uncertainty in the counting of approximately 10 per cent. To within this uncertainty the figures of Table 1 indicate that the stars produced by the DNA after the osmotic shock or the DNA deproteinized by phenol extraction and separated on the column, all had approximately the same size as the stars produced by the intact phage. Thus we can conclude that the material which is in solution after either of these methods of preparation had about the same P^{32} content, and therefore the same DNA content as the phage themselves.

Discussion.—The autoradiographic experiments indicate that all the DNA in the T2 bacteriophage is contained in one physical entity. Rubenstein, Thomas, and Hershey[13] in an accompanying paper present evidence leading to the same conclusion. Quantitative considerations of the susceptibility of the DNA-containing structure to hydrodynamic shear (Levinthal and Davison[8]) make it highly likely that this structure is a single long filament. Freese[14] and Kellenberger[15] have proposed that chromosomes are built up from DNA molecules joined end-to-end by hypothetical "linkers." If any such linkers were present in the T2 DNA they would be expected to be points of weakness since they would form a single chain bridge between the twin helices of the DNA molecules. Hershey and Burgi[9] found no evidence for any preferential breaking point in the T2 DNA molecules under shearing, nor is there presently any other good evidence for the presence of linkers.[16] It may therefore be reasonably assumed that the DNA is present as one uninterrupted, double helix. From previous analyses (see Rubenstein *et al.*[13] and Cummings and Kozloff[17]) the molecular weight is probably between 90×10^6 and 150×10^6. The DNA from phage grown in media with phosphorus of very high specific activity as well as phosphorus of very low specific activity was similar chromatographically to DNA from phage grown in unlabeled media, and no difference in the zone sedimentation behavior of any of the preparations was found, except after sufficient time had elapsed to cause double chain scission by the P^{32} decay. This suggests that in the former autoradiographic experiments[1, 2] the large pieces found, which were only about 40 per cent of the total DNA content of the phage, were probably produced by degradation of the whole molecule either through shear in placing the material in the liquid emulsion or by the radiation damage suffered by the phage when they were pelleted in the centrifuge for purification. In view of the suggested mechanism for the production of small pieces by shear degradation one would have expected that a significant fraction of the small pieces would be of a size of approximately 20 per cent of the intact phage molecule. Such pieces were not found (Levinthal and Thomas[2]) and in fact the small pieces were identified as having a molecular weight of about 12 million (Thomas[3]). This discrepancy may have resulted from an error in the counting of the stars or it may have resulted from some special characteristic of the shear degradation in the nonhomogeneous, highly viscous, liquid photographic emulsion in which the phage was suspended or to a combination of the shear degradation and radiation damage. In any case, we can conclude that the large piece previously observed was a partial breakdown product of the intact molecule while the smaller pieces observed by Meselson, Stahl, and Vinograd[4] and by Fleischman[5] were material more completely degraded under more extreme conditions of shearing.

The experimental results on the transfer of P^{32} label to intact phage after labeled phage particles had infected single cells do not appear to be affected by these findings as to the nature of the phage chromosome, nor is the interpretation that a significant fraction of the label is transferred in a large semiconservative fragment. On the other hand, the interpretations of the mechanism of recombination[1] which were based on the assumption that the 40 per cent piece was the intact chromosome are clearly in error, and the experiments do not represent any evidence against the idea that when recombination occurs it is accompanied by physical breakage of the chromosome (Meselson and Weigle[18]).

Summary.—*E. coli* heavily labeled with P^{32} was infected with T2 bacteriophage, and the progeny phage released on the lysis of the cells were purified by chromatography. The zone centrifugation and chromatographic behavior of the P^{32}-DNA obtained from this phage preparation did not differ from that of unlabeled T2 DNA, demonstrating that growth in radioactive solutions did not detectably modify the DNA. Comparison of the autoradiographs obtained from the phage or the DNA released from the phage showed that all the DNA in each particle was present in one structural entity which might be identified as a DNA molecule.

Part of this work was performed in the laboratory of Professor Francis O. Schmitt for whose helpful cooperation the authors are grateful. They are also indebted to Mr. D. Baltimore, who assisted in counting some of the autoradiographs, and Mrs. P. Kahn, who developed the chromatographic method of purifying the phage.

* This research was aided by Grants E-1469 and E-2028 from the National Institute of Allergy and Infectious Diseases.

† Supported by Post-doctoral fellowship No. GF-7617-C2 from the Division of General Medical Sciences of the U.S. Public Health Service.

[1] Levinthal, C., these Proceedings, **42**, 394 (1956).

[2] Levinthal, C., and C. A. Thomas, *Biochim. et Biophys. Acta*, **23**, 453 (1957).

[3] Thomas, C. A., *J. Gen. Physiol.*, **42**, 503 (1959).

[4] Meselson, M., F. W. Stahl, and J. Vinograd, these Proceedings, **43**, 581 (1957).

[5] Fleischman, J. B., *J. Molec. Biol.*, **2**, 226 (1960).

[6] Davison, P. F., these Proceedings, **45**, 1560 (1959).

[7] Levinthal, C., in *Conference on Genetics*, (Princeton: Josiah Macy, Jr., Foundation, 1960), p. 41.

[8] Levinthal, C., and P. F. Davison (in preparation).

[9] Hershey, A. D., and E. Burgi, *J. Molec. Biol.*, **2**, 143 (1960).

[10] Mandell, J. D., and A. D. Hershey, *Analytical Biochem.*, **1**, 66 (1960).

[11] Peterson, E. A., and H. A. Sober, *J. Am. Chem. Soc.*, **78**, 751 (1956).

[12] Stent, G. S., and C. R. Fuerst, *Advances in Biol. and Med. Phys.*, **7**, 2 (1960).

[13] Rubenstein, I., C. A. Thomas, and A. D. Hershey, these Proceedings, **47**, 1113 (1961).

[14] Freese, E., *Cold Spring Harbor Symposia Quant. Biol.*, **23**, 13 (1958).

[15] Kellenberger, E., in *Microbial Genetics* (Cambridge University Press, 1960), p. 39.

[16] For economy of hypothesis we exclude from consideration the model of double linker strands bridging DNA segments. DNA itself admirably fulfills the requirements of such linkers (such linkers would have to replicate, for instance) and it seems unnecessary to postulate an added complication without experimental justification.

[17] Cummings, D. J., and L. M. Kozloff, *Biochim. et Biophys. Acta*, **44**, 445 (1960).

[18] Meselson, M., and J. Weigle, these Proceedings, **47**, 857 (1961).

ARTICLE 5

From *Journal of Molecular Biology*, Vol. 4, 407–409, 1962. Reproduced with permission of the author and publisher.

A Minimum Estimate for the Length of the DNA of *Escherichia Coli* obtained by Autoradiography

JOHN CAIRNS

Department of Microbiology, The Australian National University, Canberra, Australia

(*Received 23 January 1961*)

E. coli labelled with [³H]thymine was lysed in EDTA and 2 M-sucrose by dialysis against Duponol. After 20 hours the DNA of some bacteria had partly untangled and was trapped on the wall of the dialysis membrane when the dialysis chamber was drained. Autoradiography showed this DNA to be several hundred microns long, indicating a molecular weight of 10^9 or more. Short fragments, which would correspond to the usual molecular weight of extracted bacterial DNA, were not seen.

1. Introduction

The nucleus of *E. coli* contains about 4×10^9 daltons of DNA (Fuerst & Stent, 1956). All this DNA replicates semiconservatively in an orderly fashion, each round of replication being completed before the next starts (Meselson & Stahl, 1958). It is dispensed semiconservatively among the descendant nuclei, only occasionally being subject to further dispersal (Forro & Wertheimer, 1960). Genetically, the bacterial nucleus is haploid (Witkin, 1951) and the genome comprises a single linkage group which can be transferred entire as a linear structure at the time of mating (Jacob & Wollman, 1958). In short, each bacterial nucleus behaves as if it contained a single molecule of DNA, by inference of some 4×10^9 molecular weight.

Against this notion of a single huge DNA molecule within each nucleus is the fact that extracted bacterial DNA has always proved to be of uniformly low molecular weight. Such uniformity of molecular weight, it has been argued, could hardly be the product of random breakage at the time of extraction. Pre-existing breakage points, such as protein links, must be built into the genome at uniformly spaced points.

The persuasiveness of this argument is now seen to be illusory. Large molecules of DNA, subject to shear breakage, break near the centres; the half molecules then repeat this process, and so on until some size is reached that will survive the shear being applied (Hershey & Burgi, 1960; Levinthal & Davison, 1961). It is not at all improbable that breakage of bacterial DNA should produce fairly uniform fragments. Indeed, judging from the fragility of T2 bacteriophage DNA (Davison, 1959), it is unlikely that this hypothetical molecule, which would be 20 to 40 times as large and some 2 mm long, could survive the stresses involved in many of the procedures used for determining molecular weight, let alone the violent explosion that must normally occur at the moment of bacterial lysis.

An attempt has therefore been made to determine the length of bacterial DNA by autoradiography (Cairns, 1961), using an extraction procedure, known to release DNA (Meselson & Stahl, 1958) but applied in this case in such a way that all explosive forces, turbulence and shear were minimal.

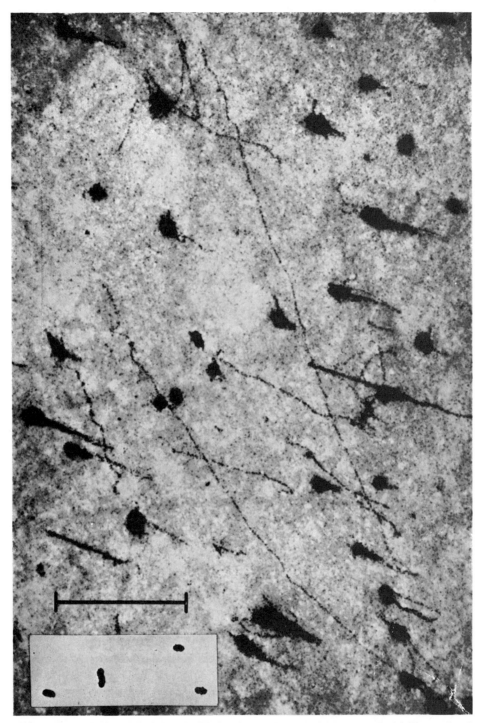

PLATE I. *E. coli* DNA, labelled with [³H]thymine (11·2 c/m-mole), extracted with Duponol and EDTA and collected on a Millipore filter at a concentration of 0·01 μg/ml. The inset photograph shows, at the same magnification, an autoradiograph of unlysed bacteria. The autoradiographic exposure was 70 days. The scale shows 100 μ.

2. Materials and Methods

Bacterium. The thymineless strain, *E. coli* B3 (Brenner).

Medium. The glucose–ammonium medium described by Hershey (1955), supplemented with 2 μg/ml. thymine.

Preparation of labelled bacteria. E. coli B3 was grown with aeration to $1\cdot5 \times 10^8$/ml., centrifuged and resuspended in an equal volume of medium containing 2 μg/ml. [³H]thymine ($11\cdot2$ c/m-mole). After 2 hr the viable count had risen to $3\cdot3 \times 10^8$/ml. This was rather less than the increase of control bacteria in cold thymine, suggesting that the presence of tritium had caused some inactivation.

Lysis of bacteria. Labelled bacteria were suspended at a concentration of 5×10^5/ml. in 2 M-sucrose, $0\cdot02$ M-NaCl, $0\cdot01$ M-EDTA pH 8, with 6 μg/ml. phenol-extracted T2 DNA. This mixture was placed in a cylindrical chamber 20 mm diameter and $2\cdot5$ mm deep, faced on one side with glass and on the other with a VM Millipore filter (50 mμ pore-size). The enclosed bacteria were then dialysed at 42°C against 1% sodium lauryl sulphate (Duponol) in sucrose–NaCl–EDTA for 45 min. The Duponol was removed by dialysis, the salt concentration raised to 4 M-NaCl for 1 hr and then lowered to $0\cdot04$ M, EDTA being present throughout. After a further 16 hr, the chamber was stood on its side, punctured in two places, and drained slowly on to filter paper. The Millipore filter was allowed to dry *in situ*.

The reasoning behind this sequence was as follows. Duponol and EDTA liberate DNA from *E. coli* in such a form that much of any attendant protein (and this includes the hypothetical linkers) can be separated by high salt concentration when the mixture is banded in CsCl (Meselson & Stahl, 1958). Here no removal of any protein freed by high salt concentration can be expected, but the considerable excess of carrier DNA and the very low concentration of bacterial DNA ($0\cdot01$ μg/ml.) should tend to prevent bacterial DNA and protein from re-uniting when the salt concentration, after being raised, is lowered again. To avoid explosion of the bacteria at the moment of lysis, lysis was carried out in 2 M-sucrose.

Autoradiography. The Millipore filters were stuck to microscope slides with a cement, overlaid with Kodak autoradiographic stripping film, AR 10, and stored at 4°C over $CaSO_4$ in an atmosphere of CO_2. After exposure the film was developed with Kodak D19b for 20 min at 16°C.

3. Results

The results can be seen in Plate I and may be summarized as follows.

(1) Although lysis has occurred (compare the inset autoradiograph of unlysed bacteria), the DNA of most bacteria has not unravelled enough to display any features of its form. This is not surprising since the only disentangling forces were Brownian movement, convection currents and finally the slow passage of the meniscus across the filter when the dialysis chamber was drained.

(2) The DNA extracted from bacteria by normal methods has a sedimentation coefficient of around 25 s (Marmur, 1961), indicating a molecular weight of 20 to 30×10^6 (Rubinstein, Thomas & Hershey, 1961) and a length of some 10 μ. No such pieces are to be seen in the autoradiograph. If bacterial DNA is truly divided into such lengths by linkers, then the extraction procedure used here has totally failed to dissolve these hypothetical links even though virtually the same procedure invariably breaks them when applied without precautions against shear; this seems highly improbable. If, on the other hand, such lengths are a casual product of turbulence and shear *per se*, it is reasonable that there should have been little fragmentation.

(3) A few pieces of DNA can be seen which have two identifiable ends and can therefore be measured. Their lengths range from 25 μ upwards.

(4) More commonly the DNA has partly untangled to display a long unbroken thread; two such threads, both about 400 μ long, are shown in the Plate. Such threads may cross others, but they show no branches. As the concentration of untangled labelled DNA must be much less than 0·01 μg/ml., they can hardly represent artificial end-to-end aggregation of smaller pieces. Until the existence in DNA of non-nucleic acid links has been demonstrated, it is probably legitimate to think of these threads as molecules. From their length, they must have a molecular weight of around 10^9. As it has not been possible to identify with certainty both ends of any of these long molecules, this must be a minimum estimate of the molecular weight of bacterial DNA.

Note added in proof.

Since this paper was written, Kleinschmidt, Lang & Kahn (1961) have produced beautiful electron micrographs showing that protoplasts of *M. lysodeikticus*, lysed at an air–water interface, release their deoxyribonucleoprotein in the form of a tangled skein which has no visible free ends. Thus two dissimilar procedures suggest that bacterial DNA may exist as a single molecule.

REFERENCES

Cairns, J. (1961). *J. Mol. Biol.* **3**, 756.

Davison, P. F. (1959). *Proc. Nat. Acad. Sci., Wash.* **45**, 1560.

Forro, F. & Wertheimer, S. A. (1960), *Biochim. biophys. Acta*, **40**, 9.

Fuerst, C. R. & Stent, G. S. (1956). *J. Gen. Physiol.* **40**, 73.

Hershey, A. D. (1955). *Virology*, **1**, 108.

Hershey, A. D. & Burgi, E. (1960). *J. Mol. Biol.* **2**, 143.

Jacob, F. & Wollman, E. L. (1958). *Symp. Soc. Exp. Biol.* **12**, 75.

Kleinschmidt, A., Lang, D. & Kahn, R. K. (1961). *Z. Naturf.* **16b**, 730.

Levinthal, C. & Davison, P. F. (1961). *J. Mol. Biol.* **3**, 674.

Marmur, J. (1961). *J. Mol. Biol.* **3**, 208.

Meselson, M. & Stahl, F. W. (1958). *Proc. Nat. Acad. Sci., Wash.* **44**, 671.

Rubinstein, I., Thomas, C. A. & Hershey, A. D. (1961). *Proc. Nat. Acad. Sci., Wash.* **47**, 1113.

Witkin, E. M. (1951). *Cold Spr. Harb. Symp. Quant. Biol.* **16**, 357.

ARTICLE **6**

From *Journal of Molecular Biology*, Vol. 6, 208–213, 1963. Reproduced with permission of the author and publisher.

The Bacterial Chromosome and its Manner of Replication as seen by Autoradiography

JOHN CAIRNS

Department of Microbiology, Australian National University, Canberra, A.C.T., Australia

(*Received 19 November 1962*)

In order to determine the form of replicating DNA, *E. coli* B3 and K12 Hfr were labelled for various periods with [³H]thymidine. Their DNA was then extracted gently and observed by autoradiography. The results and conclusions can be summarized as follows.

(1) The chromosome of *E. coli* consists of a single piece of two-stranded DNA, 700 to 900 μ long.

(2) This DNA duplicates by forming a fork. The new (daughter) limbs of the fork each contain one strand of new material and one strand of old material.

(3) Each chromosome length of DNA is probably duplicated by one fork. Thus, when the bacterial generation time is 30 min, 20 to 30 μ of DNA is duplicated each minute.

(4) Totally unexpected was the finding that the distal ends of the two daughter molecules appear to be joined during the period of replication. The reason for this is obscure. Conceivably the mechanism that, *in vivo*, winds the daughter molecules lies at the point of their union rather than, as commonly supposed, in the fork itself.

(5) The chromosomes of both B3 (F⁻) and K12 (Hfr) appear to exist as a circle which usually breaks during extraction.

1. Introduction

The semiconservative nature of DNA replication, predicted on structural grounds by Watson & Crick (1953), was demonstrated experimentally first for bacterial DNA (Meselson & Stahl, 1958), but how the two strands of the double helix come to separate during replication is not known. There must be formidable complexities to any process which, however feasible energetically (Levinthal & Crane, 1956), at once unwinds one molecule and winds two others. So a study was undertaken of the shape and form of replicating DNA.

Bacteria, despite their complexity, promised to be the most accessible source of replicating DNA. Each bacterium in an exponentially growing culture of *E. coli* makes DNA for more than 80% of the generation time (McFall & Stent, 1959; Schaechter, Bentzon & Maaløe, 1959); if, as seems likely, the bacterial chromosome is a single molecule of DNA, this molecule must be engaged in replication most of the time. A method had already been devised for extracting this DNA with little degradation (Cairns, 1962*a*) and it seemed probable that, with more care, the chromosome could be isolated intact and, caught in the act of replication, its DNA be displayed by autoradiography.

2. Materials and Methods

Bacteria. Since the chromosomes of F⁻ and Hfr bacteria differ in the type of their genetic linkage (Jacob & Wollman, 1958) and in the manner of their duplication (Nagata, 1962), two strains of *E. coli* were used, B3 (F⁻) (Brenner) and K12 3000 thy^- B_1^- (Hfr). Both strains require thymine or thymidine.

Medium. The A medium of Meselson & Weigle (1961) was used. To this was added 3 mg/ml. casein hydrolysate, which had first been largely freed of thymine by steaming with charcoal. In this medium, supplemented with 2 μg/ml. TDR,† both strains have a generation time of 30 min.

Preparation of labelled bacteria for autoradiography. The bacteria were grown with aeration to 10^8/ml., centrifuged and resuspended in an equal volume of medium containing 2 μg/ml. [³H]TDR (9 c/m-mole). In pulse-labelling experiments, incorporation of label was stopped by diluting the bacteria either 50-fold into medium containing 20 μg/ml. TDR or 250-fold into cold 0·15 M-NaCl containing 0·01 M-KCN and 0·002% bovine serum albumin. In long-term experiments, the bacteria were labelled for two generations (1 hr) so that roughly half of the DNA would be fully labelled and half would be a hybrid of labelled and unlabelled strands.

Lysis of bacteria. Only in a few minor respects has the procedure been altered from that already published (Cairns, 1962a). Labelled bacteria are lysed after dilution to a final concentration of about 10^4/ml. Since it was important in certain experiments to be sure that DNA synthesis did not continue beyond the time the bacteria were sampled, 0·01 M-KCN was added to the lysis medium (1·5 M-sucrose, 0·05 M-NaCl, 0·01 M-EDTA). Various types of cold carrier DNA (4·7 μg/ml. calf thymus, *E. coli* or T2 DNA) were used at various times without apparently influencing the results. As before, lysis was obtained by dialysis against 1% Duponol C (Dupont, Wilmington, Delaware, U.S.A.) in lysis medium for 2 hr at 37°C. The Duponol was then removed by dialysis for 18 to 24 hr against repeated changes of 0·05 M-NaCl, 0·005 M-EDTA. As before, the DNA was collected on the dialysis membrane (VM Millipore filter, Millipore Filter Corporation, Bedford, Mass., U.S.A.).

Autoradiography. As before, Kodak AR10 stripping film was used and the exposure was about 2 months.

Thymidine incorporation experiments. To determine whether incorporation was delayed following transfer to a medium containing [³H]TDR, bacteria were grown in cold medium to 5×10^8/ml., centrifuged and resuspended in medium containing 2 μg/ml. [³H]TDR (1 c/m-mole). Samples were then removed into cold 5% TCA and washed on Oxoid membrane filters (average pore diameter 0·5 to 1·0 μ, Oxo Ltd, London, England) with cold TCA and finally with 1% acetic acid. The filters were dried, placed in scintillator fluid (0·4% 2,5-diphenyloxazole, 0·01% 1,4-bis-2[5-phenyloxazolyl]-benzene in toluene) and counted in a scintillation counter.

3. Results

In interpreting autoradiographs of extracted DNA certain assumptions are necessary. These can be stated at the outset.

(1) It is not clear why some molecules of bacterial DNA choose to untangle whereas others do not. However, the few that do are assumed to be a fairly representative sample; specifically we assume that they do not belong to some special class that is being duplicated in some special way.

(2) The ratio of mass to length for this untangled DNA is taken to be at least that of DNA in the B configuration, namely 2×10^6 daltons/μ (Langridge, Wilson, Hooper, Wilkins & Hamilton, 1960). We assume that single-stranded DNA will not be found in an extended state and so, even if present, will not contribute to the tally of untangled DNA.

† Abbreviation used: TDR stands for thymidine.

(3) The density of grains along these labelled molecules is assumed to be proportional to the amount of incorporated label. Specifically we assume that if one piece of DNA has twice the grain density of another this shows that it is labelled in twice as many strands.

(a) *Pulse labelling experiments*

Simple pulse-labelling experiments could tell much about the process of DNA replication. As pointed out already, DNA synthesis in *E. coli* is virtually continuous. If, therefore, the bacterial chromosome is truly a single piece of two-stranded DNA and if, at any moment, duplication is occurring at only a single point on this molecule, then the length of DNA labelled by a short pulse will be just that fraction of the total length of the chromosome that the duration of the pulse is a fraction of the generation time; if there are several points of simultaneous duplication on the single molecule, or several molecules which are duplicated in parallel, then the length of DNA labelled will be appropriately less. Further, from such pulse experiments it should be possible to determine whether one or two new strands are being made in each region of replication.

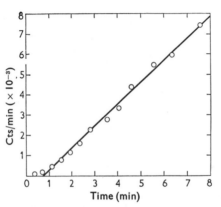

FIG. 1. Incorporation of ³H into cold TCA-insoluble material, following transfer of *E. coli* B3 to medium containing [³H]TDR.

However, first it was necessary to find out how quickly [³H]TDR gets into DNA when the bacteria are transferred from a medium containing cold TDR. The result of such an experiment with *E. coli* B3 is shown in Fig. 1 (see also Materials and Methods). A similar result was obtained for *E. coli* K12. So it seems that, under these particular conditions, the error in timing a pulse will be less than 1 minute.

Autoradiographs of *E. coli* B3 DNA were prepared following (i) a 3 minute pulse of [³H]TDR, (ii) a 3 minute pulse, followed by 15 minutes in cold TDR, and (iii) a 6 minute pulse, followed by 15 minutes in cold TDR. Examples of what was found are shown in Plate I. Similar results were obtained using *E. coli* K12.

From these experiments the following conclusions can be drawn.

(1) Immediately after a 3 minute pulse of label (Plate I(a), (b) and (c)) the labelled DNA consists of two pieces lying in fairly close association. Fifteen minutes later (Plate I(d), (e) and (f)) these pieces have moved apart and can be photographed separately. It seems therefore that *two* labelled molecules are being formed in the region of replication.

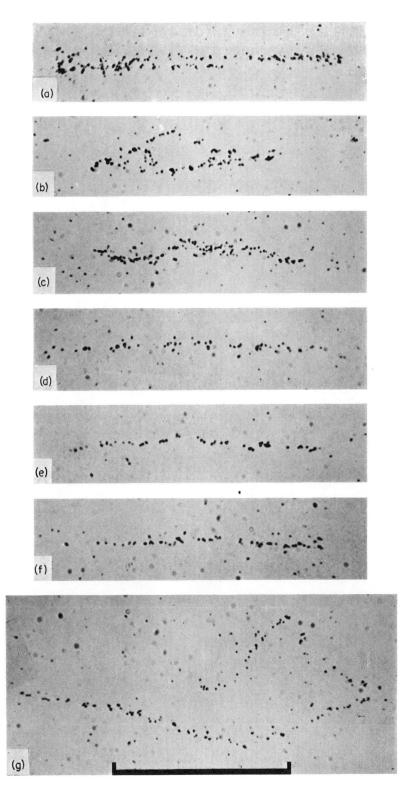

PLATE I. Autoradiographs of *E. coli* B3 DNA labelled by a pulse of [³H]TDR. Exposure time was 61 days. The scale shows 50μ.

(a), (b), (c): Immediately after a 3 min pulse; (d), (e), (f): 15 min after a 3 min pulse; (g): 15 min after a 6 min pulse.

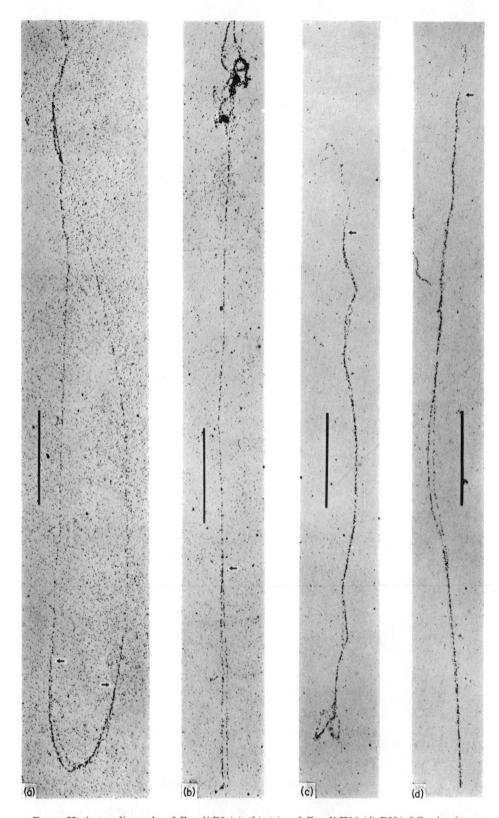

Plate II. Autoradiographs of *E. coli* B3 (a), (b), (c) and *E. coli* K12 (d) DNA following incorporation of [³H]TDR for a period of 1 hr (two generations). The arrows show the point of replication. Exposure time was 61 days. The scales show 100 μ.

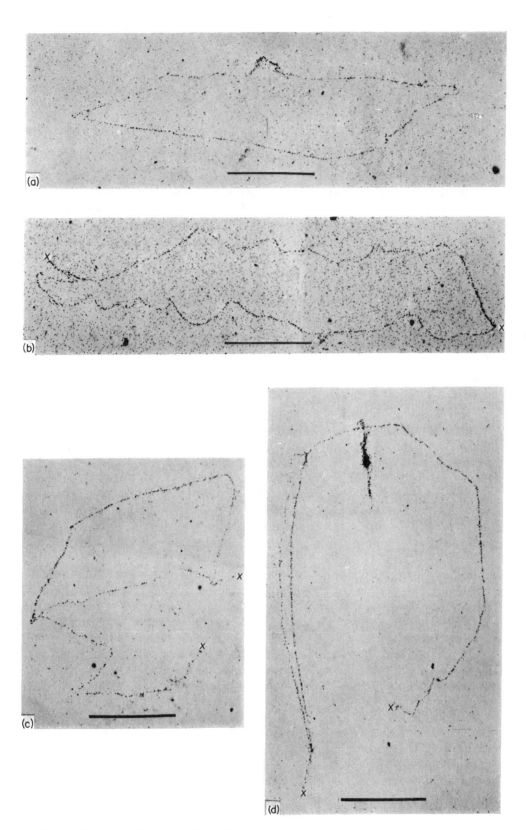

PLATE III. Autoradiographs of *E. coli* B3 (a), (b), (c) and *E. coli* K12 (d) DNA following incorporation of [³H]TDR for a period of 1 hr. In (b), (c), and (d), the postulated break is marked ×—×. Exposure time 61 days. The scales show 100 μ.

(2) There is approximately 1 grain per micron in these labelled regions. Since similarly labelled T2 and λ bacteriophage DNA, both of which are known to be largely two-stranded, show about twice this grain density (Cairns, 1962b), the two molecules being created at the point of duplication must each be labelled in one strand.

(3) A 3 minute pulse labels two pieces of DNA each 60 to 80 μ long (Plate I(a) to (f)); a 6 minute pulse labels about twice this length (Plate I(g)). In one generation time (30 minutes) the process responsible for this could cover 600 to 800 μ, or slightly more than this if the duration of the pulse is being over-estimated (see Fig. 1). There is unfortunately no precise estimate for the DNA content of the *E. coli* chromosome. When the generation time is 1 hour, each cell contains 4×10^9 daltons DNA (Hershey & Melechen, 1957). If each cell contained only one nucleus, this value would have to be divided by 1·44 (1/ln 2) to correct for continuous DNA synthesis. However, as such cells are usually multinucleate (see Schaechter, Maaløe & Kjeldgaard, 1958), this corrected value of $2·8 \times 10^9$ daltons (or 1400 μ DNA) must be too high. There is therefore no marked discrepancy between the total length of DNA that has to be duplicated ($< 1400 \mu$) and the distance traversed by the replication process in one generation (at least 600 to 800 μ). This suggests that one or at the most two regions of the chromosome are being duplicated at any moment.

These various conclusions are reinforced by the results reported in the next section.

(b) *Two-generation labelling experiments*

One conspicuous feature of the pulse experiments is not apparent from the photographs and was not listed. The proportion of the labelled DNA that had untangled, and could therefore be measured, was far lower immediately after the 3 minute pulse than 15 minutes later. Thus the replicating region of a DNA molecule is apparently less readily displayed than the rest. It was not surprising therefore that considerable search was necessary before untangled and replicating molecules were found in the two-generation experiments. Since the products of such searches are generally somewhat suspect, the guiding principles of this particular search will be given.

As pointed out earlier there are reasons for assigning the hypothetical DNA molecule of *E. coli* a length of less than 1400 μ. Therefore no molecules were accepted whose length, presumably through breakage, was much less than 700 μ. No replicating (forking) molecules were accepted unless both limbs of their fork were the same length. Lastly many extended molecules were excluded because of the complexity of their form; although perhaps interpretable in terms of a known scheme they could not be used to provide that scheme. Samples of what remained are shown in Plate II. They have been selected to illustrate what appear to be various stages of DNA replication.

First, the rough agreement between the observed length of *E. coli* DNA (up to 900 μ) and the estimated DNA content of its chromosome ($< 1400 \mu$) supports the conclusion, arising from the pulse experiments, that this chromosome contains a piece of two-stranded DNA.

Second, it seems clear that this DNA replicates by forking and that new material is formed along both limbs of the fork. This latter was shown by the pulse experiments and is confirmed here. In each of the forks shown, one limb plainly has about twice the grain density of the other and of the remainder of the molecule. The simplest hypothesis is that we are watching the conversion of a molecule of hybrid (hot–cold)

DNA into one hybrid and one fully-substituted (hot–hot) molecule. It is not surprising that, nominally after two generations of labelling, the process is seen at various stages of completion. In Plate II(a), duplication has covered about a sixth of the visible distance in what appear to be two sister molecules. In Plate II(b) and (c), duplication has gone about a third and three-quarters of the distance, respectively. In Plate II(d), duplication is almost complete, about 800 μ of DNA having been replicated. So these pictures support what seemed likely from the pulse experiments—namely, that the act of replication proceeds from one end of the molecule to the other.

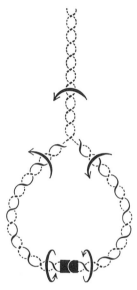

FIG. 2. The consequence of uniting the ends of the replicating fork. The arrows show the direction of rotation, as the parent molecule unwinds and the two daughter molecules are formed (modified from Delbrück & Stent, 1957).

The most conspicuous and totally unexpected feature of these pictures has been left for discussion last. In the case of each replicating molecule, the ends of the fork are joined. This complication seems to have been taken one stage further in Plate II(a); here the two limbs of the fork may be joined to each other but they also appear to be joined to their opposite numbers which are being formed from the sister molecule. Conceivably such terminal union of the new double helices (which must be alike in their base sequence) is the artificial consequence of a freedom to unite that only comes with lysis; this union may not exist inside the bacterium. Alternatively, terminal union may be the rule during the period of replication. If so, whatever unites the two ends must have the freedom to rotate so that the new helices can rotate as they are formed (Fig. 2). This uniting structure, or swivel, could in fact be the site of the mechanism that, *in vivo*, spins the parent molecule and its two daughters.

(c) *The bacterial chromosome*

The primary object of this work was to determine the form of DNA when it is replicating, not the form of the entire bacterial chromosome. The pictures presented so far give little indication of the latter as they show molecules that are either broken

(Plate II(c)) or partly tangled (Plate II(a) and (b)). It seemed possible, however' that out of all the material that had been collected some model for the shape of the whole chromosome might emerge.

In searching for such a model that could account for all the kinds of structure seen, the premise was adopted that since excess cold carrier DNA was invariably present these structures must be related to the model by breakage, if needs be, but not by end-to-end aggregation. Granted this premise, there seems to be only one model that

FIG. 3. Two stages in the duplication of a circular chromosome. (B) and (D) mark the positions of the breaks postulated to have produced the structures shown in Plate III(b) and (d), respectively.

could generate every structure merely by breakage. This model supposes that the chromosome exists as a circle. Duplication, as in Fig. 2, proceeds by elongation of a loop at the expense of the remainder of the molecule; since, however, the distal end of the molecule is also attached to the swivel, duplication creates a figure 8 each half of which ultimately constitutes a finished daughter molecule (Fig. 3). Depending on how this structure breaks at the time of extraction, it may form a rod with a terminal loop (Plate II), a rod with a subterminal loop (Plate III(c) and (d)), or a circle (Plate III(a) and (b)); in the case of a circle, the circumference may be up to twice the length of the chromosome. All structures seen, including circles of varying circumference, can be readily derived from this model whereas they do not conform to any other obvious scheme. It is, however, possible that the process of duplication may vary; that, for example, the structure shown in Plate II(a) was genuinely an exceptional case. In any event, here as elsewhere no significant difference was detected between the chromosomes of *E. coli* B3 (F⁻) and *E. coli* K12 (Hfr).

I am greatly indebted to Dr. A. D. Hershey and Professor Max Delbrück for helpful criticism and to Miss Rosemary Henry for able technical assistance.

REFERENCES

Cairns, J. (1962*a*). *J. Mol. Biol.* **4**, 407.
Cairns, J. (1962*b*). *Cold Spr. Harb. Symp. Quant. Biol.* **27**. In the press.
Delbrück, M. & Stent, G. S. (1957). In *The Chemical Basis of Heredity*, ed. by W. D. McElroy & B. Glass. Baltimore: Johns Hopkins Press.
Hershey, A. D. & Melechen, N. E. (1957). *Virology*, **3**, 207.
Jacob, F. & Wollman, E. L. (1958). *Symp. Soc. Exp. Biol.* **12**, 75.
Langridge, R., Wilson, H. R., Hooper, C. W., Wilkins, M.H.F. & Hamilton, L.D. (1960). *J. Mol. Biol.* **2**, 19.
Levinthal, C. & Crane, H. R. (1956). *Proc. Nat. Acad. Sci., Wash.* **42**, 436.
McFall, E. & Stent, G. S. (1959). *Biochim. biophys. Acta*, **34**, 580.
Meselson, M. & Stahl, F. W. (1958). *Proc. Nat. Acad. Sci., Wash.* **44**, 671.
Meselson, M. & Weigle, J. J. (1961). *Proc. Nat. Acad. Sci., Wash.* **47**, 857.
Nagata, T. (1962). *Biochem. Biophys. Res. Comm.* **8**, 348.
Schaechter, M., Bentzon, M. W. & Maaløe, O. (1959). *Nature*, **183**, 1207.
Schaechter, M., Maaløe, O. & Kjeldgaard, N. O. (1958). *J. Gen. Microbiol.* **19**, 592.
Watson, J. D. & Crick, F. H. C. (1953). *Cold Spr. Harb. Symp. Quant. Biol.* **18**, 123.

ARTICLE **7**

From the *Proceedings of the National Academy of Sciences*, Vol. 55, No. 4, 792–797, 1966. Reproduced with permission of the author.

ISOLATION OF THE DNA OF THE E. COLI CHROMOSOME IN ONE PIECE*

By C. I. Davern

COLD SPRING HARBOR LABORATORY OF QUANTITATIVE BIOLOGY,
COLD SPRING HARBOR, NEW YORK

Communicated by A. D. Hershey, February 10, 1966

The autoradiographs of H^3-labeled chromosomes of *Escherichia coli* prepared by Cairns,[1] and the explanations of the chromosome's growing mechanism by Bonhoeffer and Gierer[2] and Nagata,[3] are consistent with the notion that the DNA of the replicating chromosome is a single molecule with a single growing point. Such a molecule would have a molecular weight of $2.3–4.6 \times 10^9$ daltons,[1] depending on its stage of replication. Attempts to isolate the DNA of the *E. coli* chromosome in one piece, by methods designed to minimize shear and nuclease action, have so far failed,[4,5] although very large DNA molecules (4×10^8 daltons), corresponding roughly to half the nuclear DNA content, have been isolated from *Hemophilus influenzae* by Berns and Thomas.[5]

Such attempts to relate the isolated DNA molecule to the molecular organization of the bacterial chromosome are indirect, for they rely on comparison of an estimate of the molecular weight of the isolated molecule with an estimate of the chromosome DNA content. This paper describes a more direct method for assessing the relation between the isolated DNA molecule and the DNA of the chromosome. The method was applied to DNA extracted from *E. coli* and banded in a CsCl density gradient by means of a technique designed to avoid shear stress and minimize nuclease action. It showed that the DNA of the *E. coli* chromosome can be isolated in one piece.

The assessment is based on CsCl equilibrium density gradient analysis[6] of DNA isolated from cells that have incorporated 5-bromouracil into their DNA for a portion of the chromosome replication cycle. If each chromosome has all its DNA in one molecule, then those molecules that were replicating during the period of labeling with 5-bromouracil should be partly hybrid and, if isolated without fragmentation, should be detectably denser than normal.

The sensitivity of the method for detecting fragmentation can be gauged from the following arguments. If the DNA in the chromosome replicates by means of a single growing point, and if about one quarter to one half of each replicating molecule has become hybrid during the labeling period, then the breakage of such molecules into two fragments will produce some fragments that are hybrid in density.

Two innovations were employed in the release of DNA. First, to avoid unnecessary shear, spheroplasts were lysed on top of the CsCl solution in the centrifuge tube. Second, sodium dodecyl sarcosinate was used to promote lysis. Unlike the commonly used detergent, sodium dodecyl sulfate, the sarcosinate is soluble in concentrated CsCl.

Material and Methods.—*Bacterial cultures:* *Escherichia coli* strain 15 TAU-bar,[7] which requires thymine, uracil, arginine, proline, tryptophan, and methionine, was grown in glucose-ammonium medium[8] supplemented with 2.5 μg/ml each of L-arginine, L-proline, L-tryptophan, and L-methionine, 10 μg/ml uracil, and 0.02% Difco Bacto-peptone. Although not essential for growth, the peptone supplement was essential for renewal of growth of this strain after filtration on Millipore filters. The bacteria were grown with aeration at 37°C.

C[14]-labeled DNA: Bacteria were grown for six generations to saturation in the presence of 2 μg/ml C[14]-thymidine (sp. act. 30 mc/mM), washed, and suspended in 0.01 M ethylenediaminetetracetate (EDTA), 0.01 tris(hydroxymethyl) aminomethane (tris buffer), pH 8.5. The cells were then lysed by the sequential addition of 40 μg/ml egg white lysozyme (Worthington Biochemical Corp., Freehold, N.J.), chloroform, and 0.5% sodium dodecyl sarcosinate (Sarkosyl N.L.-97 from Geigy Industrial Chemicals, Ardsley, N.Y.). The cleared lysate, containing approximately 6 μg/ml of DNA, was thoroughly shaken to ensure fragmentation of the DNA.

H[3]-thymidine labeling of 15 TAU-bar: The cells were grown in 0.5 ml of the supplemented glucose-ammonium medium containing 4 μg/ml H[3]-thymidine (sp. act. 3.35 c/mM) for 3–4 generations to 10[8] ml, whereupon they were filtered and washed on an HA Millipore filter. Because of the small quantity of labeled cells being filtered, the filters had to be pretreated with unlabeled cells to ensure good recoveries.

Density labeling of the chromosome with 5-bromouracil: The washed H[3]-thymidine-labeled cells were suspended in growth medium containing 40 μg/ml 5-bromouracil in place of thymidine, and then incubated with aeration at 37°C. At the end of a labeling period, the culture was filtered.

Spheroplast formation:[9] The cells from the filter were suspended in 0.3 ml spheroplasting medium consisting of 20% w/w sucrose and 1 mg/ml egg white lysozyme in 0.01 M EDTA, 0.01 M tris buffer, pH 8.5, and converted to spheroplasts by incubation at 37°C for 30 min.

Lysis of spheroplasts: The following solutions were layered successively into a centrifuge tube containing 0.02 ml C[14]-DNA: (1) 4 ml 60% w/w CsCl in 0.01 M phosphate buffer, pH 6.4; (2) 0.5 ml of 40% w/w glycerol in 0.01 M EDTA, 0.01 M tris buffer, pH 8.5, containing 0.5% sodium dodecyl sarcosinate and 100 μg/ml pronase (Calbiochem, Los Angeles, Calif.); and (3) 0.5 ml 0.01 M EDTA, 0.01 M tris buffer pH 8.5, containing 100 μg/ml pronase. Immediately after preparation of the centrifuge tube in this manner, 0.02 ml of the spheroplast suspension (containing 2–4 $\times$ 10[7] spheroplasts/ml) was carefully applied at the junction between the glycerol layer and the top buffer layer. The tube was then covered with parafilm and left undisturbed for at least 16 hr before being centrifuged. During this period, detergent and pronase (which is active in the presence of the detergent) diffused into the spheroplast layer and lysed the spheroplasts. Meanwhile, CsCl diffused to form a gradient similar to the one achieved by centrifugation.

Equilibrium density gradient centrifugation: The samples were centrifuged at 35,000 rpm for at least 50 hr at 15–20°C in the SW39 rotor of the Spinco model L ultracentrifuge. At the end of the run the meniscus from each tube was sampled by dabbing with a disk of Whatman no. 1 filter paper, and the remainder of the solution was collected in 5-drop fractions directly on 1-in. squares of Whatman no. 1 filter paper in vials. The empty tubes were cut transversely into three pieces, and each piece was inserted into a separate vial. The samples were dried, and the radioactivity was measured in a scintillation spectrometer.

Results.—Log-phase bacteria, uniformly labeled with H^3-thymidine, were transferred to warm medium containing 5-bromouracil. Samples were taken immediately before, 40 min after, and 87 min after transfer, and the cells from each sample converted to spheroplasts. Three centrifuge tubes, containing the buffer-glycerol lysing layers floating on concentrated CsCl, were each carefully loaded with a 0.02-ml aliquot of spheroplasts at the buffer-glycerol boundary. Tubes 1 and 2 received the spheroplasts from the 40-min sample. Tube 3 received a mixture of spheroplasts from the 0-min and 87-min samples. The tubes were left undisturbed for 16 hr, then the contents of tube 2 were subjected to shear by five passages through a Pasteur pipet. All tubes were then centrifuged for 52 hr at 35,000 rpm.

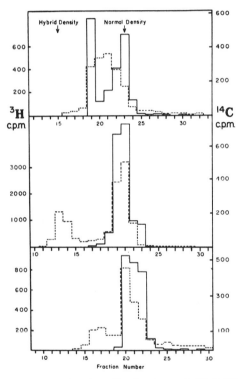

Fig. 1.—Density distributions of H^3-DNA (- - - - -) released from spheroplasts and C^{14}-DNA (———) used as a density marker. The spheroplasts were prepared from log-phase cells, uniformly labeled with H^3-thymidine, that had been incorporating 5-bromouracil for the times indicated. *Top:* (tube 1) H^3-DNA from cells labeled with 5-bromouracil for 40 min. *Middle:* (tube 2) H^3-DNA from same sample as tube 1, subjected to shear immediately before centrifugation. *Bottom:* (tube 3) H^3-DNA from a mixture of cells not labeled with 5-bromouracil and cells so labeled for 87 min.

The density distribution of the H^3-DNA derived from the lysed spheroplasts is shown in Figure 1. In the middle frame of the figure, a calibration of the density gradient is provided by the distribution of label from the deliberately sheared sample. Here the left-hand peak marks the position of hybrid DNA, and the other peak marks that of DNA of normal density. This profile also shows that 27 per cent of the spheroplast DNA present at the beginning of transfer had replicated to become hybrid during the 40 min of labeling with 5-bromouracil.

Density distribution of unsheared DNA: DNA extracted from the 40-min sample of spheroplasts formed a single broad band ranging from normal to almost hybrid density. No DNA was detectable at the position corresponding to the density of hybrid DNA (Fig. 1, *top*). DNA extracted from the mixture of spheroplasts from the 0-min and 87-min samples formed two bands, one corresponding to DNA of normal density, and the other to a density intermediate between hybrid and normal. Again no DNA of hybrid density was detectable.

Trapping of DNA by spheroplast DNA band: The anomalous distribution in tube 1 of the C^{14}-DNA used as a density marker (Fig. 1, *top*) indicates that the very small amount of unsheared DNA (about 0.002 μg) in the spheroplast

TABLE 1

RECOVERIES, AFTER CENTRIFUGATION, OF H³-DNA FROM LYSED SPHEROPLASTS AND
C¹⁴-DNA USED AS A DENSITY MARKER

	Tube 1*		Tube 2*		Tube 3*	
Location of label	H³ (%)	C¹⁴ (%)	H³ (%)	C¹⁴ (%)	H³ (%)	C¹⁴ (%)
Banded in gradient	23	98	98	98	17	97
At meniscus	46	0	1	0	69	0
Adsorbed to centrifuge-tube walls	31	2	1	2	14	3

* Loaded with spheroplasts as described in legend of Fig. 1.

DNA band can impede the free sedimentation of other DNA through it. The absence of such trapping by the smaller amount of unsheared DNA in tube 3 (Fig. 1, *bottom*) suggests that trapping depends on concentration. Results of an experiment in which a sample three times smaller than those of Figure 1 was analyzed confirm that suggestion (Fig. 2).

DNA complex of low density: Although extraction as described yielded some DNA banding in the gradient at a density expected for pure DNA, much of the DNA from the lysed spheroplasts was recovered from the meniscus and the tube wall (Table 1). Most of the DNA that ended up on the tube walls was in fact derived from DNA left behind at the meniscus after its rather inefficient sampling. Shearing of this complex permitted nearly complete recovery of DNA of normal density.

The buoyant density of the complex was measured by using 56 per cent CsCl in the CsCl layer. Under these conditions, the DNA complex banded one third of the distance from the meniscus to the tube bottom, corresponding to a density of about 1.55 gm/ml.

Discussion.—The results show that detergent and pronase release only part of the DNA from spheroplasts in the free state. Most of it remains bound to a low-density material in an extended form that is sensitive to shear.

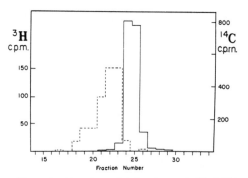

FIG. 2.—Density distribution of H³-DNA (- - - - -) released from spheroplasts and C¹⁴-DNA (———) used as a density marker. The spheroplasts were prepared from log-phase cells, uniformly labeled with H³-thymidine, that had been incorporating 5-bromouracil for 60 min.

The density distribution of the free DNA is consistent with that expected for a population of molecules heterogeneous in size, each of which has been partially replicated with 5-bromouracil as a precursor, and all or most of which have been isolated without fragmentation.

Very low concentrations of this DNA in the band can trap fragmented DNA also present in the preparation subjected to density gradient centrifugation. This phenomenon is another indication of the large size of the unsheared DNA. However, the same phenomenon may conceal the presence of any fragments of hybrid density generated by the extraction procedure. Even in the experiment shown in Figure 2, where there is no evidence of trapping of severely fragmented DNA, no guarantee can be given that there would be no trapping of the much larger fragments that might arise during extraction. On the other hand, spheroplast DNA of two density species deliberately mixed can be separated, at least in part (Fig. 1, *bottom*). This result makes it unlikely that any considerable fragmentation would be missed in these experiments. Thus, in spite of poor recoveries and banding anomalies, the data allow one to conclude that the entire DNA of the chromosome can be isolated as a single molecule, provided, of course, that the chromosome replicates by way of a single growing point. Massie and Zimm[4] concluded, on the contrary, that the chromosome of *E. coli* consists of some eight or so subunits joined together by pronase-susceptible linkers.[10]

The extraction procedure described in this paper yields most of the DNA from spheroplasts in a complex of low density. Virtually all of the DNA in the complex is sensitive to shear and must be present in extended form. The complex could originate by casual entanglement of the emerging DNA molecule with spheroplast debris, or could reflect the persistence of a chromosome-membrane association.[11-13] Whatever the nature of the complex, DNA associated with it must, like the free DNA, be unfragmented.

Owing to the banding anomalies already referred to, the method used here for assessing fragmentation becomes insensitive when the amount of DNA in the band exceeds 0.001 μg. The use of 5-bromouracil as a density label imposes further limitations because the analogue depresses the replication of DNA and may disturb its regulation.[14, 15]

Cairns' autoradiographs show that the replicating *E. coli* chromosome has the configuration of a circularized Y with fused ends.[1] Fragmentation of such a structure requires at least two or three breaks. Thus, a test for fragmentation cannot be used to assess the intactness of the chromosome, a limitation it shares with other methods.

Summary.—A method is described for extracting the DNA from *E. coli* by lysing spheroplasts on top of a CsCl solution by the combined action of pronase and sodium dodecyl sarcosinate. Density gradient analysis of DNA isolated by this method from cells that have incorporated 5-bromouracil into their DNA for about one third of a replication cycle reveals that the entire DNA complement of the chromosome can be isolated as one molecule.

The author wishes to thank Dr. H. J. F. Cairns for persuading him to tackle this project, and to acknowledge the skilled assistance of Miss Paula de Lucia.

* This work was supported by the National Science Foundation.

[1] Cairns, H. J. F., *J. Mol. Biol.*, **6**, 208 (1963); Cairns, H. J. F., in *Synthesis and Structure of Macromolecules*, Cold Spring Harbor Symposia on Quantitative Biology, vol. 28 (1963), p. 43.

[2] Bonhoeffer, F., and A. Gierer, *J. Mol. Biol.*, **7**, 534 (1963).

[3] Nagata, T., these Proceedings, **49**, 551 (1963).

[4] Massie, H. I., and B. H. Zimm, these Proceedings, **54**, 1636 (1965).

[5] Berns, K. I., and C. A. Thomas, *J. Mol. Biol.*, **11**, 476 (1965).

[6] Meselson, M., F. W. Stahl, and J. Vinograd, these Proceedings, **43**, 581 (1957).

[7] Hanawalt, P. C., *Nature*, **198**, 286 (1963).

[8] Hershey, A. D., *Virology*, **1**, 108 (1955).

[9] Spheroplasting procedure modified from methods suggested by Neu, H. C., and L. A. Heppel, *J. Biol. Chem.*, **239**, 3893 (1964); and Repaske, R., *Biochim. Biophys. Acta*, **30**, 225 (1958).

[10] Freese, E., in *Exchange of Genetic Material: Mechanisms and Consequences*, Cold Spring Harbor Symposia on Quantitative Biology, vol. 23 (1958), p. 13.

[11] Goldstein, A., and B. J. Brown, *Biochim. Biophys. Acta*, **53,** 19 (1961).

[12] Ryter, A., and F. Jacob, *Ann. Inst. Pasteur*, **107,** 384 (1964).

[13] Ganesan, A. T., and J. Lederberg, *Biochem. Biophys. Res. Commun.*, **18,** 824 (1965).

[14] Lark, K. G., T. Repko, and E. J. Hoffman, *Biochim. Biophys. Acta*, **76,** 9 (1963).

[15] Pettijohn, D., and P. Hanawalt, *J. Mol. Biol.*, **9,** 395 (1964).

Questions

ARTICLE 3 Davison

1. Why did Davison's argument not require a direct measurement of molecular weight?

2. What was the big piece in the Levinthal-Thomas experiment? The little pieces? Would you expect to get a mixture of big and little pieces of the sizes they calculated? Explain.

•3. Why did Davison calculate distribution curves?

4. What evidence did Davison provide for the lack of protein linkers?

5. What would be the relative shear sensitivity of a single-stranded DNA having molecular weight $M/2$ and a double-stranded DNA of mass M? Assume there is no internal hydrogen bonding in the single-stranded DNA.

6. How would the shear sensitivity of a DNA change if a protein was added that caused side-to-side aggregation of the DNA molecules?

7. If a solution of DNA of mass M was passed through a needle and molecules of mass $M/8$ were detected, would you expect to find molecules of mass $M/2$ and $M/4$ also?

•8. At high DNA concentration the shear sensitivity decreases. Propose an explanation.

9. What would be the molecular weight of the *first* shear degradation product of a *circular* molecule of mass M?

ARTICLE 4 Davison, Freifelder, Hede, and Levinthal

1. Why do the rays appear to come from a single point? Why is this true even for DNA molecules whose total length is 100 times the length of the phage head?

2. Estimate the fraction of *rays* that might be missed.

3. Estimate the fraction of *stars* that might be missed if there is an average of 10 rays per star. 5? 20? (Use the Poisson distribution.)

•4. Discuss the effect on the data if a fraction of the phages had not released their DNA. Suppose it was as high as 20%? What criterion did the workers use to determine whether DNA had been released?

5. What are the data indicating that all T2 phages contain the same amount of DNA? If they were not in the paper, how would that affect the conclusions of the paper?

6. Why is it necessary to perform the statistical analyses used in this paper?

7. Why was the DNA sedimented through a H_2O-D_2O gradient? Why were the rays counted by several observers?

8. Suppose each phage contained two molecules of equal size. How would this have affected star number and ray number?

9. What assumptions would have to be made to calculate M from the ray count?

ARTICLES 5 and 6 Cairns

1. Why were the cells lysed in a chamber, one side of which is a membrane filter?

2. In using the Cairns procedure, what factors would determine the exposure time you would choose?

3. Suppose two DNA molecules aggregated in such a way that their lengths overlapped to give a unit whose length was shorter than the sum of the two lengths. How would you recognize this result?

1. We now know that *E. coli* DNA is replicated bidirectionally from a single origin with two growing points. Is this inconsistent with Davern's findings? Explain.

2. Why was it necessary for Davern to use a density marker of light DNA? What ambiguity might have arisen if he had not?

3. Explain why Davern included an experiment with sheared DNA.

4. Do any of his experiments exclude protein linkers? Why do you think the idea was dropped?

5. Why did Meselson and Stahl get a different result from that of Davern? (See also Article 16, Section V.)

References

Bode, H. R., and H. J. Morowitz. 1967. "Size and Structure of the *Mycoplasma hominis* H39 Chromosome." *J. Mol. Biol.* **23**, 191–199.

Davison, P. F. 1960. "Sedimentation of Dexoyribonucleic Acid Isolated Under Low Hydrodynamic Shear." *Nature* **185**, 918–920.

Fleischman, J. B. 1960. "The Physical Properties of DNA from T2 and T4 Bacteriophage." *J. Mol. Biol.* **2**, 226–240.

Levinthal, C., and C. A. Thomas. 1957. "Molecular Autoradiography: The β-ray Counting from Single Virus Particles and DNA Molecules in Nuclear Emulsions." *Biochim. Biophys. Acta* **23**, 453–465.

MacHattie, L. A., K. I. Berns, and C. A. Thomas. 1965. "Electron Microscopy of DNA from *Hemophilus influenzae*." *J. Mol. Biol.* **11**, 648–649.

Additional Readings

Bendich, A., and H. S. Rosenkranz. 1963. "Some Thoughts on the Double-Stranded Model of Deoxyribonucleic Acid," *Prog. Nucleic Acid Res.* **1**, 232–300. Protein linkers are discussed.

Freifelder, D. 1976. *Physical Biochemistry.* W. H. Freeman and Company. Both shearing methods and autoradiography are described in detail.

Levinthal, C. and P. F. Davison. 1961. "Degradation of Deoxyribonucleic Acid under Hydrodynamic Shearing Forces." *J. Mol. Biol.* **3**, 674–683. The chemical events in the shearing process are described.

Stent, G., and R. Calendar. 1978. *Molecular Genetics,* 2nd ed. W. H. Freeman and Company. The phage life cycle is described.

IV

THE LARGE SIZE OF DNA MOLECULES

Measurement of the Molecular Weight of DNA

Beginning in the mid-1950's an enormous effort went into the determination of the molecular weight M of DNA, the values of which had both biological and physicochemical significance. Biologically, knowledge of M can reveal the number of DNA molecules per chromosome or per organism, the rates of replication and transcription, an estimate of the number of genes, and so forth. For the physical chemists admittedly the motivation was in part simply to show that M could in fact be measured for such large molecules, although a major factor was certainly the opinion (and a correct one) that for every biological molecule it is useful to have a reasonably complete description of its physical properties. By the time it became evident that DNA is an important molecule, the determination of M for proteins had become quite straightforward, and accurate values could be obtained by any one of the following techniques—light scattering, sedimentation-diffusion, sedimentation equilibrium, or osmometry. The determination of this value for DNA, however, was a special

problem because even for the smaller DNA molecules (M = 5–10 million) M is generally too large for accurate use of these techniques.

Specifically, because DNA is so long and thin, its diffusion coefficient is too small to measure accurately. Owing to its great mass, the sedimentation coefficient of DNA is so high that centrifuges could not be run slowly enough to obtain usable data by standard sedimentation equilibrium methods. Furthermore, in any measurement requiring high molarity (e.g., osmotic pressure) the required solution would not be a liquid but a semisolid gel. In addition, DNA is so asymmetric and so flexible that it could not be treated by the theories of viscosity or of sedimentation velocity. Finally, for light scattering,* if M is greater

---

*In light scattering a beam of light is passed through a solution of a macromolecule. A fraction of this light is scattered at all angles. Using a light-sensing device the intensity of the scattered light as a function of scattering angle is measured. From the intensity versus angle relation and a few other angle-independent parameters, M can be calculated.

than 10 million, (1) solutions cannot easily be rendered dust free (and macromolecules of this size usually scatter light less than dust particles), (2) the theory is inadequate since an assumption is made that the dimensions of the scattering center (i.e., the molecule) are small enough that a single molecule does not scatter the same photon twice, an assumption that is not satisfied for DNA, and (3) errors produced by the design of the light scattering instrument become increasingly greater as M increases.*

Attempts to Relate M to the Sedimentation Coefficient

For M of less than 5 million it was generally agreed (in 1957) that the most reliable technique probably was light scattering, although obtaining reproducible values of M with this technique often seemed to be an art. However, Bruno Zimm had just extended the theory and worked out a simple method for analyzing the data (the Zimm plot), and Paul Doty, Stuart Rice and their colleagues had achieved a high degree of skill in obtaining reproducibility. Light scattering is, however, time consuming and difficult, whereas the measurement of a sedimentation coefficient† s in the analytical ultracentrifuge is straightforward and simple. Therefore, Doty, Barbara McGill, and Rice (1958) decided to prepare a set of DNA samples, each having a different value of M (0.3–8 $\times 10^6$), and to measure for each very precisely both M by light scattering and s by ultracentrifugation. They expected, by doing so, to obtain an equation relating s and M, from which M could be obtained easily not only for the range originally studied but also for larger molecules. The equation Doty, McGill and Rice obtained was $s = 0.063 M^{0.37}$. Unfortunately, this equation turned out to be nearly correct only for M values of less than 5 million.

The principal cause of inaccuracy in the range of low M was that the DNA samples consisted of fragments and were therefore heterogeneous with respect to both M and s; hence all measured values of M and s were averages (and not necessarily the same kind of statistical average). Furthermore, for the higher values of M the precision of the measurement was lower, mostly for technical reasons—a fact not appreciated until several years later when low-angle light scattering instruments were developed. For example, this equation predicted a value of $M = 150$ million for phage T2 DNA rather than the value of 130 million accepted at the time (the currently accepted value is 106 million). Therefore it was suspected that, for the large values, the equation overestimated M.

Alfred Hershey and two colleagues decided to use the same principle—of obtaining an s versus M curve—but to determine M very accurately for one very large molecule only and then to measure s of both the intact molecule and its halves, quarters and so forth. This method has the advantage that, as long as the fragments are truly successively halved, at least the exponent a in the equation $s = kM^a$ will be correct even if the measured value of the standard turns out to be wrong. He selected T2 DNA because it had already been extensively studied. For a method of successive halving he chose shear degradation and for M determination, he used the ^{32}P star counting method. This work is described in Article 8 by I. Rubenstein, C. A. Thomas, Jr., and A. D. Hershey.

The equation of Rubenstein, Thomas, and Hershey has different values of k and a than that of Doty, McGill, and Rice (1958) even when it is rewritten in terms of the same parameters. This variation produced considerable concern, and various hypotheses were proposed to explain the discrepancy, hypotheses that differed according to which methods were thought to have greater validity. As will be seen in Article 13 the values of M from light scattering and star counting were both incorrect. Most important, however, was the fact that the correct form of the s-M relation is not $s = kM^a$ but $s = b + kM^a$, where b is a constant; thus in a sense the two equations arose from "forcing" two sets of data into an equation of

---

*This paragraph may be obscure for the reader without knowledge of the techniques of physical chemistry. Suffice it to say that the great mass and length of DNA molecules rendered the standard techniques quite useless.

†If a solution of macromolecules is sedimented in a centrifugal force field, the macromolecules will move in the centrifugal direction if they are denser (greater mass per unit volume) than the solvent. The value s is the ratio of the velocity of motion to the magnitude of the centrifugal force.

the wrong form. Nevertheless all was not chaos, for even though neither equation was correct, they were close enough that it was generally possible to obtain a value of M with little more than a 30% error from a simple measurement of s or of the distance of sedimentation. (It is worth calculating M from each equation for a DNA molecule whose $s = 32$. The correct value is 25 million.)

It soon became clear that a valid s-M curve could be constructed only if one used DNA samples having a single M value,* and furthermore that it was worthwhile to obtain such a relation with high precision. It was pointed out by Peter Davison that one should start with a small phage whose DNA was relatively resistant to degradation by shear and hence could be isolated without breakage. He selected $E.$ $coli$ phage T7 since he guessed from the size of the phage (as observed by electron microscopy) that it probably contained a DNA molecule having an M value of 20 to 30 million. The characterization of T7 DNA is described in Article 9 by Davison and me. The principal importance of this paper was the point that this DNA, which was easy to purify in large quantities, was probably the material of choice for both physicochemical and enzymatic studies. It is interesting that three years passed before T7 DNA was widely used.†

A New Principle in Determining *M* of DNA

The value of M reported in Article 9 was obtained in a new way. Davison had realized that the problems encountered in determining M of DNA when the DNA molecule was examined directly could be avoided if one instead measured both M for the intact phage and the weight fraction of the phage that is DNA.

(Whereas this solution now seems obvious, it was a new principle at the time, deriving from the demonstration that the phage DNA comprised a single molecule.) He set out to do this with phage T7, using the techniques of light scattering and sedimentation-diffusion for two independent measurements of M and simple chemical analysis for the weight fraction. Because of technical difficulties (principally in the measurement of the diffusion coefficient of the phage) the M value obtained was too low—i.e., 17 million for M of the DNA versus the value of 25 million now known to be correct, but the principle was established.

The problem of establishing a valid s-M equation was ignored for several years because the various people who had worked on it had been distracted by more exciting developments and because the existing equations seemed adequate. In addition, determination of M from a measurement of length by electron microscopy seemed at that time to be a simpler and more reliable procedure.

End-Group Labeling to Determine *M*

Several years after Davison's work with T7 DNA, a new method became available for determining M, i.e., end-group labeling, a procedure already in use for many years by protein chemists to determine M for proteins. The idea was the following. A pure sample of a protein had its N-terminal amino acid labeled by reaction with ^{14}C-labeled fluorodinitrobenzene (FDNB) of known specific activity. M was then calculated simply by dividing the mass of the material labeled by the number of N-termini, since there is only one N-terminus per molecule:

$$M = \frac{wq}{(dpm)}$$

where dpm and q are the number of disintegrations per minute and the specific activity of the FDNB (expressed as dpm per mole), w is the weight in grams of material labeled, and M is expressed in atomic mass units. In 1965, Charles Richardson isolated the enzyme T4 polynucleotide kinase, which could be used to label a 5′phosphoryl (5′P) terminus of DNA, there being two such termini per double helix.

---

*When all the molecules of a DNA sample have the same M value, the sample is commonly described as being "homogeneous with respect to M." Such a sample is termed *monodisperse* in polymer chemistry; *polydisperse* refers to a sample containing molecules of diverse M values.

†At the end of this paper there is a "Note added in proof" pointing out that numerical discrepancies exist between the physical properties of T7 and λ DNA. Although not recognized at the time, the error was in Cairns' measured length of λ since the molecular weight of λ DNA is now known to be 31×10^6.

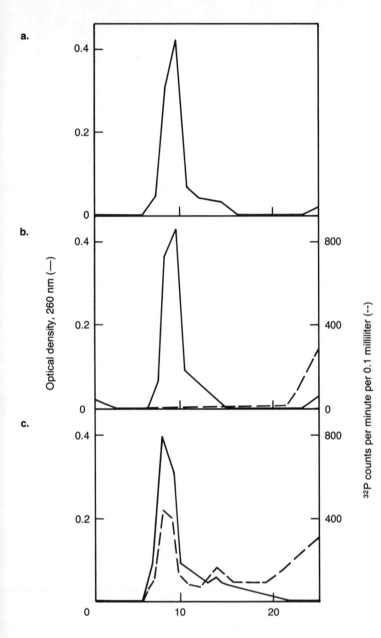

Figure IV-1
Analysis by zonal centrifugation of denatured T7 DNA (a) untreated, (b) after kinase reaction with ^{32}P-ATP but without prior phosphatase treatment, and (c) treated with phosphatase and then kinase with ^{32}P-ATP. Only in (c) does the ^{32}P cosediment with the DNA. Note the ^{32}P at the top of centrifuge tube (right end of graph); the slow sedimentation indicates that these are phosphorylated fragments. M is calculated from the ratio of the amount of ^{32}P (expressed in micrograms) to the OD$_{260}$, both values measured at the peaks of the curves. Sedimentation was from right to left. [From C. C. Richardson,] *J. Mol. Biol.* **15** (1966), 49–61.]

This enzyme can transfer a γ-^{32}P from γ-labeled ATP to a 5'OH group of DNA. Since DNA does not have a free 5'OH group, it is necessary to remove the 5'P first with alkaline phosphatase. The entire scheme is the following:

$$
\begin{array}{c}
3'\text{OH}\!\!-\!\!\!-\!\!5'\text{P} \\
5'\text{P}\!\!-\!\!\!-\!\!3'\text{OH}
\end{array}
\xrightarrow{\overset{\text{alkaline}}{\text{phosphatase}}}
\begin{array}{c}
3'\text{OH}\!\!-\!\!\!-\!\!5'\text{OH} \\
5'\text{OH}\!\!-\!\!\!-\!\!3'\text{OH}
\end{array}
$$

$$
\xrightarrow[\gamma\text{-}^{32}\text{P-ATP}]{\overset{\text{polynucleotide}}{\text{kinase}}}
\begin{array}{c}
3'\text{OH}\!\!-\!\!\!-\!\!5'^{32}\text{P} \\
5'^{32}\text{P}\!\!-\!\!\!-\!\!3'\text{OH}
\end{array}
$$

This end-group labeling procedure can be used to determine M for a DNA molecule. The first measurement of this type was carried out by Charles Richardson (1966), who also used T7 DNA. Although the technique seemed simple enough in principle, he first had to establish some criteria for ascertaining that the alkaline phosphatase had removed the 5'P from every molecule and that the polynucleotide kinase had labeled every 5'OH group. After reasonable criteria were established, a more serious problem arose: when the number of ends present in a DNA preparation was measured, frequently it was several times greater than expected so that the calculated value of M was unreasonably low. Richardson realized that this could be a result of the presence of DNA fragments. For example, if every DNA molecule were broken in half, the number of ends would be doubled; more significantly, if 1% of the molecules were broken into 100 pieces, the number of ends would also be doubled. Obviously it was necessary to purify the unbroken DNA molecules. Richardson chose the procedure of labeling T7 DNA by the kinase reaction and then purifying the labeled DNA by zonal centrifugation through a sucrose concentration gradient. The result of this purification is shown in Figure IV-1.* Note that the radioactivity is distributed

---

*In Figure IV-1 the DNA concentration in each fraction of the sucrose gradient was determined by measuring the optical density at a wavelength of 260 nm (OD$_{260}$) with the use of a spectrophotometer. The bases of DNA absorb strongly at 260 nm, and DNA concentration is obtained from the simple relation
$$\text{Concentration } (\mu g/ml) = OD_{260} \times 50$$

TABLE IV-1
Molecular weights of several *E. coli* phages and their DNA molecules

Phage	% DNA[1,2]	M_{phage}[2] (millions)	M_{phage}[3] (millions)	M_{DNA} (millions)	M_{DNA}[5] (millions)
T4	54.9	192 ± 6.6	—	105.7 ± 3.8	113.5
T5	61.7	109.2 ± 4.0	—	67.3 ± 3.1	68.5
T7	51.2	50.4 ± 1.8	48.7 ± 1.3	25.4 ± 1.0[4]	24.8

[1]Calculated from phosphorus and nitrogen content of phages determined by chemical analysis, by the phosphorus and nitrogen content of DNA (from nucleotide composition) and by the phage protein (from amino acid composition).

[2]Taken from S. B. Dubin, G. B. Bendeck, F. C. Bancroft, and D. Freifelder, 1970. *J. Mol. Biol.* **54**, 547–556.

[3]Taken from F. C. Bancroft and D. Freifelder, 1970. *J. Mol. Biol.* **54**, 537–546.

[4]Average of values calculated from the values of M_{phage} from the two preceding columns.

[5]Determined by C. W. Schmid and J. E. Hearst, 1969. *J. Mol. Biol.* **44**, 143–160. Values have been increased by 7.4% to accommodate an error (communicated to the author by J. E. Hearst) in the original paper.

in two portions. The more rapidly moving fraction has the expected ratio of ends to DNA mass and consists of labeled unbroken molecules. The very slowly moving fraction has a very high ratio and consists of labeled small fragments. Thus to obtain *M* one must calculate the ratio of ends to mass only for the unbroken molecules. In so doing Richardson obtained a value of $M = 25.4 \times 10^6$, which is almost identical to the value derived by the latest physical methods (which are described in Table IV-1 and Article 13).

From the information presented here on Richardson's work, how would you answer the following questions?

1. Suppose 10% of the 5'P were not removed by the alkaline phosphatase, how would this affect the measurement of *M*? Suppose only 80% of the 5'OH (assume 100% efficiency of conversion of 5'P to 5'OH) were labeled, how would *M* be changed? What criteria could be used to decide whether the reactions are 100% efficient?
2. What value of *M* would have been obtained if the sucrose gradient purification presented in Figure IV-1 had not been done?
3. If errors such as those mentioned in questions 1 and 2 entered the measurement, would the calculated value be a minimum, maximum, or neither?

New Methods for Determining *M* of Phages

Richardson's work suggested that the earlier value of *M* for T7 DNA obtained by Davison and Freifelder was incorrect and aroused interest once again in obtaining a good *s-M* curve. Although end-group labeling is in principle highly accurate, the analysis could not be used for larger DNA molecules because (1) the specific activity of the ATP would have to be higher than what could be easily attained, and (2) the DNA could be more easily fragmented, owing to the greater shear sensitivity of large DNA. The method of choice again proved to be to determine for several phages both *M* of the phage and the weight percent that is DNA. Although determination of *M* for large particles had been very difficult, it was made possible by two new technical developments. One of these—a modified sedimentation equilibrium procedure devised by David Yphantis (1964)—allowed *M* for very large molecules to be determined. In addition, Beckman Instruments, the manufacturer of the ultracentrifuge, improved the speed control system of the machine such that the instrument could be run at very low speeds with little speed variation. For example, in the data presented in Table IV-1, the ultracentrifuge was operated at the lowest possible speed, 807 rpm (hardly *ultra*); this was probably one of the few times that an

M determination was done with such a low centrifugal force—42 g—that one had to consider the vertical component of 1 g due to the earth's gravity.

The second technical advance was the method of optical mixing, beat-frequency spectroscopy developed at the Massachusetts Institute of Technology by George Benedeck; this allowed precise measurement of the diffusion coefficient D. With this technique a sample is illuminated with a laser beam. The particles in solution scatter the light, and because the particles are in motion, the scattered light shows a Doppler shift—i.e., the frequency of the scattered light is different from that of the incident light. This frequency shift is measured by allowing the incident and scattered beams to recombine and then determining the beat frequency.* Since the speed of the particles is a function of D, one can calculate D from the frequency shift. M is then obtained from the equation

$$ M = \frac{sRT}{D(1 - \bar{v}\rho)} $$

where s is the sedimentation coefficient, R is the universal gas constant, T is the absolute temperature, v is the partial specific volume of the molecule and ρ is the solvent density. (The partial specific volume is the volume increment produced in a solution when unit mass of solute is added to the solvent. Its measurement is tedious but generally straightforward.†)

It would have been more convenient to measure M for the DNA molecule itself with these methods. However the Yphantis procedure requires such a high DNA concentration that the resultant viscosity would necessitate running the centrifuge for an extraordinarily long period of time. In addition, like the Benedeck procedure, it too requires measurement of v, and for technical reasons, the determination of v for DNA is difficult and

generally yields a value with a large percent error.

The measurement of M of the phage by these two methods is described in two papers (Bancroft and Freifelder, 1970; Dubin et al., 1970), both of which are beyond the scope of this book. However the calculated values of M of the DNA obtained by the two methods are shown in Table IV-1.

The Determination of *M* for DNA by Sedimentation to Equilibrium in a Density Gradient

In 1957 Matthew Meselson, Franklin Stahl, and Jerome Vinograd developed the method of equilibrium centrifugation in a density gradient. With this technique, molecules were sedimented to equilibrium in a density gradient (usually of CsCl), extending both above and below the density range of the sedimenting molecules. Homogeneous high molecular weight sample material (e.g., DNA) forms a Gaussian band centered at the position in the density gradient corresponding to the density of the macromolecules. The theory showed that M is inversely proportional to the square of the width of the band at equilibrium, so that in principle M for DNA could be determined by this method.

It is assumed that the DNA would bind the positive ion (Cs^+) but not the negative ion (Cl^-) so that one would obtain M of CsDNA. This method is described in Article 10 by Meselson, Stahl, and Vinograd.

In the ensuing years the equilibrium-sedimentation method was used extensively. However, in all reported data the observed value of M, when M was greater than 10^6, appeared to be too low. In fact, from the results on a series of DNA molecules, for each of which the reported value of M was incorrect, one research group concluded that the observed value was always one-half the real value, and it was proposed that a correction factor of ½ merely be introduced into the equation used in order to account for all those factors not understood. This remedy was not especially satisfactory since the correction factor, when examined closely, appeared to vary slightly with M. The situation was ultimately

---

*The phenomenon of beating should be familiar to musicians. If two musical instruments playing the same note simultaneously are slightly out of tune, i.e., they are producing tones of close but not identical frequencies, a pulsating sound is heard—the beat—the frequency of which is the difference in the frequencies produced by the two instruments.

†For details, see D. Freifelder, *Physical Biochemistry*, 1976, W. H. Freeman and Company, Chapter 12.

resolved when the entire method was re-evaluated by Carl Schmid and John Hearst (1969) who showed that the measured value of M depends upon the DNA concentration in the particular measurement, and that the real value of M must be obtained by an extrapolation to zero concentration. The values of M determined in this way are listed in Table IV-1.

Electron Microscopy of DNA

The molecular weight of a DNA molecule can be calculated from its length if the mass per unit length is known. An obvious way to determine length is by direct observation of a DNA molecule by electron microscopy. Many groups developed techniques for visualization of DNA molecules with an electron microscope, but the molecules observed were always either aggregated with other molecules, highly stretched, or greatly tangled. In the late 1950's Albrecht Kleinschmidt (1959, 1962) developed an extraordinary method for visualizing DNA by electron microscopy. In this method the DNA is mixed with the protein, cytochrome C, and a droplet of the mixture is used to form a protein film on a water surface, which is then picked up onto a metal supporting grid. The DNA remains embedded in the film, coated with excess protein. Metal is then evaporated at a low angle onto the film.* Since the DNA projects above the film, excess metal is deposited on the DNA, which is thereby made visible (Fig. IV-2). The grid with its film is then examined with the electron microscope. The original work from Kleinschmidt's laboratory was published in German only, but Article 11 by Kleinschmidt's student Dimitrij Lang and coworkers is an excellent introduction to the method.

Lang went on to apply the Kleinschmidt technique to the problem of accurate determination of molecular length, l, and of mass per unit length, M/l. His efforts were important for two reasons: (1) The Kleinschmidt method is very simple and was becoming a routine procedure in many laboratories; (2) to determine M from a measurement of s one needed a minimum of 0.5 μg of DNA dissolved in a solu-

tion free of all substances absorbing ultraviolet light, whereas the standard amount for electron microscopy is 0.006 μg and as little as 0.0002 μg can be used. Article 12 is one of Lang's many important papers on the measurement of M.

A Valid s-M Relation

By 1970, then, four methods were available to determine M: an s-M relation, the CsCl procedure, end-group labeling, and electron microscopy. Each had advantages and disadvantages, and even today the optimal method is not yet clear. However for routine determination of M where a 10% error is allowable, any one of the methods is satisfactory. These methods and the general problem of determining M are summarized by me in Article 13.

New Methods Applicable to DNA of Very High M

The four methods for determining M were applicable only to M having a magnitude of 0.5–110×10^6. However, for bacterial DNA M is about 2–3×10^9 and for animal cells, it is 10 times greater. Molecules of this size became accessible to measurement only after Bruno Zimm developed two important improvements on the technique of viscometry, a physicochemical method for assessing molecular size. Throughout the 1960's determination of the s-M relation was a major goal. During this period many workers also attempted to obtain a $[\eta]$-M relation, where $[\eta]$ is the intrinsic viscosity determined by viscometry.* This was a worthwhile endeavor for two reasons: first, a laboratory that did not have a $30,000 ultracentrifuge could easily afford a $50 viscometer; second, s determination requires that the DNA sample be free of substances absorbing ultraviolet light (since the movement of the DNA in the centrifuge cell is measured by ul-

---

*In current work it is also common to deposit uranium oxide on the DNA by a chemical reaction.

*Intrinsic viscosity for very asymmetric molecules is for all practical purposes a measure of the length of the molecule. For DNA, length correlates with M. The symbol η refers to the viscosity measured, η_{rel} refers to the ratio of this viscosity to that of the solvent, and $[\eta]$ to the intrinsic viscosity. An instrument for measuring η is called a viscometer.

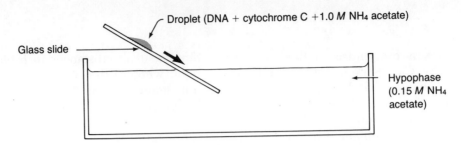

Droplet flows down and forms a protein film on air-water interface.

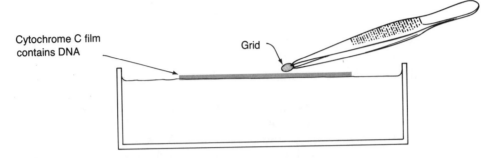

Film spreads across surface. Grid is touched to film surface surface so that support film on grid is in contact with protein film.

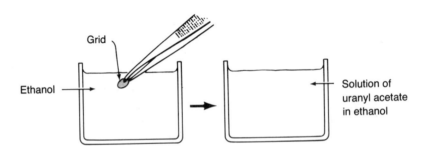

Sample is dehydrated by immersion in ethanol and then dipped into uranyl acetate solution for staining.

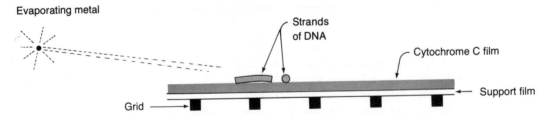

Grid is shadowed at very low angle while rotating

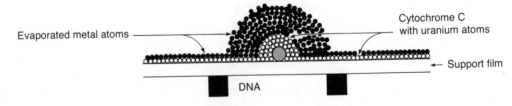

Enlarged view of a DNA strand coated with cytochrome C, stained and shadowed

Figure IV-2
Preparation of DNA for electron microscopy with the Kleinschmidt method. [From *Physical Biochemistry* by D. Freifelder. W. H. Freeman and Company. Copyright © 1976.]

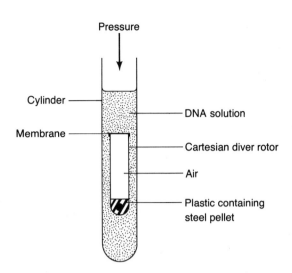

Figure IV-3
Schematic diagram of a Cartesian diver viscometer. A cylinder is partially filled with a solution whose viscosity is to be determined. The cylinder also contains an air-filled, weighted tube, which is covered by a flexible membrane. The inner tube, which is the rotor, normally floats in the sample but when air pressure of an appropriate magnitude is applied to the cylinder, the rotor (which is a Cartesian diver) moves below the liquid surface. In this way the rotor does not contact any surface of the liquid. The rotor contains steel pellets and is made to rotate by a spinning magnet situated below the cylinder. As the viscosity of the solution increases, the speed of rotation for a given magnetic torque decreases.

traviolet absorption optics), whereas the optical properties of a solution are irrelevant in viscometry.

However, viscometry does pose one major problem: the measurement of viscosity requires subjecting a solution to hydrodynamic shear forces, but DNA of high M is frequently broken by these shear forces (see page 35). To avoid this, Zimm developed a special low-shear viscometer consisting of a cylindrical Cartesian diver suspended in a DNA solution, which is in turn contained in a cylinder (Fig. IV-3). The Cartesian diver contains iron pellets and is made to rotate by a rotating magnetic field. Low shear is obtained by rotating the Cartesian diver very slowly. The speed of rotation of the Cartesian diver decreases as the viscosity of the solution increases and this speed can easily be measured; $[\eta]$ is calculated from the speed. Using this viscometer to measure $[\eta]$ for a series of DNA molecules of known M, Donald Crothers and Zimm (1965) obtained an $[\eta]$-M relation valid in the same range of M as was the s-M relation. Although there was no reason to believe that Zimm's equation would also apply for very high M—i.e., far greater than that of the largest molecule used to derive the equation—M for the $E.\ coli$ chromosome obtained in this way was within 20% of the value expected from the known DNA content per cell, thus suggesting the method was reliable.

While using this instrument, Zimm's group observed that when the magnetic field was turned off, the Cartesian diver did not gradually come to a stop; instead, it stopped rotating and then reversed direction before coming to a final stop. Zimm realized that they were observing viscoelasticity—i.e., the DNA molecules were stretched by hydrodynamic shear during the rotation of the Cartesian diver so that when the shear force was no longer applied, they sprang back or *relaxed* into the normal unstressed configuration. After this chance observation, Zimm developed a theory from which an equation relating the relaxation time to M was derived. This equation was then used to measure M for DNA molecules as large as those found in eukaryotic chromosomes.* The theoretical and early experimental papers from Zimm's laboratory are beyond the scope of this book, but much of the work done with this method has been succinctly reviewed in Article 14 by Ruth Kavenoff, Lynn C. Klotz, and Zimm.

---

*It is interesting to note that Zimm's first contribution was an instrumental improvement. This new instrumentation not only increased the sensitivity and reliability of viscosimetric measurements but also allowed a new phenomenon to be observed. His study of this new phenomenon, i.e., viscoelasticity of DNA, led to his second contribution, the development of an experimentally usable theory.

From the *Proceedings of the National Academy of Sciences*, Vol. 47, No. 8, 1113–1122, 1961. Reproduced with permission of the authors.

THE MOLECULAR WEIGHTS OF T2 BACTERIOPHAGE DNA AND ITS FIRST AND SECOND BREAKAGE PRODUCTS

BY I. RUBENSTEIN, C. A. THOMAS, JR., AND A. D. HERSHEY

BIOPHYSICS DEPARTMENT, JOHNS HOPKINS UNIVERSITY, AND DEPARTMENT OF GENETICS, CARNEGIE

INSTITUTION OF WASHINGTON

Communicated by A. D. Hershey, May 29, 1961

Recent evidence indicates that DNA molecules of uniform size and high molecular weight can be extracted from phage T2.[1,2] Controlled degradation by stirring yields fragments tentatively identified as halves and quarters of the original molecules.[3] In previous work, these materials were partially characterized by physical methods, but their absolute molecular weights remained unknown.

Our purpose in the work reported here was threefold: to determine the size and number of DNA molecules per phage particle, to check these measurements against the putative half and quarter molecules (and vice versa), and to calibrate sedimentation coefficients for DNA in terms of molecular weight in a high molecular weight range previously unmeasurable. These objectives are pertinent to the organization of the phage "chromosome," which by genetic criteria is a single structure.[4]

We show below that, contrary to earlier indications,[2,3] the particle of phage T2 contains indeed a single molecule of DNA of molecular weight at least 130 million. Some of our results are supported by those presented in the companion paper by Davison, Freifelder, Hede, and Levinthal.[5]

Our conclusions are reached by employing a method of molecular autoradiography.[6] When phage particles or DNA molecules labeled with P^{32} are embedded in a radiation-sensitive ("nuclear") emulsion, the emitted β-particles produce tracks which, emanating from a point source, give rise to characteristic figures recognizable as "stars." If it is assumed that every β-particle forms a track, counts of the average number of tracks per star measure the number of P^{32} atoms per particle and, depending on the specific radioactivity, the total phosphorus content per particle. Less accurately, the number of labeled particles can be measured from the ratio of stars to added polystyrene latex spheres.

Materials and Methods.—*Escherichia coli* strain H was grown for 2 hours with aeration at 36°C in 7 ml of a synthetic culture medium,[7] to which was added 0.2 mg/ml of bacteriological peptone and the required amount of radiophosphate (Squibb and Sons, 1,000 c/g). The total phosphorus content of the medium (about 3 μg/ml) was measured colorimetrically and the P^{32} content was determined by comparison with a standard P^{32} solution (calibrated by the National Bureau of Standards). After the bacterial mass had increased 10-fold and the number approached 10^8/ml, the culture was infected with three particles of phage T2H per bacterium. One hour later the culture was lysed by shaking with a few drops of $CHCl_3$ and digested serially for 10 minutes each with small quantities of deoxyribonuclease, ribonuclease, and pancreatin. After this treatment 10^{11}/ml of carrier phage, heavily irradiated with a germicidal lamp, were added and the lysate was filtered through a layer of analytical grade Celite on filter paper in a small Gooch crucible. A sample of the filtrate was diluted 10^6-fold in buffered peptone solution[8] for daily titration of surviving phage particles.

Five ml of the filtrate was mixed with 1.8×10^{13} purified unlabeled phage particles. An aliquot of this mixture was purified by centrifugation for assay of P^{32} and DNA contents, and for the star-size measurements of whole phage particles. The remainder of the mixture (2×10^{12} phage per ml) was shaken twice with phenol and five times with ether to yield 8 ml of DNA solution (about 0.4 mg/ml.)[9] The carrier phage provide a high concentration of DNA which serves to protect both the labeled and unlabeled molecules from breakage during extraction. Also, the

1113

large excess of unlabeled molecules prevents the aggregation of labeled molecules which would distort the autoradiographic results.

Aliquots of the DNA solution were stirred under conditions ("5,000 rpm") expected to produce mainly half molecules, or ("9,000 rpm") to produce mainly quarter molecules.[3] Stirred and unstirred samples were then passed through separate fractionating columns[9] and the resulting fractions were analyzed spectrophotometrically for DNA content and by radioassay for P^{32} content. Fractions were selected for embedding in the nuclear emulsion.

It should be noted that the procedure described avoids purification of radioactive phage particles before extraction of the DNA and relies on the column to effect final purification. This procedure prevents possible radiation damage to phage particles packed in the centrifuge, shortens the time required to get the DNA into the emulsion, and undoubtedly removes some broken or radiation-damaged DNA from the preparations.[9] We observed that the labeled DNA does in fact fragment at an appreciable daily rate, dependent on specific radioactivity, even when stored at effectively infinite dilution. In the experiments reported here, DNA was embedded in the emulsion within 24 hours or less after the time of synthesis. Less than four atoms of P^{32} per molecule of DNA decayed in the interim.

A few modifications of the autoradiographic procedures[6] were introduced mainly to minimize inadvertent mechanical breakage of the DNA. Samples were diluted with carrier DNA solutions (40 to 400 μg/ml) and volumes were measured in wide-mouth pipettes operated by screw-delivery. Solutions were mixed mechanically in tubes rotated at 10 rpm while inclined at 30° to the horizon. Five ml of melted emulsion and 1 ml of DNA (total 400 μg containing 10,000 to 50,000 labeled particles and an equal number of styrene latex spheres) were rotated for 5 minutes while submerged in water at 45°C. In a control test DNA of very high specific radioactivity was mixed with emulsion, and two drops of the mixture (the maximum amount compatible with satisfactory chromatography) were passed through the column. This test revealed that not more than 8% mechanical breakage was caused by the mixing.

Samples of the emulsion mixtures were poured into glass rings attached with paraffin to coated glass slides[6] to form a disk two to three mm thick before drying. After drying, the plates were stored for not more than two weeks and replicate samples were developed at appropriate intervals. They were viewed with an oil immersion objective at a total magnification of 575 ×. The specimens were counted by two observers (I. R. and C. A. T.) and the results were compared, usually with good agreement. Histograms shown in this paper include the results of both observers.

The number of styrene latex spheres[6] added to the emulsion was determined by counting an aqueous suspension in a calibrated Petroff-Hauser bacterial counting chamber. The ratio of spheres to stars counted in the emulsion measured the number of radioactive particles per ml and, together with the P^{32} assay, the number per phage equivalent of radioactive DNA.

Sedimentation constants were measured in an aluminum cell at 35,600 rpm, 10 μg DNA/ml, in 0.72 M NaCl, pH 6.7, at 22°C. and are reported without correction.[3]

DNA concentrations were measured spectrophotometrically in terms of the specific extinction (carefully measured for T2 DNA both before and after chromatographic purification) of 0.208 $cm^2/\mu g$ P or 0.0181 $cm^2/\mu g$ DNA. The latter figure, and all molecular weights cited in this paper, refer to the sodium nucleate with an average residue weight of 357 daltons, 8.7% P.

We also measured the specific extinction of phage particles: 0.311 $cm^2/\mu g$ P at 260 mμ. Since 98% or more of the phosphorus in phage particles is in DNA,[2] the absorbancy includes a contribution of 0.208/0.311 = 67% from the DNA and a contribution of 33% mainly due to light scattering. The absorbancy of phage particles was therefore corrected for scattering by subtracting 33% of the observed optical density.

Results.—Our principal results are derived from three independent experiments of the type just described. Their general plan is depicted schematically in Figure 1A.

Characterization of labeled DNA molecules: Our objective in this work is to characterize *unlabeled* DNA molecules. On the other hand, the autoradiographic information is obtained from *labeled* molecules. Therefore, it is important to demonstrate that the two classes of molecules are identical in the properties we seek to measure.

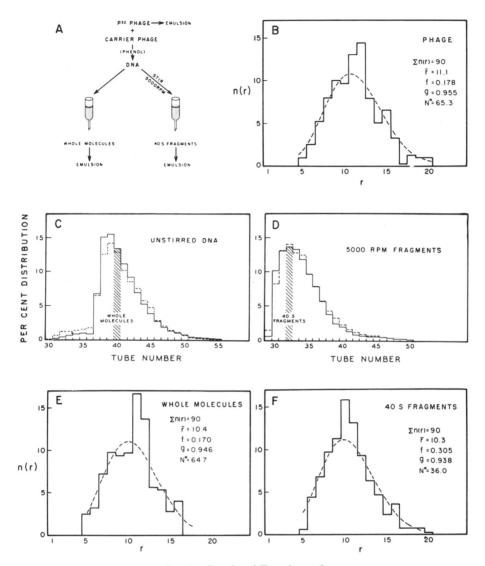

Fig. 1.—Results of Experiment 2.

A, experimental plan.   B, E, and F, frequency distributions of star sizes.
C and D, DNA chromatograms.   Solid line, optical density;   broken line, P³².
The shaded fractions were selected for autoradiography.

In Figures 1 and 2 the chromatograms of the mixed DNA preparations are shown. In all cases the elution of the P³² and optical density overlaps quite well.   This indicates that, of the molecules which pass through the column, the two classes have similar chromatographic properties and similar sensitivity to breakage by stirring.

There is little preferential retention of labeled molecules by the column.   This is evidenced by the constancy of the specific activity determinations listed in Table 1.   Here one can see that the specific activity of the mixed phage stock or of the extracted DNA is essentially the same as that of the DNA fractions collected from the column.

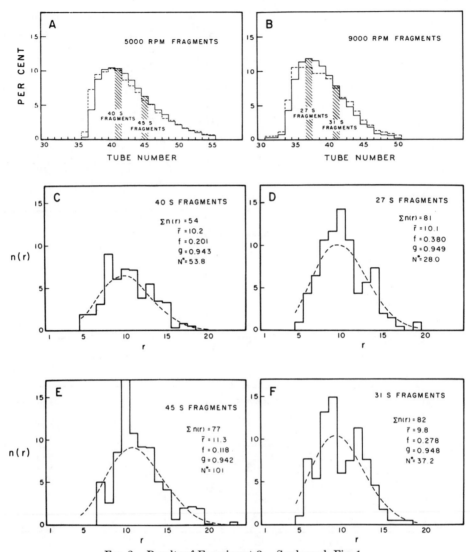

Fig. 2.—Results of Experiment 3. See legend, Fig. 1.

To summarize, labeled and unlabeled DNA respond similarly to stirring and column chromatography.

The first two entries in Table 1 show determinations of the specific radioactivity of the growth medium, the first by direct measurement, the second by employing an empirical relation between the rate of phage inactivation by "suicide" and the specific activity of the growth medium.[10, 11] The two determinations agree. The suicide experiments gave linear plots indicating that more than 90% of the viable phage had been uniformly labeled at the specific activities quoted.

Table 2 lists the sedimentation velocity measurements that were made on pertinent fractions from the column. Because of the similar response of the labeled and unlabeled DNA to stirring and column chromatography it is assumed that their sedimentation constants were also the same.

TABLE 1

SPECIFIC RADIOACTIVITY MEASUREMENTS

	Expt. 1	Expt. 2	Expt. 3
Growth medium (c/g P)	71	49	109
Same, from suicide rate of phage*	70	54	106
Labeled + carrier phage†	13,000	3,800	1,000
Extracted DNA‡	9,200	3,800	950
Unbroken DNA from column	13,000	3,800	. . .
5000 rpm, 45 S fragments	. . .	. . .	740
5000 rpm, 40 S fragments	13,000	3,700	790
9000 rpm, 31 S fragments	. . .	. . .	830
9000 rpm, 27 S fragments	. . .	. . .	750

* Equation (2) of Hershey *et al.*,[10] $\alpha N = 43,000$.
† Cpm/μg DNA, corrected for 33% scattering contribution to the absorbancy at 260 mμ. The specific radioactivities of DNA are not simply related to those of the growth medium because of the arbitrary admixture of carrier phage.
‡ Cpm/μg DNA. Sample purified before radioassay by Schmidt-Thannhauser fractionation.[7]

TABLE 2

SEDIMENTATION COEFFICIENTS OF DNA FRACTIONS

	Expt. No.	s (Svedbergs)
Whole molecules	1	61.4 (. . ., 64.0)*
45 S fragments	3	45.3 (43.9, 47.2)
40 S fragments	1	40.7 (38.7, 41.6)
"	2	40.0 (36.8, 43.3)
"	3	40.0 (38.4, 41.9)
31 S fragments	3	31.3 (29.5, 31.7)
27 S fragments	3	27.4 (23.8, 27.7)

* Figures in parentheses are sedimentation coefficients of fractions eluting from the columns imediately before and after, respectively, the fractions used for star counts. They serve as checks.

Autoradiography: The β-ray stars arising from the labeled phage particles and the DNA molecules were counted and grouped into size classes according to the number of rays per star. Some of the resulting frequency distributions are shown in Figures 1 and 2. The dotted curve in each of the histograms represents a Poisson distribution having the same mean and size as the star population. As one can see, all of the star populations follow a distribution reasonably close to the Poisson distribution. In each star histogram the following numbers are quoted: $\Sigma n(r)$, the number of stars counted in this collection; $\bar{r}$, the mean number of rays per star; f, the fraction of P^{32} atoms that decayed in the emulsion; g, the fraction of P^{32} atoms that remained at the time of embedding (used to correct all N^* values to the time of phage growth); and N^*, the calculated number of P^{32} atoms per star-forming unit found by dividing $\bar{r}$ by fg.

The autoradiographic results are summarized in Table 3. In the first column for each of the three experiments is seen N^*, the calculated number of P^{32} atoms per star-forming unit, and in brackets the total number of stars counted. In some cases these data were collected from plates that had been exposed for different lengths of time. The values of N^* were in good agreement indicating that the efficiency of β-ray detection was the same for plates stored for different lengths of time.[6, 12]

The number of P^{32} atoms per phage, N_0^*, was determined in each experiment. Hence, the ratio N^*/N_0^* in Table 3 indicates the fractional P^{32} content of the DNA molecules as compared to the P^{32} content of the phage particle. This ratio is approximately one for the whole DNA molecule, a fact which indicates that there is only one molecule per phage particle. This ratio is about one-half for the 45 S and

TABLE 3
RELATIVE MOLECULAR WEIGHTS

	Experiment 1			Experiment 2			Experiment 3		
	N^*	N^*/N_0^*	$(s/s_0)^{1.88}$	N^*	N^*/N_0^*	$(s/s_0)^{1.88}$	N^*	N^*/N_0^*	$(s/s_0)^{1.88}$
Phage	92.4 (31)	1.00	...	65.3 (90)	1.00	...	130.1 (91)	1.00	...
Whole molecules	83.4 (178)	0.90	1.00	64.7 (90)	0.99	1.00	...	...	1.00
45 S fragments	...	...	...	...	...	...	94.9 (128)	0.73†	0.54
40 S fragments	39.3 (19)	0.43	0.42	33.1 (196)	0.51	0.43	53.8 (54)	0.41	0.43
31 S fragments	...	...	...	...	...	...	37.2 (82)	0.29	0.27
27 S fragments	...	...	...	...	...	...	28.0 (81)	0.22	0.21

N^* = number of P^{32} atoms per star-forming unit measured autoradiographically, with number of stars counted in parentheses.
N^*/N_0^* = fractional phosphorus content.
$(s/s_0)^{1.88}$ = fractional molecular weight from sedimentation coefficients in Table 2 according to Burgi and Hershey.[3]
† This measurement is presumably in error.

40 S fragments, which represent "half molecules," and about one-quarter for the 31 S and 27 S fragments, which represent "quarter molecules."

Finally, the fractional molecular weight, $(s/s_0)^{1.88}$, calculated from the sedimentation coefficient is listed in the third column. In this expression s_0 is the nominal average (63) of many measurements on unbroken DNA (values range from 60 to 65). The exponent 1.88 is taken from the work of Burgi and Hershey,[3] whose estimates are $1/0.55$ for the upper portion and $1/0.51$ for the lower portion of the range of s here considered. One may see that the fractional phosphorus content determined by autoradiography corresponds with the fractional molecular weight computed from the sedimentation coefficient.

An independent measure of the number and size of the fragments can be obtained by employing polystyrene latex spheres to measure the number of stars per phage-equivalent of radioactivity. For experiments 1 and 2, respectively, this ratio was 1.52 and 1.15 for intact phage, and 1.14 and 1.36 for unbroken DNA molecules. In experiment 3, the number of stars per phage equivalent of counts ranged from 1.35 for the 45 S fragments to 5.68 for the 27 S fragments. Although these data are qualitatively clear, they lack the precision that was expected. After a consideration of many sources of error, we concluded that we were undercounting the number of polystyrene indicator particles.

Absolute molecular weights: Our conclusion that the T2 phage particle contains a single molecule of DNA reduces the problem of molecular weight measurement to that of determining the amount of DNA per phage particle. From the autoradiographic data of Table 3, five such estimates can be made. The appropriate relation is

$$M = 1.05 \times 10^8 \frac{N^*}{A_0} \tag{1}$$

where A_0 is the specific radioactivity in c/g, the DNA is assumed to contain 8.7%

P, and all β-emissions are assumed to produce countable tracks in the emulsion. The estimates of N^* (from the counts of phage particles and unbroken DNA molecules) listed in Table 3 together with the specific activities given in Table 1 yield five measures of M ranging from 123 to 140 $\times$ 10[6], with a mean of 133 $\times$ 10[6]. The measurements of N^* are evidently repeatable (as is also shown by the measurements of broken DNA) and are subject to a purely statistical error of only a few percent. The measurements do contain a known bias, since it is not certain that all of the emissions produce countable tracks in the emulsion. On this basis 133 million is a minimum estimate.

A maximum molecular weight is obtained by direct measurements of DNA per infective phage particle. A careful analysis of three phage preparations yielded an optical cross section at 260 mμ of 7.3, 7.5, and 7.6 $\times$ 10[-12] cm²/plaque-forming particle and 2.3, 2.3, and 2.5 $\times$ 10[-11] μg P per plaque-forming particle, respectively. These yield an average sodium polynucleotide molecular weight of 162 $\times$ 10[6], which is a maximum since it is unlikely that every phage particle forms a plaque.

We conclude that the sodium polynucleotide molecular weight of T2 DNA lies between 130 $\times$ 10[6] and 160 $\times$ 10[6]. The lower limit is consistent with a recent estimate of the weight of the whole phage particle based on sedimentation-diffusion measurement.[13] The phage particle weight was found to be 215 $\times$ 10[6] daltons, which correspond to a DNA molecular weight of 130 million if the phosphorus content of the dry phage is 5.2%.[14]

If the molecular weight is taken to be 130 $\times$ 10[6], the relationship between sedimentation coefficient and molecular weight is that shown in Figure 3. The best straight line corresponds to the equation

$$s = 0.00244 \, M^{0.543} \qquad (2)$$

The coefficient of this equation has a standard error of less than 10%, while the error in the exponent is less than 1%. This relation disagrees with a commonly employed extrapolation from the low molecular weight region.[15] The exponent 0.543 agrees with that arrived at by stepwise degradation by stirring.[3]

Systematic errors: The errors in the autoradiographic technique have been discussed previously[6] and only a few comments will be made here.

First, we recall that the possibility that our results are biased by molecular aggregation is ruled out by the method of measurement, in which a small minority of radioactive molecules is examined in the presence of a large excess of carrier DNA.

It is possible that there is some difficulty in the enumeration and classification of the stars shown in the histograms in Figures 1 and 2. Stars with only a few (five to six) rays can be classified accurately, but they are not always seen. Stars with 18 to 25 rays are easy to find, but there is probably some tendency to undercount the number of rays. Thus there is a tendency to overestimate the mean number when it is small, and to underestimate it when large. When the mean lies between 9 and 13 rays per star, there may be a tendency to see a narrower size distribution than actually exists.

To counteract this tendency, plates that had been exposed for various periods were examined. After long exposures, it proved feasible to count rays per star for small stars and at the same time score the total number of stars. The observed frequency of small stars was in reasonable agreement with the Poisson distribution based on counts made at earlier times.

In the extreme case, the tendencies discussed above could be expected to generate a spurious Poisson distribution when the distribution was in fact very broad. Three experiments with phage T4 performed shortly before those reported above are pertinent in this connection. In the T4 experiments, radioactive and carrier phage particles were mixed and purified by differential centrifugation, and DNA was extracted from them either by osmotic shock or by the phenol method. In each case, the DNA preparations yielded many stars equivalent in size to those produced by phage particles, but also yielded smaller stars. Figure 4 illustrates the broad,

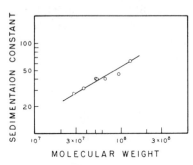

Fig. 3.—The relation between auto-radiographically determined molecular weight and sedimentation coefficient for T2 DNA and its breakage products. The straight line is given by equation (2).

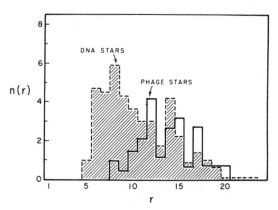

Fig. 4.—Frequency distribution of star sizes for T4 phage and DNA. The DNA was apparently partly broken during preparation.

often bimodal, distributions observed. These results certainly do not mean that the DNA of T4 differs from that of T2. Both have the same sedimentation coefficient and are broken at the same critical speed of stirring. The two phages show the same DNA content per infective particle, and are closely related by all biological criteria. For reasons that have not been determined, the T4 DNA preparations must have been partly broken before embedding. Since each break produces two fragments, very little breakage is sufficient to distort the histograms.

For present purposes, the results of these unsuccessful experiments are valuable in showing that one can detect by autoradiography heterogeneous DNA populations when they exist.

Another source of error anticipated in these experiments failed to materialize. Since DNA solutions are highly viscous, the molecules must be greatly extended in solution. In the extreme cases, such molecules might fail to produce countable stars or might produce spuriously small ones. In point of fact, some two or three per cent of the countable stars appeared to have diffuse centers as much as 10 to 20 microns across. The remainder of the DNA stars, like all of those produced by phage particles, emanated from a single point. Evidently most of the molecules are crumpled in the emulsion so as to lie within volumes not greatly exceeding one micron in diameter. This is shown most convincingly by the similarity of the star-size distributions of T2 phage particles and whole DNA molecules in the present work (Figs. 1*B* and *E*).

To summarize, after considerable experience we have not found any significant source of error in the autoradiographic method that might tend to bias the results or alter the conclusions presented in this paper.

The stability of the DNA molecule: We refer to the DNA structures isolated from phage T2 as "molecules" because of their physical and functional uniformity. The possibility remains that nonphosphodiester linkages exist uniting the polynucleotide chains. It is appropriate to summarize some results which indicate that the molecule is a stable structure.

Passing DNA through the column does not alter its sedimentation coefficient or the band width seen in a CsCl density gradient. Banding in CsCl does not alter the chromatographic behavior. Heating whole molecules to 70°C for 10 minutes in 0.1 M NaCl, 0.032 M PO$_4$, pH 7 does not alter the chromatographic behavior of T2 DNA nor diminish its intrinsic viscosity. The star population from T4 DNA (similar to that in Fig. 4) is not altered by heating under the conditions stated above. The intrinsic viscosity of T4 DNA is not altered by prolonged exposure to protease.[2] Study of the kinetics of breakage[3] indicate that the molecules are not characterized by a few weak bonds. The presumed separation of polynucleotide chains from the helical structure is accompanied by only a two-fold decrease in apparent molecular weight.[16] Thus, if unknown linkages exist, they must be at least as stable as the polynucleotide bonds.

Discussion.—Our first conclusion, that the particle of T2 contains a single molecule of DNA, seems quite unambiguous and has been reached independently by Davison *et al.*[5] Previous autoradiographic studies[6, 12] leading to a different conclusion were undoubtedly influenced by the then unknown susceptibility of the molecules to mechanical breakage,[17] though certain details of the early results remain a mystery.

The finding that there exists a single DNA molecule in T2 bacteriophage is consistent with the fact that the known genes in T2 form a single linkage group[4] and with the idea that the "chromosome" is a molecular structure. Phage T2 belongs to a class of phages having the largest known particle size. One of the smallest, phage ϕX-174, also contains a single DNA molecule, in this case single-stranded, with a molecular weight of only 1.7×10^6 (see ref. 18). One may anticipate that other phages will be found to contain a single molecule of nucleic acid. If so, it follows that different viruses will provide a continuous series of unique DNAs covering a wide range of molecular weights.

Our second conclusion, that molecules of T2 DNA break successively into halves and quarters at critical rates of shear, suggests that the molecules are linear. Recent evidence that the genetic map of phage T4 is circular[19] might suggest that the DNA molecule is ring-shaped. Ring-shaped molecules would be expected to break under shear into fragments of $^1/_3$ to $^1/_5$ the original length of the molecule.[3] Since T2 and T4 DNAs are very similar, it is unlikely that either molecule has a circular shape, at least after phenol extraction.

Our third conclusion, that the molecular weight of T2 DNA lies between 130×10^6 and 160×10^6, likewise seems unambiguous. For the molecular weight to be lower than these limits, it would be necessary for the observers to count β-disintegrations that did not occur, which is unlikely, if not impossible. Moreover, the molecular weight measured autoradiographically is supported by estimates based on the mass and composition of the phage particles.[13]

The physical-chemical methods of molecular weight determination of DNA do not provide much assistance here; indeed it was one of the objectives of this work to relate molecular weight and sedimentation coefficient. In short, the molecular weights calculated from sedimentation and intrinsic viscosity are about a factor of two too low to agree with the autoradiography. Molecular weights calculated from band widths in CsCl are also too low.[2]

Summary.—The bacteriophage T2 contains a single molecule of DNA which accounts for virtually all of the phosphorus of the virus particle.

The first and second breakage products produced by controlled stirring are, on the average, one-half and one-quarter of the total molecular weight.

The sodium salt of T2 DNA has a molecular weight of at least 130×10^6 and less than 160×10^6.

An empirical relation between molecular weight and sedimentation coefficient over a five-fold range downward from 130×10^6 is proposed.

This work was aided by grants from the Atomic Energy Commission (AT(30-1)-2119) and from the National Institutes of Health (E-3233 and C-2158).

It is a pleasure to acknowledge the contribution of Mr. T. C. Pinkerton, who performed several equilibrium sedimentations pertinent to this work.

[1] Hershey, A. D., and E. Burgi, *J. Molec. Biol.*, **2**, 143–152 (1960).

[2] Thomas, C. A., Jr., and K. I. Berns, *J. Molec. Biol.*, in press.

[3] Burgi, E., and A. D. Hershey, *J. Molec. Biol.*, in press.

[4] Streisinger, G., and V. Bruce, *Genetics*, **45**, 1289–1296 (1960).

[5] Davison, P. F., D. Freifelder, R. Hede, and C. Levinthal, these PROCEEDINGS, **47**, 1123–1129 (1961).

[6] Levinthal, C., and C. A. Thomas, Jr., *Biochim. et Biophys. Acta*, **23**, 453–465 (1957).

[7] Hershey, A. D., and N. E. Melechen, *Virology*, **3**, 207–236 (1957).

[8] Matheson, A. T., and C. A. Thomas, Jr., *Virology*, **11**, 289–291 (1960).

[9] Mandell, J. D., and A. D. Hershey, *Anal. Biochem.*, **1**, 66–77 (1960).

[10] Hershey, A. D., M. D. Kamen, J. W. Kennedy, and H. Gest, *J. Gen. Physiol.*, **34**, 305–319 (1951).

[11] Stent, G. S., and C. R. Fuerst, *J. Gen. Physiol.*, **38**, 441–458 (1955).

[12] Thomas, C. A., Jr., *J. Gen. Physiol.*, **42**, 503–523 (1959).

[13] Cummings, D. J., and L. M. Kozloff, *Biochim. et Biophys. Acta*, **44**, 445–458 (1960).

[14] Herriott, R. M., and J. L. Barlow, *J. Gen. Physiol.*, **36**, 17–28 (1952).

[15] Doty, P., J. Marmur, J. Eigner, and C. Schildkraut, these PROCEEDINGS, **46**, 461–476 (1960).

[16] Berns, K. I., and C. A. Thomas, Jr., *J. Molec. Biol.*, in press.

[17] Davison, P. F., these PROCEEDINGS, **45**, 1560–1568 (1959).

[18] Sinsheimer, R. L., *J. Molec. Biol.*, **1**, 43–53 (1959).

[19] Streisinger, G., R. S. Edgar, and G. Harrar, personal communication.

From *Journal of Molecular Biology*, Vol. 5, 643–649, 1962. Reproduced with permission of the authors and publisher.

The Physical Properties of the Deoxyribonucleic Acid from T7 Bacteriophage

Peter F. Davison and David Freifelder†

Department of Biology, Massachusetts Institute of Technology
Cambridge, Mass., U.S.A.

(*Received 24 May 1962, and in revised form 15 August 1962*)

The isolation and the physical characterization of the DNA from T7 bacteriophage are described. The DNA is shown to be homogeneous by ultracentrifugation and chromatography. It is probable that this DNA preparation yields a monodisperse solution of molecules of molecular weight 19×10^6.

1. Introduction

Physical studies of DNA should be carried out preferably on monodisperse preparations, so that the measured physical properties of the samples may be related to the characteristics of the individual molecules; until the present time, however, most investigations have been made on animal or bacterial DNA preparations containing a polydisperse mixture of molecules and the physical properties correspondingly relate to a sample of unknown and frequently irreproducible molecular weight distribution. In the search for a suitable source of a homogeneous DNA, coliphage T2 was studied; it was shown that the DNA in this virus consists of one large molecule (Davison, Freifelder, Hede & Levinthal, 1961) but the difficulties attendant upon manipulating solutions of DNA of this size (Davison, 1959) and the presence of an atypical and glucosylated base (5-hydroxymethylcytosine) caused us to turn our attention to smaller viruses with DNA of more normal base composition. The properties of such a phage, T7, have been described in the accompanying paper (Davison & Freifelder, 1962). In the present paper we describe some properties of the DNA isolated from phage T7.

2. Materials and Methods

Isolation and handling of T7 DNA. Preparation and purification of coliphage T7 are described in detail in the accompanying paper (Davison & Freifelder, 1962). The most satisfactory method for preparing T7 DNA is by phenol extraction (Mandell & Hershey, 1960). As will be discussed later, the criterion of a good preparation is taken to be the appearance of a sharp sedimenting boundary in the ultracentrifuge.

DNA was prepared only from phage stocks purified by CsCl banding. The phage at an optical density at 260 mμ (o.d.$_{260}$) of at least ten was dialysed overnight against the desired solvent (usually 0·1 to 0·01 m-phosphate buffer, pH 7·8, or 0·1 m-NaCl) and then gently agitated with an equal volume of water-saturated, colorless phenol. In weakly buffered or unbuffered solvents it was necessary to add alkali to the water saturating the phenol until the water was neutral. In the absence of this precaution the pH of the aqueous layer was found to be as low as 2 and from such phenol DNA was obtained in low yield and partly degraded. The phenol could not be stored in this neutralized state because it discolored rapidly.

† Present address: University Institute of Microbiology, Copenhagen, Denmark.

The mixture, after agitation, was briefly centrifuged to separate the phenol from the DNA solution. The phenol layer was removed by aspiration and the phenol which remained dissolved in the aqueous layer was extracted by successive ether washes. To avoid shear degradation during the ether extractions, the solutions were mixed by gentle rocking.

The ether was both redistilled and stored over aqueous $FeSO_4$. Just before use the ether was shaken with $FeSO_4$, followed by 0·1 M-EDTA and distilled water. A trace of $FeSO_4$ was often sufficient to degrade the DNA.

The removal of phenol from the DNA solution was assessed by the O.D.$_{260}$/O.D.$_{280}$ ratio which should be close to 1·8 (phenol absorbs strongly at 280 mμ). Approximately 5 mg DNA is obtainable from 1 liter of bacterial lysate.

DNA could be released by treating the phage with pH 11 buffer solutions or with EDTA; at high pH, or in the absence of divalent ions, the phage slowly disrupted. Treatment at pH 11 sometimes gave rise to a slightly degraded preparation, presumably because of hydrolytic damage. Attempts to release the DNA by treating the phage with the phase II lipopolysaccharide isolated by Jesaitis & Goebel (1955) were unsuccessful.

The DNA was stored at 4°C in solution over chloroform. By the criterion of a sharp sedimenting boundary, no degradation occurred over a period of several weeks. However, preparations left with a low O.D.$_{260}$/O.D.$_{280}$ ratio were sometimes degraded during storage. This was assumed to result from chemical effects of traces of phenol.

Ultracentrifugation. Sedimentation velocities were measured in the Kel-F cells of the Spinco model E ultracentrifuge; ultraviolet optics were used for concentrations less than 30 μg/ml. and schlieren optics for higher concentrations. Ultraviolet photographs were scanned with a Joyce–Loebl recording microdensitometer; schlieren plates, with a Gaertner microcomparator. In most experiments the solution was centrifuged at 33,450 rev./min. At a given concentration of DNA the sedimentation coefficient was not reproducible in low ionic strength solvents. In 0·2 M-NaCl or 0·15 M-NaCl–0·015 M-sodium citrate, the $S_{20,w}$ for DNA at 20 μg/ml. was 30 ± 2 s. This variability was unaffected by the addition of either 10^{-3} M-MgCl$_2$ or EDTA. In the absence of definitive experiments it was assumed that the irreproducibility resulted from instability of the sedimenting boundary. All subsequent experiments were performed in 1 M-NaCl. In the presence of a density gradient provided by the distribution of a concentrated salt solution in the centrifugal field, the sedimentation velocity was reproducible and at a concentration of 20 μg/ml. $S_{20,w}$ was $30 \pm 0·4$ s.

Viscometry. Intrinsic viscosity was determined with an Ubbelohde four-bulb, multi-gradient dilution viscometer with a capillary bore of 1 mm. The viscometer was immersed in a water bath maintained at 30°C, constant to $\pm 0·002$°C.

Chromatography. Chromatography was performed in the methylated serum albumin columns described by Hershey & Burgi (1960). The stock solutions used to create the gradient were 0·4 and 0·84 M-NaCl.

3. Results

Sedimentation properties. Figure 1 shows a densitometer trace of an ultraviolet photograph of T7 DNA sedimenting in the ultracentrifuge. This is not a self-sharpened boundary, since other molecules sedimenting with the T7 DNA can be resolved (Fig. 2(a)), and any degraded DNA added is readily discerned by the appearance of material of lower sedimentation coefficient (Fig. 2(b)).

Figure 1 shows that the T7 DNA boundary has two features, a nearly vertical rise, and a pronounced curvature at the top of the trace; the relative contribution of each depends upon the conditions of centrifugation. For DNA concentrations about 20 μg/ml. the boundary is as shown in Fig. 1. When concentrations are lower, the upper curvature becomes more pronounced; for higher concentrations it is lessened. At concentrations below 5 μg/ml. the curvature starts so low that the entire boundary is canted. The shape of the boundary is also speed dependent. As the speed is lowered, the canted boundary obtained at low DNA concentrations is sharpened but as the speed is increased, the boundaries for all DNA concentrations broaden. Solvent composition is also important. For instance, at a DNA concentration of 5 μg/ml. the boundary sedimenting in M-NaCl is sharper than the boundaries in 0·2 M-NaCl, 0·2 M-CsCl and 0·15 M-NaCl–0·015 M-sodium citrate; similar to those in 2 M-NaCl, M-CsCl or 3 M-CsCl; and broader than that in M-NaCl + 20% glycerol.

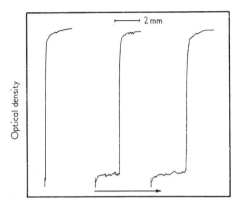

Fig. 1. Photometric traces of ultraviolet absorption photographs of sedimenting T7 DNA (concentration 20 μg/ml., solvent M-NaCl). Photographs taken at 0, 20 and 28 min after reaching speed of 33,450 rev./min. The vertical bar at the left of each trace is the meniscus. The arrow shows the direction of sedimentation. The scale indicates radial distance in the centrifuge cell.

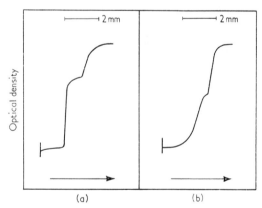

Fig. 2. Photometric traces of ultraviolet absorption photographs of: (a) a mixture of T5 and T7 bacteriophage DNA, the latter being the slower boundary; (b) a mixture of intact and sonically degraded T7 DNA. In each experiment the total DNA concentration was 20 μg/ml., solvent M-NaCl, speed 33,450 rev./min.

Figure 3 shows the reciprocal of the median sedimentation coefficient as a function of concentration for nine different T7 DNA preparations. $S_{20,w}$ extrapolated to zero concentration is 32·2 ± 0·5 s.

The sedimentation coefficient of the DNA is independent of centrifuge speed between 19,160 rev./min and 44,770 rev./min for concentrations of 12 and 18 μg/ml.

Viscosity. Samples from two separate preparations of T7 L DNA were dialysed against M-NaCl or SSC (0·15 M-NaCl–0·015 M-sodium citrate). At each concentration the viscosity plotted as a function of shear gradient showed a clear upward curvature.

Since the shear range obtainable in the viscometer is 230 to 100 sec^{-1} any extrapolation to zero shear is a highly arbitrary and subjective procedure. In these experiments a straight line was drawn as close to the points as possible. The intrinsic viscosity at zero shear estimated in this way in each experiment was 86 and 88 dl./g in M-NaCl, and 93 and 91 dl./g in SSC.

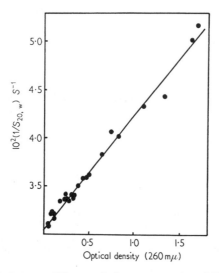

FIG. 3. The relationship between $1/S_{20,w}$ and the concentration of T7 DNA; solvent M-NaCl.

Chromatography. The DNA was eluted from the chromatographic column as a single peak at a salt concentration of 0·66 molar. This peak was not significantly narrower than that for T2 phage DNA. From several tubes, samples having an optical density greater than 0·2 were centrifuged. In each case a sharp boundary resulted with no significant variation of $S_{20,w}$ with tube number. We conclude that the DNA is chromatographically homogeneous.

Shear insensitivity. According to the calculations of Levinthal & Davison (1961), the critical shear for halving a DNA molecule of 19 million molecular weight should be approximately 50-fold greater than that needed to break a molecule of T2 DNA. Such shear rates are unlikely to be imposed in ordinary laboratory operations. T7 DNA can be degraded by being forced at high pressure through a no. 27 hypodermic needle but calculation showed that at the high flow rate obtained the flow was probably turbulent and hence the rupturing stress could not be determined. As expected, with a no. 22 needle no precautions need be taken when filling ultracentrifuge cells and the DNA may be safely pipetted.

DNA from T7 L and T7 M. In the preceding paper (Davison & Freifelder, 1962), we described two strains of T7 phage (L and M) which differed by 0·016 g/ml. in buoyant density in a CsCl gradient. Within experimental error no difference could be found with respect to nitrogen or phosphorus content. When DNA preparations from these two strains were banded together, they behaved as a single species, i.e. when equilibrium was reached, a single gaussian band resulted; this band had approximately the same variance as that of either of the DNAs banded alone. It is clear that the density difference between the two strains of phage cannot be attributed to density difference in the two DNAs. Furthermore, the sedimentation properties of the two DNAs are indistinguishable and sedimentation of a 50:50 mixture of L and M DNA gives a single boundary. At present, work is continuing to determine the differences between the two strains.

4. Discussion

We have described a DNA which we believe to be exceptionally well-suited for physical studies.

The experiments we have performed do not demonstrate the homogeneity of T7 DNA—least of all the chemical uniformity of all the molecules in solution—but on the other hand by ultracentrifugation or chromatography, we could detect no heterogeneity; from the nature of the source it is likely that this DNA comes close to being an ideally monodisperse preparation.

Having made the claim that no heterogeneity is detectable in the T7 DNA by ultracentrifugation, we must defend that statement against certain features of the sedimentation diagram which may appear to contradict it. A monodisperse preparation of any sedimenting material with a low diffusion coefficient should theoretically give rise to a densitometer trace resembling a step function, but experimentally a detectable slope will be imparted to the trace by optical imperfections and rotor wobble in the centrifuge, the movement of the boundary during the photographic exposure and the finite width of the densitometer scanning slit. When account of these factors is taken (by comparing the theoretical and measured width of the trace of the reference edge in the counterbalance) it is found that little or no boundary spread ascribable to diffusion or polydispersity is detectable in the lower part of the trace of sedimenting T7 DNA.

The upper curvature of the densitometer trace remains an unexplained anomaly, but it is one which has been observed also in T2 DNA (Hearst & Vinograd, 1961) and for several viruses, including T1, T3, T4, T5 and T7 bacteriophages (Davison & Freifelder, unpublished work) and tobacco mosaic virus (Hearst & Vinograd, 1961). On the basis of the stabilizing effect of viscosity (high DNA concentration and the glycerol–salt solvent) and density gradients (strong salt solutions), we believe the anomaly reflects boundary instability.

The ultracentrifugal and chromatographic behavior of T7 DNA is consistent with that of a monodisperse preparation. However, in the absence of molecular weight determinations by any absolute method there is at present no direct evidence that the phage contains only a single DNA molecule. By analogy with T2 and ϕX174 bacteriophages and a number of RNA viruses—all of which have been shown to contain single nucleic acid molecules—it seems likely that T7 similarly contains one molecule. This conclusion is supported by Meselson's studies (1959) which showed that the DNA of T7 is replicated as a single semi-conserved structure.

The molecular weight we have calculated from CsCl banding is 15×10^6, in fair agreement with the value of 18×10^6 found by Meselson (1959). Although these values are close to the calculated DNA content of the phage (19×10^6, Davison & Freifelder, 1962), the agreement must be accepted with reservation until account has been taken of the effects of selective solvation (Ifft, Voet & Vinograd, 1961) and of density heterogeneity. The difficulties of relating CsCl gradient bandwidths to molecular weights have been discussed by Pinkerton & Thomas (6th Annual Meeting of the American Biophysical Society, 1962).

The physical homogeneity of the DNA requires that either the phage contains one molecule of molecular weight 19×10^6 or two or more molecules of submultiples of this size; support for the idea of a single molecule comes from the sedimentation coefficients of the intact T7 DNA and its sheared fragments. These values of $S_{20,w}$ are inserted in Fig. 4 together with the empirical relationships between $S_{20,w}$ and molecular weight derived by Doty, McGill & Rice (1958) and Eigner (1960). While in reasonable agreement with these curves, our values of $S_{20,w}$ do not agree with the more recent empirical curve derived from measurements on T2 DNA shear products (Rubenstein, Thomas & Hershey, 1961). However, until more is known concerning the effects of the glucosylation on hydrodynamic properties, T2 DNA must at present be considered atypical.

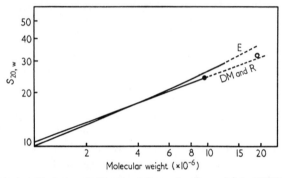

FIG. 4. The relationship between $S_{20,w}$ and the molecular weight of DNA as derived by Doty, McGill & Rice (1958) and Eigner (1960). The appropriate values for T7 DNA (○) and its first shear (●) product are superimposed. The DNA was sheared by passage through a no. 27 hypodermic needle. The sheared product had an $S_{20,w}$ of 25 s.

If it is accepted that the T7 DNA comprises a single molecule, the molecular weight and the viscosity of this DNA are *not* concordant with the empirical relationship of Eigner (1960) which would predict a viscosity 20% greater than that measured. The reason for this disparity is not known but it could result from an inadequate extrapolation to zero shear gradient in the present experiments. It may be noted that Katz (1961) recently reported a viscosity for a calf thymus DNA preparation much lower than that predicted for the molecular weight of the DNA from these same empirical relationships.

Although T7 phage is a very convenient source of monodisperse DNA we do not consider it to be a unique source; we would expect that homogeneous DNA can be obtained from any bacteriophage or DNA virus. For instance, the DNA of *Papilloma* virus (Watson & Littlefield, 1960), T1, T3 and T5 phages give sharp sedimentation boundaries.

In a polydisperse DNA, small degrees of degradation and aggregation are not detectable. On the other hand, the sharp sedimentation boundary of a homogeneous DNA enables a small percentage of superimposed heterogeneity to be detected. We would estimate that, working with 15 μg/ml. DNA (at which concentration the Johnston–Ogston effect is negligible), we could detect by a "tail" on the slow side of the sharp boundary a 10% decrease in molecular weight of 5% of the molecules. Cross-linking between 5% of the molecules is similarly manifested by a perceptible fast component. It was therefore possible to examine critically the methods used to isolate and handle DNA. Some of the results of this study are incorporated in the preparative methods described.

This work was performed in the laboratories of Professor Francis O. Schmitt, to whom the authors are grateful for his interest and help. The authors also thank Dr. M. A. Jesaitis for generously providing a sample of the phase II lipopolysaccharide, and Mrs. K. Y. Ho for her valuable technical assistance. This investigation was supported by research grant E-1469 from the National Institute of Allergy and Infectious Diseases, National Institutes of Health, United States Public Health Service. Dr. Freifelder was supported by a post-doctoral fellowship from the Division of General Medical Sciences, United States Public Health Service, 1961.

REFERENCES

Cairns, J. (1962). *Nature*, **194**, 1274.
Davison, P. F. (1959). *Proc. Nat. Acad. Sci., Wash.* **45**, 1560.
Davison, P. F. & Freifelder, D. (1962). *J. Mol. Biol.* **5**, 635.
Davison, P. F., Freifelder, D., Hede, R. & Levinthal, C. (1961). *Proc. Nat. Acad. Sci., Wash.* **47**, 1123.
Doty, P., McGill, B. B. & Rice, S. A. (1958). *Proc. Nat. Acad. Sci., Wash.* **44**, 432.
Eigner, J. (1960). Ph.D. Thesis, Harvard University.
Hearst, J. E. & Vinograd, J. (1961). *Arch. Biochem. Biophys.* **92**, 206.
Hershey, A. D. & Burgi, E. (1960). *J. Mol. Biol.* **2**, 143.
Ifft, J. B., Voet, D. H. & Vinograd, J. (1961). *J. Phys. Chem.* **65**, 1138.
Jesaitis, M. A. & Goebel, W. F. (1955). *J. Exp. Med.* **102**, 733.
Katz, S. (1961). *Nature*, **191**, 280.
Levinthal, C. & Davison, P. F. (1961). *J. Mol. Biol.* **3**, 674.
Mandell, J. D. & Hershey, A. D. (1960). *Analyt. Biochem.* **1**, 66.
Meselson, M. (1959). *The Cell Nucleus.* London: Butterworth & Co.
Rubenstein, I., Thomas, C. A. & Hershey, A. D. (1961). *Proc. Nat. Acad. Sci., Wash.* **47**, 1113.
Watson, J. D. & Littlefield, J. W. (1960). *J. Mol. Biol.* **2**, 161.

Note added in proof.

Despite the consistency of the independent estimates of the phage particle weight, the size ascribed to T7 DNA in this paper must be accepted with reservation. Cairns (1962) found that λ phage DNA has a length up to 23 μ, and hence a molecular weight 47 to 63×10^6 (according to whether the dimensions of the A or C form of sodium deoxyribonucleate are assumed). Meselson (manuscript submitted) found that the sedimentation coefficient of λ DNA is only slightly higher than that of T7 DNA: since it is unlikely that single molecules of similar structure differing twofold in weight could differ so little in sedimentation coefficient, it appears probable that some of these measurements are in error or that the viral nucleic acids are not as homologous as is usually assumed.

ARTICLE 10

From the *Proceedings of the National Academy of Sciences*, Vol. 43, No. 7, 581–588, 1957. Reproduced with permission of the authors.

EQUILIBRIUM SEDIMENTATION OF MACROMOLECULES IN DENSITY GRADIENTS*

By Matthew Meselson,† Franklin W. Stahl,‡ and Jerome Vinograd

GATES AND CRELLIN LABORATORIES OF CHEMISTRY§ AND NORMAN W. CHURCH LABORATORY OF CHEMICAL BIOLOGY

CALIFORNIA INSTITUTE OF TECHNOLOGY, PASADENA, CALIFORNIA

Communicated by Linus Pauling, May 27, 1957

I. QUALITATIVE DESCRIPTION

This communication presents a new method for the study of the molecular weight and partial specific volume of macromolecules, with some illustrations based on results with deoxyribonucleic acid (DNA) and several viruses. The method involves observation of the equilibrium distribution of macromolecular material in a density gradient itself at equilibrium. The density gradient is established by the sedimentation of a low-molecular-weight solute in a solution subject to a constant centrifugal field.

A solution of a low-molecular-weight solute is centrifuged until equilibrium is closely approached. The opposing tendencies of sedimentation and diffusion have then produced a stable concentration gradient of the low-molecular-weight solute (Fig. 1). The concentration gradient and compression of the liquid result in a continuously increasing density along the direction of centrifugal force. Consider the distribution of a small amount of a single macromolecular species in this density gradient. The initial concentration of the low-molecular-weight solute, the centrifugal field strength, and the length of the liquid column may be chosen so that the range of density at equilibrium encompasses the effective density of the macromolecular material. The centrifugal field tends to drive the macromolecules into the region where the sum of the forces acting on a given molecule is zero. (The effective density of the macromolecular material is here defined as the density of the solution in this region.) This concentrating tendency is opposed by Brownian motion, with the result that at equilibrium the macromolecules are distributed with respect to concentration in a band of width inversely related to their molecular weight.

CONCENTRATION DISTRIBUTIONS

1. *Gaussian Bands.*—It is shown in Section II that in a constant density gradient under certain attainable conditions, the concentration distribution at equilibrium of a single macromolecular species is Gaussian. The standard deviation of this Gaussian band is inversely proportional to the square root of the macromolecular weight. The band is centered about the cylindrical surface corresponding to the effective density of the macromolecular material. Figure 2 is a photometric record showing the equilibrium distribution of bacteriophage DNA in a density gradient of cesium chloride in water.

If the macromolecular material is composed of species with various molecular weights and effective densities, the observed equilibrium distribution is the sum of the separate Gaussian distributions with standard deviations and means corresponding to the molecular weights and effective densities, respectively, of the various molecular species. When heterogeneities in molecular weight and effective density are both present, it is possible, but a priori unlikely, that the separate distributions would conspire to produce a single observed band of essentially Gaussian form.[1]

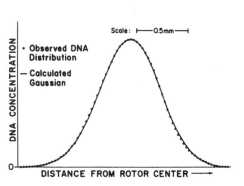

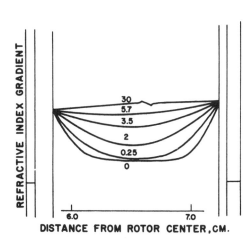

FIG. 1.—Superimposed schlieren diagrams of successive stages in the development of a cesium chloride equilibrium gradient. The numbers above each curve are the times in hours that the centrifuge has been running at 31,410 rpm. These diagrams are from an experiment in which 8 μg. of DNA was present in the cell; the banded DNA produces a diphasic pip in the final schlieren diagram.

FIG. 2.—The equilibrium distribution of DNA from bacteriophage T4. An aliquot of osmotically shocked T4r containing 3 μg. of DNA was centrifuged at 27,690 rpm for 80 hours in 7.7 molal cesium chloride at pH 8.4. Evidence for the attainment of equilibrium was provided by the essential identity of the final band shapes, whether the DNA was initially distributed uniformly in the cell or in an extremely tight band. At equilibrium the observed DNA distribution does not depart appreciably from Gaussian form, indicating essentially uniform molecular weight and effective density. The mean of the distribution corresponds to an effective density of 1.70, and the standard deviation corresponds to a molecular weight for the Cs·DNA salt of 18×10^6. Assuming the base composition for T4 reported by G. R. Wyatt and S. S. Cohen (*Nature*, **170**, 1072, 1952) and the glucose content reported by R. L. Sinsheimer (these PROCEEDINGS, **42**, 502, 1956), this corresponds to a molecular weight of 14×10^6 for the sodium deoxyribonucleate. The density gradient is essentially constant over the band and is 0.046 gm/cm.[4]. The concentration of DNA at the maximum is 20 μg/ml.

2. *Bi- or Polymodal Distribution of Banded Material.*—If the effective densities of the macromolecular species are sufficiently distinct, a distribution with more than one mode will be observed. In extreme cases this may lead to the formation of discrete bands. An example is the separation of normal DNA from DNA which contains, instead of thymine, the analogue 5-bromouracil. This unusual DNA is considerably more dense than normal DNA, and prepared mixtures of the two give rise to well-resolved bands in a cesium chloride gradient (Fig. 3). With DNA in cesium chloride, density differences of less than 0.001 gm/cm³ may be detected.

3. *Skewed Unimodal Bands.*—A skewed band indicates the presence of material heterogeneous with respect to effective density. Such bands are shown in Figures 3 and 4 for bacteriophage DNA containing 5-bromouracil and calf thymus DNA, respectively. The skewness of the former band is the result of compositional

heterogeneity; i.e., some molecules contain more 5-bromouracil (in place of thymine) than others. Effective density heterogeneity may in general be compositional or structural in origin.

4. *Symmetric Unimodal Non-Gaussian Bands.*—For the Gaussian function $y = \exp(-x^2/2\sigma^2)$ a plot of $\ln y$ against x^2 yields a straight line of slope $-1/2\sigma^2$. A plot of the logarithm of the concentration in a band against the square of the distance from the maximum provides a convenient test for heterogeneity. Downward concavity anywhere in this plot signifies heterogeneity in effective density. Such a case may be rare in view of the a priori unlikelihood that the effective densities would be distributed in just such a way as to give rise to a symmetrical band.[1] The absence of downward concavity coupled with the observed symmetry of the concentration distribution is strong presumptive evidence for density homogeneity. Under this presumption, upward concavity is evidence for heterogeneity in molecu-

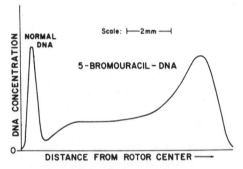

Fig. 3.—The concentration distribution of a mixture of normal and 5-bromouracil-containing DNA from bacteriophage T4. A mixture of osmotically shocked normal and 5-bromouracil-containing T4 (prepared by the method of R. M. Litman and A. B. Pardee, *Nature*, **178**, 529, 1956) was centrifuged at 44,700 rpm in 8.9 molal cesium chloride at pH 8.4. The density gradient is 0.12 gm/cm.[4]. The position of the normal DNA indicates an effective density of 1.70; the maximum effective density of the substituted DNA is 1.80.

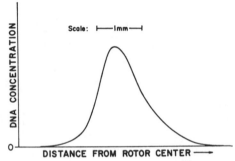

Fig. 4.—The concentration distribution of calf thymus DNA banded in cesium chloride. 2.7 μg. of calf thymus DNA (prepared by a detergent method by Dr. C. Jardetzky) was centrifuged at 44,700 rpm in 7.7 molal cesium chloride at pH 8.4. The skewness in the resultant band indicates heterogeneity in effective density.

lar weight, and the slope at any point is inversely proportional to the weight mean molecular weight of the material at the corresponding position in the band (see Sec. II).

MEASUREMENTS OF EFFECTIVE DENSITY

The mean effective density of macromolecular material (distributed in any manner within the density range of the solution) may be found from the mean of the mass distribution evaluated from the observed concentration distribution. If effective density is influenced by composition, the distribution provides a basis for the analysis of the composition of the material. This application is illustrated by the results with phage DNA containing 5-bromouracil (Fig. 3). The effective density of this DNA is found to be related to the degree of substitution of thymine by 5-bromouracil as determined chromatographically.

MOLECULAR-WEIGHT DETERMINATIONS

In the absence of density heterogeneity, both the number and weight mean molecular weights may be calculated from the observed shape of the band at equilibrium, as is shown in Section II. Molecular weights may be calculated from concentration distributions influenced by density heterogeneity; however, these should be considered minimal values. In the examples presented, only a few micrograms of the macromolecular material were present in the cell. The usefulness of the method for biological studies is further illustrated by the study of intact viruses in cesium chloride gradients. Both bacteriophage and tobacco mosaic virus[2] have been banded without loss of infectivity. Preliminary work with bovine albumin and human hemoglobin suggests the applicability of this method to smaller macromolecules. Details of the experimental procedures and results with several materials will be published elsewhere.

II. QUANTITATIVE RELATIONS

The total potential of any component at equilibrium in a closed system at constant temperature must be uniform throughout the system. In a centrifugal field this requirement results in the rigorous condition[3]

$$M_i(1 - \bar{v}_i\rho(r))\omega^2 r \, dr - \sum_K \frac{\partial \mu_i}{\partial C_k} \, dC_k = 0, \tag{1}$$

where M_i, $\bar{v}_i$, μ_i, C_i are molecular weight, partial specific volume, partial molal (Gibbs) free energy, and concentration of the ith component. The summation extends over a complete set of independently variable components. The angular velocity is given by ω, the radial co-ordinate is r, and the density of the solution at r is ρ. We shall consider a system containing water, a low-molecular-weight electrolyte XY, and a macromolecular electrolyte PX_n. The discussion will apply equally well to positive or negative polymeric ions and, with $n = 0$, to neutral polymer molecules. In the case of neutral polymers the low-molecular-weight solute may be a nonelectrolyte.

Three components are necessary and sufficient[4] to describe the composition of the system. They are chosen here as water, XY, and the neutral unsolvated molecule PX_n. Other choices are of course permissible, but this one is especially convenient. For definiteness in making certain approximations and for comparison with experiment, we shall refer to the system *water–cesium chloride–cesium deoxyribonucleate*. The total amount of polymer will be made so small as to have a negligible effect on the potentials of the salt and water. Therefore, we may first calculate the concentration distribution of XY from equation (1), ignoring the polymer, and then, again using equation (1), find the distribution of the polymer in the salt gradient. This gradient is

$$\frac{dC_{XY}}{dr} = \frac{\partial C_{XY}}{\partial a_{XY}} \cdot \frac{da_{XY}}{dr} = \frac{\partial C_{XY}}{\partial a_{XY}} \cdot \frac{a_{XY}M_{XY}(1 - \bar{v}_{XY}\rho(r))\omega^2 r}{RT}. \tag{2}$$

The sum of equation (1) has been replaced by its equivalent, $RT \, d \ln a_{XY}$, where a_{XY} is the activity. Values of a_{XY}, ρ, and $\bar{v}_{XY}$ as functions of concentration and pressure are experimentally determinable and may be found in the literature for some systems.   The salt concentration as a function of r may be found by numerical integration of expression (2).   Alternatively, it may be measured by optical methods in the centrifuge itself.

In many cases it is possible to select XY so that the density and concentration gradients are essentially constant over regions sufficiently long to encompass a polymer band.   For example, it is found, both by calculation from equation (2) and by direct observation of the refractive index gradient in the centrifuge, that the cesium chloride concentration and density gradients are essentially constant over the region of a DNA band.   For computational simplicity, we shall consider only these linear systems, so that upon choosing r_0 in the region of a band, we may write C_{XY} and ρ as

$$C_{XY}(r) = C_{XY}(r_0) + \left(\frac{dC_{XY}}{dr}\right)_{r_0} (r - r_0), \tag{3}$$

$$\rho(r) = \rho(r_0) + \left(\frac{d\rho}{dr}\right)_{r_0} (r - r_0). \tag{4}$$

Having found the distribution of XY, we may now employ equation (1) to determine the distribution of PX_n.   Making use of equation (4), we write the first term of equation (1), which represents the work per mole done against the centrifugal field moving PX_n from r to $r + dr$, as

$$M_{PX_n} \left(1 - \bar{v}_{PX_n}\rho(r_0) - \bar{v}_{PX_n}\left(\frac{d\rho}{dr}\right)_{r_0} (r - r_0)\right) \omega^2 r\, dr. \tag{5}$$

It should be emphasized that $\bar{v}_{PX_n}$ is the partial specific volume of PX_n in a solution of XY at a concentration $C_{XY}(r)$.   In order to evaluate the remaining term in equation (1), we consider it to be composed entirely of the osmotic work $RT\, d \ln C_{PX_n} + zRT\, d \ln \gamma_X C_X$, where z is the effective number of counter-ions which must be moved along with the charged polymer molecule PX_{n-z} in order to maintain electrical neutrality and γ_X is the activity coefficient of the ion X.   We thereby neglect any other change in the free energy of the polymeric component as it is moved through the solution.   This should be a valid approximation if the fractional change in the concentration of XY across a polymer band is small.   It is especially plausible for the case of DNA in cesium chloride, for which the cesium chloride concentration changes by less than one part in one hundred over the region of a band.   Also, over a small concentration range, the term $d \ln \gamma_X$ will be negligible in comparison to $d \ln C_X$.   Incorporating these approximations in the limit of low polymer concentration, we have

$$M_{PX_n}\left(1 - \bar{v}_{PX_n}\rho(r_0) - \bar{v}_{PX_n}\left(\frac{d\rho}{dr}\right)_{r_0} (r - r_0)\right)\omega^2 r\, dr -$$
$$RT\, d \ln C_{PX_n} - zRT\, d \ln \left(C_{XY}(r_0) + \left(\frac{dC_{XY}}{dr}\right)_{r_0} (r - r_0)\right) = 0. \tag{6}$$

Assuming $\bar{v}_{PX_n}$ and z to be independent of r over the region of a band, this may be integrated with respect to $(r - r_0)$ from $r = r_0$ to $r = r$, yielding

$$M_{PX_n} (1 - \bar{v}_{PX_n}\rho(r_0))\; \omega^2(r - r_0) \left(\frac{r - r_0}{2} + r_0\right) - M_{PX_n}\bar{v}_{PX_n} \left(\frac{d\rho}{dr}\right)_{r_0}$$

$$\omega^2(r - r_0)^2 \left(\frac{r - r_0}{3} + \frac{r_0}{2}\right) - RT \ln \frac{C_{PX_n}(r)}{C_{PX_n}(r_0)} - zRT \ln$$

$$\left(1 + \left(\frac{dC_{XY}}{dr}\right)_{r_0} \cdot \frac{(r - r_0)}{C_{XY}(r_0)}\right) = 0. \quad (7)$$

In many cases, including that of DNA in cesium chloride solution, the band width may be made quite small compared with the distance of the band from the center of rotation, so that $|r - r_0| \ll r_0$. Further, because of the small magnitude of the gradient of $C_{XY}(r)$, $\ln (1 + (dC_{XY}/dr)_{r_0} (r - r_0)/C_{XY}(r_0))$ may be expanded as $(dC_{XY}/dr)_{r_0} (r - r_0)/C_{XY}(r_0)$. Introducing these approximations in equation (7) and completing the square in the variable $(r - r_0)$, we obtain

$$C_{PX_n}(r) = C_{PX_n}(r_0) \left(\exp \frac{\alpha^2}{2\sigma^2}\right) \left(\exp \frac{-1}{2\sigma^2} ((r - r_0) + \alpha)^2\right), \quad (8)$$

where

$$\sigma^2 = \frac{RT}{M_{PX_n}\bar{v}_{PX_n} \left(\dfrac{d\rho}{dr}\right)_{r_0} \omega^2 r_0} \quad (9)$$

and

$$\alpha = \frac{zRT\,(dC_{XY}/dr)_{r_0}}{M_{PX_n}\bar{v}_{PX_n}\omega^2 r_0 C_{XY}(r_0)} - \frac{(1 - \bar{v}_{PX_n}\rho(r_0))}{\bar{v}_{PX_n}\,(d\rho/dr)_{r_0}}. \quad (10)$$

This is a Gaussian distribution with standard deviation σ. Equation (8) is simplified by choosing r_0 as the mean in which case $\alpha = 0$. Therefore, the density of the medium at the band center is given by the expression

$$\rho(r_0) = \frac{1}{\bar{v}_{PX_n}} \left(1 - \frac{zRT(dC_{XY}/dr)_{r_0}}{M_{PX_n}\omega^2 r_0 C_{XY}(r_0)}\right). \quad (11)$$

Thus the pull of the counter-ions displaces the origin of the Gaussian band from the region of density $1/\bar{v}_{PX_n}$ to r_0, where the density is $\rho(r_0)$. Our final result, then, for the distribution of a single polymeric species at equilibrium in a constant density gradient is

$$C_{PX_n}(r) = C_{PX_n}(r_0) \exp \frac{-(r - r_0)^2}{2\sigma^2}. \quad (12)$$

From the observed value of σ, the molecular weight is calculated as

$$M_{PX_n} = \frac{RT}{\bar{v}_{PX_n}(d\rho/dr)_{r_0}\omega^2 r_0 \sigma^2}. \quad (13)$$

The molecular weight obtained from equation (13) refers to the dry neutral molecule PX_n whether or not the species actually present is solvated or charged. Because it is much more convenient to measure the effective density $\rho(r_0)$ than to determine $\bar{v}_{PX_n}$ by the usual pycnometric method, one might inquire under what conditions it is permissible to equate the two quantities.   This may be done when

$$\left(1 - \frac{zRT(dC_{XY}/dr)_{r_0}}{M_{PX_n}\omega^2 r_0 C_{XY}(r_0)}\right) \simeq 1. \tag{14}$$

For DNA in cesium chloride solution, even if each primary phosphate carried an effective negative charge (which is surely not the case), the approximation (14) involves an error less than 10 per cent.

Having shown that a single polymeric species in a constant density gradient will be distributed at equilibrium in a band of Gaussian shape, we now turn to the more general situation of a polymer heterogeneous with respect to molecular weight although homogeneous with respect to effective density.   In the limit of low polymer concentration, interactions between polymer molecules may be presumed absent, so that the observed band will be the sum of many Gaussians with coincident origins, with each Gaussian possessing a standard deviation related to the molecular weight of the corresponding species by equation (9).

It then follows with no further assumptions that the weight and number average molecular weights of the polymeric material at r are given, respectively, by

$$M_W(r) = -2\beta \frac{d \ln C_{PX_n}}{d(r - r_0)^2}, \tag{15}$$

$$M_N(r) = \frac{-\beta C_{PX_n}(r)}{\displaystyle\int_{-\infty}^{r} (r - r_0)C_{PX_n}(r)\, dr}. \tag{16}$$

The corresponding molecular weight means for the material comprising a band are

$$M_W = \frac{-2\beta \int (d \ln C_{PX_n}/d(r - r_0)^2\, C_{PX_n}(r)\, dr}{\int C_{PX_n}(r)\, dr}, \tag{17}$$

$$M_N = \frac{-\beta \int C_{PX_n}(r)\, dr}{\int (r - r_0)^2 C_{PX_n}(r)\, dr}, \tag{18}$$

where

$$\beta = \frac{RT}{\bar{v}_{PX_n}(d\rho/dr)_{r_0}\omega^2 r_0}.$$

The integrations are to extend over the entire band and are written so as to apply to cells having straight walls whether radial or not.   In the completely general case of heterogeneity of both density and molecular weight, the molecular weights calculated from the above equations are minimal values.

Finally, we turn to the problem of estimating the time necessary for the distribution of the polymer to approach closely to its equilibrium value.   It can be shown[5] that the time required for the concentration of a single species to be within 1 per cent of its equilibrium value from the center of the Gaussian to two standard deviations may be estimated as

$$t^* = \frac{\sigma^2}{D}\left(\ln \frac{L}{\sigma} + 1.26\right), \qquad L \gg \sigma. \tag{19}$$

where σ is the standard deviation at equilibrium, D is the diffusion constant, and L is the length of the liquid column in which the polymer was evenly distributed at the start of centrifugation.   This estimate is based on the assumption that the density gradient is fully established at time zero.   The time actually required for the equilibration of XY may be estimated theoretically.[6]   In the DNA–cesium chloride system, the cesium chloride equilibrium is, in fact, approached much more quickly than that for the polymer and the estimate of equation (19) agrees with observation.

We gratefully acknowledge the technical assistance of Mr. Richard Holmquist, Mrs. Ilga Lielausis, and Mrs. Janet Morris.   Mr. Darwin Smith rendered valuable assistance with the photometry.   We are indebted to Dr. John Lamperti and to Mr. Charles Steinberg for many valuable discussions.

* Aided by grants from the National Institutes of Health and the National Foundation for Infantile Paralysis.

† Predoctoral Fellow of the National Science Foundation.

‡ Fellow in the Medical Sciences of the National Research Council.

§ Contribuion No. 2205.

[1] It should be pointed out that this possibility is subject to experimental test.   By means of a partition cell the material on either side of the mean may be isolated and rebanded.   The new band will be skewed if there was density heterogeneity in the original band.   Alternatively, one may compare the concentration distribution observed in cells of different shape.   For a single species, the concentration distribution is independent of cell shape.   However, for material heterogeneous with respect to either molecular weight or effective density or both, the observed concentration distribution is dependent on variations in the area of the cell along the radius of rotation.   This dependence is such that for material of homogeneous density the symmetry of the band is not disturbed (although its shape may be).   For a material with density heterogeneity, however, the band will show various departures from symmetry, depending on cell shape and the particular density distribution present.

[2] A. Siegel, private communication, 1957.

[3] T. Svedberg and K. O. Pedersen, *The Ultracentrifuge* (Oxford, 1940).

[4] Although three components are sufficient for a thermodynamic description of the system in the absence of the centrifugal field, more may be required in its presence.   A system subject to a centrifugal field may be regarded as being composed of a continuous sequence of phases of infinitesimal depth in the direction of the field.   A number of components sufficient to describe the chemical composition of each phase must be employed.   In the present case, more than three components would be required if any product arises with composition not describable in terms of the three chosen components and which sediments differently than any of the components involved in its formation.   This complication is explicitly excluded from the present discussion.

[5] G. M. Nazarian and M. Meselson, to be published.

[6] R. A. Pasternak, G. M. Nazarian, and J. R. Vinograd, *Nature*, 179, 92, 1957; G. M. Nazarian, doctoral dissertation, Department of Chemistry, California Institute of Technology, 1957.

ARTICLE 11

From *Journal of Molecular Biology*, Vol. 23, 163–181, 1967. Reproduced with permission of the authors and publisher.

Electron Microscopy of Size and Shape of Viral DNA in Solutions of Different Ionic Strengths

D. Lang, H. Bujard, B. Wolff and D. Russell

Division of Biology
Southwest Center for Advanced Studies
Dallas, Texas, U.S.A.

(*Received 6 September 1966*)

Diffusion-controlled adsorption of DNA from bulk solution onto a monolayer of surface-denatured cytochrome c has been used to characterize, by electron microscopy, the size and shape of single DNA molecules from bacteriophages T3 and T1 and from bovine papilloma virus. For ionic strengths, $I \geqslant 0.14$, the following DNA contour lengths and sample standard deviations are found: T3, (11.6 ± 0.4) μ; T1, (16.0 ± 1.0) μ and bovine papilloma virus (circular), (2.45 ± 0.07) μ. For DNA solutions with $I < 0.1$, the contour length of the adsorbed DNA increases up to 20% with decreasing I, presumably by a uniform, partial unwinding of the B configuration. Within the experimental error (standard deviation $= \pm 4\%$), the DNA contour lengths are found equal, whether (a) perchlorate or phenol extraction, (b) diffusion or spreading methods, or (c) ammonium acetate or sodium chloride as added salts were used. Based on experimental and operational values of standard deviation, variability of natural contour length of intact T3 or bovine papilloma virus DNA must be less than about $\pm 2\%$, if there is any. The conformation of intact T3 DNA, as measured by the distribution of end-to-end distances with the diffusion method, has been identical, at $I = 0.4$, within the error limits, with the calculated theoretical distribution for random coils, with small increasing deviations at 0.15 and 0.05 ionic strength.

1. Introduction

Individual nucleic acid molecules as shown by electron microscopy are of wide-spread recent interest (Josse & Eigner, 1966). Suitable techniques were initiated by Hall & Litt (1958), who prepared pseudoreplicas of DNA sprayed from solution onto freshly cleaved mica. Kleinschmidt & Zahn (1959) spread monomolecular surface films from DNA-containing protein solutions on water. Beer (1961) transferred DNA from solution to a glass slide covered with a thin film of a basic copolymer by slowly retracting the slide and thereby adsorbing and straightening out the DNA. Lang, Kleinschmidt & Zahn (1964a) used the diffusion-controlled adsorption of DNA on a protein surface film. Native and artificially introduced properties of nucleic acids, if represented by individual or average geometric or topologic parameters of single molecules, can be determined by the above methods. Such measurements include:

(a) *Length and diameter* (Hall & Litt, 1958; Hall & Doty, 1958; Kleinschmidt & Zahn, 1959; Inman & Jordan, 1960b; Kleinschmidt, 1960; Beer, 1961; Hall & Cavalieri, 1961; Kisselev, Gavrilova & Spirin, 1961; Bendet, Schachter & Lauffer, 1962; Inman & Baldwin, 1962; Kleinschmidt, Lang, Jacherts & Zahn, 1962; MacLean & Hall, 1962; Frank, Zarnitz

163

& Weidel, 1963; Kleinschmidt, Burton & Sinsheimer, 1963; MacHattie, Bernardi & Thomas, 1963; Ris & Chandler, 1963; Spirin, 1963; Stoeckenius, 1963; Chandler, Hyashi, Hyashi & Spiegelman, 1964; Freifelder, Kleinschmidt & Sinsheimer, 1964; Gomatos & Stoeckenius, 1964; Kleinschmidt, Dunnebacke, Spendlove, Schaffer & Whitcomb, 1964; Lang, Kleinschmidt & Zahn, 1964a,b; MacHattie & Thomas, 1964; Richardson, Inman & Kornberg, 1964; Thomas & MacHattie, 1964; MacHattie, Berns & Thomas, 1965; Bartl & Boublik, 1965; Caro, 1965; Hotta & Bassel, 1965; Kaiser & Inman, 1965; Inman, 1965; Soehner, Gentry & Randall, 1965; Kellenberger, 1965; Wells, Ohtsuka & Khorana, 1965; Kleinschmidt, Kass, Williams & Knight, 1965; Inman, Schildkraut & Kornberg, 1965; Freifelder & Kleinschmidt, 1965; Freifelder, 1966; Anderson, Hickman & Reilly, 1966; Cohen & Eisenberg, 1966; Granboulan, Huppert & Lacour, 1966; Thomas, 1966; Abelson & Thomas, 1966).

(b) *Degree of flexibility, or conformation, as determined by end-to-end distances of nucleic acid molecules* (Lang *et al.*, 1964a,b; Kleinschmidt *et al.*, 1964).

(c) *Degree of branching of DNA* (Hall & Cavalieri, 1961; Inman & Baldwin, 1962; Richardson, Schildkraut & Kornberg, 1963; Richardson *et al.*, 1964; Wells *et al.*, 1965).

(d) *Circularity of molecules and their proportion to open ones* (Stoeckenius, 1963; Kleinschmidt *et al.*, 1963,1965; Ris & Chandler, 1963; Thomas & MacHattie, 1964; MacHattie & Thomas, 1964; Freifelder *et al.*, 1964; Caro, 1965; Kaiser & Inman, 1965), *and the super-twisting of circular DNA* (Vinograd, Lebowitz, Radloff, Watson & Laipis, 1965).

(e) *Attachment of specific electron stains* (Stoeckenius, 1961; Beer & Moudrianakis, 1962; Moudrianakis & Beer, 1965; Granboulan, 1966), *or of specific biomolecules* like ribosomes (Bladen, Byrne, Levin & Nirenberg, 1965) or RNA-polymerase (Fuchs, Zillig, Hofschneider & Preuss, 1964).

(f) *Identification of partially* (Stoeckenius, 1961; Beer & Moudrianakis, 1962; Highton & Beer, 1963; Inman, 1966) *and completely separated single-stranded denatured nucleic acid molecules* (Doty, Marmur, Eigner & Schildkraut, 1960; Lang *et al.*, 1964b; Freifelder *et al.*, 1964; Freifelder & Kleinschmidt, 1965; Granboulan *et al.*, 1966).

The most frequently measured parameter is the contour length, which allows comparisons of DNA from different sources and after different treatments. Contour lengths also permit calculations of average and individual molecular weights under a reasonable assumption for the molecular weight per unit length. From X-ray diffraction analysis of wet fibers of sodium DNA, a value of about 196 dalton/Å was obtained for the B configuration, which is believed to be valid also in solution (Luzzati, Nicolaieff & Masson, 1961; Wilkins, 1963; Luzzati, Mathis, Masson & Witz, 1964; Caro, 1965). Drying of DNA from solution on specimen grids may change that figure, but no supporting evidence has been recorded so far. However, an established drying-artifact is the lateral aggregation with many branchings and the straightening of DNA into preferential orientations as seen in many previously published micrographs.

The essential feature of the spreading technique of Kleinschmidt & Zahn (1959) and of the diffusion method (Lang *et al.*, 1964a) is that nucleic acid molecules in solution are adsorbed and configurationally fixed in an insoluble, stable protein monolayer *prior* to transfer onto supporting specimen grids. Aggregation and distortion by drying of the filamentous molecules are thereby avoided. Moreover, with the diffusion method, all external forces, such as shearing forces during spreading and solution convection are excluded. The molecules are subject only to Brownian motion and to adsorption forces at the interface DNA-solution–protein monolayer. The optimum concentration of about 2×10^{-8} g DNA/ml. is sufficiently low so that the molecules are independent of each other and no extrapolation to zero concentration is necessary. This is not the case with other physical chemical methods, such as viscometry, light scattering and analytical ultracentrifugation, which require minimum amounts of 10×10^{-6} g DNA/ml. (Eigner & Doty, 1965).

One of the questions open to investigation by these methods is the effect of various agents on the contour length of nucleic acid. This paper discusses operational errors of the spreading and diffusion methods, and biological variability of intact DNA, and presents evidence for a linear expansion of duplex DNA in solutions of low ionic strength.

In addition to the determination of contour lengths, the diffusion method also permits one to observe the conformation of filamentous macromolecules as directly determined, for the first time, by the distribution of end-to-end distances on the micrographs. This is valid provided there is a correlation to the three-dimensional state in solution. It will be shown that the average DNA molecule is projected into its spacial conformation in a plane during adsorption at the interface. The data presented demonstrate that a complete characterization of viral DNA with respect to size and shape in solution can be accomplished by electron microscopy. This will help in the interpretation and calibration of hydrodynamical methods, as well as in the detection of molecular properties of nucleic acids and other linear macromolecules.

2. Materials and Methods

(a) Virus DNA

Lysates of bacteriophage T3 and T1 were obtained by infecting liter batches of exponentially growing Nutrient Broth cultures of *Escherichia coli* B (titer about 5×10^8 cells/ml.) at multiplicities of about 0·05. The cultures were shaken for one additional hour at 37°C on a rotatory shaker operating at maximum speed. The phages were concentrated and partially purified by 2 to 3 cycles of low- and high-speed centrifugation. Some samples were purified by equilibrium centrifugation in a CsCl density gradient. Finally, the pellets were resuspended in 0·15 M-NaCl, 0·015-M sodium citrate and stored at 4°C. Phage T3 was originally obtained from Dr R. Latarjet; phage T1 stems from the collection of Dr C. Bresch. T3 and T1 DNA were extracted by 5-min treatment with sodium perchlorate (4 or 5 M-NaClO₄, 0·005 M-EDTA, pH 7), according to Freifelder (1965a,b,1966). Some of the T3 DNA was extracted at 2°C by gentle mixing of phage in 1 M-sodium chloride solution, buffered with 0·01 M-sodium phosphate, 0·001 M-EDTA (pH 7·5), with freshly distilled phenol previously equilibrated with buffered sodium chloride solution. Excess phenol was removed by dialysis against buffer.

Bovine papilloma virus was isolated from wart tissue (Bujard, manuscript in preparation). Circular BPV† DNA was extracted according to Weil (1963).

(b) Spreading of DNA

The spreading procedure (Kleinschmidt & Zahn, 1959) is depicted in the left-hand part of Fig. 1, except that the salt concentration in the bulk solution varied from 0 to 0·5 M-ammonium acetate or NaCl in double-distilled water. If no salt was added, the spreading solution contained in some instances 12% (exp. 99 and 105) or 1·6% (exp. 177, 179, 180 and 181) formaldehyde (pH 6) (Fisher Sci. Co., without methanol) in order to facilitate spreading.

(c) Diffusion-controlled adsorption of DNA

The diffusion method (Lang *et al.*, 1964a) was modified by delivering the cytochrome c (Nutritional Biochemical Corp.) to the clean surface of the DNA bulk-solution in solid form dried on a needle (right-hand part of Fig. 1), instead as a solution flowing down a tilted glass slide. This is more convenient and minimizes disturbance of the bulk solution, which should be free of convection. By lowering the needle slowly into the bulk solution, the protein dissolves and spreads to a monolayer. Talc particles, previously sprinkled

† Abbreviation used: BPV, bovine papilloma virus.

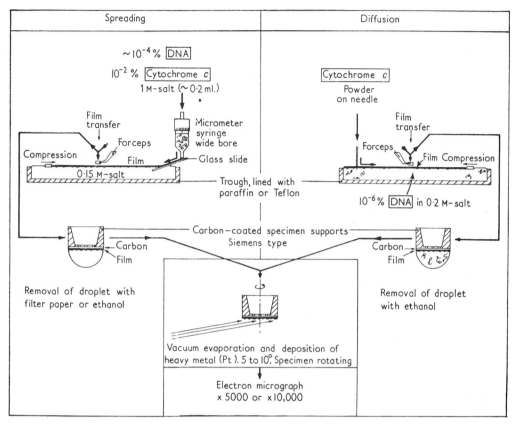

FIG. 1. Scheme of electron microscopic preparation steps using a Langmuir trough of 65 cm × 14 cm × 1·5 cm. Left: spreading; right: diffusion method.

onto the surface, make this spreading visible. When the film has covered about 70% of the surface, the needle is withdrawn. The film is then compressed (about 2%) with a Teflon-coated barrier. Choosing a DNA concentration of about 2×10^{-8} g/ml. permits a sufficient amount of DNA to reach the spread protein monolayer by diffusion within 30 min. The DNA is irreversibly adsorbed in its full molecular length, and the amount adsorbed is proportional to the square root of the time after spreading of the surface film. Salt concentrations ranged from 0·008 to 0·5 M-ammonium acetate (pH 7). In two experiments, NaCl was used. The ionic strength (I) of salt-free water was estimated with an electrical conductivity bridge (Barnstead; Boston, Mass.) calibrated with NaCl solutions. The diffusion technique, at $I < 0·1$, resulted in molecular distortion of DNA in the protein film (see Results, (a)). Therefore, in some experiments ("Diff.*" in Table 1), the interior of a Langmuir trough, 65 cm × 14 cm × 1·5 cm, was subdivided into 4 compartments by 3 Teflon bars, placed slightly below the rim of the trough. A similar method was suggested by Kleinschmidt (personal communication) and used for extraction procedures (Dunnebacke & Kleinschmidt, 1966). The first compartment contained 0·2 M-ammonium acetate on which the cytochrome c monolayer was spread and aged for 10 min without compression. By means of two equidistant barriers, the film was then moved (velocity approximately 0·5 cm/sec) to the next compartment filled with water (pH 7) or 0·02 M-ammonium acetate (pH 7). The film remained there for 5 min in order to wash out salt solution possibly carried along from the first compartment. Then the film was brought over the third compartment, which contained DNA in water or 0·02 M-ammonium acetate. The film was then stable and ready for adsorption of DNA. (The fourth compartment contained water or 0.02 M-ammonium acetate.)

(d) Electron microscopy

Supporting films were prepared in a vacuum evaporator by carbon sublimation onto freshly cleaved mica sheets. Then they were floated off on water and picked up from below by specimen grids (Siemens-type, 7 or 19 holes of $70\,\mu$ diameter). Carbon was used throughout this investigation since, unlike with Formvar, no shrinkage or expansion in the electron beam could be detected. After transfer of portions of the DNA-containing protein surface film to grids (Fig. 1), the necessary contrast was produced by small-angle (6°) vacuum deposition of platinum evaporating from two electrically heated tungsten wires, from two directions 90° apart. This is a compromise, since the DNA width is less uniform than with continuously rotating grids (Heinmets, 1949; Kleinschmidt *et al.*, 1960), but the three-dimensional shadow effect, obtained with stationary grids, is not completely lost. Micrographs (Kodak plates, contrast, $6\ \text{cm} \times 9.5\ \text{cm}$) were taken with an Elmiskop IA (Siemens & Halske) (double-condenser illumination, intermediate lens switched off, projective polepiece II or III) at linear magnifications of 4750 or 9290, at least 30 min after setting for operation. Grids were scanned without overlapping, and only long molecules distinctly visible in every part were photographed. In experiments done to determine complete length and end-to-end distance distributions (no. 132, 134, 127, 144), all molecules on randomly picked non-overlapping fields were recorded and measured. Regarding circular BPV DNA, only molecules with few or no intersections were measured. All grids were prepared at room temperature. The magnification was repeatedly calibrated by a carbon replica (E. Fullam Inc.) of an optical grating (28,800 lines/in.). From light microscopy, a figure of 28,200 lines/in. was found, with $\pm\,760$ lines/in. sample standard deviation ($\pm\,2.7\%$).

Neglecting image rotation, the position of an image point off the optical axis due to pincushion distortion equals cr^3 (Glaser, 1956), where c is a constant and r the distance from the axis without distortion. From this it was calculated by elementary geometry that a given radially oriented distance ($\ll r$) between two points is too long by a factor of $1 + 3cr^2$, or, if in tangential orientation, of $1 + cr^2$, whereas the calculated average for random orientations gave a factor of $1 + 2cr^2$. The constant c was determined for the two magnifications used by measuring the center-to-center distances of pairs of neighboring small holes in a carbon film in radial as well as tangential orientation at different distances r from the midpoint of the micrographs: $c = (1.4 \pm 0.2) \times 10^{-3}\ \text{cm}^{-2}$. With this, all measured lengths were corrected, either by applying an individual factor for each molecule or by using an average correction of -1.7% for each $6.5\ \text{cm} \times 9\ \text{cm}$ plate of randomly selected fields photographed.

(e) Length measurements

After enlarging the plates about 25 times with an optical projector (Beseler "Slide King", average pincushion distortion 0.2%) and tracing the DNA image on paper, the contour lengths and end-to-end distances were measured with map rulers (1 scale division $= 0.64\ \text{cm}$).

3. Results

(a) Appearance of DNA

Plates I and II at 32,600 magnification show two preparations by the diffusion method, at ionic strength $I = 0.20$, of phenol-extracted T3 DNA which was shadowed from two directions with platinum. No aggregation or preferential orientation is present. The average quality of the electron micrographs taken was similar to Plate II, and sufficient for length measurements if the DNA concentration was about 100 times less than in Plate II, so that the probability of two or more molecules intersecting was negligible.

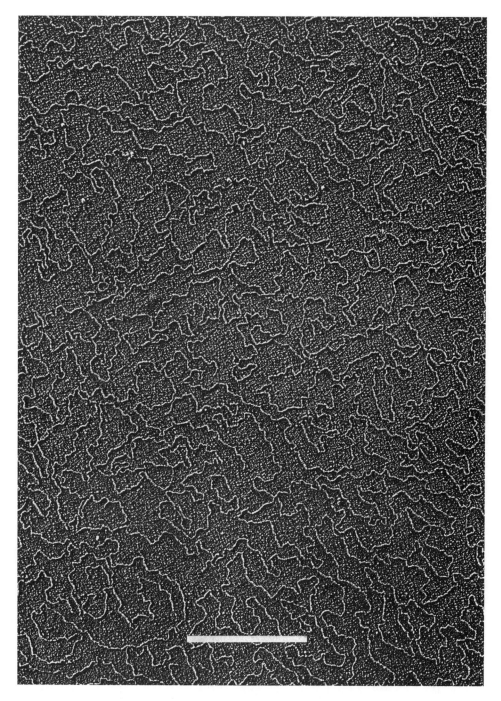

PLATE I

PLATES I AND II. DNA extracted from bacteriophage T3 by phenol, prepared on grids by the diffusion method at 0·2 ionic strength, and shadowed with platinum from two directions 90° apart. × 32,600; the bar indicates 1 μ. The DNA concentration for actual length measurements was about 100 times less than in Plate II.

PLATE II

Preparations by the diffusion method at $I < 0.05$ resulted in about 95% radial, flower-like loops of DNA originating from one spot (Kleinschmidt & Zahn, 1959; Stoeckenius, 1961), most of which were not suitable for measurement, and in collapsed loops of constant width, indicating artificial distortion. These effects disappeared, and micrographs similar to Plate I or II were obtained, when the cytochrome c film was spread and aged on 0.2 M-ammonium acetate before transfer to the compartments containing DNA in water or in solutions of ionic strength less than 0.05 (Materials and Methods, (c)).

(b) DNA size

Table 1 gives the mean contour length, L, of T3, T1 and BPV DNA, with the sample standard deviation defined by

$$s = \left[\sum_i (L_i - L)^2/(N - 1) \right]^{1/2}$$

where L_i is the length of the ith molecule and L the mean length of N molecules. The (smaller) standard deviation of the mean length may be obtained by dividing s by $N^{1/2}$. Figure 2 shows length distributions for the experiments indicated. Even though shearing forces during handling were minimized, all DNA preparations of virus stored over weeks or months at 4°C contained molecules smaller than the full length (see Inman, 1965), 21 to 42% by number or 7 to 15% by weight (Fig. 2(a), exp. 127, 132, 134 and 144). DNA extracted from T3 phage purified in a CsCl density gradient and prepared within one day for electron microscopy resulted in 11% fragments by number and 2% by weight. These percentages are upper limits, since fragments diffuse more rapidly to the protein film. If necessary, a correction leading to the true distributions of the fragments can be made (Lang et al., 1964a). The mean contour lengths were calculated only from the listed number N of molecules (Table 1) longer than 10 μ for T3 DNA and 14 μ for T1 DNA. These lower limits, 10 μ and 14 μ, mark the approximate beginnings of the frequency peaks of full-length DNA and are necessarily somewhat arbitrary. However, the effect described in the next paragraph is masked rather than enhanced by this procedure.

(i) Influence of ionic strength

With increasing I from about 0.00002 to 0.1, the contour length of T3 DNA and BPV DNA decreased to a value which remained constant between 0.1 and 0.5 ionic strength (Fig. 3(a)). A similar difference was obtained for T1 DNA in water and in 0.2 M-ammonium acetate (Table 1), with a length ratio of $18.24/16.04 = 1.137$, and for BPV DNA (Fig. 3(b)).

(ii) Molecular weight

Molecular weights of DNA can be calculated from contour lengths if the molecular weight per length unit is known. It was assumed that the DNA at ionic strengths between 0.14 and 0.50 is in the B configuration of the Watson–Crick model with 196 dalton/Å (Luzzati et al., 1961,1964; Caro, 1965; Thomas, 1966). Contour lengths at these ionic strengths were taken from Table 1, their mean values and sample standard deviations calculated, and were listed in Table 2 together with molecular weights.

(iii) Influence of experimental parameters

Table 1 indicates that for given ionic strengths there are no significant differences in DNA contour lengths under the following conditions: extraction of DNA by sodium

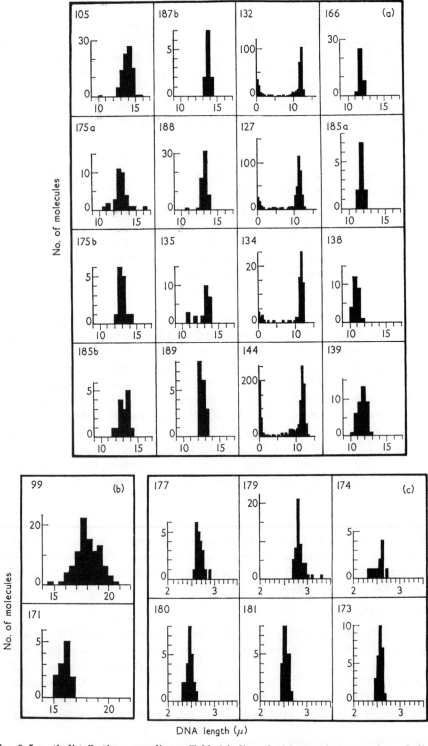

Fig. 2. Length distributions according to Table 1 indicated with experiment numbers. Ordinate, number of molecules; abscissa, DNA length in μ.

(a) T3 DNA, 0·5 μ length interval; the third column shows complete length distributions; (b) T1 DNA, 0·5 μ length interval; (c) circular BPV DNA, 0·05 μ length interval.

TABLE 1

Exp. no.	DNA from	Extracted by	Method	Salt	Bulk solution Ionic strength	pH	N (no. of molecules)	L (μ)	$\pm s$† (μ)	$\pm s$ (%)
105	T3	4 M-NaClO$_4$	Spr.	—	>0·000 03	4·9–5·2	87	13·90	0·70	5·0
175 a	T3	4 M-NaClO$_4$	Spr.	—	>0·000 02	6·8	32	13·04	0·96	7·3
175 b	T3	4 M-NaClO$_4$	Spr.	—	>0·000 02	6·8	14	13·20	0·41	3·1
185 b	T3	Phenol	Spr.	—	>0·000 02	6·2	15	13·22	0·66	5·0
187 b	T3	Phenol	Diff.*	—	0·000 03	7·6	11	13·77	0·33	2·4
188	T3	Phenol	Diff.	—	0·000 07	6·5	58	13·18	0·44	3·3
135	T3	4 M-NaClO$_4$	Diff.	NaClO$_4$	0·007 6	7·2	24	13·00	0·96	7·4
189	T3	Phenol	Diff.*	Amm. ac.‡	0·020	7·0	17	12·69	0·37	2·9
132	T3	5 M-NaClO$_4$	Diff.	Amm. ac.	0·050	7·0	223	11·96	0·54	4·5
127	T3	5 M-NaClO$_4$	Diff.	Amm. ac.	0·15	7·0	306	11·36	0·54	4·8
134	T3	4 M-NaClO$_4$	Spr.	Amm. ac.	0·15	7·0	59	11·68	0·42	3·6
166	T3	4 M-NaClO$_4$	Diff.	Amm. ac.	0·20	7·0	35	11·90	0·21	1·7
185 a	T3	Phenol	Diff.	Amm. ac.	0·20	7·0	11	11·83	0·29	2·5
138	T3	4 M-NaClO$_4$	Spr.	NaCl	0·33	5·4	26	10·91	0·40	3·7
144	T3	5 M-NaClO$_4$	Diff.	Amm. ac.	0·40	7·0	671	11·79	0·57	4·8
139	T3	4 M-NaClO$_4$	Diff.	Amm. ac.	0·50	7·0	39	11·64	0·52	4·5
99	T1	4 M-NaClO$_4$	Spr.	—	>0·000 03	5·2	94	18·24	1·05	5·8
171	T1	5 M-NaClO$_4$	Diff.	Amm. ac.	0·20	7·0	12	16·04	0·54	3·4
177	BPV	Phenol	Spr.	—	>0·000 03	6·8	21	2·70	0·087	3·2
179	BPV	Phenol	Spr.	—	>0·000 03	6·8	129	2·87	0·103	3·5
174	BPV	Phenol	Diff.	Amm. ac.	0·020	7·0	12	2·55	0·120	4·7
180	BPV	Phenol	Spr.	NaCl	0·14	6·4	24	2·48	0·074	3·0
181	BPV	Phenol	Spr.	NaCl	0·14	6·8	24	2·56	0·055	2·2
173	BPV	Phenol	Diff.	Amm. ac.	0·15	7·0	27	2·58	0·056	2·2

Under Method, Spr. and Diff. signify spreading method and diffusion method, respectively. Diff.*, see p. 166 for explanation.

† s, standard deviation.

‡ Amm. ac., ammonium acetate.

perchlorate or by phenol; spreading or diffusion method; ammonium acetate or sodium chloride in bulk solution; and pH of the bulk solution between 5 and 7.

(iv) *Accuracy of contour lengths*

The estimated operational errors involved in the presented measurements are summarized in Table 3, grouped according to error sources.

The grating constant, 28,800 lines/in., of the carbon replica used for determination of the electron optical magnification was given by the manufacturer. Light microscopy resulted in a value $(2·1 \pm 2·7)$% lower. Therefore, it seemed appropriate to list 2·7% in Table 3. The error in determining the grating constant in the central portion of the micrographs was about 1·0%. Since the magnification depends on the specimen position within the electron microscope, different specimen grids and holders were tested by comparing the objective lens focusing currents repeatedly. A standard deviation of 0·8% was found. The magnifications were standardized for a 9·60-cm diameter field of view on the plate; this diameter had to be measured twice (on plate

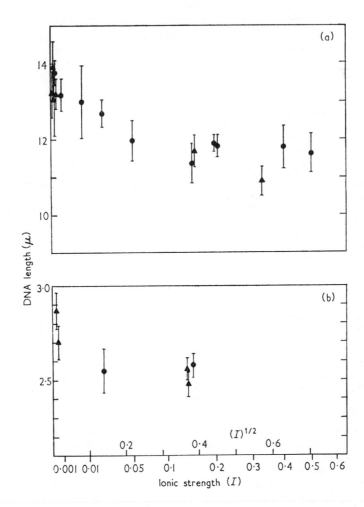

FIG. 3. Contour lengths obtained by the diffusion method (●) and by spreading (▲) as a function of ionic strength. Error limits are sample standard deviations, according to Table 1. (a) T3 DNA; (b) circular BPV DNA.

TABLE 2

Mean contour length and molecular weight above 0·14 ionic strength

DNA source	Contour length (μ)	Molecular weight (10⁶ daltons)
T3	11·6 ± 0·4	22·7
T1	16·0 ± 1·0	31·4
BPV	2·54 ± 0·07	4·98

TABLE 3

Error sources in absolute length measurements

	Error (±%)
1. Electron microscope:	
a. Grating constant	2·7
b. Its measurement on micrograph	1·0
c. Specimen position	0·8
d. Diameter of field of view (twice)	0·4
e. Pincushion distortion	1·7
2. Projector:	
a. Calibration slide	0·4
b. Measurement of its image	0·2
c. Pincushion distortion	0·5
3. Length on final image:	
a. Doubtful portions of molecules	1·0
b. Tracing on paper	0·8
c. Map ruler	0·5
d. Measurement of trace, sample standard dev.	0·8
Expected sample standard deviation on contour lengths; geometric sum:	3·9

with grating, and actual plate) with a combined error of 0·4%. In most experiments, an average correction factor of 1·017 allowing for pincushion distortion was used, regardless of the location of the molecules on the micrograph, resulting in an error of about 1·7%, as listed in Table 3. Measurements of selected and properly centered molecules, where the distortion factor was determined individually, involved a smaller error of about 0·3%.

The projector magnification was determined with 0·2% accuracy by a calibration slide with an engraved straight line of 2·60 cm ± 0·4%. The error due to pincushion distortion averaged 0·5% (Table 3), except for selected and centered molecules (0·1%).

In tracing *all* randomly photographed molecules on paper, some doubtful kinks or small loops were present, accounting for about 1·0% error. Additional errors made by tracing (0·8%) and by measuring contour lengths with map rulers (0·8%) were effected by a compromise between accuracy and reasonable speed. The mechanical and reading error of the map rulers was 0·5%.

The standard deviations in Table 3 may have positive or negative sign. The expected total sample standard deviation is then the square root of the sum of the squares (geometric sum) and amounts to 3·9%. A survey of the cited literature (Fig. 4) evidently resulted in a similar lower limit of 3 to 4% accuracy in contour lengths of viral DNA.

Spreading of DNA from 1 M-ammonium acetate on water usually resulted in variations higher than those encountered in the diffusion method. In view of the dependence of length upon ionic strength, this may be explained by a temporarily ill-defined salt gradient at the interface between water and cytochrome c during spreading, the final length of a molecule being determined by the actual local salt concentration during adsorption. This could also explain, at least in part, differences in contour length of circular DNA from Shope papilloma virus (Kleinschmidt *et al.*, 1965) and of DNA from λ-phage (Inman, 1966).

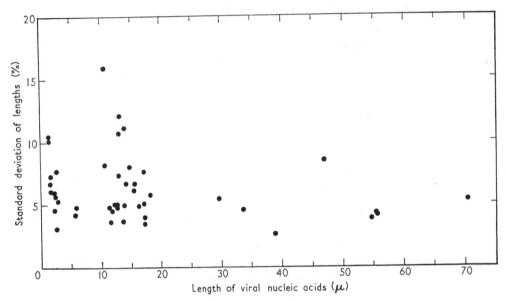

Fig. 4. Sample standard deviations for contour length measurements of viral nucleic acid as obtained from the literature cited.

(c) *Shape of DNA*

Apart from the complete distributions of contour lengths measured for T3 DNA in diffusion experiments 127, 132, 134 and 144 (see Fig. 2(a)), the end-to-end distance, a, of each adsorbed and traced molecule has been measured, since this is a function of molecular conformation or shape. Figure 5(a) to (c) shows the end-to-end distance distribution of "full-length" T3 DNA molecules greater than $10\,\mu$ at the ionic strengths 0·40, 0·15 and 0·050. The horizontal errors shown are the widths of the abscissa intervals chosen ($0\cdot5\,\mu$), the vertical ones are the statistical errors given by $\pm\,\sqrt{N_j}$, N_j being the number of molecules in the jth interval.

With regard to the conformation in solution, the inherent difficulty in drawing conclusions from the measured a-distribution in the adsorption plane is the evaluation of the mode of adsorption. In the following we calculate the theoretical a-distribution under two assumptions: first, the molecules in solution are random coils of equal contour length, and second, the end-to-end distance of the average coil is parallel-projected onto the adsorption plane. The result will be compared with experiment.

The well-known expression for the distribution of h, the end-to-end distance of not extremely extended random coils in solution, is (see e.g. Flory, 1953)

$$w(h)\mathrm{d}h = \left(\frac{3}{2\pi\overline{h^2}}\right)^{3/2} 4\pi h^2\left\{\exp\left(-\frac{3h^2}{2\overline{h^2}}\right)\right\}\mathrm{d}h \tag{1}$$

where $w(h)dh$ is the probability of finding a molecule having an end-to-end distance between h and $h + dh$; $\overline{h^2}$ denotes the mean square of h. $w(h)dh$ has now to be multiplied by the probability that h, when parallel-projected, lies between a and $a + da$, so that

$$w(a)da = 2 \int_{h=a}^{h=L} w(h) \frac{2\pi a \cdot da}{4\pi h^2 \cdot \cos\vartheta}\, dh \tag{2}$$

where ϑ is the angle between the h-vector and a vector normal to the adsorption plane and $\cos^2\vartheta = 1 - (a/h)^2$. The upper integration limit, $h = L$, can be replaced by ∞ since a full extension of the coil has a negligible probability in the case of DNA with $L > 6 \times 10^{-5}$ cm. Then we obtain the a-distribution

$$w(a)da = \frac{3}{\overline{h^2}}\, a\left\{\exp\left(-\frac{3a^2}{2\overline{h^2}}\right)\right\}da \tag{3}$$

with a maximum at $a_{\max} = (\overline{h^2}/3)^{1/2}$, a slope of $3/\overline{h^2}$ at $a = 0$ and a mean square of

$$\overline{a^2} = 2\overline{h^2}/3. \tag{4}$$

Combining equations (4) and (3) gives

$$w(a)da = \frac{2}{\overline{a^2}}\, a\left\{\exp\left(-\frac{a^2}{\overline{a^2}}\right)\right\}da \tag{5}$$

as an alternative to equation (3). Equation (5) may be compared with the experimental distribution in two ways: by inserting $\overline{a^2}$ either from the actual experiments or from earlier data on $\overline{a^2}$ as a function of the contour length L at ionic strengths I between 0·1 and 0·5. Both ways lead to essentially equal results, but here we prefer the second one because of its potential to calculate the end-to-end distance distribution for different contour lengths.

These earlier data on DNA could be expressed by a relation derived (Lang *et al.*, 1964a) from a theory of Katchalsky & Lifson (1953):

$$(\overline{h^2})^{1/2} = \frac{bL}{A}\left\{1 + \sqrt{(1 + A^3/b^2L)}\right\}, \quad L > 6\times10^{-5}\text{ cm}, 0\cdot1 < I < 0\cdot5. \tag{6}$$

The quantity b allows for electrostatic repulsion between parts of an unbranched polyelectrolyte. b is zero for an uncharged molecule and equation (6) then becomes:

$$\overline{h^2} = AL \tag{7}$$

defining a random coil where A is the Kuhn statistical element (Kuhn & Kuhn, 1943). It has to be noted that the values of A and b given earlier were based on an incorrectly derived equation, $\overline{a^2} = \overline{h^2}/3$ (Lang *et al.*, 1964a,b). Supplying the correct equation (4) to the former measurements one obtains: $A = 1\cdot85\times10^{-5}$ cm, $b = 1\cdot68\times10^{-6}$ cm,

$$(\overline{h^2})^{1/2} = 9\cdot1\times10^{-2} L\left\{1 + \sqrt{(1 + 1\cdot84\times10^{-3}/L)}\right\}\text{ (cm)} \tag{8}$$

and

$$(\overline{a^2})^{1/2} = 7\cdot4\times10^{-2} L\left\{1 + \sqrt{(1 + 1\cdot84\times10^{-3}/L)}\right\}\text{ (cm).} \tag{9}$$

Calculating $\overline{a^2}$ according to equation (9) for T3 DNA from contour lengths L as given by Table 1, equation (5) renders the theoretical distribution of end-to-end distances in the two-dimensional adsorption plane (solid lines in Fig. 5(a) to (c)).

From the agreement with the experimental points at $I = 0\cdot40$ (Fig. 5(a), exp. 144), we conclude that sufficiently long DNA behaves like a random coil with regard to its conformation at a given contour length, and that the average molecule actually parallel-projects itself during adsorption.

The agreement is less good and the maxima are slightly shifted towards higher a-values if the ionic strength decreases to 0·15 and 0·050 (Fig. 5(b) and (c)).

4. Discussion

(a) *Diffusion method*

The following picture has emerged concerning the diffusion method. In a convection-free solution of ionic strength 0·1 to 0·5, the dissolved DNA is continuously deformed and transported by random thermal collisions. Molecules diffusing to an insoluble and immobilized monomolecular film of surface-denatured cytochrome c (or other suitable basic protein) are irreversibly adsorbed—or "frozen"—on the surface film. In agreement with diffusion theory, assuming irreversible adsorption, the amount of DNA adsorbed per unit area is proportional to $\sqrt{t}$, where t is the time after spreading of the cytochrome c film (unpublished result). In order to keep negligible the number of molecules which are adsorbed and simultaneously sheared during the spreading of the protein, the condition $\sqrt{(t/t_0)} \gg 1$ should be observed, where t_0 is the spreading time (usually a few seconds). At typical concentrations of about 2×10^{-8} g/ml., the DNA molecules are practically independent of each other, but they can still be conveniently counted or measured by electron microscopy; extrapolation to zero concentration is not necessary. The protein film with its positively charged groups has the essential functions to take negatively charged DNA out of solution by adsorption at a defined salt concentration and to preserve the two-dimensional shape of DNA through the consecutive procedures of film transfer onto electron microscopic specimen grids and drying. Owing to its occurrence as a duplex molecule, DNA is relatively stiff compared to simplex polymers; but in the native form, it is extremely long and therefore sufficiently flexible to collapse in one dimension during adsorption and to appear in micrographs as two-dimensional random coils. The theoretical treatment of this transition, assuming parallel-projection of the average random-coil conformation, leads to the same end-to-end distance distribution as directly measured, and may therefore be applied to such questions as the probability of a linear molecule appearing circular or the interpretation of data from hydrodynamical methods.

DNA can also be adsorbed from salt-free water if the cytochrome c film is preformed and stabilized on a salt solution.

(b) *Error limits and biological variability*

The sample standard deviation estimated from errors in the individual experimental procedures for length measurements is about $s_0 = \pm 4\%$ (Table 3), which agrees well with the sample standard deviations found experimentally (Fig. 4 and Table 1). However, by selecting molecules on the fluorescent screen for unequivocal measurability and with additional restrictions, such as examining one grid only and making detailed distortion corrections for each molecule, one obtains experimental sample standard deviations as low as 1·7% (Table 1).

The measured standard deviations depend on the definition of what is called an intact molecule and a fragment. It seemed appropriate to define an intact molecule as being beyond the "beginning" of the peak in the contour length distribution. For simplicity, this limit was set at 10 μ for T3- and 14 μ for T1 DNA, irrespective of the exact location of the peaks. This error vanishes with circular DNA.

The above considerations are relevant to the following: Is there a biological variability of length or molecular weight of viral genomes? The observed, the estimated operational and the biological sample standard deviations, denoted by s, s_o and s_x, assuming a Gaussian distribution of native lengths, are related by:

$$s^2 = s_o^2 + s_x^2. \tag{10}$$

The significance of a determination of s_x according to equation (10) is dependent on the accuracies of s and s_o, which are low ($s = (4 \cdot 2 \pm 1 \cdot 6)\%$ for T3/T1 DNA and $s = (3 \cdot 1 \pm 0 \cdot 9)\%$ for BPV DNA as calculated from Table 1, last column). From a more detailed calculation based on the data for selected molecules, one may conclude from electron microscopy that a biological variability in molecular weights of T3, T1 and BPV DNA is less than 1 to 2%, if present at all.

The precision of length measurements could be increased by additional efforts (Reisner, 1965; Bahr & Zeitler, 1965) in order to reduce some of the errors exemplified in Table 3. However, at present it seems to be difficult to bring the operational standard deviation below 2%.

(c) Influence of ionic strength

Four effects of 0·00002 to 0·10 ionic strength solutions have been observed.

(i) Instability of the cytochrome c film

No films stable at low compressions were obtained on water; this is probably the cause of the second effect.

(ii) Flower-like appearance of DNA

This effect disappears with films preformed and aged on a 0·2 M-salt solution (see Results, (a)). From this one can conclude that the flower-like DNA conformation is caused by an unstable cytochrome c film, due to low salt concentrations in the bulk solution, and bears no correlation to the packing pattern within the virus particles. This is also obvious when the DNA is released by spreading on water, via a bacteriophage tail, one end being still in the phage head, the other end forming a flower-like structure, as seen in micrographs (e.g., Caro, 1965, Fig. 7). The frequency of this artifact is lower, but more variable, using the spreading method on solutions of low ionic strength (less than 0·05 or, most commonly, on water) because of an ill-defined salt gradient below the protein monolayer during its formation. This cannot be avoided, since the DNA/cytochrome c spreading-solution requires a higher ionic strength than the bulk solution (Fig. 1, left-hand part) in order to obtain suitable surface films (Kleinschmidt & Zahn, 1959). A systematic study of the qualifications of other protein substances has not been made.

(iii) Increase of contour length

Three trivial explanations of this effect (Fig. 3) can be ruled out. First, variations of the instrumental magnification were eliminated by means of repeated recalibrations and by alternating measurements at low- and high-ionic strengths. Secondly, the suspicion that the cytochrome c film forced the elongation of DNA at low ionic strengths cannot be true, since the effect was quantitatively reproducible also in diffusion experiments, indicated as Diff.* in Table 1, in which the cytochrome c film was spread and aged on solution of high ionic strength, then washed in water and shifted to the compartment containing DNA in solution of low ionic strength (water or 0·020 M-ammonium acetate). Moreover, this DNA did not show flower-like forms, in contrast to non-aged cytochrome c films spread on water. Thirdly, "microkinks" in the DNA molecule beyond a resolution of about 100 Å (Freifelder & Kleinschmidt, 1965), which would simulate a DNA elongation if straightened out in solution of low ionic strength, should cause a drastic reduction in coil volume. This is not the case; the Kuhn statistical element A is practically constant at low and high ionic strength, and with a value of A about 1800 Å it is 18 times larger than 100 Å. Production of microkinks during and after adsorption is unlikely, at least with diffusion experiments using aged and stabilized cytochrome c films.

Therefore, we conclude that the linear expansion of DNA in solution of low ionic strength is real and not an artifact introduced by the protein film during or after adsorption.

The idea of linear expansion may be supported by measurements of Scruggs & Ross (1964), who found a constant intrinsic viscosity of T4 DNA in NaCl above 1 M and an increase of 60% below 1 M. There are probably several reasons for this large increase; in addition to changes in hydration and coil volume, one of them might be linear expansion.

However, the sedimentation coefficient of DNA is not altered between 0·01 and 1·0 ionic strength (Studier, 1965). Sedimentation measurements at ionic strengths lower than 0·01 are obscured by charge effects (Pedersen, 1958).

Rowen (1953) found by light-scattering an increase in apparent coil volume of thymus DNA in KCl solution below about 0·7 ionic strength. Since actual coil volumes are only slightly changed at 0·05 to 0·4 ionic strength (Fig. 5(a)), linear expansion may also have contributed here.

When ionic strengths are below 0·1, the repulsive forces between electric charges of equal sign in the DNA molecule increase. Apparently, this causes the linear extension. Earlier work by Thomas (1954), Lawley (1956), Cavalieri, Rosoff & Rosenberg (1956), Inman & Jordan (1960*a*), indicates that DNA is denatured in salt-free water. Since denaturation is accompanied by unwinding of the double helix, elongation might be the first step in a denaturation process in which a uniform partial unwinding is taking place. Indeed, T3 DNA has been found to have up to $22\,\mu$ contour length after spreading with cytochrome c on salt-free water, if the protein film was not compressed and in the state of a two-dimensional "gas" or liquid (Sobotka & Trurnit, 1961). At the spreading front of the film, the ionic strength due to ions from the spreading solution (1 M-NaCl or 1 M-ammonium acetate) carried along by the film is minimal and the linear extension maximal. Total unwinding would correspond to a doubling of the duplex length.

If it is true that the number of helical twists of the DNA duplex depends on the ionic strength, then circular DNA is forced to change its number of supertwists (Vinograd *et al.*, 1965) at different ionic strengths. If one assumes n helical and n_s supertwists, positive or negative numbers meaning right- or left-handedness, then the sum $n + n_s$ is a topological invariant and $dn + dn_s = 0$, such that a torque tending to unwind (dn negative) the closed double helix is relaxed by right-hand supertwists (dn_s positive). We hope to find this effect during future electron microscopic investigations.

One wonders whether local reduction of ionic strength within the cell has a biological significance for DNA function such as facilitating intercalation (Lerman, 1964), transcription and replication.

(iv) *Increase in coil volume*

The slight spacial expansion of DNA as measured by the shift of the maximum in the end-to-end distance distributions in going from 0·4 to 0·15 and to 0·05 ionic strength (Fig. 5), is not large enough to be very significant. This agrees with expectation, since the DNA coil is a very loose one, with a large average distance between charged groups. As pointed out earlier (Lang *et al.*, 1964*a*), spacial or coil expansion of DNA, in contrast to linear expansion, is not effected by electrostatic repulsion of adjacent charges along the molecule because of its intrinsic stiffness over short distances; it is caused by charges on DNA segments which are far apart by contour length, but close by distance.

(d) *Size and shape*

Of the three DNA types used here, only T3 DNA appears to have been measured (Bendet *et al.*, 1962). They adsorbed and oriented T3 DNA from 0·1 M-ammonium acetate solution onto collodion supports and found a length of (14·0 ± 1·6) μ (sample s.D.) which is higher than our value at this ionic strength.

The length of circular BPV DNA is the same as that of the circular DNA from Shope papilloma virus (Kleinschmidt *et al.*, 1965).

It was concluded in Results, section (c) from the directly measured distribution of end-to-end distances on micrographs that undegraded, native DNA of a given molecular weight in solution of 0·4 ionic strength has the conformation of a random coil. The underlying assumption is the parallel projection of the average coil, which is subjected only to Brownian motion, onto the protein film during adsorption. Figure 5 is not strong evidence in favor of both a random coil and parallel projection, but it is difficult to understand how a non-random coil which projects its average conformation in other than a parallel manner, can have an end-to-end distance distribution identical with that of a parallel-projected random coil.

In the case of DNA populations containing circular molecules, as seen in micrographs, equation (5) may be used to decide whether these molecules are really circular or whether two ends are located very near to each other by chance. At a resolution δ, small compared with $(\overline{a^2})^{1/2}$, the probability of finding a molecule which appears circular by chance, is given by integrating equation (5) over a from 0 to δ and amounts to

$$1 - \exp\left(-\delta^2/\overline{a^2}\right) \approx \delta^2/\overline{a^2}.$$

With $\delta = 100$ Å on a low-magnification micrograph and $(\overline{a^2})^{1/2} = 2\,\mu$, we expect only 0·01% "circular" molecules. Significantly higher frequencies indicate genuine circularity.

In contrast to the appearance of DNA as a random coil with regard to the distribution of end-to-end distances at a given contour length at $I = 0·4$, homologous DNA, with different contour lengths, does not obey equation (7) for random coils; instead, equation (6) holds. Thus, with regard to the mean square end-to-end distance as a function of the contour length, DNA behaves as an intermediate between a random coil and a rod (Lang *et al.*, 1964*a*); or, in other words, the Kuhn statistical element A in equation (7) is not constant but a function of the electric charge density along the molecule and increases with contour length.

In conclusion, electron microscopy utilizing diffusion-controlled adsorption provides quantitative information about the following molecular parameters relating

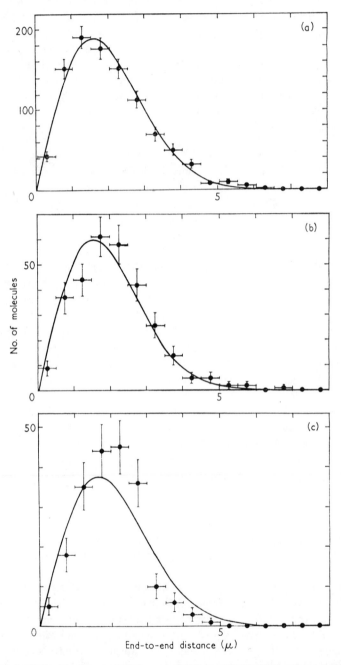

FIG. 5. Distribution of end-to-end distances of full-length T3 DNA after adsorption to the cytochrome c film. Ionic strengths of the DNA solution in ammonium acetate are: (a) 0·40, (b) 0·15 and (c) 0·050. (————) Theoretical distribution for random coils parallel-projected onto a plane according to equations (5) and (9).

to size and shape of DNA in solution: individual contour length and end-to-end distance as well as their distributions and average values.

We are grateful to Dr R. Hausmann and Miss N. Reusse of this Laboratory for gifts of concentrated phage T3 and T1 solutions, and to Drs J. Jagger and A. K. Kleinschmidt for critical discussions of the manuscript.

This work was supported in part by research grants GB–4388 from the National Science Foundation and GM 13234 from the National Institutes of Health, U.S. Public Health Service.

REFERENCES

Abelson, J. & Thomas, C. A., Jr. (1966). *J. Mol. Biol.* **18**, 262.
Anderson, D. L., Hickman, D. D. & Reilly, B. E. (1966). *J. Bact.* **91**, 2081.
Bahr, G. F. & Zeitler, E. (1965). *Lab. Invest.* **14**, 880.
Bartl, P. & Boublik, M. (1965). *Biochim. biophys. Acta*, **103**, 678.
Beer, M. (1961). *J. Mol. Biol.* **3**, 363.
Beer, M. & Moudrianakis, E. N. (1962). *Proc. Nat. Acad. Sci., Wash.* **53**, 564.
Bendet, I., Schachter, E. & Lauffer, M. A. (1962). *J. Mol. Biol.* **5**, 76.
Bladen, H. A., Byrne, R., Levin, J. G. & Nirenberg, M. (1965). *J. Mol. Biol.* **11**, 78.
Caro, L. G. (1965). *Virology*, **25**, 226.
Cavalieri, L. F., Rosoff, M. & Rosenberg, B. H. (1956). *J. Amer. Chem. Soc.* **78**, 5239.
Chandler, B., Hyashi, M., Hyashi, M. N. & Spiegelman, S. (1964). *Science*, **143**, 47.
Cohen, G. & Eisenberg, H. (1966). *Biopolymers*, **4**, 429.
Doty, P., Marmur, J., Eigner, J. & Schildkraut, C. L. (1960). *Proc. Nat. Acad. Sci., Wash.* **46**, 461.
Dunnebacke, T. H. & Kleinschmidt, A. K. (1966). *Z. Naturf.*, in the press.
Eigner, J. & Doty, P. (1965). *J. Mol. Biol.* **12**, 549.
Flory, P. J. (1953). *Principles of Polymer Chemistry*, chapter 10. Ithaca, N.Y.: Cornell University Press.
Frank, H., Zarnitz, M. L. & Weidel, W. (1963). *Z. Naturf.* **18b**, 281.
Freifelder, D. (1965a). *Biochem. Biophys. Res. Comm.* **18**, 141.
Freifelder, D. (1965b). *Biochim. biophys. Acta*, **108**, 318.
Freifelder, D. (1966). *Virology*, **28**, 742.
Freifelder, D. & Kleinschmidt, A. K. (1965). *J. Mol. Biol.* **14**, 271.
Freifelder, D., Kleinschmidt, A. K. & Sinsheimer, R. L. (1964). *Science*, **146**, 254.
Fuchs, E., Zillig, W., Hofschneider, P. H. & Preuss, A. (1964). *J. Mol. Biol.* **10**, 546.
Glaser, W. (1956). In *Encyclopedia of Physics*, ed. by S. Flügge, vol. 33, p. 243. Berlin: Springer-Verlag.
Gomatos, P. J. & Stoeckenius, W. (1964). *Proc. Nat. Acad. Sci., Wash.* **52**, 1449.
Granboulan, N. (1966). *C. R. Acad. Sci. Paris*, Série D, **262**, 403.
Granboulan, N., Huppert, J. & Lacour, F. (1966). *J. Mol. Biol.* **16**, 571.
Hall, C. E. & Cavalieri, L. F. (1961). *J. Biophys. Biochem. Cytol.* **10**, 347.
Hall, C. E. & Doty, P. (1958). *J. Amer. Chem. Soc.* **80**, 1269.
Hall, C. E. & Litt, M. (1958). *J. Biophys. Biochem. Cytol.* **4**, 1.
Heinmets, F. (1949). *J. Appl. Phys.* **20**, 384.
Highton, P. J. & Beer, M. (1963). *J. Mol. Biol.* **7**, 70.
Hotta, Y. & Bassel, A. (1965). *Proc. Nat. Acad. Sci., Wash.* **53**, 356.
Inman, R. B. (1965). *J. Mol. Biol.* **13**, 947.
Inman, R. B. (1966). *J. Mol. Biol.* **18**, 464.
Inman, R. B. & Baldwin, R. L. (1962). *J. Mol. Biol.* **5**, 172.
Inman, R. B. & Jordan, D. O. (1960a). *Biochim. biophys. Acta*, **42**, 421.
Inman, R. B. & Jordan, D. O. (1960b). *Biochim. biophys. Acta*, **43**, 206.
Inman, R. B., Schildkraut, C. L. & Kornberg, A. (1965). *J. Mol. Biol.* **11**, 285.
Josse, J. & Eigner, J. (1966). *Ann. Rev. Biochemistry*, **35**, part II, 789.
Kaiser, A. D. & Inman, R. B. (1965). *J. Mol. Biol.* **13**, 78.
Katchalsky, A. & Lifson, S. (1953). *J. Polymer Sci.* **11**, 409.

Kellenberger, E. (1965). *Path. Microbiol.* **28**, 540.

Kisselev, N. A., Gavrilova, L. P. & Spirin, A. S. (1961). *J. Mol. Biol.* **3**, 778.

Kleinschmidt, A. K. (1960). In *III. Internat. Kongr. Grenzflächenaktive Stoffe, Köln,* vol. 2, p. 138. Mainz: Univers. Druckerei.

Kleinschmidt, A. K., Burton, A. & Sinsheimer, R. L. (1963). *Science,* **142**, 961.

Kleinschmidt, A. K., Dunnebacke, T. H., Spendlove, R. S., Schaffer, F. L. & Whitcomb, R. F. (1964). *J. Mol. Biol.* **10**, 282.

Kleinschmidt, A. K., Kass, S. J., Williams, R. C. & Knight, C. A. (1965). *J. Mol. Biol.* **13**, 749.

Kleinschmidt, A. K., Lang, D., Jacherts, D. & Zahn, R. K. (1962). *Biochim. biophys. Acta,* **61**, 857.

Kleinschmidt, A., Lang, D. & Zahn, R. K. (1960). *Naturwiss.* **47**, 16.

Kleinschmidt, A. K. & Zahn, R. K. (1959). *Z. Natur.* **14b**, 770.

Kuhn, W. & Kuhn, H. (1943). *Helv. chim. Acta,* **26**, 1394.

Lang, D., Kleinschmidt, A. K. & Zahn, R. K. (1964a). *Biochim. biophys. Acta,* **88**, 142.

Lang, D., Kleinschmidt, A. K. & Zahn, R. K. (1964b). *Biophysik,* **2**, 73.

Lawley, P. D. (1956). *Biochim. biophys. Acta,* **14**, 481.

Lerman, L. S. (1964). *J. Cell Comp. Physiol.* **64**, Suppl. 1, 1.

Luzzati, V., Mathis, A., Masson, F. & Witz, J. (1964). *J. Mol. Biol.* **10**, 28.

Luzzati, V., Nicolaieff, A. & Masson, F. (1961). *J. Mol. Biol.* **3**, 185.

MacHattie, L. A., Bernardi, G. & Thomas, C. A., Jr. (1963). *Science,* **141**, 59.

MacHattie, L. A., Berns, K. I. & Thomas, C. A., Jr. (1965). *J. Mol. Biol.* **11**, 648.

MacHattie, L. A. & Thomas, C. A., Jr. (1964). *Science,* **144**, 1142.

MacLean, E. C. & Hall, C. E. (1962). *J. Mol. Biol.* **4**, 173.

Moudrianakis, E. N. & Beer, M. (1965). *Proc. Nat. Acad. Sci., Wash.* **53**, 564.

Pedersen, K. O. (1958). *J. Phys. Chem.* **62**, 1282.

Reisner, J. H. (1965). *Lab. Invest.* **14**, 875.

Richardson, C. C., Inman, R. B. & Kornberg, A. (1964). *J. Mol. Biol.* **9**, 46.

Richardson, C. C., Schildkraut, C. L. & Kornberg, A. (1963). *Cold Spr. Harb. Symp. Quant. Biol.* **28**, 9.

Ris, H. & Chandler, B. L. (1963). *Cold Spr. Harb. Symp. Quant. Biol.* **28**, 1.

Rowen, J. W. (1953). *Biochim. biophys. Acta,* **10**, 391.

Scruggs, R. L. & Ross, P. D. (1964). *Biopolymers,* **2**, 593.

Sobotka, H. & Trurnit, H. J. (1961). In *Analytical Methods of Protein Chemistry,* ed. by P. Alexander & R. J. Block, vol. 3, p. 212. Oxford: Pergamon Press.

Soehner, R. L., Gentry, G. A. & Randall, C. C. (1965). *Virology,* **26**, 394.

Spirin, A. S. (1963). In *Proc. Internat. Congr. Biochemistry,* ed. by V. A. Engelhardt, vol. 1, p. 66. Oxford: Pergamon Press.

Stoeckenius, W. (1961). *J. Biophys. Biochem. Cytology,* **11**, 297.

Stoeckenius, W. (1963). In Weil, R. & Vinograd, J. (1963). *Proc. Nat. Acad. Sci., Wash.* **50**, 730.

Studier, F. W. (1965). *J. Mol. Biol.* **11**, 373.

Thomas, C. A., Jr. (1966). *J. Gen. Physiol.* **49**, 143.

Thomas, C. A., Jr. & MacHattie, L. A. (1964). *Proc. Nat. Acad. Sci., Wash.* **52**, 1197.

Thomas, R. (1954). *Biochim. biophys. Acta,* **14**, 231.

Vinograd, J., Lebowitz, J., Radloff, R., Watson, R. & Laipis, P. (1965). *Proc. Nat. Acad. Sci., Wash.* **53**, 1104.

Weil, R. (1963). *Proc. Nat. Acad. Sci., Wash.* **49**, 480.

Wells, R. D., Ohtsuka, E. & Khorana, H. G. (1965). *J. Mol. Biol.* **14**, 221.

Wilkins, M. H. F. (1963). *Science,* **140**, 941.

From *Journal of Molecular Biology*, Vol. 54, 557–565, 1970. Reproduced with permission of the author and publisher.

Molecular Weights of Coliphages and Coliphage DNA

III. Contour Length and Molecular Weight of DNA from Bacteriophages T4, T5 and T7, and from Bovine Papilloma Virus

D. Lang

Division of Biology
The University of Texas at Dallas
P. O. Box 30365, Dallas, Texas 75230, U.S.A.

(*Received 23 March 1970*)

DNA has been prepared for electron microscopy at ionic strength 0·20. The lengths of DNA molecules from the following viruses were found to be: bovine papilloma virus, $(2·49 \pm 0·05)\mu$ (circular DNA); T7, $(12·15 \pm 0·25)\mu$; T5, $(36·2 \pm 0·8)\mu$; and T4, $(52·0 \pm 1·0)\mu$. The limits given are estimated systematic errors of unknown sign plus sample standard deviations. In units of T7 DNA, the relative lengths and their maximum errors are: bovine papilloma virus, $0·205 \pm 0·003$; T5, $2·98 \pm 0·06$; and T4, $4·28 \pm 0·06$; these values should equal relative molecular weights except for T4 DNA (4·71). Based on the mean value of $25·1 \times 10^6$ daltons, obtained from three recent and independent determinations of T7 DNA molecular weight, the molar linear density of duplex DNA as prepared by standardized electron microscopy was found to be $2·07 \times 10^{10}$ for DNA with common bases, $2·28 \times 10^{10}$ for T4 DNA, and $2·23 \times 10^{10}$ daltons/cm for T2 DNA. Consequently, molecular weights of the following viral DNA's are: bovine papilloma virus, $5·15 \times 10^6$; T5, $74·9 \times 10^6$; T4, 119×10^6; and T2, 116×10^6 daltons.

1. Introduction

The determination of molecular weight of viral DNA by electron microscopy employs the relationship between apparent contour length, L, and molecular weight, M:

$$M = M'L \text{ (daltons)}.$$

The molar linear density, M' (dalton/cm), is calculated either according to the Watson–Crick B configuration of double-stranded DNA, assuming that this configuration prevails during preparation for electron microscopy, or, M' is obtained by calibration with DNA of independently determined molecular weight. In both cases, the accuracy of a molecular-weight determination by electron microscopy is limited by errors not only of length measurements, but also of values for M' or for the molecular weight of the calibrating DNA.

This paper is part of such a calibration by known DNA's. Recently, Bancroft & Freifelder (1970) and Dubin, Benedek, Bancroft & Freifelder (1970) made absolute determinations of molecular weights of DNA from *Escherichia coli* bacteriophages T4, T5 and T7. Freifelder kindly sent a sample of each original stock solution of purified phage for DNA extraction and length measurements. This was done in this laboratory with a mixture of all three phages, and also separately. The former procedure obviously permitted more precise determinations of length ratios and hence

557

molecular-weight ratios, since systematic magnification errors cancelled when the three DNA species were prepared simultaneously on the same grid. Addition of circular bovine papilloma virus DNA widened the range of molecular weights under study.

Grids for electron microscopy were prepared by the diffusion method of Lang, Kleinschmidt & Zahn (1964) as modified by Lang, Bujard, Wolff & Russell (1967). The resulting lengths of the three DNA species will be given in absolute units and also in terms of their proportions, followed by calculation of molecular-weight ratios and an error analysis. The molar linear density of duplex DNA, prepared for electron microscopy, will be evaluated using recently published molecular weights of T7 DNA.

2. Materials and Methods

(a) *Virus*

Bacteriophages T4, T5 and T7 were a gift from D. Freifelder. The stock solutions, each 0·3 ml., were 0·01 M-Tris buffer, 0·1 M-NaCl, 0·01 M-$MgSO_4$ and 0·001 M-$CaCl_2$, pH 7, containing phages with optical densities (260 nm, 1 cm path) of 12 for T4 and T5, and 36 for T7. Bancroft & Freifelder (1970) and Dubin *et al.* (1970) used phages from the same stock solution for independent DNA molecular-weight measurements with the following results: T4, 106×10^6 daltons; T5, $67·3 \times 10^6$ daltons; and T7, $25·5 \times 10^6$ daltons.

(b) *DNA*

First, the three DNA species were released simultaneously from the phages by mixing 0·10 ml. T4, 0·05 ml. T5 and 0·01 ml. T7 of the above stock solutions and then adding 1·44 ml. 5 M-$NaClO_4$, pH 7·8 (Freifelder, 1965). Using this solution, one preparation for electron microscopy was made at once and a second one 24 hr later.

Second, each DNA species was released separately by adding 0·01 ml. stock solution to 0·1 ml. 5 M-$NaClO_4$.

Bovine papilloma virus DNA extracted by Bujard (1967) and T3 DNA were characterized earlier (Lang *et al.*, 1967; Bujard, 1968).

(c) *Electron microscopy*

The diffusion method was applied as described by Lang *et al.* (1967). To 0·05 ml. of the above DNA mixture, or to each 0·11 ml. of the separately released DNA, were added 30 ml. 0·20 M-ammonium acetate, 10^{-3} M-EDTA, pH 6·5. The solutions were poured into Teflon-coated dishes (29 ml., see Lang & Mitani, 1970), and left there 20 min for temperature equilibration. After this, the surface was cleaned with a Teflon-coated bar and sprinkled with a few talc particles. Then cytochrome *c* powder (Nutritional Biochemicals Corp.) was spread from a glass needle. After a diffusion time of 15 min, portions of the surface film with adsorbed DNA were transferred to 7-hole Siemens-type grids, dried with ethanol, and shadowed with platinum. Carbon films served as specimen supports on the grids.

Micrographs were taken on 6·5 cm × 9 cm Kodak Electron Image Plates with a Siemens Elmiskop 1A at nominal magnifications of 5000 and 10,000, after lens currents and high tension had been switched on for at least 30 min. In order to reduce that part of the magnification error which arises from grid to grid by differences in specimen position with respect to the objective lens and which amounts to ± 0·8% sample S.D. (Lang *et al.*, 1967), images of molecules were focussed by adjusting the vertical specimen position without changing the objective-lens current. For each set of 12 plates, the magnification was determined by micrographing a selected area of a carbon replica made from a cross-lined optical grating with 54,800 lines/inch. The replica was mounted on Siemens-type grids by E. F. Fullam, Inc. During the 16 days when micrographs were made, the sample S.D. of 9 independent magnification measurements was ± 0·4% which reflects an excellent stability of the equipment.

The plates, including the ones with the grating image, were optically enlarged 23·7 times by projection. The images were traced on paper and measured with a Minerva curvimeter (map measurer). For each molecule the length was corrected for pincushion distortion as described earlier (Lang *et al.*, 1967); on the average, the correction was 0·5%.

3. Results

The length distribution of the two mixed-DNA preparations, Figure 1(a), shows three major peaks, corresponding to T7-, T5- and T4 DNA in the order of increasing length. The peak at 2·5 μ is produced by added bovine papilloma virus DNA.

Discrimination between intact linear molecules and fragments is not possible without arbitrariness. Suitably, one may define a peak region as the abscissa value under the peak, plus and minus three times the sample S.D., and one may then assume that all molecules within peak regions are intact. Since the sample S.D. of the length of intact viral DNA is usually ±4% or less, a peak region would then be approximately the mean length ± 12%. In this way one obtains 31% fragments by number and 19% by weight from the distribution in Figure 1(a).

No significant differences between corresponding mean lengths of simultaneously and separately prepared DNA species were found. Therefore, the measurements of Figure 1(a) were combined with the results of the separate determinations, which were

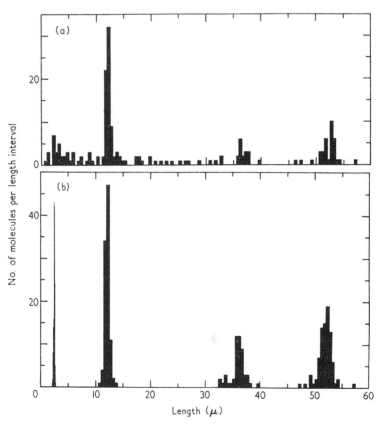

Fig. 1. (a) Length spectrum of simultaneously prepared DNA from bovine papilloma, T7, T5, and T4 viruses, showing four major frequency peaks of full-length molecules and interspersed fragments. Length intervals are 0·5 μ.

(b) Length spectrum obtained from the values of (a) *plus* the lengths from the separately prepared DNA's. Such pooling is justified since there were no significant differences in lengths or accuracy. Fragments between peaks have been omitted (see text). Length intervals are 0·05 μ for the 2·49 μ peak and 0·5 μ elsewhere.

TABLE 1

Result of length measurements

DNA from	Number of molecules	L (μ)	±Sample s.d. (μ)	±s.d. of the mean (μ)	Total error ±(s.d. of the mean+1·7%) (μ)	Relative length (in units of T7 DNA)
BPV†	151	2·49	0·08	0·01	0·05	0·205 ±0·003
T7	100	12·15	0·40	0·04	0·25	1·00
T5	51	36·2	1·3	0·18	0·8	2·98±0·05
T4	86	52·0	1·3	0·14	1·0	4·28±0·06

† BPV, bovine papilloma virus.

made twice for T4- and T5 DNA and once for T7- and bovine papilloma virus DNA. The combined length spectrum, showing peak regions only, is given in Figure 1(b). The mean lengths of the four DNA species are listed in Table 1, together with the experimental sample S.D. and S.D. of the mean.

The sample S.D. of about $\pm 3\%$ characterizes the precision of the method used. The S.D. of the mean of about $\pm 0.2\%$ shows into what range the effect of random errors has been confined by measuring more than one molecule.

Systematic errors, not included in these deviations, are estimated in Table 2.

TABLE 2

Systematic errors of length measurements

	Error (%)
1. Grating constant of cross-lined carbon replica	1·3
2. Tracing of DNA molecules on paper	0·8
3. Measurement of trace with curvimeter	0·8
Arithmetic sum	2·9
Geometric sum	1·7

The lines of the grating replica (length standard) could not be resolved by visible light. Therefore, two identifiable particles on the replica, found by light microscopy to be $7.49\ \mu$ apart, were micrographed by electron microscopy at a magnification of 5000. By measuring the distances between the then visible grating lines, a grating constant of $(5.41 \pm 0.05)\ 10^4$ lines/inch ($\pm$ S.D. of the mean) was found, 1.3% less than the manufacturer's value of 5.4800×10^4 lines/inch. Hence, $\pm 1.3\%$ was listed in Table 2 as possible error. The next systematic error in Table 2 is an interpretation error which depends on the quality of the preparation and on the degree to which the DNA is one-dimensionally wrinkled. This error is usually $\pm 0.8\%$ or less, as well as the last one, which is a personal error. Adding up, the maximum systematic error is $\pm 2.9\%$. But the signs of the component errors are unknown, and a more likely value is the geometric sum $\pm 1.7\%$.

The total error of lengths may then be the S.D. of the mean plus 1.7%, as given in column 6 of Table 1. With regard to the length ratios in column 7, systematic errors cancel each other. The remaining random errors, obtained by geometric addition of standard deviations of the two means involved, were multiplied by three in order to find the maximum error (column 7). Hence, the true value of a length ratio is most likely within the limits indicated in Table 1.

Another way of estimating the accuracy is to look at the history of independent length determinations, as shown in Figure 2. Measurements of T3 DNA from the same stock solution (Fig. 2, squares), and more recently of T7 DNA (circles), were done in this laboratory by several workers over a period of almost four years using the same electron microscope but different length standards. There is no trend with time and, as expected, the lengths of T3- and T7 DNA are equal. This T7 DNA was phenol-extracted from a phage strain given by B. Gomez and originally obtained from M. S. Meselson. The mean value of all determinations, $12.20 \pm 0.40\ \mu$ sample S.D. $\pm 0.07\ \mu$ S.D. of the mean, agrees with the length of T7 DNA in Table 1 ($12.15\ \mu$).

37

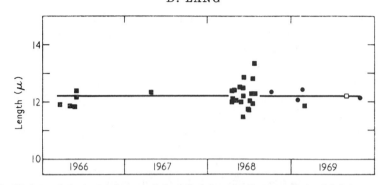

FIG. 2. History of electron microscopic length determinations, made in this laboratory by the diffusion method at 0·2 ionic strength, for T3 DNA (■) and T7 DNA (●). Both have probably equal length. The symbol (□) refers to 0·15 ionic strength and presence of formaldehyde (Lang & Mitani, 1970). The values scatter randomly with respect to the horizontal line representing the mean length of 12·20 μ.

4. Discussion

Figure 3 shows a survey of length measurements. It includes this and other work and also data on related viruses which are believed to have similar DNA lengths. Since the apparent DNA length depends on the ionic strength of the solutions from

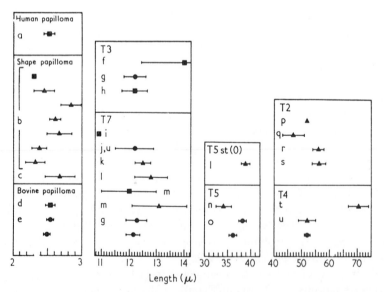

FIG. 3. Comparison of DNA lengths found by myself (values at the bottom) and by others for the viruses indicated and at the following ionic strengths: <0·1, (▲); 0·1 to <0·2, (■); and ≥0·2, (●). In each box, earlier determinations are entered above later ones. Errors are sample s.d.'s and letters stand for references. a: Crawford, Follett & Crawford, 1966; b: Kleinschmidt, Kass, Williams & Knight, 1965; c: Kellenberger, 1965; d: Lang et al., 1967; e: Bujard, 1970; f: Bendet, Schachter & Lauffer, 1962; g: Fig. 2; h: Lang & Mitani, 1970; i: Inman, Schildkraut & Kornberg, 1965; j: Freifelder & Kleinschmidt, 1965; k: Bode & Morowitz, 1967; l: Abelson & Thomas, 1966; m: Misra, Sinha & DasGupta, 1969; n: Frank, Zarnitz & Weidel, 1963; o: Bujard, 1969; p: Cairns, 1961; q: Kleinschmidt, Lang, Jacherts & Zahn, 1962; r: Thomas & MacHattie, 1964; s: nonglucosylated T2 DNA, Thomas, 1966; t: Chandler, 1963; u: Kleinschmidt, 1967.

which DNA is adsorbed to the cytochrome c interface (Lang *et al.*, 1967; Inman, 1967), different symbols are used in Figure 3 for different ionic strengths. The indicated errors are sample standard deviations. As can be seen, agreement and precision are generally best for results obtained at the higher ionic strengths, indicating that the diffusion method at ionic strength 0·2 was a suitable choice for standardized length determination of DNA.

As pointed out in the Introduction, utilization of length measurements for determining molecular weights requires knowledge of M', the molar linear density of DNA in a situation corresponding to the one in which it was measured.

One cannot expect that the B configuration applies necessarily to DNA prepared on grids. Conditions and steps are involved all of which may distort molecular configuration: ionic strength of the DNA solution, binding to cytochrome c and adsorption to the interface, mechanical instability of the protein film, adsorption to specimen grids, drying in ethanol, shrinkage or expansion of specimen supports in the electron beam. Indeed, DNA with two nicks on opposite strands may break, even if the nicks are 20 nucleotide pairs apart (Lang & Mitani, 1970), and lengths between 8 and 22 μ of otherwise intact T3 DNA can be brought about willfully under extreme conditions. However, with the standardized method used here, results are reproducible within a few per cent, and earlier indirect evidence suggested that M' is not more than 15% above the value calculated for the DNA B configuration (Lang & Coates, 1968).

For the B configuration with 3·46 Å length per pair of complementary nucleotides having the common bases adenine, thymine, guanine and cytosine (Langridge,

TABLE 3

A comparison of measured molecular weights of NaDNA (M), of their ratios and of molar linear densities

	Schmid & Hearst,[†] 1969	Leighton & Rubenstein, 1969	Bancroft & Freifelder, 1970; Dubin *et al.*, 1970	This paper[‡]
M_{T7} (10^6 daltons)	24·8	25	25·5	—
M_{T5}	68·5	83	67·3	—
M_{T4}	113·5	—	106	—
M_{T2}	—	132	—	—
M_{T5}/M_{T7}	2·76	3·32	2·63	2·98
M_{T4}/M_{T7}	4·58	—	4·14	4·71
M_{T2}/M_{T7}	—	5·28	—	(4·62)[§]
M_{T4}/M_{T5}	1·66	—	1·58	1·58
M_{T2}/M_{T5}	—	1·59	—	(1·55)[§]
M'_{T7} (10^{10} daltons/cm)	2·04	2·06	2·10	—
M'_{T5}	1·89	2·29	1·86	—
M'_{T4}[‖]	1·98	—	1·85	—
M'_{T2}[‖]	—	(2·35)[§]	—	—
Mean value of M'_T	1·97	2·23	1·94	

[†] Molecular weights by Schmid & Hearst are increased by 7·4% (Freifelder, personal communication).

[‡] From length ratio.

[§] Values in parentheses were calculated assuming that T4- and T2 DNA have equal length.

[‖] For comparison, reduced to non-glucosylated cytosine.

Wilson, Hooper, Wilkins & Hamilton, 1960), one obtains $1 \cdot 913 \times 10^{10}$ daltons/cm for NaDNA using International Atomic Weights of 1966 based on carbon isotope ^{12}C. The molecular weights of the two possible nucleotide pairs differ by only $0 \cdot 3 \%$; M' should then be practically independent of base composition. In T4- and T2 DNA, supposedly without effect on configuration (Thomas, 1966), cytosine is replaced by hydroxymethylcytosine (Wyatt & Cohen, 1953), which is 100% glucosylated in T4 and $76 \cdot 5 \%$ in T2 (Sinsheimer, 1960). M' for these DNA's, again in B configuration, is therefore $2 \cdot 104 \times 10^{10}$ (T4 DNA) and $2 \cdot 064 \times 10^{10}$ daltons/cm (T2 DNA). The numbers given in this paragraph are based on the assumption that the dimensions of the B configuration do not depend on base composition. The alternative is discussed by Freifelder (1970).

A comparison of measured molecular weights of sodium salts of DNA (M), of their ratios, and of molar linear densities, is given in Table 3. Listed are most recent M-values from three laboratories using different methods: Schmid & Hearst (1969), CsCl density-gradient equilibrium centrifugation; Leighton & Rubenstein (1969), auto-radiography of single ^{32}P-labeled DNA molecules: Bancroft & Freifelder (1970), Dubin *et al.* (1970), high-speed equilibrium centrifugation, sedimentation-diffusion measurements, and phosphorus–nitrogen determinations. The last column shows molecular weight ratios calculated from length ratios (Table 1) using M'-values of the preceding paragraph.

Good agreement exists for M_{T7}, but not for M_{T5} and M_{T4}, M_{T2}. For the M ratios, good agreement exists only for ratios *not* containing M_{T7}. The following reasoning might then be adopted.

Since three recent and independent determinations of M_{T7} are agreeing well, and since there is no indication that the T7 DNA's used might be different, the mean molecular weight of $(25 \cdot 1 \pm 0 \cdot 4) \times 10^6$ daltons (Table 3) is considered to be correct. This leads to an apparent molar linear density of $M'_{T7} = (25 \cdot 1 \pm 0 \cdot 4) \times 10^6 / [(1 \cdot 215 \pm 0 \cdot 025) \times 10^{-3}] = (2 \cdot 07 \pm 0 \cdot 04) \times 10^{10}$ daltons/cm. Assuming independence of M' on base composition this value should apply to other duplex DNA's when prepared for electron microscopy by the diffusion method from $0 \cdot 2$ ionic strength as described, or by a similar new method reported by Lang & Mitani (1970). From the lengths in Table 1, and using $M'_{T4} = (2 \cdot 07 \times 2 \cdot 104 / 1 \cdot 913) \times 10^{10} = 2 \cdot 28 \times 10^{10}$ and $M'_{T2} = (2 \cdot 07 \times 2 \cdot 064 / 1 \cdot 913) \times 10^{10} = 2 \cdot 23 \times 10^{10}$ daltons/cm, one may thus calculate molecular weights as shown in Table 4.

TABLE 4

Molecular weights calculated from length measurements and $M_{T7} = 25 \cdot 1 \times 10^6$ daltons

DNA from:	BPV†	T5	T4	T2
M (10^6 daltons)	5·15	74·9	119	116

† BPV, bovine papilloma virus.

This paper was prepared with the technical assistance of Mr L. Lewis, Jr. and Miss B. Bruton and I would like to thank them for their help. I also thank Dr D. Freifelder who sent us bacteriophage T4, T5 and T7; also Drs H. Bujard and B. Gomez who provided bovine papilloma virus DNA and another strain of T7 bacteriophage. This work was supported by National Science Foundation research grant GB6837 and by U.S. Public Health Service career development award GM-34,964 and divisional grant GM-13,234.

REFERENCES

Abelson, J. & Thomas, C. A., Jr. (1966). *J. Mol. Biol.* **18**, 262.
Bancroft, F. C. & Freifelder, D. (1970). *J. Mol. Biol.* **54**, 537.
Bendet, I., Schachter, E. & Lauffer, M. A. (1962). *J. Mol. Biol.* **5**, 76.
Bode, H. R. & Morowitz, H. J. (1967). *J. Mol. Biol.* **23**, 191.
Bujard, H. (1967). *J. Virology*, **1**, 1135.
Bujard, H. (1968). *J. Mol. Biol.* **33**, 503.
Bujard, H. (1969). *Proc. Nat. Acad. Sci., Wash.* **62**, 1167.
Bujard, H. (1970). *J. Mol. Biol.* **49**, 125.
Cairns, J. (1961). *J. Mol. Biol.* **3**, 756.
Chandler, B. (1963). *M. Sc. Thesis*, University of Wisconsin.
Crawford, L. V., Follett, E. A. C. & Crawford, E. M. (1966). *J. Microscopie*, **5**, 597.
Dubin, S. B., Benedek, G. B., Bancroft, F. C. & Freifelder, D. (1970). *J. Mol. Biol.* **54**, 547.
Frank, H., Zarnitz, M. L. & Weidel, W. (1963). *Z. Naturf.* **18b**, 281.
Freifelder, D. (1965). *Biochem. Biophys. Res. Comm.* **18**, 141.
Freifelder, D. (1970). *J. Mol. Biol.* **54**, 567.
Freifelder, D. & Kleinschmidt, A. K. (1965). *J. Mol. Biol.* **14**, 271.
Inman, R. B. (1967). *J. Mol. Biol.* **25**, 209.
Inman, R. B., Schildkraut, C. L. & Kornberg, A. (1965). *J. Mol. Biol.* **11**, 285.
Kellenberger, E. (1965). *Path. Microbiol.* **28**, 540.
Kleinschmidt, A. K. (1967). *Naturwiss.* **54**, 417.
Kleinschmidt, A. K., Kass, S. J., Williams, R. C. & Knight, C. A. (1965). *J. Mol. Biol.* **13**, 749.
Kleinschmidt, A. K., Lang, D., Jacherts, D. & Zahn, R. K. (1962). *Biochim. biophys. Acta*, **61**, 857.
Lang, D., Bujard, H., Wolff, B. & Russell, D. (1967). *J. Mol. Biol.* **23**, 163.
Lang, D. & Coates, P. (1968). *J. Mol. Biol.* **36**, 1968.
Lang, D., Kleinschmidt, A. K. & Zahn, R. K. (1964). *Biochim. biophys. Acta*, **88**, 142.
Lang, D. & Mitani, M. (1970). *Biopolymers*, **9**, 373.
Langridge, R., Wilson, H. R., Hooper, C. W., Wilkins, M. H. F. & Hamilton, L. D. (1960). *J. Mol. Biol.* **2**, 19.
Leighton, S. B. & Rubenstein, I. (1969). *J. Mol. Biol.* **46**, 313.
Misra, D. N., Sinha, R. K. & DasGupta, N. N. (1969). *Virology*, **39**, 183.
Schmid, C. W. & Hearst, J. E. (1969). *J. Mol. Biol.* **44**, 143.
Sinsheimer, R. L. (1960). In *The Nucleic Acids,* ed. by E. Chargaff & J. N. Davidson, vol. 3, p. 187, New York and London: Academic Press.
Thomas, C. A., Jr. (1966). *J. Gen. Physiol.* **49**, No. 6, Part 2, 143.
Thomas, C. A., Jr. & MacHattie, L. A. (1964). *Proc. Nat. Acad. Sci., Wash.* **52**, 1297.
Wyatt, G. R. & Cohen, S. S. (1953). *Biochem. J.* **55**, 774.

ARTICLE **13**

From *Journal of Molecular Biology*, Vol. 54, 567–577,
1970. Reproduced with permission of the publisher.

Molecular Weights of Coliphages and Coliphage DNA

IV. Molecular Weights of DNA from bacteriophages T4, T5 and T7 and the General Problem of Determination of *M*

DAVID FREIFELDER

Graduate Department of Biochemistry
Brandeis University, Waltham, Mass. 02154, U.S.A.

(*Received 23 March 1970, and in revised form 3 August 1970*)

Molecular weights (M) of the DNA from *Escherichia coli* bacteriophages T4, T5 and T7 are obtained by averaging the values from several methods. From these values and data for molecular length (l) under specific conditions of preparation for electron microscopy, M/l is derived. The possibility that M/l is not constant for all DNA molecules is discussed. Using sedimentation data an equation relating $S_{20,w}^0$ and M for bulk sedimentation and a second equation for sedimentation in sucrose gradients are presented. The general problems of measurement of M, M/l and $S_{20,w}^0$ and the current state of affairs are discussed.

1. Introduction

For the past 15 years considerable effort has gone into measuring DNA molecular weights—especially bacteriophage and viral DNA. However, the results have been disappointing in that the measurements have shown a great deal of variation and there have been no criteria for determining the reliability of the values obtained. Nonetheless, a set of values has through the years grown to be "accepted" and a few of them, that is, those for *Escherichia coli* phage λ and T4 have even become known as "standards". Unfortunately these so-called standard values were never obtained from absolute measurements on purified material. The result of this is that almost the entire set of accepted values is not correct (although certainly not in error by more than 25%). Schmid & Hearst (1969) were the first to report reliable values for the molecular weights of T4, T5 and T7 DNA measured by an absolute method, i.e. equilibrium centrifugation. These values were considerably lower than the accepted ones (Thomas & MacHattie, 1967), but were consistent with values calculated from the hydrodynamic model (worm-like coil) of Hearst and co-workers (Hearst & Stockmayer, 1962; Harris & Hearst, 1966; Gray, Bloomfield & Hearst, 1967; Hearst, Harris & Beals, 1966; Hearst, Beals & Harris, 1968; Hearst, Schmid & Rinehart, 1968). In the first two papers of this series (Bancroft & Freifelder, 1970; Dubin, Benedek, Bancroft & Freifelder, 1970) absolute determinations of the molecular weights of T4-, T5- and T7 DNA's have also been carried out and have yielded values in good agreement with those of Schmid & Hearst (1969). Hence a good set of standards may be available for the first time.

In the third paper of this series (Lang, 1970), electron micrographic lengths of these three DNA's are reported for a particular set of conditions. These data were obtained in the hope that the mass per length of DNA under specified conditions could be derived so that electron microscopy could become the method of choice for determination of M. However, it will be seen that the situation concerning mass per unit length for DNA in aqueous solution is not yet settled so that there probably is not yet a method for determination of DNA molecular weights other than tedious absolute methods.

In this paper the general problem of determination of molecular weight will be discussed. Second, using the molecular weight values obtained here, a useful relation between sedimentation coefficient, $S_{20,w}^{0}$, and molecular weight, M, will be derived. Third, a more reliable value of the molecular weight of λ DNA will be derived from end-group labeling data in the literature and sedimentation data using the above-mentioned S *versus* M curve. This value is probably more nearly correct than literature values and can be used until a precise value is obtained by absolute methods. Fourth, the commonly used S–M relation for centrifugation in sucrose gradients will be revised. Fifth, a brief compilation of corrected molecular weights and a list of some corrected values for several biological constants all of which were previously based upon λ and T4 standards will be given.

2. Molecular Weights of T4, T5 and T7 DNA

In papers I and II of this series, values for the molecular weights of phages T4, T5 and T7 have been obtained by equilibrium centrifugation (Bancroft & Freifelder, 1970) and sedimentation-diffusion (Dubin *et al.*, 1970). For both measurements it was necessary to have accurate values for the partial specific volume, $\bar{v}$, so these were measured pycnometrically. The most difficult part in the determination of $\bar{v}$ is to obtain the concentration of material accurately; in this work this was accomplished by chemical measurement of the nitrogen and phosphorus concentrations and calculation therefrom of the DNA and protein concentrations from the nucleotide and amino-acid compositions. Values of $\bar{v}$ were obtained with about 1% precision. The molecular weights so determined for these phages are not based on the use of any molecular weight standards and are therefore absolute determinations. The molecular weights of the phage DNA were then obtained from the DNA and protein contents. It should be noticed that there is excellent agreement in the values of T7 DNA obtained by the two methods, $25 \cdot 3 \pm 0 \cdot 8$ million daltons by equilibrium centrifugation and $25 \cdot 8 \pm 0 \cdot 9$ million by sedimentation-diffusion, as shown in Table 1. It must be remembered that the two determinations are not independent since they share the value of $\bar{v}$ which is in fact the principal source of error. Hence they will be averaged to $25 \cdot 55$ million and considered to be the result of a single measurement. The Table also lists the values for T4 and T5 DNA obtained from the sedimentation-diffusion experiments and a calculated value for non-glucosylated T4 which we will use in the discussion below of mass per unit length. These values compare favorably to those obtained by CsCl density-gradient equilibrium centrifugation by Schmid & Hearst (1969) as indicated in Table 1. The Schmid–Hearst values have all been increased by 7·4% since in their paper Schmid & Hearst used calculated rather than measured values for the effective density gradient and later found by direct measurement of the density gradient that their values of M should be so increased (Hearst, personal communication), which brings the various sets of values to within their experimental errors. Richardson (1966) has also measured the molecular weight of T7 DNA by labeling of the 5′-phosphoryl termini (also an absolute method) and obtained a value

TABLE 1

Molecular weights of T4, T5, T7 and φX174 DNA

Phage	Bancroft & Freifelder (1970)	Dubin et al. (1970)	Schmid & Hearst (1969)	Leighton & Rubenstein (1969)	Richardson (1966)	Sinsheimer (1959)	Mean[†]
		M Determined by various observers ($\times 10^6$ daltons)					
T4	—	106	113·5	130(T2 phage)	—	—	110
T5+	—	67·3	68·5	83	—	—	67·9
T7	25·3	25·8	24·8	25	26·2	—	25·2
φX174	—	—	—	—	—	3·4	3·4

[†] For T4 and T5 the mean is the average of the values of Dubin et al. and Schmid & Hearst. For T7, the values of Bancroft & Freifelder and Dubin et al. have been averaged to give 25·55, as described in the text and then averaged with that of Schmid & Hearst.

of 26·2 million. This value is an upper limit since it is difficult to know whether the phosphatase or kinase reaction goes to completion. For the purposes of this paper we are taking for the molecular weight of T7 DNA the mean of the average value obtained by sedimentation equilibrium centrifugation and sedimentation-diffusion and that by CsCl isopycnic equilibrium centrifugation, i.e. 25·2±0·8 million daltons. For T4 and T5 DNA the data of Dubin *et al.* (1970) and Schmid & Hearst (1969) have been averaged: T4, 110 million and T5, 67·9 million.

3. Molecular Weights by Autoradiography—A Critique

Recently Leighton & Rubenstein (1969) have reported values of M for the DNA of coliphages T2, T5 and T7 determined by molecular autoradiography. This method involves counting tracks from ^{32}P-labeled DNA or phage embedded in a photographic emulsion. From the number of tracks emanating from a single point (this configuration is called a "star"), and the specific activity, the phosphorus content per DNA molecule is obtained; from this value and the mean residue weight, M can be calculated. The values obtained were 132, 83 and 25 million for T2, T5 and T7 DNA, respectively. The first two are significantly larger than those of Schmid & Hearst (1969), and Dubin *et al.* (1970), whereas there is good agreement for T7. The wide differences for T2 and T5 are of some concern since in principle the autoradiographic method, as well as those of the latter two groups of workers, are absolute. There are four possible sources of error with the autoradiographic technique: (1) *Measuring specific activity.* Here there should be no problem since specific activity can be determined with accuracy. (2) *Biasing of track counting by the observer.* Most users of this technique have felt that there is a tendency to miss tracks so that if there is a bias, it would be to underestimate. However, this need not be the case, since the observer is often called upon to decide whether very short tracks (especially those in the vertical direction which can be seen only by continuously focusing the microscope up and down) are indeed the result of ^{32}P decay. (3) *Inability to count stars with a small or large number of rays.* Because of a background of cosmic rays and other tracks in the emulsion, the convention, adopted by Levinthal & Thomas (1957) and all subsequent workers, has been to count only stars with five or more rays. However, stars with more than 20 rays are virtually impossible to count accurately. This counting procedure results in a biased distribution as is made clear by the peaking of the mean above the Poisson distribution, as discussed in detail by Kahn (1964), although it has been estimated that this results in a mean value only 2·5% high for average star size of ten (Levinthal & Thomas, 1957). (4) *Purity of the DNA preparation.* Fragments or concatenated forms of phage DNA can bias the distribution. The concatenated forms are very important since if they are large enough, they can increase the value of the mean even if in relatively small amounts. We believe that this fourth point is the major source of the high values of Leighton & Rubenstein (1969) for T2 and T5 for the following reasons. The T7 phage they used was purified by differential centrifugation followed by isopycnic centrifugation in CsCl density gradients, which is a procedure yielding phage of high purity. However, the T2 and T5 preparations were purified only by two cycles of high- and low speed centrifugation, a procedure which is usually considered inadequate since it normally yields phage samples of low purity as evidenced by the translucent and multicomponent appearance of the phage pellet and the viscosity of the resuspended phage (Freifelder, unpublished results). Some of the impurities are "cell debris" which may or may not contain phosphorus; of greater significance is the free DNA whose molecular weight is often very high either because of aggregation or because it consists in part of concatenated replication intermediates. It should be realized that purification by sedimentation strongly selects for large molecules of sedimentable size and against fragments. These large molecules can bias the distribution strongly even though their concentration may be relatively low, thereby increasing the value of the mean. Unfortunately, one cannot argue vigorously that this particular artifact has in fact given rise to the high values of M for the two DNA's (T2 and T5) which were inadequately purified; it would have helped if Leighton & Rubenstein (1969) had provided the ray distributions in their report. Since this may be the case, it is recommended that the autoradiographic experiments be repeated using an osmotic shock-resistant mutant of T2 or T4 phage and purification by CsCl banding (stabilizing the phages by addition of Mg^{2+} and Ca^{2+} during banding).

4. Electron Micrographic Determination of DNA Molecular Weight
Evaluation of Mass per Unit Length (M/l)

Two problems have beset the use of the protein monolayer electron microscopic technique in molecular-weight measurement–length variation and lack of knowledge of M/l. Inman (1967) and Lang, Bujard, Wolff & Russell (1967) have shown that length depends upon the ionic strength and composition of the medium supporting the protein film (the hypophase). However, at ionic strengths greater than 0·1 the length is constant, so that variation can be minimized by working in this range. Addition of low concentrations of formaldehyde also seems to reduce this variation. Using the diffusion method instead of the spreading method for preparing the cytochrome c films containing adsorbed DNA may also improve reproducibility, but this has not been studied systematically.

As described in paper III of this series (Lang, 1970) it is possible to obtain reproducibility of $\pm 2\%$ in length so that it should be possible to determine the molecular weight of any DNA molecule fairly accurately if M/l were known. Clearly since length varies with conditions, a value of M/l for defined conditions must be used. (In the past a value for M/l derived from X-ray diffraction of DNA fibers has been used by most workers for whatever ionic strength they happen to have chosen for the hypophase; this is certainly not justified.) In particular, we will consider those conditions used by Lang (1970). Using the lengths from that paper and our values of M, then M/l for non-glucosylated T4, T5 and T7 DNA are $1·93 \pm 0·04$, $1·88 \pm 0·08$ and $2·07 \pm 0·08$ million daltons per micron, respectively. It should be noted that M/l for T7 is somewhat greater than for T4 and T5. It seems unlikely that the value for T7 is incorrect in view of the agreement for M between the methods discussed above and it is equally unlikely that the measured length of T7 DNA is in error since in Lang's measurements, T4, T5 and T7 DNA were all present on the same grid, so that even if the absolute length were in error, the relative lengths would be correct. However, two points should be made concerning M/l for T4 and T5. For T4 we have used a value of M from which the glucose mass has been subtracted but l for the normal glucosylated form. Thomas (1966) has reported that l for T2 is unaffected by the absence of glucose. However, in that experiment the hypophase was water in which case l is often variable from one experiment to the other so that the agreement in the values of l might be fortuitous. Furthermore, the degree of glucosylation of T2 is less than that of T4 (Lehman & Pratt, 1960). Although there is no reason to believe that glucose should affect l, we intend to investigate this experimentally. For T5 the peculiarity is that all the values of M are reproducible; M seems to be too low for its sedimentation coefficient, as will be discussed in a later section. Using a value of M *calculated* from $S^0_{20,w}$ and the equation of the following section yields a value of M/l nearer that of T7. However, we feel that the anomaly in the S–M relation for T5 DNA lies in S, not in M. This should be investigated further.

One important consideration that should not be ignored *a priori* is that M/l is not constant, a discouraging possibility since it might eliminate the usefulness of electron microscopy in precise determination of M. The idea is not unreasonable though for the following reason. It is known (Printz & von Hippel, 1965) that hydrogen bonds in DNA are continually being made and broken. Associated with breakage there must be a partial unwinding of the DNA and therefore an increase in length and decrease in M/l. A high A+T DNA should, on the average, have a greater proportion of broken H bonds than a high G+C content DNA so that it might be expected that a high A+T DNA would have a lower M/l than a high G+C DNA. The A+T contents of T4, T5 and T7 DNA's are 66, 61 and 48%, respectively so that the agreement in M/l for T4 (glucoseless) and T5 and the higher value for T7 is in agreement with these considerations.

If there is indeed a dependence of M/l on A + T content, one might simply assume that $M/l = \%(A + T) \times (M/l)_{A+T} + \%(G + C) \times (M/l)_{G+C}$. Then from two measurements, $(M/l)_{A+T}$ and $(M/l)_{G+C}$ could be determined. From the values of M/l for T4 (glucoseless) and T7 there would result $(M/l)_{A+T} = 1\cdot54 \times 10^6/\mu$ and $(M/l)_{G+C} = 2\cdot42 \times 10^6/\mu$. These numbers seem more different than one might expect. If the maximum error is allowed, they could be 1·82 and 2·16, respectively. Actually, the simple assumption made above is probably not valid since it is very likely that base stacking and base sequence would have important effects. For instance the alternating double-strand copolymer d(A−T) probably has a different M/l from dA:dT.

At present the dependence of M/l on base composition must be considered an open question subject to experimental testing. Hence, electron microscopy should probably not be used as a means of determining *very accurate* DNA molecular weights until this question is settled. Nonetheless it is certainly useful if a 10% error is tolerable. (The error is probably less than that.) We are currently engaged in a study of M/l using phage DNA's having different G + C content.

5. Equation Relating Sedimentation Coefficient ($S^0_{20,w}$) and Molecular Weight (M)

For more than a decade, numerous workers have attempted to establish an empirical relationship between $S^0_{20,w}$ and M, since the relatively simple measurement of $S^0_{20,w}$ could then replace tedious absolute determinations of M. Unfortunately, there have been major difficulties: first, there have not been until very recently reliable values of M. Second, when relatively accurate values of M of polydisperse samples were obtained by light scattering, the corresponding S was either a modal or median S which cannot be related in a simple way to the molecular-weight averages yielded by light scattering. Third, at first there did not exist a good hydrodynamic model for DNA so that the mathematical form of the S–M relationship was not clear—in fact even after a series of papers appeared in which the hydrodynamic model was defined (i.e. the worm-like coil), these papers passed unnoticed. Fourth, the principal attempts were to fit the data to a linear relation between log S and log M whereas Hearst and co-workers (Hearst & Stockmayer,

1962; Gray et al. 1967) have shown that the equation takes the form, $S - b = k\,M^a$, where b, K and a are constants. In the simple model of Hearst & Stockmayer (1962), $a = 0.5$. When excluded volume effects are included in the theory, a is slightly less than 0.5, although the magnitude of the difference could not be calculated from first principles (Gray et al., 1967). Crothers & Zimm (1965) were the first to attempt to obtain a curve of this type experimentally and argued on the basis of their data that $b = 2.7$ and $a = 0.445$ and not 0.5. Unfortunately, they did not have available the correct value for the molecular weight of T2 DNA and used the commonly accepted value of 130×10^6 and also insisted that their curve include the light-scattering data of Eigner & Doty (1965) for polydisperse samples. (See below for discussion of light-scattering work.) Furthermore, their value of $S_{20,w}^0$ for T7 (i.e. 33.7 s) is probably too high since three other laboratories have obtained 32.2 s (Davison & Freifelder, 1962), 32.0 s (Studier, 1965), and 31.8 s (Gray & Hearst, 1968). (As will be seen below their $S_{20,w}^0$ for T2 was also high.) For the purposes of this paper we have chosen to average the three lower values to yield 32.0 ± 0.2 s.

A potential difficulty with all of the attempts to obtain an S–M relationship valid for $M > 50 \times 10^6$ daltons, has been the use of glucosylated T-even DNA. Davison & Freifelder (1964) showed that the degree of glucosylation of the 5-hydroxymethylcytosine of the T-even DNA's has a small but significant effect on $S_{20,w}^0$. Similarly, Rosenblum & Cox (1966) detected a difference in the S-value of T2 and T4 DNA. Clearly the $S_{20,w}^0$ and M for the non-glucosylated DNA would be a more suitable choice for construction of an empirical S–M curve. Gray & Hearst (1968) have stated that the difference in S is accounted for only by the mass difference but this conclusion was based only upon the incorrect Crothers–Zimm (1965) equation. Actually, as we shall see below, this conclusion is probably correct. At present, however, this problem is only academic since the measurement of $S_{20,w}^0$ of the T-even DNA's seems to be beset with a variety of difficulties, as has been discussed by Hearst & Vinograd (1961) and Rosenblum & Schumaker (1963). In fact, reported values have a 10% range and seem to fall into two classes. The most carefully determined values ("carefully" referring to the number of measurements and the precautions taken to assure precision) for T4 are 62.1 ± 0.7 (Gray & Hearst, 1968), and 61.5 ± 1.5 (Rosenblum & Cox, 1966), which average to 61.8 ± 0.4. The higher values 63.5 (Studier, 1965) and 64.5 (Crothers & Zimm, 1965) for T2 and 66.6 for T4 (Rosenblum & Schumaker, 1963), can probably be ignored since these investigations did not carry out as many determinations of S and did not use the low concentrations and low speeds as those which yielded the lower values. In one case at least (Crothers & Zimm, 1965) their value of $S_{20,w}^0$ for T7 DNA was also too high (and by the same factor) suggesting that there may be an error in the buoyancy or viscosity corrections. It should be realized that there may also be a small systematic error in these S measurements resulting from the fact that the value of $\bar{v}$ which is used is that of calf thymus DNA (Hearst, 1962). Since there is a dependence of buoyant density on base composition, it might be expected that $\bar{v}$ would show a similar dependence. However, this probably does not amount to more than a 1 to 2% effect.

There are two careful determinations of $S_{20,w}^0$ for T5$^+$; 51.8 ± 0.6 s (Gray & Hearst, 1968), and 51.9 ± 1.0 s (Leighton & Rubenstein, 1969). These average to 51.8 ± 0.8. A potential difficulty arises in using T5 as a standard for constructing an S–M relation; that is, it possesses a large number of single-strand breaks (Davison, Freifelder & Holloway, 1964; Abelson & Thomas, 1966; Jacquemin-Sablon & Richardson, 1970) which could affect $S^0_{20,w}$ by increasing flexibility of the molecule. However, Hays & Zimm (1970) have clearly demonstrated that single-strand breaks have no effect on the sedimentation properties of native DNA.

It would be valuable to have values of $S^0_{20,w}$ and M other than those for T4, T5 and T7 DNA for constructing an $S-M$ relationship. However, no such values exist; in every case in the literature M has been determined either by sedimentation using one of the various incorrect $S-M$ relationships or by electron microscopy using conditions for which the mass per unit length is not known. The only case for which there has been an absolute determination of M is for the DNA of the single-strand phage, ϕX174 (Sinsheimer, 1959) in which case Schmid & Hearst (1969) have suggested that the measurement might have a small error. Nonetheless, we shall include the $S^0_{20,w}$ for the linear form of the double-strand replicative form, i.e. $S^0_{20,w} = 14\cdot3$ s, and $M = 3\cdot4 \pm 0\cdot2 \times 10^6$ (Sinsheimer, 1959; Burton & Sinsheimer, 1962). Eigner & Doty (1965) have reviewed a large number of measurements of M of polydisperse DNA samples (e.g. calf thymus, salmon sperm, and bacterial DNA's) determined by light scattering. However, Schmid, Rinehart & Hearst (personal communication and data to be published) have shown that in no case has reliable light scattering been performed on double-strand DNA because of the high values of $P(\theta)^{-1}$ used to extrapolate to zero-angle molecular weights. These data will therefore not be used.

We now attempt to construct a measurable $S-M$ relation of the form $S - b = kM^a$. The mean values of M used are: T4, 108 million; T5$^+$, 67·9 million; T7, 25·2 million; T4 (glucose-free), calculated to be 101 million, and ϕX174, 3·4 million. In the first trial a was chosen to be 0·5 in accord with the original Hearst–Stockmayer (1962) theory, but it was not possible to draw a satisfactory straight line through three points.

Instead of plotting S *versus* $M^{1/2}$, $\log (S - b)$ has been plotted against $\log M$ and an attempt was made to determine values for b and a. The problem which arose immediately was that if the values for T7 were considered to be exact, then for no reasonable value of b was it possible to draw a line through both the T4 and T5 points. If the T4 and T7 values are considered to be good, then a curve can be found which passes through the ϕX174 replicative form DNA point. No curve can be found which passes through the ϕX174, T7 and T5. In every case S for T5 seems to be too high for its M. This may indicate that the measured S of T5 is too high because of the single-strand breaks, which is unlikely, or that M is too low. In a previous section it was suggested that the error is in S. This may be a result of a singularity in T5 at the inner terminus of the FST region (i.e. the region beyond which injection cannot proceed without protein synthesis–Lanni, 1968).

By plotting curves for $b = 1\cdot6-5\cdot0$ in intervals of 0·2 and using the ϕX174, T7, and T4 values, the best curve is that obtained by choosing $b = 2\cdot8$ for which $a = 0\cdot479$ (Fig. 1). These values of b and a give 1100 Å for the Kuhn statistical length and a calculated diameter (the b of Gray, Bloomfield & Hearst, 1967) of 38 Å. Hence with all of the reservations discussed above, an $S-M$ relationship which is probably more nearly correct than any other one currently available in the literature is $S^0_{20,w} = 2\cdot8 + 0\cdot00834\ M^{0\cdot479}$

This equation is probably not valid for molecular weights below 10^6 daltons where the worm-like coil model probably does not hold but should be valid for very high values of M. The Figure also shows four points from the Eigner–Doty (1965) equation. The values of S appear to be somewhat high. This is not of great concern because of the considerations above of the average involved in obtaining these numbers.

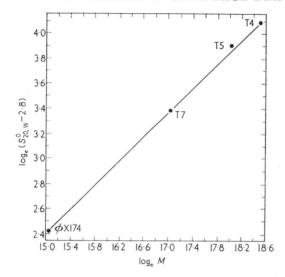

Fig. 1. Determination of constants in *S–M* relation. Plot of $\log_e(S^0_{20,w}-2.8)$ *versus* $\log_e M$. $S^0_{20,w}$ is in Svedbergs, M in daltons. The constant a determined from the slope of the least-squares line is 0.479; k is 0.00834.

6. The *S–M* Relationship in Sucrose Gradients

Burgi & Hershey (1963) showed that in zone sedimentation of DNA through a sucrose gradient of specified concentration range, the distance travelled is proportional to the sedimentation coefficient. Using values of M available at the time, they derived the relationship, $S_1/S_2 = (M_1/M_2)^{0.35}$. The considerations of Hearst & Stockmayer (1962) and Gray *et al.* (1967) would suggest that a more realistic form of the equation might be $(S_1-c)/(S_2-c) = (M_1/M_2)^d$, where c and d are constants. However, due to the increased buoyancy and viscosity of the gradient, it has been shown (Schmid, personal communication) that $c = 0$ is a good approximation. In addition there is a striking peculiarity of the sucrose gradient technique which suggests that sedimentation in this system must be considered at present to be strictly an empirical phenomenon. This odd property is that relative values of S in sucrose are not the same as relative values obtained by sedimentation in the analytical ultracentrifuge (Leighton & Rubenstein, 1969). This is probably due to convection at the walls resulting from the fact that the centrifuge tube is not sector-shaped. In any case, the exponent, 0.35, was determined from values of M which are probably incorrect (i.e. 83×10^6 and 132×10^6 for T5 and T7, respectively). If it is assumed that an equation of this form is an adequate description of sedimentation in sucrose gradients, at least in some narrow range of M, then by use of the values of M employed in the present series of papers, the equation, $S_1/S_2 = (M_1/M_2)^{0.38}$, results.

7. Molecular Weight of λ DNA

Escherichia coli phage λ DNA has sometimes been used as a standard to determine relative molecular weights. The values for M most often used (32 to 33×10^6 daltons) are those determined from electron microscopic measurements (MacHattie & Thomas, 1964; Caro, 1965) which, as we have seen, are not valid. Weiss & Richardson (1966) and

Live (1969) have recently measured M for λ by end-group labeling with polynucleotide kinase and obtained a value of 31×10^6. Live (personal communication) has pointed out that this is an upper limit since the enzymic reactions are carried out at DNA concentrations so high that there is undoubtedly some end-to-end aggregation *via* the λ cohesive ends or that even if the phosphatase and kinase reactions had gone to completion, there would still have been unlabeled termini.

Davis & Davidson (personal communication) have measured the relative length of λ DNA and double-strand ϕX174 DNA by electron microscopy of a sample in which both DNA's were present on the same grid. They obtained a length ratio of $9 \cdot 0 \pm 0 \cdot 1$ which yields a value of $30 \cdot 6 \pm 0 \cdot 4 \times 10^6$ daltons for the M of λ. The $S^0_{20,\mathrm{w}}$ for λ DNA is $34 \cdot 4 \pm 0 \cdot 3$ (unpublished results) which from the new S–M relationship yields $29 \cdot 6 \pm 1 \cdot 5 \times 10^6$ daltons for the M of λ. Hence for the present, $30 \pm 1 \times 10^6$ daltons will be taken as a better value than those currently in the literature. A detailed study of this is currently in progress.

There has been one attempt reported in the literature at absolute measurement of M for λ—that of Dyson & Van Holde (1967) using equilibrium centrifugation of whole phage. However, their use of a calculated rather than a measured value for $\bar{v}$ introduces too much uncertainty for consideration here. The value of $\bar{v}$ was calculated from the amino-acid composition and the percentage DNA—a common method for which there is neither experimental not theoretical justification. In fact as shown in paper I of this series (Bancroft & Freifelder, 1970), $\bar{v}$ for T5 is not that which would be obtained by these considerations.

8. Recommendations for Future Measurements

At this point one might ask—what is a good method of determining M?

The methods given in papers I and II of this series, i.e. Yphantis equilibrium centrifugation and sedimentation-diffusion, are capable of high precision but are very tedious and require large amounts of material (approx. 10 mg) to determine $\bar{v}$ and percentage DNA. The laser scattering diffusion measurements in addition require highly specialized equipment. Both of these methods are probably to be avoided as general procedures although they are extremely good for measuring viral particle weights.

The equilibrium centrifugation method of Schmid & Hearst (1969) seems to be the most straightforward and useful as long as one has available an ultracentrifuge equipped with a double-beam scanner. These are becoming more common. The method has the advantage of using only a few micrograms of DNA. At very high molecular weights ($> 100 \times 10^6$) the system may fail because it may not be possible to work at sufficiently low concentrations to make the virial corrections accurately.

Electron microscopy probably provides reliable data as long as the ionic conditions and the method of preparing the film is the same as that for which the mass per unit length was determined. If the A + T content departs very much from 50%, there may be a systematic error.

In the range 20 to 100×10^6 daltons, good values can probably still be obtained from the sedimentation coefficient if it is determined accurately. It must be emphasized though that the ionic conditions of sedimentation employed should not depart significantly from the range $0 \cdot 2$ to $1 \cdot 0$. It is probably best also to work at 20 to 25°C and at near neutral pH. Strong salt solution should be avoided in aluminum cells.

Much more work is needed to improve the *S–M* equation; specifically more accurate values in the range of 40 to 80×10^6 are needed. These could be provided by using many of the easily grown phages of *Pseudomonas aeruginosa*. Such work is in progress.

This work was supported by grant no. GM-14358 from the U.S. Public Health Service, no. E509 from the American Cancer Society, and contract (AT-30-1)-3797 from the Atomic Energy Commission. I was supported by a Career Development Award (GM-7617) from the U.S. Public Health Service. This is publication no. 744 from the Graduate Department of Biochemistry, Brandeis University.

REFERENCES

Abelson, J. & Thomas, C. A. (1966). *J. Mol. Biol.* **18**, 262.
Bancroft, F. C. & Freifelder, D. (1970). *J. Mol. Biol.* **54**, 537.
Burgi, E. A. & Hershey, A. D. (1963). *Biophys. J.* **3**, 309.
Burton, A. & Sinsheimer, R. L. (1962). *J. Mol. Biol.* **14**, 327.
Caro, L. (1965). *Virology*, **25**, 226.
Crothers, D. & Zimm, B. (1965). *J. Mol. Biol.* **12**, 525.
Davison, P. F. & Freifelder, D. (1962). *J. Mol. Biol.* **5**, 643.
Davison, P. F. & Freifelder, D. (1964). *Biopolymers*, **2**, 15.
Davison, P. F., Freifelder, D. & Holloway, B. W. (1964). *J. Mol. Biol.* **8**, 1.
Dubin, S. B., Benedek, G. B., Bancroft, F. C. & Freifelder, D. (1970). *J. Mol. Biol.* **54**, 547.
Dyson, R. D. & Van Holde, K. E. (1967). *Virology*, **33**, 359.
Eigner, J. & Doty, P. (1965). *J. Mol. Biol.* **12**, 549.
Gray, H. B., Bloomfield, V. A. & Hearst, J. E. (1967). *J. Chem. Phys.* **46**, 1493.
Gray, H. B. & Hearst, J. E. (1968). *J. Mol. Biol.* **35**, 111.
Harris, R. A. & Hearst, J. E. (1966). *J. Chem. Phys.* **44**, 2595.
Hays, J. B. & Zimm, B. H. (1970). *J. Mol. Biol.* **48**, 297.
Hearst, J. E. (1962). *J. Mol. Biol.* **4**, 415.
Hearst, J. E., Beals, E. & Harris, R. A. (1968). *J. Chem. Phys.* **48**, 5371.
Hearst, J. E., Harris, R. A. & Beals, E. A. (1966). *J. Chem. Phys.* **45**, 3106.
Hearst, J. E., Schmid, C. & Rinehart, F. (1968). *Macromolecules.* **1**, 491.
Hearst, J. E. & Stockmayer, W. (1962). *J. Chem. Phys.* **37**, 1425.
Hearst, J. E. & Vinograd, J. I. (1961). *Arch. Biochem. Biophys.* **92**, 206.
Inman, R. B. (1967). *J. Mol. Biol.* **25**, 209.
Jacquemin-Sablon, A. & Richardson, C. C. (1970). *J. Mol. Biol.* **47**, 477.
Kahn, P. L. (1964). *J. Mol. Biol.* **8**, 392.
Lang, D. (1970). *J. Mol. Biol.* **54**, 557.
Lang, D., Bujard, H., Wolff, B. & Russell, D. (1967). *J. Mol. Biol.* **23**, 163.
Lanni, Y. T. (1968). *Bact. Rev.* **32**, 227.
Lehman, I. R. & Pratt, E. A. (1960). *J. Biol. Chem.* **235**, 3254.
Leighton, S. B. & Rubenstein, I. (1969). *J. Mol. Biol.* **46**, 313.
Levinthal, C. & Thomas, C. A. (1957). *Biochim. biophys. Acta*, **23**, 453.
Live, T. (1969). Ph.D. Thesis, Harvard University.
MacHattie, L. A. & Thomas, C. A. (1964). *Science*, **144**, 1142,
Printz, M. P. & von Hippel, P. H. (1965). *Proc. Nat. Acad. Sci., Wash.* **53**, 363.
Richardson, C. C. (1966). *J. Mol. Biol.* **15**, 49.
Rosenblum, J. & Cox, E. C. (1966). *Biopolymers*, **4**, 747.
Rosenblum, J. & Schumaker, V. N. (1963). *Biochemistry*, **2**, 1206.
Schmid, C. W. & Hearst, J. E. (1969). *J. Mol. Biol.* **44**, 143.
Sinsheimer, R. L. (1959). *J. Mol. Biol.* **1**, 43.
Studier, F. W. (1965). *J. Mol. Biol.* **11**, 373.
Thomas, C. A. (1966). *J. Gen. Physiol.* **49**, 143.
Thomas, C. A. & MacHattie, L. A. (1967). *Ann. Rev. Biochem.* **36**, 485.
Weiss, B. & Richardson, C. C. (1966). *Cold Spr. Harb. Symp. Quant. Biol.* **31**, 471.

38

ARTICLE **14**

From *Cold Spring Harbor Symposium on Quantitative Biology*, Vol. 38, 1–8, 1973. Reproduced with permission of the authors and publisher.

On the Nature of Chromosome-sized DNA Molecules

Ruth Kavenoff, Lynn C. Klotz,* and Bruno H. Zimm

Department of Chemistry, University of California, San Diego, La Jolla, California 92037

To learn how eukaryotic DNA is organized, we have been studying its structure. Our work has been devoted primarily to the question of whether a single DNA molecule runs the entire length of a chromosome. To answer this question, we have been measuring the size of the largest DNA molecules in gentle lysates of Drosophila cells by means of a special viscoelastic technique (Chapman et al., 1969). Here we will review published results (Kavenoff and Zimm, 1973) which provide evidence that there are DNA molecules which run the entire length of a chromosome and then we will consider the extent of our knowledge of such molecules and, for comparison, the structure of the genome of some diplornaviruses.

Evidence for Chromosome-sized DNA Molecules from Drosophila

If there is only one DNA molecule per chromosome, it should have a molecular weight equal to the DNA content of its chromosome. Although Drosophila chromosomes are quite small, relative to most other chromosomes in higher eukaryotes, they contain more than ten times as much DNA as bacterial chromosomes. With conventional techniques it is difficult to measure intact bacterial DNA molecules accurately, and it is still more difficult to measure even larger DNA molecules. For example, sedimentation measurements are not reliable with DNA molecules significantly larger than those of bacteria, primarily because of a recently recognized source of artifacts, i.e., the sensitivity of the sedimentation coefficient to centrifuge speed (Rubenstein and Leighton, 1971; Kavenoff, 1972; Zimm, in prep.; Levin and Hutchinson, 1973). As a consequence of this problem, it is probably not feasible to use sedimentation to measure DNA molecules much larger than bacterial DNAs, and any attempt to do so would only *underestimate* the true size of the DNAs. On the other hand, the success of the viscoelastic technique in measuring the size of bacterial molecules (Klotz and Zimm, 1972a) encouraged us to apply the technique to Drosophila DNAs.

The viscoelastic technique principally measures the size of the largest DNA molecules in a solution. Figure 1 illustrates the property being measured, i.e., the viscoelastic relaxation of solutions of DNA molecules which have been stretched (in an instrument similar to a Couette concentric-cylinder viscometer) and then allowed to recoil toward the random coils characteristic of chain molecules at rest. This recoil decays exponentially, after some initial transients, with a decay time (τ) which can be obtained as the long-time slope of an appropriate semi-logarithmic plot of the recoil. τ varies as $M^{1.67}$, and is, therefore, a sensitive measure of molecular size. In fact, when we measure the limiting slope at long times of the recoil plot, we are selectively determining the molecular weights of the largest DNA molecules only, even though these largest molecules may be but a small fraction of the total. This quality is one of the important features of the viscoelastic method which makes it suitable for determining the size of chromosomal DNA in cell lysates; it also becomes a limitation, in that the investigation of smaller molecules which may also be present in the same solution is difficult. (More extensive discussion of the technique and interpretation are given in Klotz and Zimm, 1972a, b; Kavenoff and Zimm, 1973.)

For this work, Drosophila chromosomes are almost ideal. For example, consider the chromosomes of wild-type *Drosophila melanogaster* (see Table 1); there are only four chromosomes in the haploid set, with two large ones of almost equal size. In addition, there are other stocks of *D. melanogaster* which have significant variations in their chromosomes; we can think of some of these stocks as mutants with respect to the phenotype

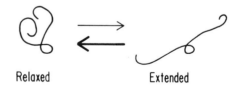

Relaxed   Extended

Figure 1. Schematic diagram of the viscoelastic relaxation of a single chain of DNA. The DNA is first stretched out and then allowed to return toward the random coil characteristic of chain molecules at rest. Relaxation of solutions of DNA is measured in a modified Couette viscometer called a viscoelastometer. (For details see Kavenoff and Zimm, 1973; Klotz and Zimm, 1972b.)

---

* Present address: Department of Biochemistry and Molecular Biology, Harvard University, Cambridge, Massachusetts 02138.

1

Chamber of Viscoelastometer

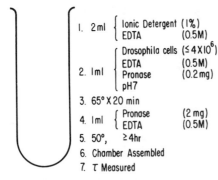

1. 2 ml { Ionic Detergent (1%)
         { EDTA   (0.5M)

2. 1 ml { Drosophila cells (≤4×10⁶)
        { EDTA   (0.5M)
        { Pronase   (0.2 mg)
        { pH7

3. 65° X 20 min

4. 1 ml { Pronase   (2 mg)
        { EDTA   (0.5M)

5. 50°, ≥4hr

6. Chamber Assembled

7. τ Measured

Figure 2. Preparation of Drosophila lysates. (For further details see Kavenoff and Zimm, 1973.)

of their largest chromosomes. We used such "mutants" to examine the relation of DNA size to chromosome structure. Likewise, we also characterized DNA from other species of Drosophila with interesting variations in chromosome structure.

The preparation of lysates of Drosophila cells is outlined in Figure 2. The cells were lysed in the chamber of our measuring instrument without further transfer in order to minimize breakage of the DNA by mechanical shear. To minimize nuclease action, high temperature was used with solutions containing detergents, chelating agents, high salt concentration, and proteolytic enzymes. The procedure was developed with cultured cells using the *D. melanogaster* cell line established by Imogene Schneider. Subsequently, we were able to obtain cell suspension of other Drosophila by disrupting pupae in a Dounce homogenizer.

The results of viscoelastic measurements of some Drosophila DNAs are summarized in Table 1. To begin with, we have results for the wild-type and two mutants of *D. melanogaster*. In the inversion mutant, the largest chromosome was identical in size to the largest chromosome in wild-type, but it carried a large pericentric inversion that changed the position of the centromere (see Table 1). In the translocation mutant, the largest chromosome

Table 1. Comparison of Chromosome Size with Results of Viscoelastic Measurements of Drosophila Lysates

Drosophila	Chromosomes	τ, hr $\left(\begin{smallmatrix}\text{Number of}\\\text{Measurements}\end{smallmatrix}\right)$	Molecular Weight of Largest DNA	DNA Content of Largest Chromosome*
melanogaster				
wild-type		1.67 ± 0.23 (44)	$41 \pm 3 \times 10^9$	43×10^9
inversion		1.77 ± 0.30 (9)	$42 \pm 4 \times 10^9$	43×10^9
translocation		2.98 ± 0.54 (10)	$58 \pm 6 \times 10^9$	59×10^9
hydei				
wild-type		1.62 ± 0.27 (8)	$40 \pm 4 \times 10^9$	
deletion		0.69 ± 0.17 (8)	$24 \pm 4 \times 10^9$	
virilis		2.12 ± 0.28 (5)	$47 \pm 4 \times 10^9$	
americana		5.00 ± 1.1 (3)	$79 \pm 10 \times 10^9$	

For each Drosophila, the idealized appearance of the chromosomes at metaphase is shown in the second column, with the largest chromosomes emphasized. The third column lists the values of τ, corrected to 50°C and 0.2 M Na⁺, with the average deviation of the measurements. The fourth column gives the values of M calculated from the values of τ; *the precision measure reflects average deviation only* (see text). (For further details see Kavenoff and Zimm, 1973.)
* Based on values of Rudkin (1965).

was larger in size than the largest chromosome in wild-type, because it carried 60% of the X chromosome attached to one end as a result of a translocation (see Table 1). Now if we consider the values obtained for τ, we see that all of them were greater than one hour; these values are much larger than the values of τ for intact bacterial DNA molecules (which are only about a minute). So the largest molecules in the Drosophila lysates are definitely much larger than the DNA molecules in bacterial chromosomes. For *D. melanogaster*, the values for τ were the same in both the wild-type and in the inversion mutant and markedly larger in the translocation mutant. These results indicate that the size of the largest DNA depends on the size of the largest chromosome and *not* on the morphology of the largest chromosome. We will return to this point.

If we compare the values for molecular weight calculated from τ's with the values from Rudkin's measurements of DNA content (Rudkin, 1965 and pers. commun.), we see that in each case the molecular weights of the largest DNA molecules approximate the DNA content of the largest chromosome. From this comparison, it is clear that the number of large DNA molecules per chromosome must be small, probably 1 or 2.

We can exclude the possibility that there are two DNA molecules per chromosome with one DNA molecule per chromosome arm, because with *D. melanogaster* we obtained the same results for both the inversion mutant and the wild-type. The inversion in the largest chromosome changed the arm ratio from the value of 1:1 (in wild-type) to about 7:1 without affecting the total DNA content of the chromosome. If each arm contained one DNA molecule, the inverted chromosomes would contain DNA molecules 75% larger than those of wild-type, and the values of τ for the inversion mutant would have been about 150% larger than the values for the wild-type. Since we obtained identical values of τ for both the inversion mutant and the wild-type, and since we also obtained larger values for the translocation mutant with larger chromosomes, we concluded that the largest DNA molecules must span a chromosome with *no* discontinuity at the centromere.

Similar results showing a dependence of the size of the largest DNA molecules on the DNA content of the largest chromosome were also obtained with other species of Drosophila. Thus with *D. hydei*, we could compare measurements of wild-type, in which the largest chromosome is the X, with a mutant strain in which the size of the X chromosome was greatly reduced by a deletion of most of one arm. We could also compare measurements of two very closely related species, *D. virilis* and *D. americana*.

The large metacentric chromosomes in *D. americana* presumably arose from the acrocentric chromosome in *D. virilis* (or a common ancestor) by Robertsonian fusion, as indicated by the fact that the individual arms of *D. americana* pair with the individual chromosomes of *D. virilis* in hybrids (Patterson and Stone, 1952). Although we do not have data for these species comparable to Rudkin's data for the DNA contents of the individual chromosomes of *D. melanogaster*, we do have values for the total DNA contents of these species, and from these we can say that our molecular weights are quite reasonable (for details see Kavenoff and Zimm, 1973).

Thus in every case we have examined, there is a good correlation between the size of the largest chromosome and the size of the largest DNA molecules as measured by τ. Since the sizes of the largest DNA molecules approximate the total DNA contents of the chromosomes, and the values span a fourfold range in molecular weight, we believe these molecules are not breakage products but represent the actual size of the DNA molecules in chromosomes. Therefore, we call them "chromosome-sized."

It is tempting to conclude from these data that there is only one chromosome-sized DNA molecule per chromosome. However, the fact is that our values for M rest on a long extrapolation of an empirical relation between τ and M. This is shown in Figure 3. The nature of the extrapolating function is not much in doubt; straightline log–log plots of hydrodynamic properties against size are well established for random-coil molecules, of which very large DNA is almost certainly a typical example. The close theoretical relation between the retardation time and the intrinsic viscosity, and the well-known relation between intrinsic viscosity and molecular weight, which is linear on a log–log scale for molecular weights above 30×10^6 daltons, support this statement. The difficulty comes in establishing the precise values of the constants in the linear relation. The values of these constants can only be found at present from plots, such as Figure 3, of experimental data on viral and bacterial DNA samples. Unfortunately, the uncertainty of these measurements, especially the molecular weights, is great. Another complication is the possible effects of proteins and detergents on the DNA conformation; these have not been thoroughly studied. At present, we can only estimate (on the basis of Figure 3) that the probable error of the absolute values of the molecular weights of our chromosome-sized DNA is about ±30%.

The *relative* values of the molecular weights of the various Drosophila DNA molecules are much more precise, however, because the ratios of the

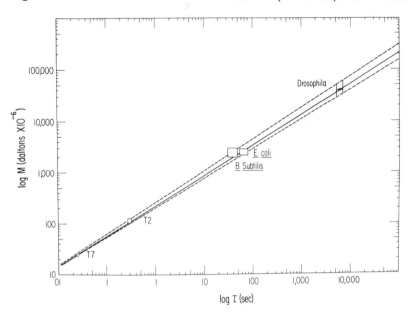

Figure 3. Empirical relation between τ and M. All data for τ corrected to 50°C and 0.2 M Na^+. Values for T7, T2, *Bacillus subtilis*, and *Escherichia coli* based on data in Klotz and Zimm (1972a); τ values for wild-type *D. melanogaster* are also plotted.

molecular weights depend only on the slope of the line, which is obviously more accurately determined than the molecular weights themselves. For this reason, we have emphasized the variations from one strain to another in the above discussion.

We have been trying to obtain an independent measure of chromosome-sized DNA molecules by radioautography. Preliminary experiments indicate that this approach is complicated by the deleterious biological effects of [³H]thymidine. Nevertheless, radioautography of lysates of *D. melanogaster* cultured cells labeled for one-third of a generation with [³H]thymidine produced a number of very long radioautograms. The longest observed are 1.2 cm, corresponding to a molecular weight somewhere between 2.4 and 3.2 × 10¹⁰ daltons (Dennis and Wake, 1966). One of these is shown in Figure 4. This does not really allow us to resolve the question of the number of DNA molecules per chromosome but it does at least lend some confidence to our viscoelastic measurements.

Fortunately, we can cite completely different

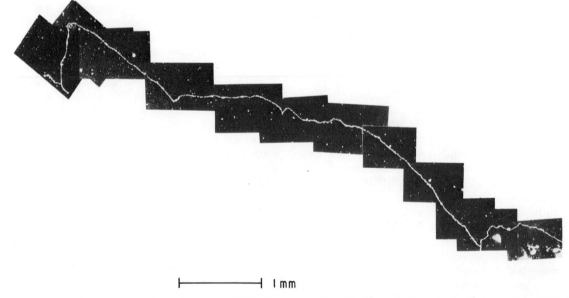

├─────────────────┤ 1 mm

Figure 4. Radioautograph of *D. melanogaster* DNA. Cultured cells of wild-type *D. melanogaster* were labeled with [³H]thymidine (500 μCi/ml) for 24 hr, washed, diluted 10,000-fold with unlabeled cells, and lyzed as described in Figure 2, except with tenfold lower levels of pronase. Small aliquots were carefully applied to lysozyme-coated slides; the slides were then held at an angle of 45° to the vertical while the drops slowly ran down the slides. The slides were subsequently dipped in Kodak NTB2 liquid emulsion, exposed five months, developed photographically, and then examined by dark-field photomicroscopy. The contour length, measured with a map measurer, was 1.2 cm.

evidence which, together with our results, leads to the conclusion that there is only one chromosome-sized DNA molecule per chromosome. That the haploid genome of *D. melanogaster* contains a significant fraction of unique, single-copy sequences was deduced from measurements of renaturation rate constants by Laird (1971) and also by Wu et al. (1972). Consequently, there should be only one chromosome-sized DNA molecule per chromosome. If there were more than one, we could not reconcile our data with the existence of both unique sequences and simple, linear, genetic maps.

Confessions of Ignorance

Our knowledge of the properties of the chromosome-sized DNA molecules from Drosophila is very incomplete. Nevertheless, it may be worthwhile to outline what we do and don't know about them.

We know very little about the biochemical nature of chromosome-sized DNA molecules. By DNA molecules we mean pieces of native DNA, without specifying whether or not the pieces contain single-stranded nicks or short gaps; such interruptions should not affect τ, because they do not affect the viscosity (Hays and Zimm, 1970). Also, we cannot exclude the possibility that some protein is still attached to the DNA at the time of measurement. However, the stability and reproducibility of the measurements (see Kavenoff and Zimm, 1973), after the first few hours of exposure to pronase, suggest that the attached protein, if any, either is completely resistant to pronase or has no effect on the configurations of the DNA molecule and hence no effect on τ. Likewise, we cannot exclude the possibility that the chromosome-sized DNA molecules contain pronase-resistant links of material other than DNA, for example, protein in a protease-resistant form, or RNA.

We cannot be sure that the DNA molecules are all linear. Although we can probably exclude any model that requires all the DNA to consist of structures with a dozen or more branches, we cannot be sure about the possibility of a few branches in some of the molecules, nor can we rule out the possibility that some, or even all, of the molecules are circular. If such were the case, however, the structures would have to be similar in all the chromosomes examined; otherwise, the relations among the different species and strains could not have been as we found them. For this reason, we think that the structures are most probably linear. (For discussion of the effect of circularity, see Klotz and Zimm, 1972a.) If the DNA is linear, the question remains as to what the ends are like.

Our results do not exclude the possibility that there may be some molecules of nuclear DNA which are much smaller than chromosome size, since, with our technique, it is difficult to measure such molecules in the presence of larger ones.

It is difficult to reconcile our data showing a good correlation between the size of the DNA molecules and the size of individual chromosomes with the idea that all of the nuclear DNA is in one or two (superchromosomal) pieces. We estimate that in our Drosophila lysates there could not have been more than 10% of the DNA in the form of linear genome-sized DNA molecules; otherwise, there should have been clear evidence in the recoil plots. However, we cannot be sure that the chromosome-sized DNA molecules which we detected did not arise through some specific fragmentation of larger molecules. In fact, precisely such fragmentation has been found to occur in the case of the genomes of some diplornaviruses.

Evidence for Genome-sized Circular Molecules from Diplornaviruses

Diplornaviruses, as their name implies, have genomes composed of double-stranded RNA. They have about 30 million daltons of RNA per particle and they infect a wide range of eukaryotes. Two of the best known are reovirus which grows in mammals and cytoplasmic polyhedrosis virus (CPV) which grows in silkworms.

The structure of these genomes has been the subject of dispute for several years. The problem has been whether the double-stranded RNA consists of one molecule or whether it consists of ten specific pieces (for review, see Shatkin, 1971). The concept of a segmented genome consisting of ten specific pieces has been widely accepted, largely on the basis of extensive physico-chemical and chemical analyses characterizing the ten specific fragments, all less than 3×10^6 daltons molecular weight (Bellamy et al., 1967; Kalmakoff et al., 1969; Fujii-Kawata et al., 1970; Millward and Graham, 1970; Millward and Nonoyama, 1970; Banerjee and Grece, 1971; Banerjee and Shatkin, 1971; Lewandowski and Millward, 1971; Furuichi and Miura, 1972; Lewandowski and Leppla, 1972). However, this overlooked some early EM evidence for longer linear molecules approaching the size of the genome (Granboulon and Niveleau, 1967; Dunnebacke and Kleinschmidt, 1967; Vasquez and Kleinschmidt, 1968; Nishimura and Hosaka, 1969) and it also presented a paradox: How could each virus particle have ten different "chromosomes" and survive? In fact, some diplornaviruses have remarkable infectivity; reovirus, for example, has been prepared such that every virus particle was infectious (Spendlove et al., 1970).

To investigate this problem, one of us (R. K.) in collaboration with John A. Mudd prepared lysates

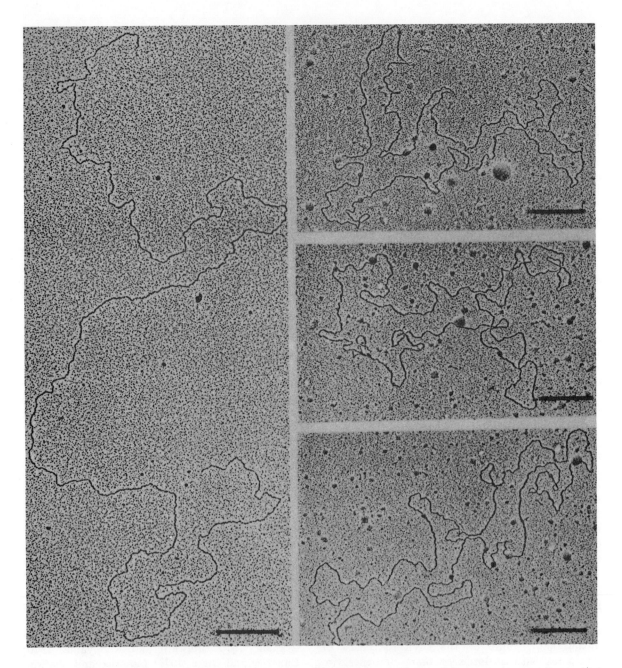

Figure 5. Electron micrographs of CPV RNA. Polyhedra from CPV-infected silkworm larvae guts (generously provided by James Kalmakoff) were purified by velocity and isopycnic sedimentation, pelleted, and resuspended in PB (0.01 M sodium phosphate buffer, pH 6.8). Equal volumes of polyhedra (about 170 μg RNA per ml) and nuclease-free pronase (10 mg/ml, Calbiochem) were combined, made 0.5 % in Sarkosyl (10 %, Sigma), and incubated as follows: 0°C for 30 min, 25°C for 30 min, and then 75°C for 15 min. The lysate was dialyzed against PB containing 1 mM EDTA at 20°C for 24 hr, and then against PB at 4°C for 72 hr. A portion of the lysate was mixed with one-tenth volume of 5 mg/ml cytochrome c (Sigma) containing 5 mM Tris and 1 mM EDTA at pH 7, and made 1 M in ammonium acetate. Aliquots were spread (Kleinschmidt, 1968) over a hypophase of 0.1 M ammonium acetate and 0.01 % formaldehyde. The monolayer was transferred to carbon-coated grids, stained with 0.1 % aqueous uranyl acetate, and rotary shadowed at low angle with platinum-palladium. Grids were examined in a Phillips EM 200 electron microscope. Bars represent 0.5 μ. Of 213 unselected molecules, 29 were open circles with a contour length of 14.7 $\pm$ 0.6 μ, and 46 were linear with a contour length of about 15 μ.

from reovirus using essentially the same methods which we had used for Drosophila. Both viscosity and sedimentation measurements indicated that the lysates contained double-stranded RNAs which were significantly larger than the pieces detected previously by sedimentation; in fact, our sedimentation data could only be explained by the existence of genome-sized circular forms of reovirus RNA. Subsequently, John A. Mudd and Susan Gillies observed circular forms of CPV RNA in the electron microscope (Kavenoff, Mudd, and Gillies in prep.). Electron micrographs of three open circles and one genome-sized linear molecule of CPV RNA are shown in Figure 5. Measurements of 29 open circles indicated a contour length of nearly 15 μ, corresponding to about 35×10^6 daltons of double-stranded RNA (Granboulon and Franklin, 1966; Boedtker, 1971).

Apparently, therefore, each circle contains all ten pieces of double-stranded RNA and also a considerable fraction which may be spacer RNA. Although many details remain to be resolved, it seems clear that the genomes of diplornaviruses are single molecules containing a number of sites that are very sensitive to RNase, so that degrading RNase (which apparently contaminates most virus preparations) with pronase prevents the specific fragmentation of the circular genomes.

We must emphasize that we have no reason to believe that a similar situation will be found in the case of Drosophila, but it is a possibility.

General Nature of Chromosomal DNA Molecules

Returning to the nature of eukaryotic DNA, we must caution that despite the evidence that there is only one DNA molecule per chromosome in Drosophila, there may be other cases in which this is not so. To illustrate the point, one need only consider the extent of chromosome polymorphism within a single fly. Some Drosophila neuroblast cells have so-called "giant" chromosomes which contain two or four times as much DNA as do simple chromosomes. In fact, the giant chromosomes even undergo mitosis with twice the usual DNA content per chromatid (Gay et al., 1970). Clearly, we cannot say whether our results apply to the DNA in the giant chromosomes. Likewise, we certainly cannot say how our results relate to the structure of the DNA in the polytene chromosomes of the salivary cells.

Conclusions

The results for Drosophila DNAs lead to the conclusion that *in its simplest form one chromosome contains one long molecule of DNA*. The qualification, "in its simplest form," is necessary because of the generality of the term "chromosome"; here we refer to the chromosomes of normal cells from anaphase through G_1, and possibly to the chromatids of metaphase chromosomes. The conclusion may not apply to all cases. Nevertheless, it should have a sufficient range of validity to be useful in studying the organization of eukaryotic DNA and the behavior of chromosomal genes.

Acknowledgments

We are indebted to Dan Lindsley for suggesting the study of *D. melanogaster* strains with chromosome aberrations and helping us to select and obtain the strains, to Ellen Rasch for performing microspectrophotometric measurements on some of our strains, to George Rudkin for providing advice and unpublished data, and to Imogene Schneider for providing her cell line of *D. melanogaster* and advice. We also thank Frank Newhook and Rob Allen for the use of their photomicroscopy equipment.

This work was supported by the Cancer Research Funds of the University of California and NIH grant GM 11916. Some of this work was conducted during the tenure (by R. K.) of a postdoctoral fellowship from the International Agency for Research on Cancer of the World Health Organization.

References

BANERJEE, A. K. and M. A. GRECE. 1971. An identical 3'-terminal sequence in the ten reovirus genome segments. *Biochem. Biophys. Res. Comm.* **45**: 1518.

BANERJEE, A. K. and A. J. SHATKIN. 1971. Guanosine-5'-diphosphate at the 5' termini of reovirus RNA: Evidence for a segmented genome within the virion. *J. Mol. Biol.* **61**: 643.

BELLAMY, A. R., L. SHAPIRO, J. T. AUGUST, and W. K. JOKLIK. 1967. Studies on reovirus RNA. I. Characterization of reovirus genome RNA. *J. Mol. Biol.* **29**: 1.

BOEDTKER, H. 1971. Conformation independent molecular weight determinations of RNA by gel electrophoresis. *Biochim. Biophys. Acta* **240**: 448.

CHAPMAN, R. E., L. C. KLOTZ, D. S. Thompson, and B. H. ZIMM. 1969. An instrument for measuring retardation times of deoxyribonucleic acid solutions. *Macromolecules* **2**: 637.

DENNIS, E. S. and R. G. WAKE. 1966. Autoradiography of the *Bacillus subtilis* chromosome. *J. Mol. Biol.* **15**: 435.

DUNNEBACKE, T. H. and A. K. KLEINSCHMIDT. 1967. Ribonucleic acid from reovirus as seen in protein monolayers by electron microscopy. *Z. Naturforsch.* **22b**: 159.

FUJII-KAWATA, I., K. MIURA, and M. FUKE. 1970. Segments of genome of viruses containing double-stranded ribonucleic acid. *J. Mol. Biol.* **51**: 247.

FURUICHI, Y. and K. MIURA. 1972. The 3'-termini of the genome RNA segments of silkworm cytoplasmic polyhedrosis virus. *J. Mol. Biol.* **64**: 619.

GAY, H., C. C. DAS, K. FORWARD, and B. P. KAUFMANN. 1970. DNA content of mitotically active condensed chromosomes of *Drosophila melanogaster*. *Chromosoma* **32**: 213.

GRANBOULON, N. and R. M. FRANKLIN. 1966. Electron microscopy of viral RNA, replicative form and replicative intermediate of the bacteriophage R17. *J. Mol. Biol.* **22**: 173.

GRANBOULON, N. and A. NIVELEAU. 1967. Etude au microscope electronique du RNA du reovirus. *J. Microscopie* **6**: 23.

HAYS, J. B. and B. H. ZIMM. 1970. Flexibility and stiffness in nicked DNA. *J. Mol. Biol.* **48**: 297.

KALMAKOFF, J., L. J. LEWANDOWSKI, and D. R. BLACK. 1969. Comparison of the ribonucleic acid subunits of reovirus, cytoplasmic polyhedrosis virus, and wound tumor virus. *J. Virol.* **4**: 851.

KAVENOFF, R. 1972. Characterization of the *Bacillus subtilis* W23 genome by sedimentation. *J. Mol. Biol.* **72**: 801.

KAVENOFF, R. and B. H. ZIMM. 1973. Chromosome-sized DNA molecules from *Drosophila*. *Chromosoma* **41**: 1.

KLEINSCHMIDT, A. K. 1968. Monolayer techniques in electron microscopy, p. 361. In *Methods in enzymology*, vol. 12, ed. L. Grossman and K. Moldave. Academic Press, New York.

KLOTZ, L. C. and B. H. ZIMM. 1972a. Size of DNA determined by viscoelastic measurements: Results on bacteriophages, *Bacillus subtilis* and *Escherichia coli*. *J. Mol. Biol.* **72**: 779.

———. 1972b. Retardation times of deoxyribonucleic acid solutions. II. Improvements in apparatus and theory. *Macromolecules* **5**: 471.

LAIRD, C. D. 1971. Chromatid structure: Relationship between DNA content and nucleotide sequence diversity. *Chromosoma* **32**: 378.

LEVIN, D. and F. HUTCHINSON. 1973. Neutral sucrose sedimentation of very large DNA from *Bacillus subtilis*. I. Effect of random double-strand breaks formed by gamma ray irradiation of the cells. *J. Mol. Biol.* **75**: 455.

LEWANDOWSKI, L. J. and S. H. LEPPLA. 1972. Comparison of the 3′ termini of discrete segments of the double-stranded ribonucleic acid genomes of cytoplasmic polyhedrosis virus, wound tumor virus and reovirus. *J. Virol.* **10**: 965.

LEWANDOWSKI, L. J. and S. MILLWARD. 1971. Characterization of the genome of cytoplasmic polyhedrosis virus. *J. Virol.* **7**: 434.

MILLWARD, S. and A. F. GRAHAM. 1970. Structural studies on reovirus: Discontinuities in the genome. *Proc. Nat. Acad. Sci.* **65**: 422.

MILLWARD, S. and M. NONOYAMA. 1970. Segmented structure of the reovirus genome. *Cold Spring Harbor Symp. Quant. Biol.* **35**: 773.

NISHIMURA, A. and Y. HOSAKA. 1969. Electron microscopic study on RNA of cytoplasmic polyhedrosis virus of the silkworm. *Virology* **38**: 550.

PATTERSON, J. T. and W. S. STONE. 1952. *Evolution in the genus* Drosophila. Macmillan, New York.

RUBENSTEIN, I. and S. B. LEIGHTON. 1971. The influence of rotor speed on the sedimentation behavior in sucrose gradients of high molecular weight DNA's. *Biophys. Soc. Abstr.* 209a.

RUDKIN, G. T. 1965. Photometric measurements of individual methaphase chromosomes. *In Vitro* **1**: 12.

SHATKIN, A. J. 1971. Viruses with segmented ribonucleic acid genomes: Multiplication of influenza versus reovirus. *Bact. Revs.* **35**: 250.

SPENDLOVE, R. S., M. E. McCLAIN, and E. H. LENNETTE. 1970. Enhancement of reovirus infectivity by extracellular removal or alteration of the virus capsid by proteolytic enzymes. *J. Gen. Virol.* **8**: 83.

VASQUEZ, C. and A. K. KLEINSCHMIDT. 1968. Electron microscopy of RNA strands released from individual reovirus particles. *J. Mol. Biol.* **34**: 137.

WU, J., J. HURN, and J. BONNER. 1972. Size and distribution of the repetitive segments of the *Drosophila* genome. *J. Mol. Biol.* **64**: 211.

Questions

ARTICLE 8 Rubenstein, Thomas, and Hershey

1. Note how this paper differs from Article 4 by Davison et al. in that here it is necessary to know the specific activity of the ^{32}P and the detection efficiency very accurately. Calculate how much ^{31}P would have to be in the stock solution to produce a 10% error.

2. The value of 130 million is roughly 15% too high. What systematic errors could do this?

3. Note that for the whole molecules, all molecules are the same length, whereas for halves the size varies although the mean value is 1/2. This means that the weight average ($\Sigma N_i M_i^2/\Sigma N_i M_i$) and the number average ($\Sigma N_i M_i/\Sigma N_i$) values of M are not the same (where N_i and M_i refer to the number of molecules having each value of M). Which average of M was obtained in the star count and in the s determination? Does this mean that the equation obtained would be different if one had a collection of precise wholes, halves, quarters, and so on?

ARTICLE 9 Davison and Freifelder

1. In the chromatography experiment explain the reasoning behind taking samples from several tubes for sedimentation analysis.

2. What is the argument that T7 phage cannot contain two molecules of equal size? Two molecules of different size (e.g., with a 10:1 weight ratio)?

ARTICLE 10 Meselson, Stahl, and Vinograd

1. The centrifuge cell extends roughly from 5.7 to 7.3 cm from the axis of rotation. What is the smallest value of M that could be measured by this method? How long would it take for a DNA having this M to come to equilibrium?

2. If the centrifugal velocity increases, how will the band position change?

•3. What might be the explanation for the appearance of a non-Gaussian (in particular, too broad at the base) but symmetric band? A non-Gaussian, asymmetric band?

4. Suppose a ^{12}C-labeled DNA, having M of 25×10^6 has a density of 1.700 g/ml. This sample is mixed with another sample of the same DNA labeled with ^{13}C. What will be the density and band width of the ^{13}C DNA? Assume a speed of 45,000 rpm.

•5. For high molecular weight DNA, concentration, c, effects are usually important when c exceeds about 5 μg/ml. If a centrifuge cell contains 1 μg DNA in 0.8 ml, what is the approximate concentration at the peak of the band at equilibrium if the speed is 20,000 rpm? 40,000 rpm? Why should the concentration dependence be less for lower M? Will concentration dependence be affected by speed?

6. If the mass per unit length of DNA is 2 million per micron, and a DNA molecule occupies a sphere whose diameter is 20% of the length of the molecule, how many T4 DNA molecules are side by side at the half height of the band at 40,000 rpm?

7. What criterion should be used to insure that equilibrium is reached?

8. It is known that one half of the T4 DNA molecule has a higher density than the other. Will this affect the band width? Will it affect the orientation of the molecules at equilibrium?

•1. Would you expect M/l from electron microscopy to be the same as M/l determined by Watson and Crick? Discuss.

•2. Since the technique usually involves allowing a droplet of solution to spread rapidly over a surface to form a film, it might be argued that the DNA is stretched. What data in these papers can be used to argue that stretching does not occur? Could the stretching break the DNA, if it occurred?

ARTICLE 13 Freifelder

1. It is pointed out that M/l may not be constant for all linear DNA molecules. Would you expect M/l to be constant for all circular DNA molecules? Why?

2. If you had 1 ml of a pure sample of a DNA at 50 μg/ml, which method would you select to obtain M to 20% accuracy? For 2% accuracy? Suppose 10% of the molecules were fragmented, which method would you choose to get 5% accuracy? Would it affect your decision if you knew that M was about 2 million? About 100 million? Explain the reasons for your decisions.

ARTICLE 14 Kavenoff, Klotz, and Zimm

1. How would you prove that each eukaryotic chromosome contains a single DNA molecule?

2. If during a viscosity measurement, η suddenly decreased, what has probably happened?

3. What factors should determine the degree of extension of a DNA molecule?

4. What factors contribute to the viscoelasticity of DNA?

5. Generally the cells were lysed within the viscometer sample chamber. Why?

•6. Design an experiment to determine the accuracy of a viscoelasticity measurement of M having a magnitude of $2-3 \times 10^9$.

References

Bancroft, F. C., and D. Freifelder. 1970. "Molecular Weights of Coliphages and Coliphage DNA. I. Measurement of the Molecular Weight of Bacteriophage T7 by High-Speed Equilibrium Centrifugation." *J. Mol. Biol.* **54**, 537–546.

Crothers, D. M., and B. H. Zimm. 1965. "Viscosity and Sedimentation of the DNA from Bacteriophages T2 and T7 and the Relation to Molecular Weight." *J. Mol. Biol.* **12**, 525–536.

Doty, P., B. B. McGill, and S. A. Rice. 1958. "The Properties of Sonic Fragments of Deoxyribose Nucleic Acid." *Proc. Nat. Acad. Sci.* **84**, 432–438. The first *s-M* curve was derived here.

Dubin, S. B., G. B. Benedeck, F. C. Bancroft, and D. Freifelder. 1970. "Molecular Weights of Coliphages and Coliphage DNA. II. Measurement of Diffusion Coefficients Using Optical Mixing Spectroscopy, and Measurement of Sedimentation Coefficients." *J. Mol. Biol.* **54**, 547–556.

Kleinschmidt, A. K., D. Lang, D. Jacherts, and R. K. Zahn. 1962. "Darstellung und Längenmessungen des Gesamten Desoxyribonucleinsäure-inhalten von T2-Bakteriophagen." *Biochim. Biophys. Acta* **61**, 857–864. The method is described and a classical micrograph is shown.

Kleinschmidt, A. K., and R. K. Zahn. 1959. "Über Desoxyribonucleinsäure Molekeln in Protein-Mischfilmen." *Z. für Naturforschung* **14b,** 770–779. This is the paper presenting the discovery of the method.

Klotz, L. C., and B. H. Zimm. 1972. "Size of DNA Determined by Viscoelastic Measurements: Results on Bacteriophages, *Bacillus subtilis,* and *Escherichia coli.*" *J. Mol. Biol.* **72,** 779–800.

Richardson, C. C. 1966. "The 5'-Terminal Nucleotides of T7 Bacteriophage Deoxyribonucleic Acid." *J. Mol. Biol.* **15,** 49–61.

Schmid, C., and J. E. Hearst. 1969. "Molecular Weights of Homogeneous Coliphage DNA's from Density-Gradient Sedimentation Equilibrium." *J. Mol. Biol.* **44,** 143–160.

Yphantis, D. 1964. "Equilibrium Ultracentrifugation of Dilute Solutions." *Biochem.* **3,** 297–317.

Additional Readings

Eigner, J., and P. Doty. 1965. "The Native, Denatured, and Renatured States of Deoxyribonucleic Acid." *J. Mol. Biol.* **12,** 549–580. An important discussion of the s-M relation.

Freifelder, D. 1976. *Physical Biochemistry,* Chap. 13. Freeman. A detailed description of Zimm's instruments.

Kavenoff, R., and B. H. Zimm. 1973. "Chromosome-sized DNA Molecules from *Drosophila.*" *Chromosoma* **41,** 1–27.

SEPARATION OF THE DNA STRANDS

DNA Denaturation

The DNA Replication Model

As we saw earlier, James Watson and Francis Crick showed that DNA consists of two single polynucleotide strands that are intertwined to form a double-stranded helix and whose nucleotide bases are hydrogen-bonded according to the base pairing rule: adenine to thymine and guanine to cytosine. Such a structure lends itself easily to models for DNA replication because the base pairing rule requires that each strand must specify the base sequence of the other strand. Thus, if each strand functions as the template for synthesis of a complementary strand, two new strands, one of each type, are formed. Watson and Crick proposed a simple replication scheme consisting of simultaneous separation of the double helix to form two single strands and copying of each with the newly synthesized strand winding about the parent strand to reform a double helix. They describe this model in Article 15.

Demonstration of Semi-Conservative Replication

In the Watson-Crick model, after one round of replication the two daughter molecules each contain one parental strand. This is called *semi-conservative replication*. One can, however, devise a scheme of *conservative* replication as well. In such a model the base pairing rule would be obeyed but the newly synthesized strands would remain paired to the parental strands for only a brief interval.* The two models are shown in Fig. V-1. As the replication site moved along, the daughter strands

---

*Another type of conservative replication, in which no separation of parental strands occurs is formally possible. In this model the polymerizing enzyme would recognize the base and carry out synthesis in the groove of the helix (see Fig. A-1, Appendix A). This idea was never seriously considered since base recognition was thought to require the hydrogen bonding groups, which are inaccessible to an enzyme in the groove.

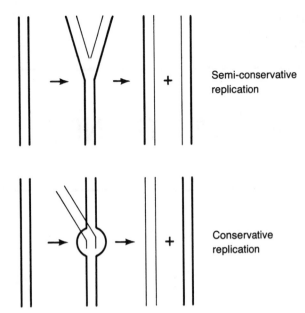

Figure V-1

Two ways in which DNA can replicate. In semi-conservative replication, the parental strands (heavy lines) separate, each daughter molecule receiving one parental and one newly synthesized daughter strand (thin line). In conservative replication the parental strands remain entwined so that, after one round of replication, one of the daughter molecules contains only the parental strands, and the other only daughter strands.

Semi-conservative replication

Conservative replication

that replication is semi-conservative. To extend this statement to the Watson-Crick model, one must assume not only that unwinding is possible but also that the DNA Meselson and Stahl had isolated and analyzed was indeed double-stranded. At the time neither assumption seemed to need proof since the elegance of the Watson-Crick model made all alternatives seem unnecessary. As we will see, however, these assumptions were in fact put to a test several years later.

Equivalence of DNA Denaturation and Strand Separation

In an effort to understand the elements of the replication process, physical biochemists attempted to duplicate the unwinding process in the test tube. The basic procedure was to treat DNA with a physical or chemical agent that would break hydrogen bonds, in hopes of observing the separation of the strands.

The principal method used to study this problem was to obtain *melting curves*, whose significance can be explained as follows. DNA absorbs ultraviolet light having a wavelength of 260 nm,* owing to the absorption by the bases. However, the absorbance per mole of bases (the molar extinction coefficient) of a DNA molecule is 60% less than that of the free bases (i.e., those not contained in a double-stranded DNA molecule). The term *hypochromic* is used to describe this absorption property of DNA. An explanation of the cause of hypochromicity is beyond the scope of this book, but for purposes of this discussion we can regard it simply as a result of the very close proximity and parallel arrangements of the planes of the bases. One early observation was that, when DNA is heated, the absorbance or *optical density* (the optical density, OD, is the absorbance of a layer 1 cm thick), increases with rising temperature. A graph showing OD versus temperature is called a melting curve (Fig. V-2), because it was believed that the increase in OD reflected disruption or "melting"

would detach from the parental strands and associate with one another. Hence the parental DNA would always retain its original strands. Both the conservative and semi-conservative models were put to a test by Matthew Meselson and Franklin Stahl—who, with Jerome Vinograd, had invented the powerful technique of equilibrium sedimentation in a density gradient in order to carry out this important experiment (see Article 10). They argued that DNA replicates semi-conservatively in the bacterium, *Escherichia coli*, as they explain in Article 16.

The Watson-Crick model describes a mode of replication of a double helix. Implicit in this model is the idea that the two strands of the double helix could unwind. In the Meselson-Stahl experiment the results were interpreted in terms of two possible modes of replication—the semi-conservative one in which total unwinding occurs, and the conservative in which only partial unwinding is necessary. The simple conclusion of this experiment was

---

*The abbreviation nm is for nanometer or 10^{-9} meters. In the papers reprinted here wavelength is expressed in millimicrons (mμ) or angstrom units (Å). The unit nm became the standard one by international agreement only recently. 1 nm = 1 mμ = 10 Å.

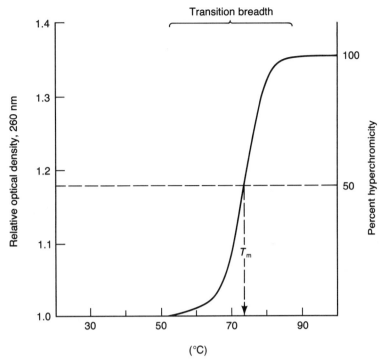

Figure V-2

A typical melting curve for DNA. The DNA is heated to the temperature indicated, and the optical density (which, for DNA, is customarily measured at the wavelength 260 nm) is measured. Relative optical density is the ratio of the optical density at the temperature indicated to that at 25°C. The melting temperature T_m is the temperature at which 50% of the maximum OD increase is reached.

of the ordered DNA structure. A melting curve is usually described by the following parameters. As the OD increases, the DNA is said to become *hyperchromic*. The maximum OD reached is usually 37% greater than that before heating; it is common to refer to this value as 100% hyperchromicity and similarly an increase of 18.5% is called 50% hyperchromicity. The temperature at which 50% hyperchromicity is reached is called the *melting temperature* T_m. The transition breadth is the temperature span required to go from 0% to 100% hyperchromicity.

Evidence that the increase in OD is a result of a major structural change in the DNA was provided by measurements of the variation of several physical parameters with temperature. For instance, the *optical activity* of the DNA decreases, indicating that heated DNA is less helical; the *intrinsic viscosity* (a measure of both mass and rigidity) decreases, suggesting that heated DNA is more flexible; and the radius of gyration (a measure of the size of a sphere in which a molecule can be placed) decreases, indicating that some sort of collapse of the molecule occurs. Since the melting transition is accompanied by a loss of ordered structure, the entire process has been called *denaturation*, a general term used to describe loss of order in a polymer; DNA heated to the point of maximum collapse is denatured DNA, and unheated DNA having the normal, double-stranded structure is called *native* DNA.

Denaturation of both proteins and DNA can be effected by the same agents (heat) and the same reagents (acid, alkali, and urea). Since these reagents and agents were known to destabilize hydrogen bonds and since it was believed that the two DNA strands are held together by hydrogen bonds, it was presumed that in the molecular collapse accompanying DNA denaturation the two polynucleotide strands separated. Such a separation should have been detectable by a halving of M after exposure to a denaturing reagent and many people looked for this event. Several groups examined this by the light scattering method by measuring M in the presence and absence of 7 M urea. Halving of M was observed, but unfortunately it was overlooked that the equations used to calculate M by light scattering were derived theoretically under the assumption that the system contains only two components, water and the macromolecule DNA; the low molarity of salts in the solution, e.g. 0.1 M NaCl, was usually ignored, an omission that was hardly appropriate for 7 M urea. Paul Doty and his students at Harvard University pointed out this oversight so emphatically that a portion of the scientific community not familiar with the light scattering method missed the point of this criticism and drew the unwarranted conclusion that strand separation was impossible. (In later years it was said by some that this misunderstanding may have delayed progress in the DNA field for several years). Other denaturants were then used, but for a long time no evidence could be obtained to support the idea that strand separation accompanies denaturation. It was clear though that answering the question of whether the strands actually could be separated was imperative if DNA replication was to be understood.

Finally a glimmer of light appeared in a paper by Meselson, Stahl and Vinograd (Article 10); these workers found that after centrifugation to equilibrium in CsCl, the band width (which is theoretically proportional to 1/M) of a boiled DNA sample was greater than that of unheated DNA. This finding was appreciated by some and thought to be a result of separation of the individual polynucleotide strands but still the discrepancy between this and the light scattering data remained. An understanding of the problem finally came from the work of Joseph Eigner in Doty's laboratory when it was realized that when boiled DNA is returned to room temperature (the temperature at which the physical measurements were usually made), hydrogen bonds can reform. This reformation follows the base pairing rule but, because of the large number of similar short base sequences in DNA, the original DNA structure is not restored—in fact, the hydrogen bonding can be either within a strand or between many strands to form a large aggregate. Eigner showed that this hydrogen bonding occurs only at high ionic strength, since in dilute salt solutions the charged phosphate groups repel one another and keep the bases apart. Furthermore, interstrand aggre-

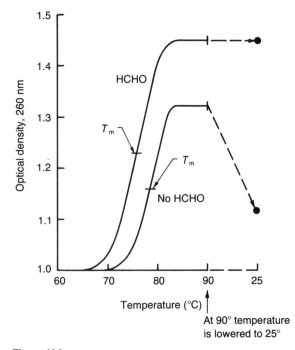

Figure V-3
The effect of formaldehyde (HCHO) on DNA melting curves. This graph shows the usual plot of optical density at 260 nm versus temperature. In this hypothetical example the melting temperature T_m in the absence of HCHO is 78°. After the solution is heated to 90° the temperature is immediately lowered to 25° and the optical density drops as indicated by the lower dashed arrow. When HCHO is present, the curve is changed in three ways: (1) T_m is lowered, in this example, by 2°; (2) the maximum increase in optical density is 45%; (3) when, after the solution is heated to 90°, the temperature is lowered to 25°, the optical density does not decrease (upper dashed arrow).

gation requires that the DNA concentration be in excess of 10 μg/ml. Thus in the CsCl experiment of Meselson, Stahl, and Vinograd, halving of M was observed since the DNA concentration was 1 μg/ml, even though the salt concentration was 7.7 M. In light scattering and several other physical techniques, the DNA concentration used was usually 1000 μg/ml or greater and in a mid-range of ionic strength. Eigner's work thus provided experimental conditions for re-examination of the question of strand separation without the aggregation problem. This too was done in Doty's laboratory by Julius Marmur and his co-workers. They mixed two denatured and distinguishable DNA samples and allowed controlled aggregation (renaturation) to occur. The result was the formation of double-stranded DNA having properties derived from each sample. They called this hybrid

DNA. This work is reported in Articles 17, 18, and 19.

At the same time it was independently observed by Charles Thomas and Lawrence Grossman that when DNA at any concentration or ionic strength is heated in the presence of formaldehyde, its physical state at the high temperature is unchanged when restored to low temperature. In particular, they studied melting curves in the presence and absence of formaldehyde. Recall that it had been shown (in Doty's laboratory) that if DNA is heated and the OD is measured at the high temperature, the OD increase is 37%; but if the OD of the same sample is measured after returning it to room temperature, the increase is only 12%. Thomas and Grossman each noted that if DNA is heated in formaldehyde and then cooled, the OD remains high (Fig. V-3). Since formaldehyde reacts with the amino groups of the bases, it was concluded that formaldehyde probably acts by preventing reformation of hydrogen bonds. Hence if heating causes strand separation, addition of formaldehyde should maintain the single-strand state, as

numerous physical experiments did in fact demonstrate. In one series of experiments the use of formaldehyde allowed Peter Davison and me to demonstrate that separation of the conserved subunits studied by Meselson and Stahl occurred in the upper region of a melting curve. We prepared hybrid (^{14}N-^{15}N) *E. coli* DNA and heated it in 1% formaldehyde to various temperatures. For each sample the OD was measured and the sample was centrifuged to equilibrium in CsCl. Strand separation was detected by the appearance of the heavy and light bands (Fig. V-4). We had hoped to find

Figure V-4
(a) A plot of the optical density (which is proportional to DNA concentration) of *E. coli* DNA, labeled in one strand with ^{14}N and in the other with ^{15}N, against DNA density in a CsCl density gradient. T2 DNA was added as a density standard in order to align the other bands. The panels from top to bottom are samples heated in 1% formaldehyde having 30%, 87%, and 100% of full hyperchromicity. The dashed curve in the uppermost panel indicates the position of unheated DNA. The dashed curves in the lower panels indicate the approximate areas assigned to the ^{14}N, hybrid (^{14}N-^{15}N), and ^{15}N bands (from left to right). Note that at 30% of full hyperchromicity the DNA has increased in density (indicating the presence of single-stranded regions) but no molecules show complete separation of strands. At 100% hyperchromicity (the top of the melting curve of Fig. V-2) strand separation is still incomplete.
(b) Summary of data from many such experiments as in part A. This curve shows that, with temperatures higher than that producing the onset of 100% hyperchromicity, strand separation is ultimately complete. [Redrawn from D. Freifelder and P. F. Davison, *Biophys. J.* **2** (1962), 249–256.]

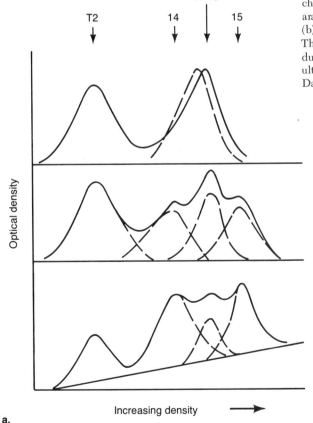

a.

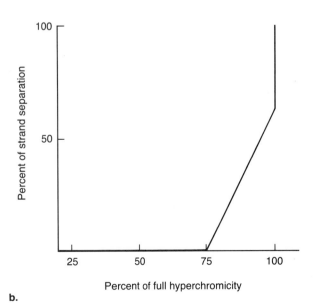

b.

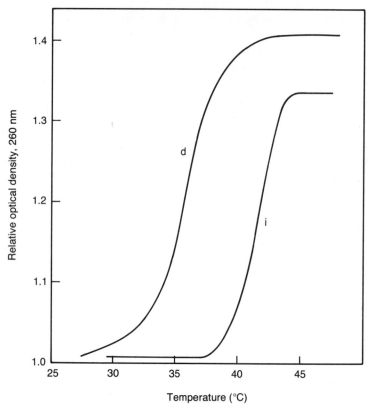

Figure V-5
Two melting curves of phage T2 DNA, the d-curve and the i-curve. The d-curve is a standard melting curve as in Fig. V-2, in which OD is measured at the indicated temperature. In the i-curve the DNA has been heated to the indicated temperature and then cooled to 25°C before measuring OD_{260}. [Reprinted from E. P. Geiduschek, *J. Mol. Biol.* **4** (1962), 467–487.]

the strands separating at a unique temperature on the upper flat portion of the melting curve, but since the *E. coli* DNA was heterogeneous—consisting of hundreds of fragments of the *E. coli* chromosome caused by hydrodynamic shear degradation (see Article 9)—this was not observed. The situation was clarified when we used a homogeneous phage DNA and measured *s* as a function of temperature, in the presence of formaldehyde. It was found that as base pairs were disrupted, *s* increased owing to the increased flexibility of the DNA. At a unique temperature *s* decreased abruptly to a value corresponding to that expected for single-stranded DNA. This work is reported in Article 20.

At the same time Peter Geiduschek was examining strand separation from a different point of view. He reasoned that if denaturation meant strand separation, then if denaturation were incomplete in the sense that a single

hydrogen bond remained unbroken, lowering the temperature should result in complete reversibility of the melting process—i.e., the OD and all other physical properties should return to the native values. To test this idea he introduced a cross-link between the two strands using nitrous acid and in so doing prepared "reversible" DNA which is described in one of the papers reprinted in this section (Article 21). In a second and more extensive study (which is too long and complicated to be reprinted here) Geiduschek (1962) presented a new method for studying the melting of DNA. A DNA sample was heated and the OD was measured. The heated sample was then cooled to 25°C and the OD remeasured. This was done over a wide range of temperatures to obtain a pair of melting curves—one obtained at the high temperature (a d-curve) and the other at 25° (an i-curve). A typical pair of curves is shown in Fig. V-5. The i-curve (a measure of

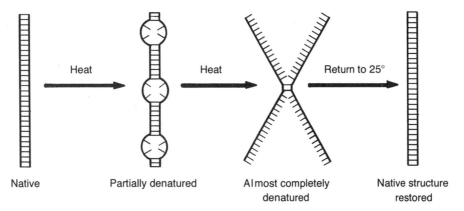

Native   Partially denatured   Almost completely denatured   Native structure restored

Figure V-6
The physical state of a DNA molecule that has been heated almost to the point of complete denaturation and then cooled. It is important to note that strand separation does not always start at the ends of the molecule. The first regions to open are usually rich in AT pairs.

irreversibility) indicates that, if the DNA is homogeneous, the OD does not remain high until a temperature is reached that is in the upper region of the d-curve. The interpretation was that, as DNA is heated, hydrogen bonds are broken and the DNA becomes hyperchromic. However, if the strands are not separated into two distinct elements, the DNA "zips up" (as does the reversible DNA described in Article 21) again when the temperature is restored to 25°, a temperature at which hydrogen bonds are normally unbroken (Fig. V-6). Hence the i-curve shows when irreversible denaturation, i.e., strand separation, sets in and is essentially a measure of the amount of free, single-stranded DNA.*

At this point it is worth restating the three basic approaches used to prove that strand separation occurs. In one approach (Articles 17, 18, and 19), the argument is that strand separation must have occurred because, when controlled aggregation (renaturation) is allowed, the DNA subunits (presumably single strands) have formed new associations—i.e., hybrid DNA's. In the second (Article 20), it is

*One might ask why, with a homogeneous DNA such as in Fig. V-4, the i-curve does not show that strand separation occurs well out onto the plateau region of the d-curve. The reason is that phage DNA frequently contains naturally occurring single-strand breaks (see page 286). Hence if a DNA molecule contains a single-strand break, a fragment can fall off before complete strand separation occurs, if all hydrogen bonds holding that fragment to the structure are broken.

shown that when formaldehyde is present to prevent reassociation, halving of M can be detected by an abrupt decrease in the sedimentation coefficient. In the third (Article 21 and Figs. V-5 and V-6), it is shown that if any link persists between the strands, irreversible denaturation is impossible; this is reflected in the melting curves in such a way that supports the belief that the percent hyperchromicity reflects the degree of unwinding of the helix.

The work just described provided sufficient evidence to convince almost everyone that the double helix could be unwound to yield two physically separated single strands. However, just at this time a series of papers from the laboratory of Liebe Cavalieri were issued. In them it was argued that the *conserved* subunits of the Meselson-Stahl work were double-strand helices and that *E. coli* DNA was four-stranded, consisting of two Watson-Crick double helices held together in a side-to-side aggregate by hypothetical bonds of an unknown type termed "biunial." Two types of experiments were given to support the biunial theory. In the first, the mass-to-length ratio was determined by light scattering and found to be twice that determined by Watson and Crick by X-ray diffraction. In the second, the decrease in molecular weight M as a function of time of digestion of boiled DNA by the enzyme DNase II, was studied, M being measured by light scattering. If boiled DNA is single-stranded, M should decrease continually with time; if it is double-stranded, there should be a lag since the number of single-strand breaks (produced linearly with time) would have to reach a number such that two would be opposite one another (one in each strand) to form a double-strand break before the molecule would break. The data showed a lag. Through no fault of the experimenters, the interpretation of the data from both kinds of experiments was incorrect for two reasons. First, the theory of light scattering was not correct for asymmetric molecules with M greater than 2×10^6, and in addition deficiencies in instrumental design produced incorrect values (see page 76). Second, the boiled DNA consisted of intermolecular aggregates. Four years passed before these problems were appreciated; in the meantime the scientific community was baffled by Cavalieri's results even though no one really believed them.

The importance of Cavalieri's experiments was mainly the disclosure that there was no real evidence either that DNA *in solution* had the Watson-Crick double-stranded structure or that the conserved subunits observed by Meselson and Stahl were single strands. It was finally realized that it was more important to prove these two points than to explain Cavalieri's results. John Cairns, Peter Davison, Alfred Hershey, and Cyrus Levinthal then got together to decide how to do so and concluded that proof of the double-stranded nature of *E. coli* DNA could be obtained only by measuring the mass per unit length (M/l) of a molecule that replicates by the Meselson-Stahl scheme and whose subunits are separable by boiling. To avoid the ambiguity of measuring average values of M and l of populations of molecules whose M and l values are not all the same (e.g., as in the fragmented DNA isolated from bacteria), they decided to use an organism from which DNA could be isolated without fragmentation and whose M was quite small—this meant a phage. *E. coli* phage λ seemed to be the object of choice since Meselson and Jean Weigle had found in the course of studying genetic recombination in λ that λ DNA is replicated semi-conservatively and Hershey had measured M for λ, using a technique that was untested but seemed to be valid. Only the measurement of l remained, and this was provided by the ingenious autoradiographic technique developed by Cairns and described in Article 22.

Cairns' paper satisfied almost everyone that DNA was double-stranded, and Cavalieri's work then seemed of little consequence. However, it is interesting to note that Cairns' result would have caused even more confusion if the molecular weight of λ DNA had been correctly known. In the paper, Cairns quotes Hershey as determining a value of 46 million for M of λ DNA. We now know that 31 million is correct. If this value is compared to the 23μ maximum length observed by Cairns, M/l would have been 1.3 million/μ, and not 2 million/μ as expected from the Watson-Crick data or 4 million/μ as Cavalieri might have predicted. Note that Cairns chose to use the length of the longest molecule rather than a mean value because it was assumed that DNA was so rigid that it could not be stretched (although kinking during drying might produce apparent shrinkage). We now know that stretching (accompanied by unwinding) is possible and that the mean value (which Cairns did not report) would probably have been more suitable. Hence the Cairns work in fact did not prove that DNA was double-stranded in solution, and it was not until 9 years later that M/l for a semi-conservatively replicating DNA was evaluated precisely. Although none of the work was done with the M/l problem in mind, during this time Meselson showed that the DNA of *E. coli* phage T7 is replicated semi-conservatively, several groups determined the molecular weight of T7 DNA, and Dimitrij Lang measured its length by electron microscopy (as described in Article 12, Section IV).

Fortunately, proof that DNA is double helical in solution did not rest solely on the work of Cairns, because other workers had concentrated on proving that the *conserved* subunits observed by Meselson and Stahl were single strands. This was done in two ways. First, I showed that the sedimentation coefficient of T7 DNA heated in HCHO depends on NaCl concentration (Fig. V-7). This indicates single-strandedness of the heated DNA because of the following facts: Both single-stranded and double-stranded DNA are highly charged owing to the negatively charged phosphate groups along the sugar-phosphate backbone. These phosphates mutually repel one another and therefore cause the single strand to be highly extended. In 1 M NaCl these charges are neutralized by the Na$^+$ ion. Hence single-stranded DNA, which has free

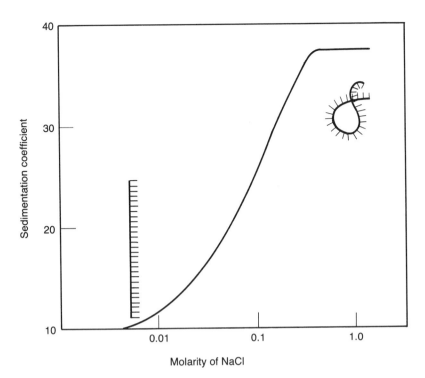

Figure V-7
Sedimentation coefficient of formaldehyde-stabilized single strands of *E. coli* phage T7 DNA, as a function of NaCl concentration. The approximate shape of a single strand at low (0.01 *M*) and high (0.3 *M*) molarity is shown. At low (0.01 *M*) molarity the negative charges on the strand repel one another and the strand is extended; at high (0.3 *M*) molarity the charges are neutralized and the molecule collapses to form a *random coil*. Since the random coil is more compact, it encounters less friction as it moves through liquid and therefore sediments more rapidly.

rotation about the phosphodiester bonds, should become less extended in 1 *M* NaCl (and under conditions preventing hydrogen bonding) and approach the random coil configuration. Since a long thin molecule sediments more slowly than a ball having the same mass (because it encounters greater resistance), single-stranded DNA should sediment more slowly in dilute than in concentrated salt solutions. Double-stranded DNA is always rigid so that its sedimentation rate is reasonably independent of salt concentration. Hence this experiment supported the view that the product of heating is single-stranded DNA. The second line of evidence was the electron-microscopy demonstration by Kleinschmidt and me that DNA heated in the presence of formaldehyde was much more flexible and kinky than unheated DNA (Fig. V-8).

Perhaps the most conclusive experiment showing that DNA is double-stranded appeared in a paper by Robert Baldwin and Eric Shooter (1963). As in the other two experiments, Baldwin and Shooter studied denaturation, but their method differed in two important ways. First they used alkaline pH for denaturation, making use of the fact that, at high pH, the charged groups contained in the hydrogen bonds are titrated. Second, they used DNA isolated from *E. coli* bacteria that were grown in a medium containing

5-bromouracil (BU). This base contains a bromine atom in place of the methyl group of thymine, and because the bromine and the methyl group have nearly the same dimensions, BU can substitute for thymine in DNA. Since the bromine atom is about 3½ times the mass of a methyl group, BU can be used as a density label in much the same way that

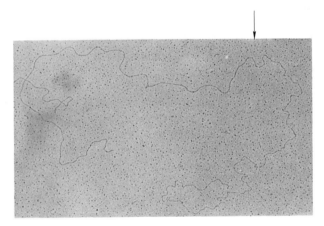

Figure V-8
An electron micrograph showing the different appearance of double- and single-stranded DNA. An *E. coli* phage λ DNA molecule was treated in such a way that a single strand was removed from a portion of the DNA, leaving a molecule that is double-stranded in one region (left side of the photograph) and single-stranded in another (indicated by the arrow at the right side of the photograph). Note that the single-stranded region is both thinner and kinkier than the double-stranded region.

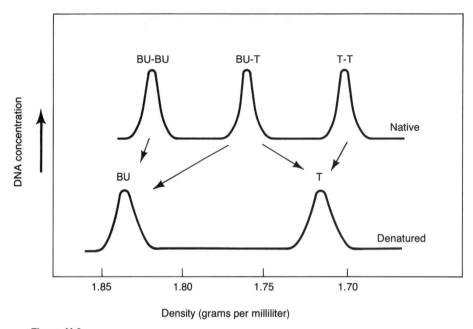

Figure V-9
A plot of the DNA concentration as a function of density for three DNA samples, one containing 5-bromouracil (BU) in both strands, one containing thymine (T) in both strands, and a hybrid with one strand of each type, all centrifuged to equilibrium in CsCl. The upper part shows a sample of unheated, native DNA; the lower part shows the result of denaturation of this sample either by heat or alkali treatment. The arrows indicate how each DNA type moves upon denaturation.

Meselson and Stahl increased the density of DNA by substituting ^{15}N for ^{14}N. As shown in Fig. V-9, the densities of normal DNA, fully BU-labeled DNA, and hybrid (i.e., obtained by growth for one generation in a medium containing BU) DNA are distinct enough that all species can be easily separated in a CsCl density gradient. The denaturation of DNA can also be observed in a CsCl density gradient since, as shown by Meselson, Stahl, and Vinograd (Article 10), denatured DNA has a higher density than native DNA (see also Fig. V-9). Hybrid DNA can, of course, be distinguished from fully heavy or fully light DNA in that it will form two bands when denatured.

Earlier in Baldwin's laboratory it had been shown that fully BU-labeled DNA denatures at 1.5 pH units lower than fully thymine-containing DNA. Baldwin and Shooter made use of this fact to prove that the biunial model is incorrect. The classical Watson-Crick model states that DNA is double-stranded; the semiconservative replication model further states that the conserved subunits observed by Meselson and Stahl are the individual single polynucleotide strands. In the biunial model, however, it is assumed that DNA is four-stranded, consisting of two double-strand helices (each of which replicates conservatively) held together by biunial bonds (Fig. V-10). In this theory, then, it is maintained that, after one round of replication in BU medium, the four-stranded DNA consists of a fully light (thymine-containing) double helix bonded to a fully heavy (BU) double helix. If denaturation, as proposed by Cavalieri, involves both breakage of biunial bonds and collapse of the double helix *without separation of the single strands*, then denaturation of hybrid DNA should occur in two steps at two distinct pH values; if the Watson-Crick model is correct, the denaturation pH should be between that for fully light and fully heavy DNA (Fig. V-11). Baldwin and Shooter found only one pH transition for hybrid DNA, located halfway between the transitions for light and heavy DNA, and thus eliminated the biunial theory. Hence it could be concluded decisively that the subunits of hybrid DNA observed by Meselson and Stahl are single strands.

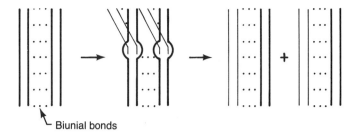

Biunial bonds

Figure V-10
A model for conservative replication of two
double-strand DNA helices held together
by biunial bonds (dashed lines).

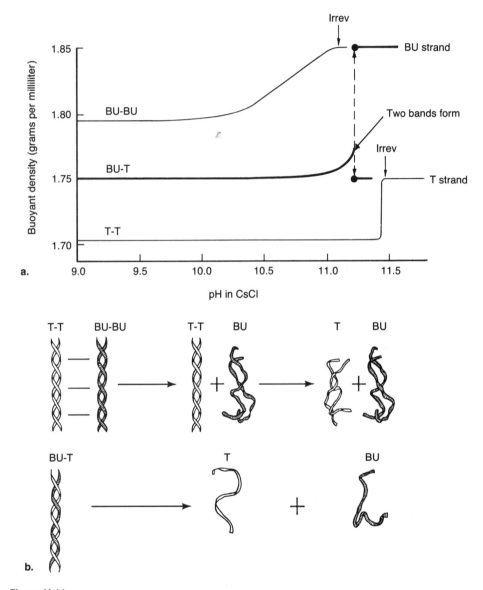

a.

b.

Figure V-11
(a) A plot of the density of three DNA samples labeled as in Figure 5-9 as a function of pH.
The increase in density indicates the separation of base pairs in local regions. The label "ir-
rev" refers to the pH at which denaturation is irreversible in the sense that the density does
not return to that at neutral pH if the alkali-treated DNA is neutralized.
(b) Two models for BU-containing hybrid DNA. In the upper row the four-stranded model is
shown. Here the two double helices melt separately, one at the melting pH value for heavy
DNA, and the other at that for light DNA. In the lower, or the two-stranded model, the helix
melts at an intermediate pH. [After R. L. Baldwin and E. M. Shooter, *J. Mol. Biol.* **7** (1963),
511–526.]

From *Nature*, Vol. 171, 964–969, 1953. Reproduced with permission of the authors and publisher.

GENETICAL IMPLICATIONS OF THE STRUCTURE OF DEOXYRIBONUCLEIC ACID

By J. D. WATSON and F. H. C. CRICK

Medical Research Council Unit for the Study of the
Molecular Structure of Biological Systems, Cavendish
Laboratory, Cambridge

THE importance of deoxyribonucleic acid (DNA) within living cells is undisputed. It is found in all dividing cells, largely if not entirely in the nucleus, where it is an essential constituent of the chromosomes. Many lines of evidence indicate that it is the carrier of a part of (if not all) the genetic specificity of the chromosomes and thus of the gene itself.

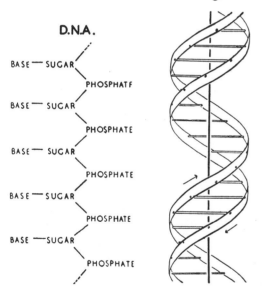

Fig. 1. Chemical formula of a single chain of deoxyribonucleic acid

Fig. 2. This figure is purely diagrammatic. The two ribbons symbolize the two phosphate-sugar chains, and the horizontal rods the pairs of bases holding the chains together. The vertical line marks the fibre axis

Until now, however, no evidence has been presented to show how it might carry out the essential operation required of a genetic material, that of exact self-duplication.

We have recently proposed a structure[1] for the salt of deoxyribonucleic acid which, if correct, immediately suggests a mechanism for its self-duplication. X-ray evidence obtained by the workers at King's College, London[2], and presented at the same time, gives qualitative support to our structure and is incompatible with all previously proposed structures[3]. Though the structure will not be completely proved until a more extensive comparison has been made with the X-ray data, we now feel sufficient confidence in its general correctness to discuss its genetical implications. In doing so we are assuming that fibres of the salt of deoxyribonucleic acid are not artefacts arising in the method of preparation, since it has been shown by Wilkins and his co-workers that similar X-ray patterns are obtained from both the isolated fibres and certain intact biological materials such as sperm head and bacteriophage particles[2,4].

The chemical formula of deoxyribonucleic acid is now well established. The molecule is a very long chain, the backbone of which consists of a regular alternation of sugar and phosphate groups, as shown in Fig. 1. To each sugar is attached a nitrogenous base, which can be of four different types. (We have considered 5-methyl cytosine to be equivalent to cytosine, since either can fit equally well into our structure.) Two of the possible bases—adenine and guanine—are purines, and the other two—thymine and cytosine—are pyrimidines. So far as is known, the sequence of bases along the chain is irregular. The monomer unit, consisting of phosphate, sugar and base, is known as a nucleotide.

The first feature of our structure which is of biological interest is that it consists not of one chain, but of two. These two chains are both coiled around a common fibre axis, as is shown diagrammatically in Fig. 2. It has often been assumed that since there was only one chain in the chemical formula there would only be one in the structural unit. However, the density, taken with the X-ray evidence[2], suggests very strongly that there are two.

The other biologically important feature is the manner in which the two chains are held together. This is done by hydrogen bonds between the bases, as shown schematically in Fig. 3. The bases are joined together in pairs, a single base from one chain being hydrogen-bonded to a single base from the other. The important point is that only certain pairs of bases will fit into the structure. One member of a pair must be a purine and the other a pyrimidine in order to bridge between the two chains. If a pair consisted of two purines, for example, there would not be room for it.

We believe that the bases will be present almost entirely in their most probable tautomeric forms. If this is true, the conditions for forming hydrogen bonds are more restrictive, and the only pairs of bases possible are :

adenine with thymine ;
guanine with cytosine.

The way in which these are joined together is shown in Figs. 4 and 5. A given pair can be either way round. Adenine, for example, can occur on either chain ; but when it does, its partner on the other chain must always be thymine.

This pairing is strongly supported by the recent analytical results[5], which show that for all sources of deoxyribonucleic acid examined the amount of adenine is close to the amount of thymine, and the amount of guanine close to the amount of cytosine, although the cross-ratio (the ratio of adenine to guanine) can vary from one source to another. Indeed, if the sequence of bases on one chain is irregular, it is difficult to explain these analytical results except by the sort of pairing we have suggested.

The phosphate-sugar backbone of our model is completely regular, but any sequence of the pairs of

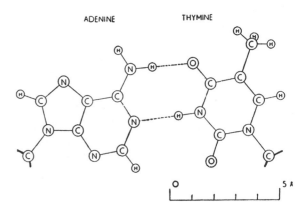

Fig. 4. Pairing of adenine and thymine. Hydrogen bonds are shown dotted. One carbon atom of each sugar is shown

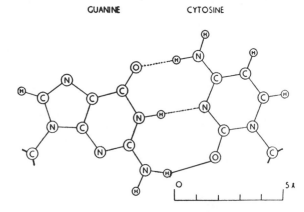

Fig. 5. Pairing of guanine and cytosine. Hydrogen bonds are shown dotted. One carbon atom of each sugar is shown

bases can fit into the structure. It follows that in a long molecule many different permutations are possible, and it therefore seems likely that the precise sequence of the bases is the code which carries the genetical information. If the actual order of the bases on one of the pair of chains were given, one could write down the exact order of the bases on the other one, because of the specific pairing. Thus one chain is, as it were, the complement of the other, and it is this feature which suggests how the deoxyribonucleic acid molecule might duplicate itself.

Previous discussions of self-duplication have usually involved the concept of a template, or mould. Either

```
        \                                                     /
         \ SUGAR—BASE  ----------  BASE—SUGAR /
PHOSPHATE /                                    \ PHOSPHATE
         /                                      \
         \ SUGAR—BASE  ----------  BASE — SUGAR /
PHOSPHATE /                                    \ PHOSPHATE
         /                                      \
         \ SUGAR—BASE  ----------  BASE—SUGAR /
PHOSPHATE /                                    \ PHOSPHATE
         /                                      \
         \ SUGAR—BASE  ----------  BASE—SUGAR /
PHOSPHATE /                                    \ PHOSPHATE
         /                                      \
         \ SUGAR—BASE  ----------  BASE—SUGAR /
PHOSPHATE /                                    \ PHOSPHATE
         /                                      \
```

Fig. 3. Chemical formula of a pair of deoxyribonucleic acid chains. The hydrogen bonding is symbolized by dotted lines

127   128

the template was supposed to copy itself directly or it was to produce a 'negative', which in its turn was to act as a template and produce the original 'positive' once again. In no case has it been explained in detail how it would do this in terms of atoms and molecules.

Now our model for deoxyribonucleic acid is, in effect, a *pair* of templates, each of which is complementary to the other. We imagine that prior to duplication the hydrogen bonds are broken, and the two chains unwind and separate. Each chain then acts as a template for the formation on to itself of a new companion chain, so that eventually we shall have *two* pairs of chains, where we only had one before. Moreover, the sequence of the pairs of bases will have been duplicated exactly.

A study of our model suggests that this duplication could be done most simply if the single chain (or the relevant portion of it) takes up the helical configuration. We imagine that at this stage in the life of the cell, free nucleotides, strictly polynucleotide precursors, are available in quantity. From time to time the base of a free nucleotide will join up by hydrogen bonds to one of the bases on the chain already formed. We now postulate that the polymerization of these monomers to form a new chain is only possible if the resulting chain can form the proposed structure. This is plausible, because steric reasons would not allow nucleotides 'crystallized' on to the first chain to approach one another in such a way that they could be joined together into a new chain, unless they were those nucleotides which were necessary to form our structure. Whether a special enzyme is required to carry out the polymerization, or whether the single helical chain already formed acts effectively as an enzyme, remains to be seen.

Since the two chains in our model are intertwined, it is essential for them to untwist if they are to separate. As they make one complete turn around each other in 34 A., there will be about 150 turns per million molecular weight, so that whatever the precise structure of the chromosome a considerable amount of uncoiling would be necessary. It is well known from microscopic observation that much coiling and uncoiling occurs during mitosis, and though this is on a much larger scale it probably reflects similar processes on a molecular level. Although it is difficult at the moment to see how these processes occur without everything getting tangled, we do not feel that this objection will be insuperable.

Our structure, as described[1], is an open one. There is room between the pair of polynucleotide chains (see Fig. 2) for a polypeptide chain to wind around the same helical axis. It may be significant that the distance between adjacent phosphorus atoms, 7·1 A., is close to the repeat of a fully extended polypeptide chain. We think it probable that in the sperm head, and in artificial nucleoproteins, the polypeptide chain occupies this position. The relative weakness of the second layer-line in the published X-ray pictures[3a,4] is crudely compatible with such an idea. The function of the protein might well be to control the coiling and uncoiling, to assist in holding a single polynucleotide chain in a helical configuration, or some other non-specific function.

Our model suggests possible explanations for a number of other phenomena. For example, spontaneous mutation may be due to a base occasionally occurring in one of its less likely tautomeric forms. Again, the pairing between homologous chromosomes at meiosis may depend on pairing between specific bases. We shall discuss these ideas in detail elsewhere.

For the moment, the general scheme we have proposed for the reproduction of deoxyribonucleic acid must be regarded as speculative. Even if it is correct, it is clear from what we have said that much remains to be discovered before the picture of genetic duplication can be described in detail. What are the polynucleotide precursors ? What makes the pair of chains unwind and separate ? What is the precise role of the protein ? Is the chromosome one long pair of deoxyribonucleic acid chains, or does it consist of patches of the acid joined together by protein ?

Despite these uncertainties we feel that our proposed structure for deoxyribonucleic acid may help to solve one of the fundamental biological problems—the molecular basis of the template needed for genetic replication. The hypothesis we are suggesting is that the template is the pattern of bases formed by one chain of the deoxyribonucleic acid and that the gene contains a complementary pair of such templates.

One of us (J.D.W.) has been aided by a fellowship from the National Foundation for Infantile Paralysis (U.S.A.).

[1] Watson, J. D., and Crick, F. H. C., *Nature*, **171**, 737 (1953).

[2] Wilkins, M. H. F., Stokes, A. R., and Wilson, H. R., *Nature*, **171**, 738 (1953). Franklin, R. E., and Gosling, R. G., *Nature*, **171**, 740 (1953).

[3] (*a*) Astbury, W. T., Symp. No. 1 Soc. Exp. Biol., 66 (1947). (*b*) Furberg, S., *Acta Chem. Scand.*, **6**, 634 (1952). (*c*) Pauling, L., and Corey, R. B., *Nature*, **171**, 346 (1953); *Proc. U.S. Nat. Acad. Sci.*, **39**, 84 (1953). (*d*) Fraser, R. D. B. (in preparation).

[4] Wilkins, M. H. F., and Randall, J. T., *Biochim. et Biophys. Acta*, **10**, 192 (1953).

[5] Chargaff, E., for references see Zamenhof, S., Brawerman, G., and Chargaff, E., *Biochim. et Biophys. Acta*, **9**, 402 (1952). Wyatt, G. R., *J. Gen. Physiol.*, **36**, 201 (1952).

Printed in Great Britain by Fisher, Knight & Co., Ltd., St. Albans.

129

130

ARTICLE 16

From the *Proceedings of the National Academy of Sciences*, Vol. 44, No. 7, 671–682, 1958. Reproduced with permission of the authors.

THE REPLICATION OF DNA IN ESCHERICHIA COLI*

By Matthew Meselson and Franklin W. Stahl

GATES AND CRELLIN LABORATORIES OF CHEMISTRY,† AND NORMAN W. CHURCH LABORATORY OF CHEMICAL BIOLOGY, CALIFORNIA INSTITUTE OF TECHNOLOGY, PASADENA, CALIFORNIA

Communicated by Max Delbrück, May 14, 1958

Introduction.—Studies of bacterial transformation and bacteriaphage infection[1-5] strongly indicate that deoxyribonucleic acid (DNA) can carry and transmit hereditary information and can direct its own replication. Hypotheses for the mechanism of DNA replication differ in the predictions they make concerning the distribution among progeny molecules of atoms derived from parental molecules.[6]

Radioisotopic labels have been employed in experiments bearing on the distribution of parental atoms among progeny molecules in several organisms.[6-9] We anticipated that a label which imparts to the DNA molecule an increased density might permit an analysis of this distribution by sedimentation techniques. To this end, a method was developed for the detection of small density differences among

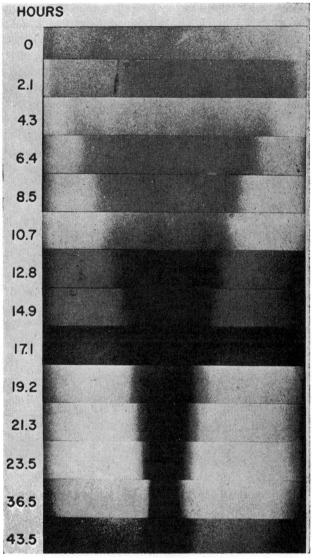

Fig. 1.—Ultraviolet absorption photographs showing successive stages in the banding of DNA from *E. coli.* An aliquot of bacterial lysate containing approximately 10[8] lysed cells was centrifuged at 31,410 rpm in a CsCl solution as described in the text. Distance from the axis of rotation increases toward the right. The number beside each photograph gives the time elapsed after reaching 31,410 rpm.

macromolecules.[10] By use of this method, we have observed the distribution of the heavy nitrogen isotope N[15] among molecules of DNA following the transfer of a uniformly N[15]-labeled, exponentially growing bacterial population to a growth medium containing the ordinary nitrogen isotope N[14].

Density-Gradient Centrifugation.—A small amount of DNA in a concentrated solution of cesium chloride is centrifuged until equilibrium is closely approached.

The opposing processes of sedimentation and diffusion have then produced a stable concentration gradient of the cesium chloride. The concentration and pressure gradients result in a continuous increase of density along the direction of centrifugal force. The macromolecules of DNA present in this density gradient are driven by the centrifugal field into the region where the solution density is equal to their own buoyant density.[11] This concentrating tendency is opposed by diffusion, with the result that at equilibrium a single species of DNA is distributed over a band whose width is inversely related to the molecular weight of that species (Fig. 1).

If several different density species of DNA are present, each will form a band at the position where the density of the CsCl solution is equal to the buoyant density of that species. In this way DNA labeled with heavy nitrogen (N[15]) may be

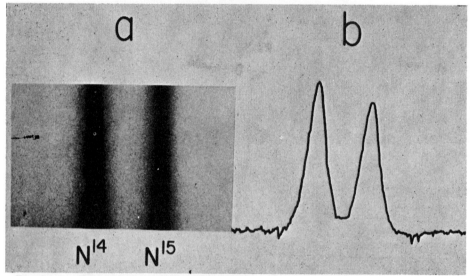

Fig. 2—*a:* The resolution of N[14] DNA from N[15] DNA by density-gradient centrifugation. A mixture of N[14] and N[15] bacterial lysates, each containing about 10[8] lysed cells, was centrifuged in CsCl solution as described in the text. The photograph was taken after 24 hours of centrifugation at 44,770 rpm. *b:* A microdensitometer tracing showing the DNA distribution in the region of the two bands of Fig. 2a. The separation between the peaks corresponds to a difference in buoyant density of 0.014 gm. cm.$^{-3}$

resolved from unlabeled DNA. Figure 2 shows the two bands formed as a result of centrifuging a mixture of approximately equal amounts of N[14] and N[15] *Escherichia coli* DNA.

In this paper reference will be made to the apparent molecular weight of DNA samples determined by means of density-gradient centrifugation. A discussion has been given[10] of the considerations upon which such determinations are based, as well as of several possible sources of error.[12]

Experimental.—*Escherichia coli* B was grown at 36° C. with aeration in a glucose salts medium containing ammonium chloride as the sole nitrogen source.[13] The growth of the bacterial population was followed by microscopic cell counts and by colony assays (Fig. 3).

Bacteria uniformly labeled with N[15] were prepared by growing washed cells for

14 generations (to a titer of 2×10^8/ml) in medium containing 100 μg/ml of $N^{15}H_4Cl$ of 96.5 per cent isotopic purity.   An abrupt change to N^{14} medium was then accomplished by adding to the growing culture a tenfold excess of $N^{14}H_4Cl$, along with ribosides of adenine and uracil in experiment 1 and ribosides of adenine, guanine, uracil, and cytosine in experiment 2, to give a concentration of 10 μg/ml of each riboside.   During subsequent growth the bacterial titer was kept between

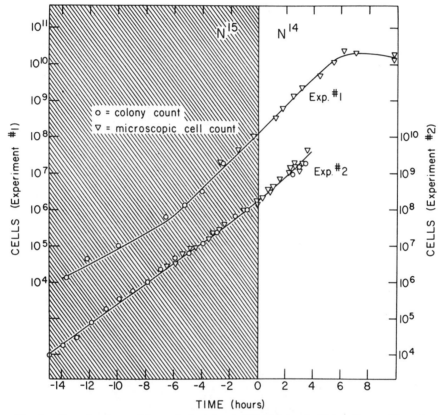

Fig. 3.—Growth of bacterial populations first in N^{15} and then in N^{14} medium.   The values on the ordinates give the actual titers of the cultures up to the time of addition of N^{14}. Thereafter, during the period when samples were being withdrawn for density-gradient centrifugation, the actual titer was kept between 1 and 2×10^8 by additions of fresh medium.   The values on the ordinates during this later period have been corrected for the withdrawals and additions.   During the period of sampling for density-gradient centrifugation, the generation time was 0.81 hours in Experiment 1 and 0.85 hours in Experiment 2.

1 and 2×10^8/ml by appropriate additions of fresh N^{14} medium containing ribosides.

Samples containing about 4×10^9 bacteria were withdrawn from the culture just before the addition of N^{14} and afterward at intervals for several generations. Each sample was immediately chilled and centrifuged in the cold for 5 minutes at $1,800 \times g$.   After resuspension in 0.40 ml. of a cold solution 0.01 M in NaCl and 0.01 M in ethylenediaminetetra-acetate (EDTA) at pH 6, the cells were lysed by the addition of 0.10 ml. of 15 per cent sodium dodecyl sulfate and stored in the cold.

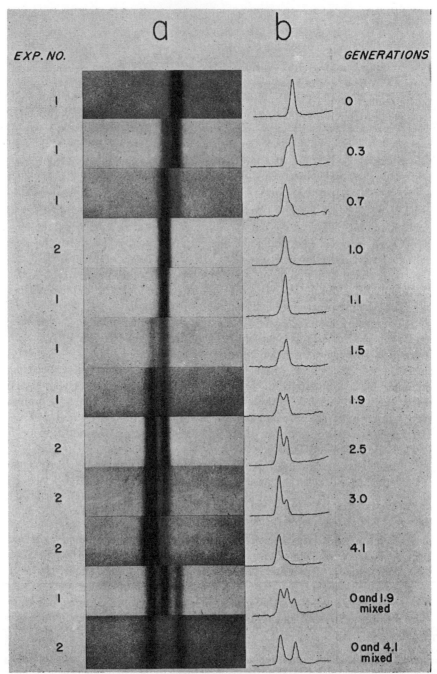

Fig. 4—*a:* Ultraviolet absorption photographs showing DNA bands resulting from density-gradient centrifugation of lysates of bacteria sampled at various times after the addition of an excess of N^{14} substrates to a growing N^{15}-labeled culture. Each photograph was taken after 20 hours of centrifugation at 44,770 rpm under the conditions described in the text. The density of the CsCl solution increases to the right. Regions of equal density occupy the same horizontal position on each photograph. The time of sampling is measured from the time of the addition of N^{14} in units of the generation time. The generation times for Experiments 1 and 2 were estimated from the measurements of bacterial growth presented in Fig. 3. *b:* Microdensitometer tracings of the DNA bands shown in the adjacent photographs. The microdensitometer pen displacement above the base line is directly proportional to the concentration of DNA. The degree of labeling of a species of DNA corresponds to the relative position of its band between the bands of fully labeled and unlabeled DNA shown in the lowermost frame, which serves as a density reference. A test of the conclusion that the DNA in the band of intermediate density is just half-labeled is provided by the frame showing the mixture of generations 0 and 1.9. When allowance is made for the relative amounts of DNA in the three peaks, the peak of intermediate density is found to be centered at 50 ± 2 per cent of the distance between the N^{14} and N^{15} peaks.

For density-gradient centrifugation, 0.010 ml. of the dodecyl sulfate lysate was added to 0.70 ml. of CsCl solution buffered at pH 8.5 with 0.01 M tris(hydroxymethyl)aminomethane. The density of the resulting solution was 1.71 gm. cm.$^{-3}$ This was centrifuged at 140,000$\times$ g. (44,770 rpm) in a Spinco model E ultracentrifuge at 25° for 20 hours, at which time the DNA had essentially attained sedimentation equilibrium. Bands of DNA were then found in the region of density 1.71 gm. cm.$^{-3}$, well isolated from all other macromolecular components of the bacterial lysate. Ultraviolet absorption photographs taken during the course of each centrifugation were scanned with a recording microdensitometer (Fig. 4).

The buoyant density of a DNA molecule may be expected to vary directly with the fraction of N^{15} label it contains. The density gradient is constant in the region between fully labeled and unlabeled DNA bands. Therefore, the degree of labeling of a partially labeled species of DNA may be determined directly from the relative position of its band between the band of fully labeled DNA and the band of unlabeled DNA. The error in this procedure for the determination of the degree of labeling is estimated to be about 2 per cent.

Results.—Figure 4 shows the results of density-gradient centrifugation of lysates of bacteria sampled at various times after the addition of an excess of N^{14}-containing substrates to a growing N^{15}-labeled culture.

It may be seen in Figure 4 that, until one generation time has elapsed, half-labeled molecules accumulate, while fully labeled DNA is depleted. One generation time after the addition of N^{14}, these half-labeled or "hybrid" molecules alone are observed. Subsequently, only half-labeled DNA and completely unlabeled DNA are found. When two generation times have elapsed after the addition of N^{14}, half-labeled and unlabeled DNA are present in equal amounts.

Discussion.—These results permit the following conclusions to be drawn regarding DNA replication under the conditions of the present experiment.

1. *The nitrogen of a DNA molecule is divided equally between two subunits which remain intact through many generations.*

The observation that parental nitrogen is found only in half-labeled molecules at all times after the passage of one generation time demonstrates the existence in each DNA molecule of two subunits containing equal amounts of nitrogen. The finding that at the second generation half-labeled and unlabeled molecules are found in equal amounts shows that the number of surviving parental subunits is twice the number of parent molecules initially present. That is, the subunits are conserved.

2. *Following replication, each daughter molecule has received one parental subunit.*

The finding that all DNA molecules are half-labeled one generation time after the addition of N^{14} shows that each daughter molecule receives one parental subunit.[14] If the parental subunits had segregated in any other way among the daughter molecules, there would have been found at the first generation some fully labeled and some unlabeled DNA molecules, representing those daughters which received two or no parental subunits, respectively.

3. *The replicative act results in a molecular doubling.*

This statement is a corollary of conclusions 1 and 2 above, according to which each parent molecule passes on two subunits to progeny molecules and each progeny molecule receives just one parental subunit.   It follows that each single molecular reproductive act results in a doubling of the number of molecules entering into that act.

The above conclusions are represented schematically in Figure 5.

The Watson-Crick Model.—A molecular structure for DNA has been proposed by Watson and Crick.[15] It has undergone preliminary refinement[16] without alteration of its main features and is supported by physical and chemical studies.[17] The structure consists of two polynucleotide chains wound helically about a common axis.   The nitrogen base (adenine, guanine, thymine, or cytosine) at each level

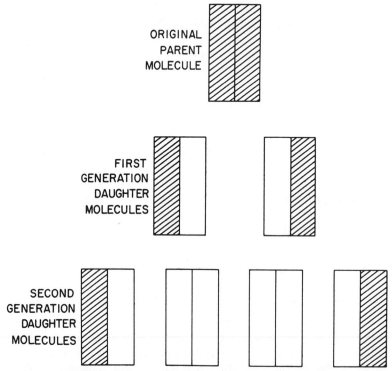

ORIGINAL
PARENT
MOLECULE

FIRST
GENERATION
DAUGHTER
MOLECULES

SECOND
GENERATION
DAUGHTER
MOLECULES

Fig. 5.—Schematic representation of the conclusions drawn in the text from the data presented in Fig. 4. The nitrogen of each DNA molecule is divided equally between two subunits. Following duplication, each daughter molecule receives one of these. The subunits are conserved through successive duplications.

on one chain is hydrogen-bonded to the base at the same level on the other chain. Structural requirements allow the occurrence of only the hydrogen-bonded base pairs adenine-thymine and guanine-cytosine, resulting in a detailed complementariness between the two chains.   This suggested to Watson and Crick[18] a definite and structurally plausible hypothesis for the duplication of the DNA molecule. According to this idea, the two chains separate, exposing the hydrogen-bonding sites of the bases.   Then, in accord with the base-pairing restrictions, each chain serves as a template for the synthesis of its complement.   Accordingly, each daughter molecule contains one of the parental chains paired with a newly synthesized chain (Fig. 6).

The results of the present experiment are in exact accord with the expectations of the Watson-Crick model for DNA duplication. However, it must be emphasized that it has not been shown that the molecular subunits found in the present experiment are single polynucleotide chains or even that the DNA molecules studied here correspond to single DNA molecules possessing the structure proposed by Watson and Crick. However, some information has been obtained about the molecules and their subunits; it is summarized below.

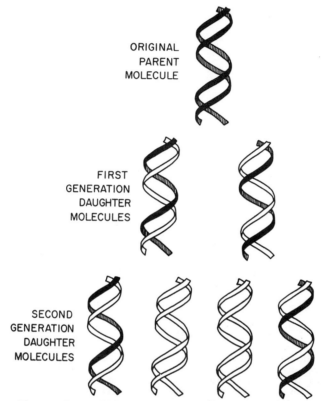

ORIGINAL
PARENT
MOLECULE

FIRST
GENERATION
DAUGHTER
MOLECULES

SECOND
GENERATION
DAUGHTER
MOLECULES

FIG. 6.—Illustration of the mechanism of DNA duplication proposed by Watson and Crick. Each daughter molecule contains one of the parental chains (*black*) paired with one new chain (*white*). Upon continued duplication, the two original parent chains remain intact, so that there will always be found two molecules each with one parental chain.

The DNA molecules derived from *E. coli* by detergent-induced lysis have a buoyant density in CsCl of 1.71 gm. cm.$^{-3}$, in the region of densities found for T2 and T4 bacteriophage DNA, and for purified calf-thymus and salmon-sperm DNA. A highly viscous and elastic solution of N^{14} DNA was prepared from a dodecyl sulfate lysate of *E. coli* by the method of Simmons[19] followed by deproteinization with chloroform. Further purification was accomplished by two cycles of preparative density-gradient centrifugation in CsCl solution. This purified bacterial DNA was found to have the same buoyant density and apparent molecular weight, 7×10^6, as the DNA of the whole bacterial lysates (Figs. 7, 8).

Heat Denaturation.—It has been found that DNA from *E. coli* differs importantly from purified salmon-sperm DNA in its behavior upon heat denaturation.

Exposure to elevated temperatures is known to bring about an abrupt collapse of the relatively rigid and extended native DNA molecule and to make available for acid-base titration a large fraction of the functional groups presumed to be blocked by hydrogen-bond formation in the native structure.[19, 20, 21, 22] Rice and Doty[22] have reported that this collapse is not accompanied by a reduction in molecular weight as determined from light-scattering. These findings are corroborated by density-gradient centrifugation of salmon-sperm DNA.[23] When this material is

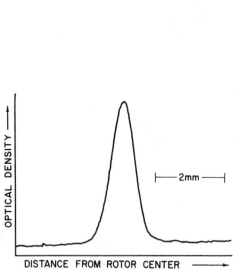

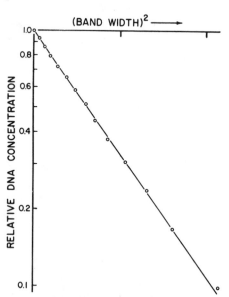

FIG. 7.—Microdensitometer tracing of an ultraviolet absorption photograph showing the optical density in the region of a band of N^{14} *E. coli* DNA at equilibrium. About 2 μg. of DNA purified as described in the text was centrifuged at 31,410 rpm at 25° in 7.75 molal CsCl at pH 8.4. The density gradient is essentially constant over the region of the band and is 0.057 gm./cm.[4]. The position of the maximum indicates a buoyant density of 1.71 gm. cm.$^{-3}$ In this tracing the optical density above the base line is directly proportional to the concentration of DNA in the rotating centrifuge cell. The concentration of DNA at the maximum is about 50 μg./ml.

FIG. 8.—The square of the width of the band of Fig. 7 plotted against the logarithm of the relative concentration of DNA. The divisions along the abscissa set off intervals of 1 mm.[2]. In the absence of density heterogeneity, the slope at any point of such a plot is directly proportional to the weight average molecular weight of the DNA located at the corresponding position in the band. Linearity of this plot indicates monodispersity of the banded DNA. The value of the the slope corresponds to an apparent molecular weight for the Cs·DNA salt of 9.4×10^6, corresponding to a molecular weight of 7.1×10^6 for the sodium salt.

kept at 100° for 30 minutes either under the conditions employed by Rice and Doty or in the CsCl centrifuging medium, there results a density increase of 0.014 gm. cm.$^{-3}$ with no change in apparent molecular weight. The same results are obtained if the salmon-sperm DNA is pre-treated at pH 6 with EDTA and sodium dodecyl sulfate. Along with the density increase, heating brings about a sharp reduction in the time required for band formation in the CsCl gradient. In the absence of an increase in molecular weight, the decrease in banding time must be ascribed[10] to an increase in the diffusion coefficient, indicating an extensive collapse of the native structure.

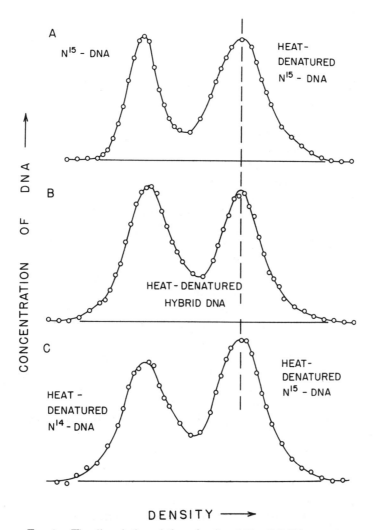

Fig. 9.—The dissociation of the subunits of *E. coli* DNA upon heat denaturation. Each smooth curve connects points obtained by microdensitometry of an ultraviolet absorption photograph taken after 20 hours of centrifugation in CsCl solution at 44,770 rpm. The baseline density has been removed by subtraction. *A:* A mixture of heated and unheated N[15] bacterial lysates. Heated lysate alone gives one band in the position indicated. Unheated lysate was added to this experiment for comparison. Heating has brought about a density increase of 0.016 gm. cm.$^{-3}$ and a reduction of about half in the apparent molecular weight of the DNA. *B:* Heated lysate of N[15] bacteria grown for one generation in N[14] growth medium. Before heat denaturation, the hybrid DNA contained in this lysate forms only one band, as may be seen in Fig. 4. *C:* A mixture of heated N[14] and heated N[15] bacterial lysates. The density difference is 0.015 gm. cm.$^{-3}$

The decrease in banding time and a density increase close to that found upon heating salmon-sperm DNA are observed (Fig. 9, *A*) when a bacterial lysate containing uniformly labeled N[15] or N[14] *E. coli* DNA is kept at 100° C. for 30 minutes in the CsCl centrifuging medium; but the apparent molecular weight of the heated bacterial DNA is reduced to approximately half that of the unheated material.

Half-labeled DNA contained in a detergent lysate of N^{15} *E. coli* cells grown for one generation in N^{14} medium was heated at 100° C. for 30 minutes in the CsCl centrifuging medium. This treatment results in the loss of the original half-labeled material and in the appearance in equal amounts of two new density species, each with approximately half the initial apparent molecular weight (Fig. 9, *B*). The density difference between the two species is 0.015 gm. cm.$^{-3}$, close to the increment produced by the N^{15} labeling of the unheated DNA.

This behavior suggests that heating the hybrid molecule brings about the dissociation of the N^{15}-containing subunit from the N^{14} subunit. This possibility was tested by a density-gradient examination of a mixture of heated N^{15} DNA and heated N^{14} DNA (Fig. 9, *C*). The close resemblance between the products of heating hybrid DNA (Fig. 9 *B*) and the mixture of products obtained from heating N^{14} and N^{15} DNA separately (Fig. 9, *C*) leads to the conclusion that the two molecular subunits have indeed dissociated upon heating. Since the apparent molecular weight of the subunits so obtained is found to be close to half that of the intact molecule, it may be further concluded that the subunits of the DNA molecule which are conserved at duplication are single, continuous structures. The scheme for DNA duplication proposed by Delbrück[24] is thereby ruled out.

To recapitulate, both salmon-sperm and *E. coli* DNA heated under similar conditions collapse and undergo a similar density increase, but the salmon DNA retains its initial molecular weight, while the bacterial DNA dissociates into the two subunits which are conserved during duplication. These findings allow two different interpretations. On the one hand, if we assume that salmon DNA contains subunits analogous to those found in *E. coli* DNA, then we must suppose that the subunits of salmon DNA are bound together more tightly than those of the bacterial DNA. On the other hand, if we assume that the molecules of salmon DNA do not contain these subunits, then we must concede that the bacterial DNA molecule is a more complex structure than is the molecule of salmon DNA. The latter interpretation challenges the sufficiency of the Watson-Crick DNA model to explain the observed distribution of parental nitrogen atoms among progeny molecules.

Conclusion.—The structure for DNA proposed by Watson and Crick brought forth a number of proposals as to how such a molecule might replicate. These proposals[6] make specific predictions concerning the distribution of parental atoms among progeny molecules. The results presented here give a detailed answer to the question of this distribution and simultaneously direct our attention to other problems whose solution must be the next step in progress toward a complete understanding of the molecular basis of DNA duplication. What are the molecular structures of the subunits of *E. coli* DNA which are passed on intact to each daughter molecule? What is the relationship of these subunits to each other in a DNA molecule? What is the mechanism of the synthesis and dissociation of the subunits in vivo?

Summary.—By means of density-gradient centrifugation, we have observed the distribution of N^{15} among molecules of bacterial DNA following the transfer of a uniformly N^{15}-substituted exponentially growing *E. coli* population to N^{14} medium.

We find that the nitrogen of a DNA molecule is divided equally between two physically continuous subunits; that, following duplication, each daughter molecule receives one of these; and that the subunits are conserved through many duplications.

* Aided by grants from the National Foundation for Infantile Paralysis and the National Institutes of Health.

† Contribution No. 2344.

[1] R. D. Hotchkiss, in *The Nucleic Acids*, ed. E. Chargaff and J. N. Davidson (New York: Academic Press, 1955), p. 435; and in *Enzymes: Units of Biological Structure and Function*, ed. O. H. Gaebler (New York: Academic Press, 1956), p. 119.

[2] S. H. Goodgal and R. M. Herriott, in *The Chemical Basis of Heredity*, ed. W. D. McElroy and B. Glass (Baltimore: Johns Hopkins Press, 1957), p. 336.

[3] S. Zamenhof, in *The Chemical Basis of Heredity*, ed. W. D. McElroy and B. Glass (Baltimore: Johns Hopkins Press, 1957), p. 351.

[4] A. D. Hershey and M. Chase, *J. Gen. Physiol.*, **36**, 39, 1952.

[5] A. D. Hershey, *Virology*, **1**, 108, 1955; **4**, 237, 1957.

[6] M. Delbrück and G. S. Stent, in *The Chemical Basis of Heredity*, ed. W. D. McElroy and B. Glass (Baltimore: Johns Hopkins Press, 1957), p. 699.

[7] C. Levinthal, these PROCEEDINGS, **42**, 394, 1956.

[8] J. H. Taylor, P. S. Woods, and W. L. Huges, these PROCEEDINGS, **43**, 122, 1957.

[9] R. B. Painter, F. Forro, Jr., and W. L. Hughes, *Nature*, **181**, 328, 1958.

[10] M. S. Meselson, F. W. Stahl, and J. Vinograd, these PROCEEDINGS, **43**, 581, 1957.

[11] The buoyant density of a molecule is the density of the solution at the position in the centrifuge cell where the sum of the forces acting on the molecule is zero.

[12] Our attention has been called by Professor H. K. Schachman to a source of error in apparent molecular weights determined by density-gradient centrifugation which was not discussed by Meselson, Stahl, and Vinograd. In evaluating the dependence of the free energy of the DNA component upon the concentration of CsCl, the effect of solvation was neglected. It can be shown that solvation may introduce an error into the apparent molecular weight if either CsCl or water is bound preferentially. A method for estimating the error due to such selective solvation will be presented elsewhere.

[13] In addition to NH_4Cl, this medium consists of 0.049 M Na_2HPO_4, 0.022 M KH_2PO_4, 0.05 M NaCl, 0.01 M glucose, 10^{-3} M $MgSO_4$, and 3×10^{-6} M $FeCl_3$.

[14] This result also shows that the generation time is very nearly the same for all DNA molecules in the population. This raises the questions of whether in any one nucleus all DNA molecules are controlled by the same clock and, if so, whether this clock regulates nuclear and cellular division as well.

[15] F. H. C. Crick and J. D. Watson, *Proc. Roy. Soc. London*, A, **223**, 80, 1954.

[16] R. Langridge, W. E. Seeds, H. R. Wilson, C. W. Hooper, M. H. F. Wilkins, and L. D. Hamilton, *J. Biophys. and Biochem. Cytol.*, **3**, 767, 1957.

[17] For reviews see D. O. Jordan, in *The Nucleic Acids*, ed. E. Chargaff and J. D. Davidson (New York: Academic Press, 1955), **1**, 447; and F. H. C. Crick, in *The Chemical Basis of Heredity*, ed. W. D. McElroy and B. Glass (Baltimore: Johns Hopkins Press, 1957), p. 532.

[18] J. D. Watson and F. H. C. Crick, *Nature*, **171**, 964, 1953.

[19] C. E. Hall and M. Litt, *J. Biophys. and Biochem. Cytol.*, **4**, 1, 1958.

[20] R. Thomas, *Biochim. et Biophys. Acta*, **14**, 231, 1954.

[21] P. D. Lawley, *Biochim. et Biophys. Acta*, **21**, 481, 1956.

[22] S. A. Rice and P. Doty, *J. Am. Chem. Soc.*, **79**, 3937, 1957.

[23] Kindly supplied by Dr. Michael Litt. The preparation of this DNA is described by Hall and Litt (*J. Biophys. and Biochem Cytol.*, **4**, 1, 1958).

[24] M. Delbrück, these PROCEEDINGS, **40**, 783, 1955.

From the *Proceedings of the National Academy of Sciences*, Vol. 46, No. 4, 453–461, 1960. Reproduced with permission of the authors.

STRAND SEPARATION AND SPECIFIC RECOMBINATION IN DEOXYRIBONUCLEIC ACIDS: BIOLOGICAL STUDIES

BY J. MARMUR AND D. LANE

CONANT LABORATORY, DEPARTMENT OF CHEMISTRY, HARVARD UNIVERSITY

Communicated by Paul Doty, February 25, 1960

It is clear that the correlation between the structure of deoxyribonucleic acid (DNA) and its function as a genetic determinant could be greatly increased if a means could be found of separating and reforming the two complementary strands. In this and the succeeding paper[1] some success along these lines is reported. This paper will deal with the evidence provided by employing the transforming activity of DNA from *Diplococcus pneumoniae* while the succeeding paper[1] will summarize physical chemical evidence for strand separation and reunion.

Bacterial transformation offers a unique approach to this problem since the activity of DNA isolated from genetically marked strains can be assayed after being subjected to various treatments. We have concentrated on thermal treatment to accomplish our goal for several reasons. First, the accumulated experience in this Laboratory[2-5] has shown that exposure of DNA to carefully controlled temperatures for certain periods of time can denature the DNA molecules with minimal damage to their chemical structure; this is a prerequisite to strand separation and furthermore the ease of precise control of the temperature offers an obvious means of providing nearly reversible conditions which would optimize the chances of reuniting the DNA strands. Moreover, it has been shown in one case that thermal treatment did lead to strand separation[6] and there is considerable evidence that transforming activity falls sharply in the region of thermal denaturation.[5, 7, 8] By utilizing these observations and by giving particular attention to the *rate* of cooling from the elevated temperature we have been able to demonstrate a pattern of inactivation and restoration of biological activity consistent with strand separation and specific reunion. It appears that this reunion can take place between complementary strands or between strands of two closely related molecules from mutant strains of *D. pneumoniae*.

Experimental Details.—The strains of *D. pneumoniae* and the transformation techniques employed have been described previously.[9-11] The isolation of DNA from *D. pneumoniae* and other bacterial species was carried out by a procedure described elsewhere.[12] The samples of pneumococcal DNA had sedimentation constants, $s_{20,w}$ measured in 0.15 M NaCl plus 0.015 M sodium citrate in the range of 24 to 26 S (corresponding to molecular weights of 8–10 million).

In most experiments pneumococcal DNA was dissolved in 0.15 M NaCl plus 0.015 M sodium citrate (hereafter referred to as standard saline-citrate) and heated in small tubes or flasks by immersion in boiling water. When *fast cooling* was employed, samples were transferred to tubes precooled in ice water. In cases where *slow cooling* was desired, samples were placed in an insulated water bath (six liters) at the elevated temperature, and allowed to cool with the heaters turned off. In a typical experiment the drop in temperature took the following time course: 89°—0 (minutes), 80°—15, 70°—40, 60°—75 and 50°—130. The relationship between concentration and transforming activity of DNA which has been heated at 100

RELATIONSHIP BETWEEN CONCENTRATION AND TRANSFORMATION
FOR NATIVE AND THERMALLY TREATED DNA

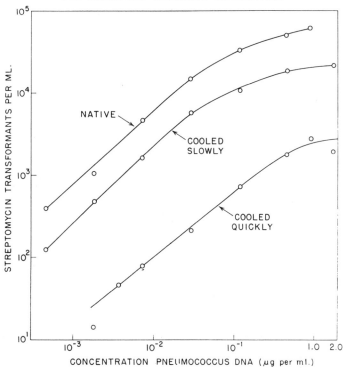

Fig. 1.—Relationship between transformation and concentration for native and thermally denatured DNA (cooled quickly and slowly).

Pneumococcal DNA was heated at 100° for 10 minutes in standard saline-citrate. Hot concentrated saline-citrate was added to a final concentration of 0.3 M NaCl and 0.03 M Na citrate and a DNA concentration of 20 μg/ml. The mixture was subdivided into two portions, one cooled quickly, the other slowly from 90° to room temperature. The native DNA was an unheated aliquot of the same preparation.

and then cooled quickly and slowly, as well as native unheated DNA is shown in Figure 1. All assays were carried out in the linear portion of the transformation-concentration curve.

Results.—Thermal denaturation and marker inactivation: The thermal denaturation of DNA can be followed by changes in the absorbance at 2600 A. observed either as a function of the temperature of the solution itself or the temperature to which it has been heated prior to cooling to room temperature (25°). The difference between these two temperature profiles results from reformation of base pairs as the temperature is lowered; such reformation has been thought to be nonspecific, involving short regions of chains that are not exactly complementary.[1, 3] The transforming activity, as measured by the ability of the heated pneumococcus DNA to effect transformation of wild-type pneumococcus with respect to three unlinked resistance markers: streptomycin, erythromycin, and bryamycin (which behaves as a high-level micrococcin resistance marker), was followed as a function of temperature. The results together with the absorbance-temperature profiles are shown in Figure 2. From this it is clear that different markers have different ther-

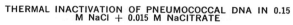

THERMAL INACTIVATION OF PNEUMOCOCCAL DNA IN 0.15
M NaCl + 0.015 M NaCITRATE

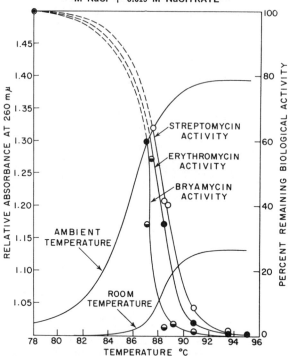

Fig. 2.—Loss in transforming activity and denaturation of
DNA as a function of temperature.

Pneumococcal DNA was heated at 20 μg per ml in standard saline-citrate. The absorbance at 260 mμ was recorded
for a series of temperatures both at the ambient temperature
(after correction for thermal expansion) and after cooling (in
air) to room temperature. After each cooling a sample was
removed, diluted, and assayed for its ability to transform
with respect to streptomycin, erythromycin, and bryamycin.

mal resistance in agreement with the results of others.[7, 8] This difference in sensitivity is most likely related to differences in base composition among the individual
DNA molecules in which the markers are located.[5] It is important to note, however,
that the thermal inactivations though differing one from another, are much more
closely related to the absorbance-temperature curve determined at room temperature than to the one obtained at the ambient temperatures. Hence, the possibility
exists that some helical regions melted out at the elevated temperature and have
reformed on cooling with return of biological activity.

A stronger suggestion that complementary chains have reunited to a very limited
extent is seen in the fact that all three markers retain some residual activity even
though all the molecules have been denatured as judged by the optical density increase having reached its maximum. In a separate experiment, transforming DNA
was heated at 120° in an autoclave for 10 minutes, cooled in air, and the streptomycin activity assayed: 0.05 per cent of the original activity remained. Since this
very elevated temperature exceeded that required to denature even pure guanine-
cytosine regions (110°),[4, 5] it is highly improbable that any DNA molecules escaped

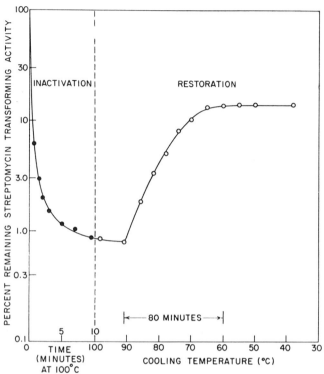

Fig. 3.—Thermal inactivation and restoration of the transforming activity of DNA.

Pneumococcal DNA at 20 μg/ml in standard saline-citrate was preheated for 1½ minutes at 85.5° (no loss in biological activity), then transferred to a boiling water bath at 0 minute. At the times shown, samples were removed to an equal volume of 1.5 M NaCl plus 0.15 M Na citrate in an ice-water bath. After 10 minutes exposure at 100°, an equal volume of hot (100°) 1.5 M NaCl plus 0.15 M Na citrate was mixed with the DNA solution, the mixture transferred to a large water bath and then cooled slowly. During the gradual descent of the temperature, samples were removed (right of the dashed vertical line) to pre-chilled tubes in an ice-water bath.

denaturation. Thus the residual biological activity must be due to a low level of transformation by denatured DNA molecules or to a small amount of specific recombination to yield two stranded helices. The former possibility is not in accord with the finding that denatured DNA is not taken up by transformable cells.[13] Thus, specific reformation of the separated strands is indicated.

As a consequence a systematic investigation was undertaken of the variables that might affect the recombination so that what has been at best a marginal effect could be maximized. The important variables have been found to be the rate of cooling of the heated solution, the concentration of DNA and the ionic strength.

The effect of slow cooling: The transforming activity of thermally denatured pneumococcal DNA was found to be very dependent on the rate of cooling of the heated solution. Thus when pneumococcal DNA solutions are heated at 100° in ctandard saline-citrate for various periods of time and a series of aliquots removed, sooled quickly and slowly from 90° to room temperature, it is found that the slow

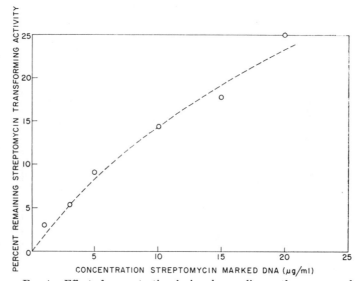

Fig. 4.—Effect of concentration during slow cooling on the recovery of transforming activity of thermally denatured DNA.

Pneumococcal DNA at 20 μg/ml in standard saline-citrate was heated for 10 minutes at 100°, transferred to a large bath at 89° and immediately diluted into pre-heated (89°C) tubes containing standard saline-citrate to give the final concentrations shown on the graph. After slow cooling all samples were diluted to the same concentration for assay. Circles are experimental points; dashed line is theoretical curve.[1]

cooling leads to an activity 35-fold greater than the fast cooling. This ratio is insensitive to the time of exposure at 100° between 5 and 20 minutes.

The transition from the activity characteristic of fast cooling to that characteristic of slow cooling can be seen in Figure 3. The transforming activity is plotted for quickly cooled aliquots taken from a solution during the exposure at 100° and then at various times during the slow cooling (at the right of the vertical dashed line) as the temperature of the bath dropped. Thus it is seen that restoration of the activity begins at about 90° and continues until about 65°. While the level of restored activity is about 15 per cent, this varies from sample to sample and in some cases has been found to be as high as 50 per cent. It is of interest to note that when fast cooled samples resulting in low biological activity are reheated and slowly cooled the transforming activity is restored to almost that of an aliquot that had been slow cooled originally. Thus the low activity of a fast cooled sample represents a repression rather than a destruction of its essential ability to transform. Essentially similar results have been obtained with thermally denatured transforming DNA isolated from *Streptococcus salivarius* and *Bacillus subtilis* on fast and slow cooling (Marmur, Lane, and Levine, unpublished results).

Effect of concentration of DNA during cooling: An important factor in determining the level of stored activity on slow cooling was found to be the concentration of the DNA itself during the cooling period. In contrast, there is practically no dependence of the activity on concentration for fast cooling. To illustrate the former a solution of pneumococcal DNA at a concentration of 20 μg/ml in standard saline-citrate was heated at 100° for 10 minutes and then diluted into hot solvent to give

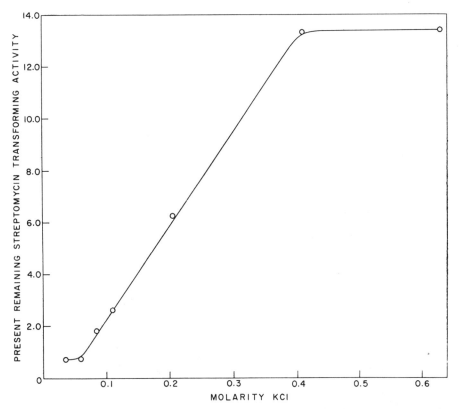

Fig. 5.—Effect of ionic strength on the recovery of transforming activity of slowly cooled denatured DNA.

Pneumococcal DNA at 200 μg/ml in standard saline-citrate was heated at 100° for 10 minutes, diluted 20-fold into KCl of various ionic strength and at 86°. The solutions were cooled slowly (the contribution of saline-citrate to the final ionic strength is included in the graph).

a series of concentration from 1 to 20 μg/ml. When these were slowly cooled the specific activities were found to increase nearly linearly from 3 per cent for 1 μg/ml to 25 per cent at 20 μg/ml (Fig. 4). As shown in the following article, this behavior is consistent with a bimolecular reaction.

Effect of salt concentration: Once the DNA strands have separated, the negatively charged phosphate groups would tend to prevent the reunion of the strands to form the helix. Increasing ionic strength would be expected to favor recombination by repressing this repulsive effect through shielding. This is shown to be the case in Figure 5. DNA was heated at a high concentration in standard saline-citrate at 100° for 10 minutes and then diluted into prewarmed tubes containing various concentrations of KCl. The solutions were cooled slowly, then assayed. There is a linear increase in biological activity between 0.05 M KCl and 0.4 M KCl. At higher concentrations of KCl (0.4 M) the biological activity reaches a maximum representing 14 per cent of the activity of the native material. Essentially similar results are obtained when the cooling is carried out in various concentrations of saline-citrate.

The effect of pH on strand reunion is broad with maximum at neutrality.

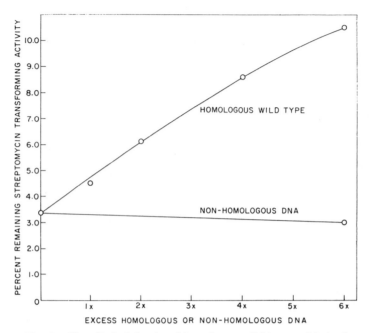

Fig. 6.—The effect of denatured homologous wild-type and heterologous DNA on the transforming activity.

Streptomycin marked pneumococcal DNA at 5 μg/ml was heated for 10 minutes at 97° in standard saline-citrate together with wild-type (homologous) pneumococcal DNA in the relative concentrations shown. Heterologous DNA and streptomycin marked DNA were also heated together.   All the mixtures were slow cooled.   Circles are experimental points; dashed line is theoretical curve.[1]

Effect of homologous and heterologous DNA: If the two complementary strands of the DNA molecule separate completely, it should be possible, by adding an excess of denatured homologous wild-type DNA during the slow cooling to reform hybrid molecules in which one strand carries a specific genetic marker and the other strand does not carry the marker.   DNA isolated from streptomycin resistant pneumococci and various concentrations of DNA isolated from streptomycin sensitive cells were heated together in standard saline-citrate at 100° for 10 minutes and then cooled slowly.   The resultant increase in biological activity is shown in Figure 6.   Thus, it would appear that hybrid molecules formed in the slow cooling are active in carrying out transformations.   Since transformants are selected and counted in a medium containing streptomycin, it is not known at the present time whether a cell transformed by a hybrid molecule results in a pure clone of streptomycin resistant cells or a mixed population of streptomycin sensitive and resistant cells.   This question is now being studied using linked markers.

If genetically marked DNA is first denatured and then slow cooled in the presence of homologous DNA in the *native* state there is no increase of the biological activity over that of the control.

If during the slow cooling, denatured heterologous DNA is added at five to six times excess, there is again no increase in transforming activity over that of the control.   The heterologous DNA samples tested have been isolated from *Salmonella typhimurium* (LT-2), *Micrococcus lysodeikticus*, *Micrococcus pyogenes* var. *aureus*, *Streptococcus* D, and calf thymus. These preparations vary in base composition, although that of *Streptococcus* D is identical to the per cent guanine plus cytosine of *D. pneumoniae* (Marmur, unpublished).

Discussion.—Previous studies on the thermal denaturation of DNA have done little to correlate its molecular and biological properties. The relationship between optical density, which is a measure of the helical content, and transforming activity, which is a fine probe of the properties of individual molecules, has led to the discovery that strands of thermally denatured DNA can reunite during slow cooling to form helical structures which are biologically active.

Beyond the demonstration that the biological activity can be restored to denatured DNA by a process of slow cooling which permits restoration of complementary-paired helical regions,[1] the most important observation appears to be the additional increment in biological activity conferred by the presence of homologous wild-type DNA during the cooling process. This is clearly a consequence of forming hybrids between strands of the marked and wild-type DNA and strongly supports the view that the essential genetic information in DNA is carried independently by each strand. Essentially similar conclusions have been arrived at by Pratt and Stent[14] working with bromouracil induced mutants of phage and by Tessman[15] working with nitrous acid mutants of phage and the bacterial virus ϕX 174 which contains single-stranded DNA.[16] The duplicate set of genetic information in the native DNA molecule of the mutant strain is disseminated among a larger number of renatured molecules in which at least one strand carries the information. Closely related DNA from *Streptococcus* D (which can be transformed by pneumococcal DNA[17]) having the same over-all base composition as pneumococcus does not have the same effect as the added homologous denatured pneumococcal DNA during slow cooling. This points to the necessity for a close homology that is required for the interacting strands to form double stranded structures.

The question might arise as to how representative the biological results are of the whole molecular population. It has already been established that a genetic marker is associated with a DNA molecule,[18–20] occupying a small portion of it.[10] The number of DNA molecules in a pneumococcal cell is of the order of several hundred[18] if the weight average molecular weight is 10 million. Since essentially similar results have been obtained with both the streptomycin and bryamycin markers, and since they represent molecules which are relatively rich in guanine-cytosine and adenine-thymine respectively, it is safe to conclude that most of the molecules exhibit the same interactions as the molecules carrying the biological markers being assayed.

Summary.—DNA from *D. pneumoniae* when heated to temperatures where all the molecules are denatured still retains transforming activity. The activity is increased by slowly cooling the denatured DNA as well as by increasing the ionic strength and concentration of DNA during the cooling period. The reunion of the strands is specific and has been shown to occur between populations of denatured molecules of mutant strains of *D. pneumoniae*. The reunion does not occur between the DNA strands of pneumococcus and that from *Streptococcus* D or from organisms with widely varying base ratios. Hybrid molecules in which one strand contains the mutant marker and the other strand derives from wild-type DNA are active in carrying out transformations with respect to the mutant property.

The authors would like to thank Dr. Paul Doty for valuable suggestions and discussions as well as Mr. C. Schildkraut and Mr. J. Eigner for their help and stimulation during the course of many of the experiments. We would like to thank Drs. M. Roger and R. D. Hotchkiss for numerous helpful discussions. This work was supported by a grant from the United States Public Health Service (C-2170).

[1] Doty, P., J. Marmur, J. Eigner, and C. Schildkraut, these PROCEEDINGS, **46**, 461 (1960).

[2] Rice, S. A., and P. Doty, *J. Am. Chem. Soc.*, **79**, 3937 (1957).

[3] Doty, P., H. Boedtker, J. R. Fresco, R. Haselkorn, and M. Litt, these PROCEEDINGS, **45**, 482 (1959).

[4] Marmur, J., and P. Doty, *Nature*, **183**, 1427 (1959).

[5] Doty, P., J. Marmur, and N. Sueoka, in *Brookhaven Symposia in Biology*, **12**, 1 (1959).

[6] Meselson, M., and F. Stahl, these PROCEEDINGS, **44**, 671 (1958).

[7] Zamenhof, S., H. E. Alexander, and G. Leidy, *J. Exptl. Med.*, **98**, 373 (1953); Zamehhof, S., in *The Chemical Basis of Heredity* (Baltimore: Johns Hopkins Press, 1957), p. 373.

[8] Roger, M., and R. D. Hotchkiss (personal communication).

[9] Fox, M. S., and R. D. Hotchkiss, *Nature*, **179**, 1322 (1957).

[10] Litt, M., J. Marmur, H. Ephrussi-Taylor, and P. Doty, these PROCEEDINGS, **44**, 144 (1958).

[11] Hotchkiss, R. D., and A. Evans, *Cold Spring Harbor Symp. Quant. Biol.*, **23**, 85 (1958).

[12] Marmur, J. (in preparation.)

[13] Lerman, L. S., and L. J. Tolmach, *Biochim. Biophys. Acta*, **33**, 371 (1959).

[14] Pratt, D., and G. S. Stent, these PROCEEDINGS, **45**, 1507 (1959).

[15] Tessman, I., *Virology*, **9**, 375 (1959).

[16] Sinsheimer, R. L., *J. Mol. Biol.*, **1**, 45 (1959).

[17] Bracco, R. M., M. R. Krauss, A. S. Roe, and C. M. MacLeod, *J. Exptl. Med.*, **106**, 247 (1957).

[18] Fox, M. S., *Biochem. Biophys. Acta*, **26**, 83 (1957); Lerman, L .S., and L. J. Tolmach, *Biochim. Biophys. Acta*, **26**, 68 (1957).

[19] Goodgal, S. H., and R. M. Herriott, in *The Chemical Basis of Heredity* (Baltimore: Johns Hopkins Press, 1957), p. 336.

[20] Schaeffer, P., *Symp. Soc. Exptl. Biol.*, **12**, 60 (1958).

From the Proceedings of the National Academy of Sciences, Vol. 46, No. 4, 461–476, 1960. Reproduced with
permission of the authors.

*STRAND SEPARATION AND SPECIFIC RECOMBINATION IN
DEOXYRIBONUCLEIC ACIDS: PHYSICAL CHEMICAL STUDIES*

BY P. DOTY, J. MARMUR, J. EIGNER, AND C. SCHILDKRAUT*

CONANT LABORATORY, DEPARTMENT OF CHEMISTRY, HARVARD UNIVERSITY

Communicated February 25, 1960

The separation and reunification of the complementary molecular strands of
DNA, so clearly indicated by the restoration of biological activity,[1] can be demonstrated by physical chemical techniques as well. Such studies permit a more
quantitative description of the phenomenon, make possible the inclusion of DNA
samples that do not participate in bacterial transformation and lead to a better
insight into the controlling features of the reaction. This, in turn, should provide
a better basis for understanding the possibilities and limitations that DNA has

in vivo for the macromolecular reactions that are the counterpart of its genetic function.

In summarizing here our current work in this direction we begin by showing how strand separation and recombination is reflected in three physical chemical methods in the order in which we took them up.   To proceed from these to more quantitative studies it became necessary to find reliable, routine means for determining the molecular weight of DNA in the various forms with which we were confronted, so that the complicating problems of aggregation and depolymerization could be assessed and minimized.   Success along these lines made it possible to follow the molecular weight changes accompanying the processes being studied and to examine the effect of molecular weight thereon.   In the final section we present the results of a study of recombination between strands which differ either in isotopic label or in species of origin.   In both cases "hybrid" or "heterozygous" reformed molecules can be demonstrated.

OBSERVATION OF STRAND RECOMBINATION

Absorbance-Temperature Curves.—It has been established that when DNA solutions are slowly heated a dramatic macromolecular change occurs in a very restricted temperature range.   The change is a cooperative melting out of the one-dimensional helical structure yielding disorganized, coiled polynucleotide chains.   At 0.2 molar sodium ions the midpoint of this transition, T_m, lies in the interval of 80 to 100° depending on the guanine-cytosine content of the DNA.[2]   The change, which involves a 40 per cent increase in absorbance, can be easily and accurately followed by measuring the absorbance at 260 mμ as a function of temperature.[3]   When the solution is cooled to 25° the absorbance decreases until it is about 12 per cent above that of the original solution at room temperature.   This appears to be due to the formation of short, imperfect, intrachain helical regions in which a major portion of the bases are paired.   Upon reheating such solutions, in the case of calf thymus and other mammalian DNA, the absorbance increases gradually without a region of rapid rise since short, imperfect helices have a broad range of transition temperatures.[3]

When DNA of *Diplococcus pneumoniae* was thermally denatured and then reheated it was found, in contrast to the general behavior just described, that a distinct maximum occurred in the middle of the temperature range; beyond this there was a sharp rise coincident with the latter half of the original curve for native DNA.   This behavior suggested that a typical helix-coil transition was taking place involving the denaturation of long, perfect helices containing about half of the bases in the DNA sample.

The meaning of this became more clear when the absorbance-temperature curves of quickly and slowly cooled samples were measured.   Such results are shown in Figure 1.   The rate of cooling of these samples is the same as that described in reference 1   The absorbance of the slowly cooled sample is seen to rise to a plateau and then continue to the abrupt region coincident with the thermal denaturation of the native DNA.   This sample had retained a large degree of its original transforming activity as described in the foregoing paper.[1]   From this it is concluded that the slowly cooled sample contained substantial amounts of the Watson–Crick helix.   The curve for the quickly cooled sample, which had very low transforming

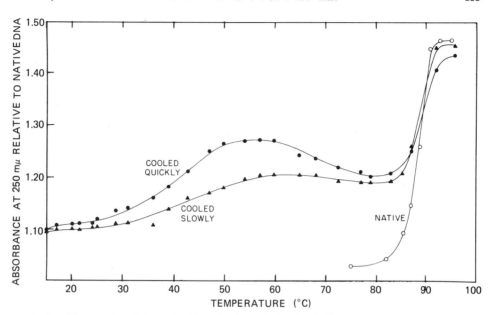

Fig. 1.—Thermal transitions of native, slowly cooled and quickly cooled *D. pneumoniae* DNA. *D. pneumoniae* DNA was heated at 100° for 10 min in standard saline-citrate. Hot concentrated saline-citrate was then added to a final concentration of 0.27 *M* NaCl plus 0.02 *M* Na citrate and subdivided into two portions, one cooled quickly in ice-water, the other cooled slowly. The native DNA was an aliquot of the same sample. The absorbance at 260 mμ, corrected for thermal expansion, was recorded after each solution was exposed for 15 min to the indicated temperature. The ordinate gives this reading relative to the absorbance of the native sample at 25°.

activity, can now be understood. At room temperature it is devoid of long, complementary helical regions, but during the gradual heating period short, imperfect helical regions melt out, releasing chains that then have time to develop long, complementary helices: these then melt in the vicinity of the characteristic temperature, T_m. Thus the display of the sharply melting region by the quickly cooled sample is seen to be an artifact in that this is not evidence that long, complementary helices exist in the same DNA sample at room temperature. This is consistent with its near absence of biological activity. We propose to call this form of DNA *denatured* and refer to the slowly cooled DNA with substantial amounts of reformed complementary helical regions as *renatured*.

Density Gradient Ultracentrifugation.—When a very dilute solution of DNA (∼ 2 γ/ml) in about 7.8 molal cesium chloride is centrifuged at high speed in an analytical ultracentrifuge for 30–40 hours it is found that a density gradient has been set up and the DNA has migrated to a band at a position corresponding to its effective density. This is the basis of avaluable new technique[4] which permits the very accurate assessment of the density of DNA and, through the band profile, reflects in a combined fashion the molecular weight and the density heterogeneity of the DNA sample.[5] It has recently been shown that the density of DNA varies linearly with the guanine-cytosine content[6, 7] and that the distribution of guanine-cytosine among the molecules making up a DNA sample is relatively narrow, particularly for bacterial DNA.[6-8] It has been also shown that the density of DNA increases about 0.016 units upon denaturation.

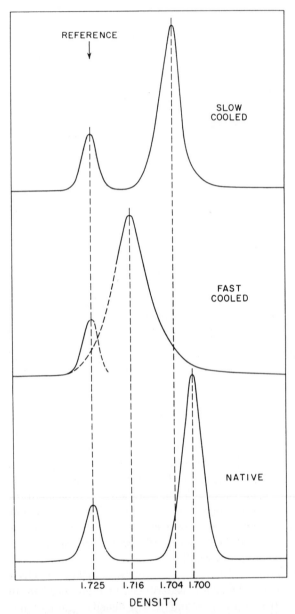

Fig. 2.—Molecular reconstitution in *D. pneumoniae* DNA.

Equilibrium concentration distributions of DNA samples banded in CsCl gradient. See text for a description of samples. Centrifugation at 44,770 rpm. The ordinate represents the DNA concentration as a function of the distance from the axis of rotation. Density increases towards the left. Conditions of heating and cooling are as described for Figure 1 of the preceding paper.[1]

In this situation it was obvious that the density of denatured and renatured pneumococcal DNA should be determined because our view of renaturation, brought about by slowly cooling thermally denatured DNA, predicts that such material should have a density close to the original native DNA. The photometric traces of the ultraviolet photographs of the appropriate density gradient ultracentrifugations are shown in Figure 2. At the bottom of the figure one sees the band for native pneumococcal DNA with a density of 1.700 together with a small band of DNA from *Pseudomonas aeruginosa* which is present in each case to provide a reference density. The trace for the quickly cooled, denatured pneumococcal DNA is shown in the middle exhibiting the expected higher density of 1.716. At the top is seen the tracing for the renatured DNA: it has a density of 1.704. Thus the renatured DNA is found to have a density close to that of native DNA supporting the view that it consists predominately of reformed, complementary helices of the same character, and density, as the native DNA.

Although the molecular weights of these three forms will be discussed at a later point it may be of interest to mention here that the pronounced widening of the band for the denatured DNA reflects an approximately five-fold reduction in molecular weight; hence only a part of this can be due to strand separation alone. The near equivalence of the profile for the renatured DNA to the native DNA involves, therefore, not only a recombination of two strands but, under the conditions employed, some aggregation which by chance increases the molecular weight to about that of the original native DNA.

Electron Microscopy.—The evidence presented in the two previous sections clearly predicts that electron microscopy should reveal in the renatured DNA the long, cylindrical threads of 20 A. diameter that are characteristic of native DNA: these should be absent in the denatured (quickly cooled) DNA. Accordingly, pneumococcal DNA was heated in saline-citrate at 100° for 10 minutes, diluted with more concentrated saline-citrate to a final concentration of 18 γ/ml in 0.30 M NaCl + 0.03 M Na citrate, and cooled in two parts, one quickly and one slowly. Both solutions were then dialysed into 0.05 M ammonium carbonate plus 0.1 M ammonium acetate. Thereafter Professor C. E. Hall sprayed the solutions on freshly cleaved mica and, proceeding in a manner previously described[9, 10] obtained electron micrographs shown in Figure 3 (slowly cooled) and Figure 4 (quickly cooled). The former shows the structure characteristic of native DNA. The only difference between this and micrographs of native DNA is the more frequent occurrence of irregular patches at the ends of cylindrical threads. These are probably regions of denatured DNA arising from incomplete recombination or the inequality in length of the two strands that are paired. The micrograph for the quickly cooled sample (Fig. 4) shows only irregularly coiled molecules with clustered regions. These results are very similar to those found for ribonucleic acid.[11] Thus the identification of renatured DNA as being for the most part similar to native DNA and denatured DNA as being an irregular chain with considerable base pairing in short regions seems to be complete.

Dependence of Molecular Recombination on Concentration.—In the foregoing report[1] it is shown that the transforming activity of renatured pneumococcal DNA increases with the concentration at which it is slowly cooled (see Fig. 4 of ref. 1). This suggests that the extent of molecular recombination is quite low at concentrations considerably below those employed in the experiments described in the foregoing sections, that is, $\sim$ 20 γ/ml. In order to find out if the biological activity reflected the state of the DNA sample itself we heated and slowly cooled a pneumococcal DNA sample in the usual manner at a concentration of 1 γ/ml. The density of this sample was found to be 1.715. This value matches that of quickly cooled pneumococcal DNA (1.716) prepared either at the usual concentration, 20 γ/ml, or at this very low concentration, 1 γ/ml. Thus cooling at what we have described as a slow rate, that is from 90 to 60° in 80 minutes, does not lead to substantial recombination of strands when the concentration is quite low.

With the physical exhibition of the molecular recombination, as well as the biological, shown to be quite concentration dependent it seemed permissible to attempt to account for the concentration dependence observed (Fig. 4 of ref. 1) by assuming a bimolecular reaction, $A + A \leftrightarrows A_2$, which reaches effective equilibrium at a particular temperature that lies within the region where slow cooling takes place. If α represents the mole fraction of strands that are combined in the helical form, the equilibrium constant $K = \alpha/[(1 - \alpha)^2 c]$ where c represents the DNA concentration. This can be applied to the data in Figure 4 of reference 1 by

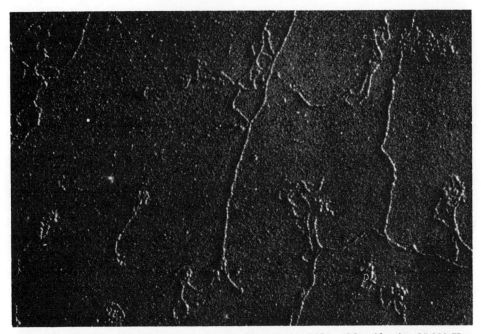

Fig. 3.—Electron micrograph of renatured *D. pneumoniae* DNA.   Magnification 95,000 X.
Shadow casting is from the right at a shadow to height ratio of 10:1.   See text for heating and cooling conditions.

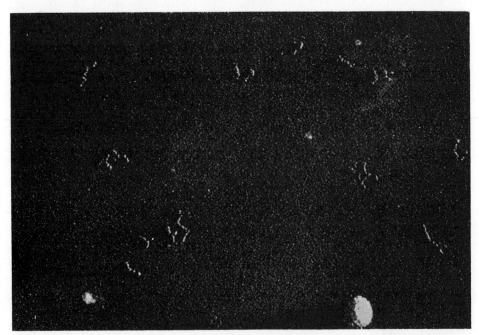

Fig. 4.—Electron micrograph of denatured *D. pneumoniae* DNA.   Magnification 95,000 X.
The polystyrene reference sphere at the lower right of the micrograph has a diameter of 880 Å.
Shadow casting is from the right at a shadow to height ratio of 10:1.   See text for heating and cooling conditions.

assuming that α is equal to the fractional activity of the renatured DNA. By choosing one point to permit an evaluation of the product Kc, α can be calculated as a function of c. The dashed line drawn in Figure 4 of reference 1 is the result calculated in this way. Since it fits the experimental data there is reason to conclude that the molecular recombination is indeed bimolecular as the character of the Watson–Crick structure for DNA would demand. Moreover, the success of this simple approach suggests that properly designed experiments which yield K as a function of temperature should lead to the determination of the heat and entropy of the recombination as well as the free energy change involved in this process at the temperature of cellular division.

Dependence of Molecular Recombination on the Source of DNA.—With the view that molecular recombination in DNA consists simply of two complementary strands coming together with the fraction of strands complexed under control of an equilibrium constant it follows that the manifestation of molecular recombination depends critically on the concentration of complementary strands. Now the concentration of complementary strands is not identical with the concentration of DNA, when DNA from different sources is considered. Take the extreme cases, of typical bacterial DNA and the DNA of mammalian cells. Here the DNA content per cell differs by the order of a thousand fold. If we take the average molecular weight to be the same in both cases and if it is assumed that each DNA molecule in a given cell is different, it follows that at the same weight concentration, the concentration of complementary pairs will be a thousand fold greater for the bacterial than for the animal DNA. From this it follows that molecular recombination will not be expected for animal DNA under the conditions that are just sufficient to permit its occurrence in bacterial DNA.

The expectation that animal DNA will not show complementary reformation when bacterial DNA will do so has been borne out in two kinds of experiments.[12] In one set the absorbance temperature profiles for a number of denatured DNA samples have been determined. The majority of bacterial DNA samples showed some of the character seen in Figure 1, that is a sharply rising region at high temperature approximately coincident with the curve for native DNA. By contrast calf thymus and salmon sperm DNA showed essentially a continuous and gradual rise with temperature. This shows that on the time scale of the heating curve there was some molecular reconstitution in most of the bacterial samples but not in the cellular DNA.

In another experiment samples of thermally denatured DNA were cooled quickly and then brought to 80° and the absorbance followed as a function of time in 0.30 M NaCl plus 0.03 M sodium citrate. The pneumococcal DNA fell from 1.40 times the absorbance of the native DNA at 25° to 1.08 whereas the calf thymus showed no drop at all. This again demonstrates the inability of calf thymus DNA to reform the helical configuration under the conditions which do permit pneumococcal DNA to do so. Consequently, the great difference in concentration of complementary strands in the denatured DNA appears to offer the explanation.

THE MACROMOLECULAR PROPERTIES OF NATIVE, DENATURED, AND RENATURED DNA

A means of making reasonably accurate molecular weight assignments to DNA in each of the forms under study is necessary in order to examine quantitatively the phenomenon of molecular recombination in DNA and to enable the complicating features of aggregation and depolymerization to be understood and eliminated in so far as possible. Although light scattering measurements were the first method to satisfactorily establish the molecular weight and shape of DNA in solution,[13, 14] its accuracy and applicability in routine use, particularly for molecular weights greater than 5 million, has been found wanting elsewhere[15] as well as in this Laboratory. Consequently, a systematic investigation of sedimentation constants, evaluated by extrapolation to zero concentration, and intrinsic viscosities evaluated by extrapolation to zero gradients, has been undertaken[16] and molecular weights have been derived by a careful application of the Mandelkern–Flory equation:[17]

$$ M = \left[\frac{s^\circ [\eta]^{1/3} \eta_0 N}{\beta(1 - \bar{v}\rho)} \right]^{3/2} $$

In this equation, s° = the sedimentation constant at zero concentration, $[\eta]$ = the intrinsic viscosity at zero gradient, η_0 = the solvent viscosity, $N = 6.03 \times 10^{23}$, $(1 - \bar{v}\rho)$ = the buoyancy factor and β = a constant which has a value which ranges from 2×10^6 to about 4×10^6 as the permeability of a coiled molecule or the axial ratio of a rigid ellipsoid increase. For most flexible polymers the value of β has been found to be about 2.6×10^6.

Native DNA.—An investigation[18] of DNA samples in which the molecular weight had been varied by ultrasonic radiation and the molecular weights measured by light scattering showed β to increase monotonically from 2.56 at 300,000 to 3.29 at 8,000,000. Other light scattering experiments, and an independent study[19, 20] in which molecular weights were determined by sedimentation-diffusion, have shown good agreement with this result. Such an empirical evaluation of β, based as it is on absolute methods, permits us to use the Mandelkern–Flory equation to derive weight average molecular weights from sedimentation-viscosity data.

TABLE 1

VALUES OF s° AND $[\eta]$ FOR SELECTED MOLECULAR WEIGHTS OF NATIVE DNA

Molecular Weight	$s^\circ_{20, w}$ in HMP	$[\eta]$ in HMP	$s^\circ_{20, w}$ in SSC	$[\eta]$ in SSC
1,000,000	8.8 S	9.1 dl/gm	10.0 S	8.3 dl/gm
2,000,000	11.5	21.0	13.1	18.7
4,000,000	15.4	45.5	17.5	38.0
10,000,000	22.9	120	26.4	82.5
16,000,000	28.2	190	32.8	119

For a number of samples s° and $[\eta]$ have been measured in a solvent of lower ionic strength, 0.011 M in Na^+ ions, as well as in the more common solvent that is 0.195 M in Na^+ ions. The former solvent contains 0.0025 M Na_2HPO_4, 0.0050 M NaH_2PO_4 and 0.001 M sodium ethylenediaminetetraacetate and has a pH of 6.8. For convenience this is referred to as HMP (hundredth molar phosphate). The more common solvent consists of 0.15 M NaCl and 0.015 M $Na_3(C_6H_5O_7)$ with a pH of about 7. This will be referred to as SSC (standard saline-citrate). On a double logarithmic plot of s° against M the data in these two solvents fall on two parallel lines exhibiting slight upward curvature. Similarly, log $[\eta]$ against log M yields lines with downward curvature. Table 1 summarizes the smoothed $s^0_{20,w}$ and $[\eta]$ values in these two solvents for selected molecular weights.

Denatured DNA and the Problem of Aggregation.—When DNA is thermally denatured and cooled, a large fraction of the bases become paired.[3] While most of this pair formation can occur within each chain, some interchain bonding can be expected and this would increase with DNA concentration and molecular weight. Consequently, the determination of the molecular weight of denatured DNA requires proof that aggregation has been eliminated. Early light scattering studies showed no significant change in molecular weight on thermal denaturation,[21] and thereby suggested that the two strands had not separated. Other reports claimed a decrease to one-half the molecular weight but they were not convincing.[22] Moreover, the $s°$ and $[\eta]$ values for such denatured DNA did not yield a molecular weight in agreement with the light scattering values by a wide margin. Upon further investigation of this dilemma we have been able to show that significant aggregation occurs upon cooling denatured DNA in SSC at concentrations in excess of 40–100 γ/ml. Since light scattering and viscosity measurements require concentrations in this range or higher we were able to conclude that aggregation had occurred in most samples previously studied.

In order to eliminate this aggregation in the useful range of concentration we have gone to a solvent of lower ionic strength, HMP. Although there is some reformation of base pairs (about 25 per cent) in this solvent at room temperature no aggregation is evident by any of the methods employed and a satisfactory dependence of $s°$ and $[\eta]$ on molecular weight has been established by using the Mandelkern-Flory equation with a value of 2.6×10^6 for β. The results of such studies can be fitted with the usual empirical relations: $s° = K_s M^{a_s}$ and $[\eta] = KM^a$.

Now it has been shown that the guanine-cytosine bond is stronger than the adenine-thymine bond[2, 8] and as a result it is to be expected that the amount of base pairs in denatured DNA will increase with the guanine content. Indeed this has been shown to be the case with RNA samples of different composition.[3] Consequently it is possible that the dependence of $s°$ and $[\eta]$ on molecular weight for denatured DNA will depend on the composition of the DNA. This has been found to be the case and it is illustrated in Table 2 where the constants for the

TABLE 2

CONSTANTS FOR THE EMPIRICAL SEDIMENTATION AND VISCOSITY RELATIONS FOR DENATURED DNA IN PHOSPHATE-VERSENE SOLUTION 0 011 M IN NA$^+$ (HMP)

Source of DNA	K_s	a_s	K	a
D. pneumoniae	0.054_5	0.35_4	$3.4_6 \times 10^{-5}$	0.93_3
E. coli K-12	0.055_8	0.36_1	$3.1_1 \times 10^{-5}$	0.91_2

empirical equations are listed for denatured DNA (quickly cooled) from *D. pneumoniae* and *Escherichia coli* (K-12).

These results indicate a rather highly swollen chain configuration for DNA, despite some base pair interaction. For a given chain length the molecular dimensions that can be deduced from this are in the range expected for typical polyelectrolytes. These dimensions are much larger than those for RNA of the same chain length and in the same solvent: this is probably due to the additional contraction of the RNA coil due to additional hydrogen bonding made possible by the 2-OH of the ribose ring.

With these results at hand it is possible to assess the molecular weight as a function of time of heating at a given temperature and to demonstrate whether or not strand separation has occurred at the early stages of heating.

Strand Separation and Depolymerization.—In the low ionic strength solvent (HMP) the midpoint of the thermal denaturation profile, T_m, for *D. pneumoniae* is shifted downward from 85° in SSC to 64° and the transition is complete at 70°. For *E. coli* DNA the corresponding values are 5° higher. A temperature 15° above the T_m is well beyond the melting out region for DNA in this solvent, indeed by about the same amount as 100° is for the higher ionic strength solvent (SSC). Consequently 79° and 84° were chosen as the temperatures at which to follow the thermal degradation of these two DNA samples. Aliquots were removed periodically from a large amount of the solution that had been brought to

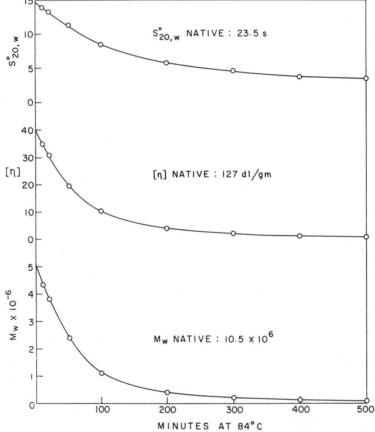

Fig. 5.—Thermal degradation of *E. coli* DNA in HMP at 84°. Molecular weights derived from Mandelkern-Flory equation[17] using $\beta = 2.6 \times 10^6$.

this temperature, and cooled quickly. The $s°$ and $[\eta]$ were measured at 25° and the molecular weight calculated as indicated above. The results for *E. coli* DNA are shown in Figure 5. It is seen that the molecular weight falls continuously but rather slowly through the period of exposure to elevated temperature. The extrapolation back to zero time is obvious and the molecular weight value there is 5.0 million. The molecular weight of the native DNA was 10.5 million ($s^\circ_{20,w}$ = 23.5 S and [η] = 126). In this case we have, therefore, a clear example of the molecular weight decreasing by a factor of two upon thermal denaturation, and with aggregation eliminated and depolymerization taken into account it can be said that strand separation did occur in the very early stages of the exposure to the elevated temperature.

The behavior of pneumococcal DNA in a similar experiment was roughly the same except that in this case the molecular weight fell from 8.2 million for the native DNA to 2.9 million for the denatured DNA at zero time. This change by a factor of somewhat more than 2, in this case 2.83, probably reflects an occasional enzymatically induced single chain scission in the original DNA preparation.

From the data obtained during the first 100 minutes of the above experiments the rate of thermal depolymerization of the polynucleotide chains of DNA can be evaluated. In practical units this can be expressed as 0.165 scission per 10 million molecular weight per 10 minutes at 79°. Similar experiments at 100° yields 1.41 in the same units. These and other results lead to a value of about 20 kcal for the activation energy for hydrolytic scission of individual strands of DNA, a value similar to that found in a quite different study of ribonucleic acid.[23] With this evaluation of the rate of bond scission it is possible to assess the extent of fragmentation induced by the exposure to 100° for 10 minutes followed by fast or slow cooling. For fast cooling the value stated above is the answer desired since the units have been chosen for convenience in this situation. Since pneumococcal DNA is often in the range of 10 million molecular weight each native molecule will receive on the average between one and two scissions as a result of the heating to 100°. Slow cooling will involve a longer exposure to temperatures where degradation is significant: in the case of our particular cooling rate the depolymerization would be about doubled.

Renatured DNA.—Renatured DNA was prepared by heating pneumococcal DNA of molecular weight 8.2 million at 100° for 10 minutes in standard saline-citrate at a concentration of 20 γ/ml and cooling slowly in double the saline-citrate concentration. In standard saline-citrate this material had a $S^\circ_{20,w}$ of 23.5 S and an [η] of 21.6 from which a molecular weight of 6.0 million was deduced. An aliquot of the same heated solution, quickly cooled, had a molecular weight of 2.0 million. When dialyzed into the lower ionic strength solvent (HMP), the renatured DNA had a $s^\circ_{20,w}$ of 15.4 S and an [η] of 43. A molecular weight of 4.0 million was obtained using these results. The denatured DNA showed no change in molecular weight when similarly transferred to the lower ionic strength solvent.

From these results it appears that a small amount of aggregation occurs when renaturation is carried out at 20 γ/ml in 0.3 M NaCl plus 0.03 M Na citrate and that the molecular weight falls by about the amount expected from the depolymerization rates reported above. In the lower ionic strength solvent, where the aggregation appears to have been eliminated, it is interesting to note that the s° and [η] values are essentially equal to those given in Table 1 for 4 million molecular weight native DNA. This supplies further evidence of the close structural similarity of the renatured to the native DNA.

The results have an important bearing on the transforming activity that can be expected for renatured pneumococcal DNA. The loss in activity to be expected from the molecular weight reduction can be obtained directly from an earlier study[24] of the molecular weight dependence of transforming activity. For a reduction of 10 to 4 million the activity should fall to 70 per cent of its original value. However, the activity would be expected to be lowered still further due to the fact that the complementary strands that recombine will generally be of different length as a result of the mild thermal depolymerization. If there was no selection according to chain length in recombination there would be a loss of about 50 per cent of the nucleotides from helical regions due to this effect and a drop to 35 per cent would be expected for the maximum activity. However, there is probably some preference for complementary chains of more nearly equal molecular weight to recombine and the evidence from the absorbance-temperature profile and DNA density in CsCl indicates that about 75 per cent of the nucleotides do reform in the helical configuration. Consequently the maximum biological activity of renatured transforming DNA under the conditions employed may be placed at about 50 per cent. The highest observed[1] has been approximately 50 per cent.

RECOMBINATION OF DNA STRANDS OF DIFFERENT DENSITY AND FROM DIFFERENT SPECIES

In this final section we return to the technique of density gradient ultracentrifugation and ask if by its use recombination between strands of different density can be observed. Two cases are of interest. In one we mix two DNA samples from the same species, one of which is denser by virtue of having N^{15} instead of N^{14}. In the other we mix N^{15} E. coli DNA with N^{14} DNA from other bacteria to see if renatured molecules of intermediate density can be observed. Such "heterozygous" molecules may be expected from closely related bacteria but not from distantly related ones.

From what has been stated in the previous section concerning aggregation and strands of unequal length in the renatured molecules, it is not surprising to find that the optimal resolution of hybrid molecules has proved somewhat elusive. Thus, while maximum helix development in the renatured sample is favored by high DNA concentrations, so is aggregation. Obviously aggregation of renatured molecules of different density will smear out the hybrid band one seeks to resolve. Likewise the inequality of chain lengths in the hybrid molecules will spread out their densities with a corresponding broadening of the hybrid band.

Against this background we present the results obtained with a mixture of N^{14} and N^{15} E. coli DNA each at a concentration of 5 γ/ml. This concentration is low enough to prevent aggregation but at the price of only limited recombination to the helical form. The results are collected in Figure 6. At the top (A) is shown the tracing for the quickly cooled, denatured DNA. The bands are seen to be completely separated without any hybrid formation. Thus, the reforming of base pairs has been entirely intrachain.

If an aliquot of the same heated mixture is slowly cooled the results are as shown in the second tracing (B). Here the peaks have remained at the same densities as in the denatured DNA but about one-third of the total DNA has shifted to a lighter density distributed about the value of 1.724. While this is indicative of hybrid formation it is far from proof since the densities of the peaks are essentially those of denatured N^{14} and N^{15} DNA and do not approach that of renatured material which should be about 0.008–0.012 density units lighter than the denatured materials.

In this situation it is necessary to provide a control consisting of the mixture of the two separately cooled DNA samples, that is, samples that have had the same thermal history as the slowly cooled mixture (*B*) but have been prevented from forming hybrids. The tracing of this control is shown in the Figure as (*C*). This provides a key to our interpretation since peaks corresponding to the renatured molecules are clearly evident at the right of each of the denatured peaks.

If the control *C* is now subtracted from the trace for the slowly cooled mixture (*B*) the migration of material to the density corresponding to renatured hybrids should be evident. This is shown at the bottom as (*D*). A shift of material from extremities to the density of the hybrid is clearly evident. About 20 per cent of the total DNA is involved: this is about that expected from (*C*) since there it can be seen that about 40 per cent of the material is in the bands corresponding to the renatured form.

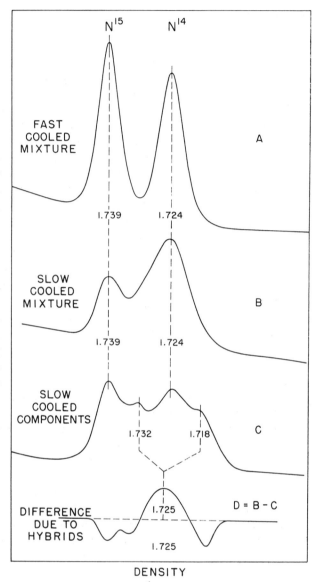

FIG. 6.—Hybrid formation in *E. coli* B DNA. Equilibrium concentration distributions of DNA samples banded in CsCl gradient. See text for a description of samples. Centrifugation at 44,770 rpm. The ordinate represents the concentration of DNA in the centrifuge cell. The area of each band is proportional to the amount of DNA it contains. N^{15} and N^{14} labeled *E. coli* B DNA were prepared by growing wild-type cells in a synthetic medium containing $N^{15}H_4Cl$ or $N^{14}H_4Cl$, respectively, as the sole nitrogen source. The DNA was isolated by a procedure to be published (J. Marmur, in preparation).

By using 4 times the concentrations employed in the above experiment the denatured species can be essentially eliminated but the additional spreading introduced by aggregation results in only a single band of somewhat irregular shape, and a proper correction cannot be made.

Despite this difficulty we have made some preliminary observations on mixtures of DNA from two different species. In these cases the peak densities appear to be sufficient for interpretation.

The results of Baron et al.[25] showing that E. coli, Shigella and Salmonella are closely related genetically, as well as the report of the similarity of their base compositions[26] prompted an experiment to see whether N^{15} labeled DNA strands from E. coli would recombine with N^{14} DNA strands from the other enteric organisms.

N^{15} E. coli B DNA and N^{14} Shigella dysenteriae (obtained from Dr. S. E. Luria) DNA were mixed, heated as usual and slowly cooled. The densities of the native samples were 1.725 and 1.710 respectively. If we assume 50 per cent of the resultant mixture is composed of hybrid molecules in which helical regions have formed to the extent of 75 per cent of the native helical content we would expect that, as was the case for E. coli and D. pneumoniae, the density of the resultant hybrid band should not be the mean of the native values but 0.004 density units heavier. A band centered at a density of 1.722 was indeed found and thus hybrid, or heterozygous molecules appear to have formed.

A similar prediction can be made for N^{15} E. coli B DNA and N^{14} Salmonella typhimurium(LT-2) (obtained from Dr. M. Demerec) since the base compositions are the same. However, a density of 1.729 was observed. Thus it appears that heterozygous renatured molecules have not formed. Since the observed density is midway between that predicted for denatured molecules and renatured, homogenous molecules, it is possible that occasional heterozygous regions have combined preventing a full development of homogeneous renatured molecules. Perhaps under optimal conditions interspecies molecular hybridization of the DNA would be observed in this case as well.

In another experiment, N^{14} D. pneumoniae DNA was heated and slowly cooled with N^{14} Serratia marcescens DNA. The density of the native samples are 1.700 and 1.717 respectively. The resulting trace showed two clearly separated bands of reformed D. pneumoniae of density 1.704 and reformed Serratia of density 1.721. There was no evidence of a component of intermediate density indicating that neither aggregation nor interspecies hybridization had occurred.

This serves as an ideal control to show that molecules having approximately the same density difference (0.017 density unit) but not the genetic similarity discussed above, show a strong preference for renaturing only with members of their homologous species. Only in the case where similarity in base composition has been correlated with genetic interaction has it been possible to demonstrate the appearance of heterozygous molecules.

Discussion.—Our aim in this paper has been to present a series of preliminary reports of closely related work on strand separation and molecular recombination in DNA. Consequently, the discussion will be postponed until the fuller accounts are published. Nevertheless, perhaps two points may be emphasized briefly.

The demonstration that strand separation actually occurs in a matter of a few minutes or less eliminates some of the objections[27] pertaining to the unwinding of DNA that have been put forward as a criticism of the Watson–Crick mechanism[28] for DNA duplication. What has been observed here favors the more recent calculations of the time required for this process.[29, 30] Also, although it is not as yet apparent whether the reformation of denatured DNA during slow cooling has a biological counterpart, the demonstration that it does occur suggests a mechanism for genetic exchange in closely related DNA molecules.

The second point is the recognition that the routine reduction of DNA into single strands and their specific reformation into essentially native molecules provides a means of creating entirely new DNA molecules. The study of the progeny of such heterozygous molecules will be of obvious interest. A particular instance is that of a heterozygous molecule each strand of which contains a different marker. The progeny from cells having a single nucleus may be pure clones, mixed clones or transformants with markers normally linked. Any of these results would, of course, illuminate the molecular aspects of the replication act. If linkage can be found, not only will crossing-over be demonstrated but the possibility of making *in vitro* new forms of viable DNA not previously existent will be assured. It seems likely that heterozygous DNA molecules will find other uses as well. For example, active hybrid molecules containing a greater variety of alterations in one strand. New possibilities of testing whether or not homologous ribonucleic acid has sequences in common with DNA can now be explored. The formation of molecular hybrids between closely related organisms should be a useful tool for plotting homologies in base sequences where no genetic exchanges have yet been demonstrated.

Summary.—When solutions of bacterial DNA are denatured by heating and then cooled, two different molecular states can be obtained in essentially pure form depending on the choice of conditions, that is, rate of cooling, DNA concentration, and ionic strength. One state corresponding to fast cooling consists of single stranded DNA having about half the molecular weight of the original DNA. The other state corresponding to slow cooling consists of recombined strands united by complementary base pairing over most of their length. This form has as much as 50 per cent of its original transforming activity and is called renatured. The quickly cooled, single stranded form is essentially inactive and is called denatured. These two forms are clearly identified by differences in (1) absorbance-temperature curves, (2) density, (3) appearance in electron micrographs and (4) hydrodynamic properties. The recombination depends on concentration in the manner expected for an equilibrium between two independent strands and a bimolecular complex.

Molecular weight determinations of the native and renatured forms have been based on an extension of an earlier calibration of sedimentation and intrinsic viscosity in terms of light scattering measurements. For the denatured form molecular weights were determined by the use of the Mandelkern-Flory relation. By these means thermal degradation was accurately assessed. To avoid aggregation it was necessary to substantially lower the cation concentration over that previously used. In this way denatured DNA has been characterized as a single-chain, unaggregated polymer and strand separation demonstrated.

Density gradient experiments on N^{14} and N^{15} *E. coli* DNA have shown the the existence of hybrids in the DNA renatured from the mixture. Similarly, hybrids have been shown to form between the strands of bacteria that are closely related genetically. Thus the possibility of forming by renaturation heterozygous DNA molecules with different genetic markers or chemical modifications in the two strands seems assured.

The authors would like to thank Miss D. Lane for valuable assistance and Professor A. Peterlin, Professor J. D. Watson, and Dr. N. Sueoka for helpful advice and discussions.   We are very grateful to Professor C. E. Hall of the Massachusetts Institute of Technology for the electron micrographs and to Drs. S. E. Luria and M. Demerec for supplying us with strains of *Shigella dysenteriae* and *Salmonella typhimurium* (LT-2).   This work was supported by a grant from the United States Public Health Service (C-2170).

* Predoctoral Fellow of the National Science Foundation.

[1] Marmur, J., and D. Lane, these Proceedings, **46**, 453 (1960).

[2] Marmur, J., and P. Doty, *Nature*, **183**, 1427 (1959).

[3] Doty, P., H. Boedtker, J. R. Fresco, R. Haselkorn, and M. Litt, these Proceedings, **45**, 482 (1959).

[4] Meselson, M., F. W. Stahl, and J. Vinograd, these Proceedings, **43**, 581 (1957).

[5] Sueoka, N., these Proceedings, **45**, 1480 (1959).

[6] Sueoka, N., J. Marmur, and P. Doty, *Nature*, **183**, 1429 (1959).

[7] Rolfe, R., and M. Meselson, these Proceedings, **45**, 1039 (1959).

[8] Doty, P., J. Marmur, and N. Sueoka, *Brookhaven Symposia in Biology*, **12**, 1 (1959).

[9] Hall, C. E., and M. Litt, *J. Biophys. and Biochem. Cytology*, **4**, 1 (1958).

[10] Hall, C. E., and P. Doty, *J. Am. Chem. Soc.*, **80**, 1269 (1958).

[11] Hall, C. E., and H. Boedtker (unpublished results).

[12] Wright, G., senior thesis, Harvard University, 1960.

[13] Doty, P., and B. H. Bunce, *J. Am. Chem. Soc.*, **74**, 5029 (1952).

[14] Reichmann, M. E., S. A. Rice, C. A. Thomas, and P. Doty, *J. Am. Chem. Soc.*, **76**, 3407 (1954).

[15] Butler, J. A. V., D. J. R. Laurence, A. B. Robins, and K. V. Shooter, *Proc. Roy. Soc.*, **A250**, 1 (1959).

[16] Eigner, J., doctoral thesis, Harvard University, 1960.

[17] Mandelkern, L., H. A. Sheraga, W. R. Krigbaum, and P. J. Flory, *J. Chem. Phys.*, **20**, 1392 (1952).

[18] Doty, P., B. B. McGill, and S. A. Rice, these Proceedings, **44**, 432 (1958).

[19] Kawade, Y., and I. Watanabe, *Biochim. Biophys. Acta*, **19**, 513 (1956).

[20] Iso, K., and Watanabe, I., *J. Chem. Soc. Japan, Pure Chem. Section*, **78**, 1268 (1957), in Japanese.

[21] Rice, S. A., and P. Doty, *J. Am. Chem. Soc.*, **79**, 3937 (1957).

[22] Alexander, P., and K. A. Stacey, *Biochem. J.*, **60**, 194 (1955).

[23] Bacher, J. E., and W. Kauzmann, *J. Amer. Chem. Soc.*, **74**, 3779 (1952).

[24] Litt, M., J. Marmur, H. Ephrussi-Taylor, and P. Doty, these Proceedings, **44**, 144 (1958).

[25] Baron, L. S., W. F. Carey, and W. M. Spilman, these Proceedings, **45**, 1752 (1959).

[26] Lee, K. Y., R. Wahl, and E. Barbu, *Ann. Inst. Pasteur*, **91**, 212 (1956).

[27] Delbrück, M., and G. S. Stent, in *The Chemical Basis of Heredity*, ed. W. D. McElroy and B. Glass (Baltimore: Johns Hopkins Press, 1957), p. 699.

[28] Watson, J. D., and F. H. C. Crick, *Nature*, **171**, 964 (1953).

[29] Levinthal, C., and H. R. Crane, these Proceedings, **42**, 436 (1956).

[30] Longuet-Higgins, H. C., and B. H. Zimm, *J. Mol. Biol.* (in press).

From *Journal of Molecular Biology*, Vol. 3, 595–617, 1961. Reproduced with permission of the authors and publisher.

The Formation of Hybrid DNA Molecules and their use in Studies of DNA Homologies

Carl L. Schildkraut†, Julius Marmur‡ and Paul Doty

Department of Chemistry, Harvard University, Cambridge 38, Massachusetts, U.S.A.

(*Received 29 March 1961*)

Heavy-isotope-labeled DNA has been used to study strand separation and recombination. The rate at which ^{14}N–^{15}N biologically half-labeled DNA can be made to separate into subunits corresponds very closely to what has been predicted for the rate of unwinding of the strands of a double helix.

The use of a phosphodiesterase from *E. coli* (Lehman, 1960) which selectively attacks single-stranded DNA has made it possible to remove unmatched single chain ends from renatured DNA and reduce the remaining differences between renatured and native DNA.

A mixture of heavy-isotope-labeled and normal bacterial DNA was taken through a heating and annealing cycle, treated with the phosphodiesterase and examined by cesium chloride density-gradient centrifugation. Three bands were observed, corresponding to heavy renatured, hybrid, and light renatured DNA. As would be expected for random pairing of complementary strands, the amount of the hybrid was double that of either the heavy or the light component. It has thus been demonstrated that the strands which unite in renaturation are not the same strands that were united in the native DNA but instead are complementary strands originating in different bacterial cells.

The formation of hybrids has been possible only where the heavy and normal DNA samples have a similar overall base composition. It has also been shown for DNA samples isolated from bacteria of different genera in a case where genetic exchange by conjugation has been demonstrated. The evidence for the parallelism between genetic compatibility and the formation of DNA hybrids *in vitro* has led to the proposal that organisms yielding DNA which forms hybrid molecules are genetically and taxonomically related.

1. Introduction

It has been shown recently that it is possible to separate the two strands of DNA molecules and then to reunite these strands so as to restore to a large degree the original helical structure and biological activity (Marmur & Lane, 1960; Doty, Marmur, Eigner & Schildkraut, 1960). The optimum conditions for bringing about maximum restoration, that is, renaturation, have been presented in the preceding paper (Marmur & Doty, 1961).

The work presented in this paper is aimed at the detailed examination of the process of strand separation and recombination, the formation of hybrid DNA molecules with

---

† Present address: Department of Biochemistry, Stanford University, Palo Alto, California.

‡ Present address: Graduate Department of Biochemistry, Brandeis University, Waltham, Massachusetts.

each of the strands coming from a different source, and the use of such hybrids in detecting overlap in nucleotide sequences in the two samples. If complementarity is the condition for maximum renaturation then it is proper to enquire to what extent deviations from complementarity are required to interfere with the renaturation. The most modest deviation is probably that which occurs between a point mutant and wild-type DNA. In a following paper (Marmur, Lane & Doty, 1961) this kind of recombination is studied with bacterial transformation as the criterion and it is concluded that such small differences do not interfere with renaturation.

The method of study in the present report was density-gradient centrifugation (Meselson, Stahl & Vinograd, 1957) coupled with the use of heavy-isotope-labeled DNA. This proves to be the best technique for following the interaction between homologous and heterologous DNA. Native DNA samples of different GC (guanine-cytosine) contents can be observed separately when banded in the density gradient (Sueoka, Marmur & Doty, 1959; Rolfe & Meselson, 1959). Two native samples having the same GC content can also be observed and their interaction studied if one of them is labeled with heavy isotopes such as ^{15}N, ^{13}C and deuterium (Meselson & Stahl, 1958; Davern & Meselson, 1960; Marmur & Schildkraut, 1961a).

A double helix resulting from the union of a heavy-labeled and non-labeled strand will have an intermediate density. It can be clearly distinguished from the original molecules if they are separated by a sufficient distance in the density gradient. The amount of hybrid molecules formed between different pairs of DNA samples expected to be homologous should vary with the degree of homology.

In order to examine the possibilities of renaturation occurring between DNA from different species that may be genetically related, more quantitative work was necessary to obtain a proper base line for comparison. The removal of non-renatured regions by means of a new phosphodiesterase, specific to single-stranded DNA (Lehman, 1960) provided a means of eliminating artifacts and improving resolution so that hybrid DNA molecules could be clearly and quantitatively resolved. As a result it was possible to examine in some detail the renaturation that was possible between pairs of DNA samples from different sources. This appears to offer a new means of estimating the extent of overlap in DNA sequence between DNA samples from genetically related species.

2. Materials and Methods

(a) *Bacterial strains*. The organisms used in this study together with the base composition and buoyant density of their DNA are listed in Table 1.

(b) *Preparation of DNA*. The isolation procedure of Marmur (1961) was used to extract DNA from cells grown in the exponential phase in Difco brain heart infusion medium. When heavy-isotope-labeled DNA was required the cells were grown in a synthetic medium containing $^{15}NH_4Cl$ as the only nitrogen source and D_2O of greater than 99% purity (Marmur & Schildkraut, 1961a).

(c) *Solvents*. In order to eliminate spurious effects due to divalent metal ion contamination, and to inhibit possible attack by nucleases, a chelating agent was always present in the DNA solutions. Citrate ions have proved most convenient in this respect and the standard saline solution (0·15 M-NaCl was used most often) contained 0·015 M Na-citrate. This solvent, standard saline-citrate, will be designated by the abbreviation SSC. Reference will also be made to various multiples of SSC. For example, 2 × SSC is 0·30 M-NaCl and 0·030 M-Na citrate.

(d) *Heating and Annealing*. Solutions containing equal amounts (by weight) of heavy-isotope-labeled and normal DNA in 1·9 × SSC were prepared. In the usual preparation of renatured DNA, 1 ml. was placed in a 2 ml. glass stoppered volumetric flask, immersed in a bath of boiling water for 10 min and immediately transferred to a bath thermostated at 68°C. After 2 hr the temperature was lowered in approximately 5°C steps at intervals of 15 min until 25°C was reached. Although all results reported here were obtained by this cooling procedure, it has been found that the salt concentration and the rate of cooling from 68°C to room temperature can be varied somewhat without noticeable change in the final results.

TABLE 1

Bacteria from which DNA was isolated for use in studies of hybrid formation

Species	Strain	Source	%GC†	Density (g/cm³)
Bacillus brevis	9999	ATCC‡		1·704
Bacillus macerans	7069	ATCC		1·713
Bacillus megaterium		Univ. of Penn.	38	1·697
Bacillus natto	MB-275	A Demain		1·703
Bacillus subtilis	168	Yale University	42	1·703
Clostridium perfringens	87b	M. Mandel	31	1·691
Diplococcus pneumoniae	R-36A	R. Hotchkiss	39	1·701
Erwinia carotovora	8061	ATCC	54	1·709
Escherichia coli	B	S. Luria	50	1·710
Escherichia coli	44 B	M. Mandel		1·710
Escherichia coli	C-600	R. Appleyard		
Escherichia coli	K12 (W678)	J. Lederberg	50	1·710
Escherichia coli	TAU⁻	S. Cohen	50	1·710
Escherichia coli	W-3110	M. Yarmolinsky		1·710
Escherichia coli	I	A. N. Belozersky	52	
Escherichia coli	II-IV-4	A. N. Belozersky	67	
Escherichia freundii	17	H. Blechman		1·710
Escherichia freundii	5610-52	M. Mandel		
Salmonella arizona	PC145	Walter Reed Hosp.		1·712
Salmonella ballerup	ETS107	Walter Reed Hosp.		
Salmonella typhimurium	LT-2	M. Demerec	50, 54	1·712
Salmonella typhimurium	ETS9	Walter Reed Hosp.		
Salmonella typhosa	643	Walter Reed Hosp.	53	1·711
Shigella dysenteriae	15	S. Luria	53	1·710

† The GC contents of the DNA samples were all obtained from Belozersky & Spirin (1960).
‡ American Type Culture Collection.

(e) *CsCl.* Optical grade CsCl was obtained from the Maywood Chemical Co., Maywood, N.J. A concentrated stock solution was prepared by dissolving 130 g CsCl in 70 ml. of 0·02 M-tris buffer (2-amino-2-hydroxymethylpropane-1:3-diol) pH 8·5. The final solution was passed through a medium grade sintered glass filter to remove large amounts of solid material that seemed to contaminate the solid CsCl. When solutions in which the DNA had been annealed at low concentrations had to be brought up to the correct density for centrifugation, the solid CsCl was added directly. These solutions were not subsequently filtered.

(f) *Density-gradient centrifugation.* The technique described by Meselson et al. (1957) was followed. CsCl was used to bring the density of the DNA solution being examined to values between 1·71 g/cm³ and 1·75 g/cm³, depending on the specific sample involved. This could be done most easily by mixing the DNA solution with the concentrated CsCl stock solution. If the DNA solution was not concentrated enough, as was generally the case, this procedure would diminish the final concentration of the DNA below that required for accurate observation. To avoid this problem it was possible to use solid CsCl to give the proper density. To 1·03 g

CsCl was added 0·80 ml. of the slow cooled DNA solution (still in the same saline citrate solution) and 0·01 ml. of a stock solution of 50 μg/ml. of a DNA sample whose density is well established and can be used as a standard. DNA from *Cl. perfringens* is generally used since its density is the lowest observed thus far. This places it in a position where its band cannot coincide with any heavy-isotope-labeled DNA samples. The final adjustment of density was made by the addition of a small amount of water or solid CsCl. The measurement of the density at various stages of the adjustment is facilitated by the use of the linear relation between refractive index and density (Meselson, 1958, personal communication; Ifft, Voet & Vinograd, 1961)

$$\rho^{25\cdot0°C} = 10\cdot8601\, n_{D}^{25\cdot0°C} - 13\cdot4974.$$

Approximately 0·75 ml. of the final CsCl solution was placed in a cell containing a plastic (Kel-F) centerpiece and centrifuged in a Spinco model E analytical ultracentrifuge at 44,770 rev/min at 25°C. After 20 hr of centrifugation, ultraviolet absorption photographs were taken on Kodak commercial film. It was clear that equilibrium had been closely approached in 20 hr since there was no difference in calculated densities when photographs taken after 48 hr were used. Moreover, the variances did not decrease by more than 5% between 20 and 48 hr. Tracings were made with a Joyce-Loebl double-beam recording microdensitometer with an effective slit width of 50 microns in the film dimension.

Densities were calculated by using the position of the standard DNA as a reference. The CsCl density gradient was obtained from the data of Ifft *et al.* (1961).

(g) E. Coli *phosphodiesterase*. This enzyme, which preferentially hydrolyses single-stranded DNA, was kindly supplied by Dr. L. Grossman and had been prepared in highly purified form by the method of Lehman (1961, personal communication). The enzymatic hydrolysis was carried out by the following modification (L. Grossman, 1961, personal communication) of the method of Lehman (1960). The annealed sample, usually at a total DNA concentration of 10 μg/ml., was dialysed against two changes of 0·067 M-glycine buffer, pH 9·2. To 0·7 ml. of the dialysed DNA solution was added 1·6 μmoles of $MgCl_2$, 2·4 μmoles of 2-mercaptoethanol, and approximately 25 to 75 units of crystallized *E. coli* phosphodiesterase. The final mixture (about 0·8 ml.) was incubated at 37°C for 3 hr. Solid CsCl and the standard DNA were then added and the solution was made up to the proper density.

3. Strand Separation

In order to form hybrid molecules by renaturation it is first necessary to examine in some detail the conditions necessary for strand separation. A very useful material for these studies has been the ^{14}N-^{15}N-half-labeled or biological "hybrid" DNA molecules first described by Meselson & Stahl (1958). DNA having one strand ^{15}N-labeled and the other strand normal was isolated from *E. coli* B according to the method of Marmur (1961). It was found that the biologically formed "hybrid" DNA banded at a density of 1·717 g/cm³ in the CsCl gradient, exactly between that of fully ^{15}N-labeled and normal DNA from *E. coli*. The band profile is shown in the top tracing of Fig. 1.

(a) *Methods for inducing strand separation*

In terms of the Watson-Crick model (1953) for DNA, the dissociation of the two strands obviously requires first the rupture of the hydrogen bonds uniting the base pairs and second the uncoiling and diffusing apart of the two strands. The rupture of the hydrogen bonds has been shown to occur in a relatively narrow temperature range at elevated temperatures but the diffusion apart of the strands did not automatically follow (Rice & Doty, 1957). Later it was found that lower ionic strengths and mcre dilute solutions did permit the diffusing apart to occur (Doty *et al.*, 1960;

Eigner & Doty, 1961) and prevented the non-specific re-association from occurring at lower temperatures where numerous hydrogen bonds re-form.

Meanwhile it had been found by Meselson & Stahl (1958) that the DNAs containing ^{14}N and ^{15}N subunits, each of which replicate in a conservative manner, produced two bands on heating in CsCl for 30 min. Thus they demonstrated that the two molecular subunits dissociated on heating. Meselson & Stahl left the question of the molecular structure of the subunits open to further investigation. This prompted us to isolate DNA from cells grown according to the procedure described by Meselson & Stahl (1958), and to see how the hybrid DNA reacted to the various methods that produce strand separation. The evidence that these methods do produce strand separation can be found in the papers to which we will refer. We are concerned here only with their effect on the biological hybrids and the detailed investigation of this effect in order to learn more about strand separation.

First we sought to reproduce the Meselson-Stahl results. They had observed the two-band pattern after heating a cell lysate in CsCl for 30 min. We found a similar pattern was produced by heating a sample of our purified biological hybrid at 100 °C for only 10 min at a concentration of 20 μg/ml. in SSC. The tracings of the bands before and after heat treatment can be seen at the top and bottom frames of Fig. 1. The apparent density of the native, hybrid DNA is 1·717 g/cm³; that of the two denatured bands is greater by 0·007 and 0·023 g/cm³ respectively. The average of these is, of course, greater than that of the native hybrid DNA by the 0·015 g/cm³ characteristic of denatured DNA. As a check, it was demonstrated that the same two-band pattern results when ^{14}N and ^{15}N DNA samples are separately denatured and then mixed.

Three other methods which have recently been shown to bring about strand separation have also caused the disappearance of the native hybrid band and the formation of two heavier bands. This additional information, combined with the demonstration of Doty et al. (1960) that the above heating procedure produces strand separation in bacterial DNA samples, now leads us to conclude that Meselson & Stahl were indeed observing the separation of single strands. We shall return to this point in the discussion, and now proceed to summarize the experimental observations on the biological hybrids.

By the addition of 8 M-urea the temperature for thermal denaturation can be reduced by nearly 20 °C (Rice & Doty, 1957). Consequently, by using 0·01 M-salt and 8 M-urea it was found possible to bring about strand separation by heating DNA only to 65 °C (Eigner & Doty, 1961). When this procedure was applied to the biological hybrid DNA, two bands formed in the CsCl density gradient, exactly as was the case after heating at 100 °C for 30 min in CsCl or for 10 min in SSC.

Formamide, being a more potent hydrogen bonding agent, and being miscible with water in all proportions, offers a substantial improvement over urea. Marmur & Ts'o (1961) have shown that in 95% formamide strand separation readily occurs at room temperature. Again, when this is applied to the biological hybrids the typical two-band pattern is produced.

Acid and base titration, by virtue of their ability to disrupt hydrogen bonding, offer another route to strand separation. It has now been shown (Cox, Marmur & Doty, 1961) that strand separation occurs when the pH is lowered to 2·5 or raised to 12·0. These two pH values lie just outside the region in which the hypochromic shift (at 260 mμ) indicates that the helix-coil transition is complete. At pH values on the other side of the hypochromic shift, strand separation has not occurred. When the biological hybrids are exposed to conditions in which the pH is below 2·5 or above 12·0, two bands again result in the CsCl gradient.

Thus it is seen that several different procedures lead to the formation of the same two-band pattern in the CsCl density gradient, and that all of them are a consequence of the prior breaking of hydrogen bonds uniting the two DNA strands. While the present work is not aimed at establishing how many subunits make up DNA molecules, it does appear that the extension to the biological hybrid DNA of the consistent finding that strand separation occurs whenever hydrogen bonds are cooperatively broken, shows that only one type of bonding is uniting the subunits. Since this can only be identified with that assumed in the Watson-Crick structure, there seems to be no alternative to equating the subunits, observed as two bands, with the two strands in the native DNA molecule.

(b) Temperature dependence of strand separation

With the demonstration that the two-band pattern of the biological hybrid DNA is indicative of strand separation we now attempt to use this material to obtain further details about the process of strand separation. The studies of Marmur & Doty (1959) demonstrated that it should be possible to melt out the molecules having a high AT (adenine-thymine) content leaving GC (guanine-cytosine)-rich molecules undenatured. This can be observed to occur with the biological hybrids by exposing them to temperatures only slightly above the temperature, T_m, of the midpoint of the absorbance rise. T_m is the same for all E. coli DNA samples, including the biological hybrids (Marmur & Doty, 1961). The band profile obtained after heating the latter at T_m for 20 min is shown as the second tracing in Fig. 1. Since there is no difference from the profile of the native sample, it is concluded that all the molecules remain in the helical form. To melt selectively the AT-rich molecules, it is necessary to expose them to temperatures a few degrees above T_m. The results of doing this are shown as the third tracing of Fig. 1, where it is seen that approximately 25% of the molecules undergo strand separation. The molecules showing the greater resistance to heat denaturation are seen to have a higher density, in agreement with the hypothesis that they should be GC-rich.

(c) Kinetics of strand separation

Since a small but detectable thermal depolymerization always accompanies strand separation it is desirable to know the minimum time at the elevated temperature required to obtain complete separation, and thus avoid this undesirable side effect. In addition, since a few theoretical calculations of the rate of uncoiling of the DNA molecule have been reported (Kuhn, 1957; Longuet-Higgins & Zimm, 1960), some comparison with experimental findings seems to be in order. The rate of separation into subunits can be observed experimentally by following the disappearance of the hybrid band or the appearance of the two heavier bands. The results presented in Plate I and Fig. 2 show that the complete disappearance of material of density 1·717 g/cm³ takes slightly over 60 sec at 100 °C. This is very close indeed to the time calculated by Kuhn (1957) for the unwinding of a double helix of comparable length in which no bonds are holding the strands together. As is evident from the absorbance-temperature profile, the rise in absorbance is complete at 95 °C. When the temperature of a DNA solution is raised to 100°C few hydrogen bonds should exist between strands and Kuhn's model seems applicable to such a molecule.

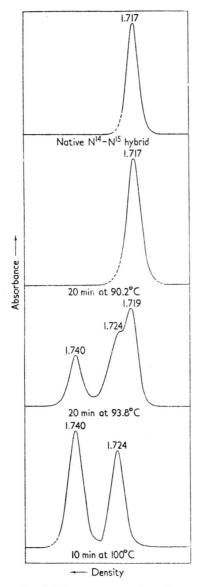

FIG. 1. The temperature dependence of strand separation. Microdensitometer tracings of ultraviolet absorption photographs of samples equilibrated in a CsCl density gradient. Each sample was heated in SSC at a concentration of 20 μg/ml.

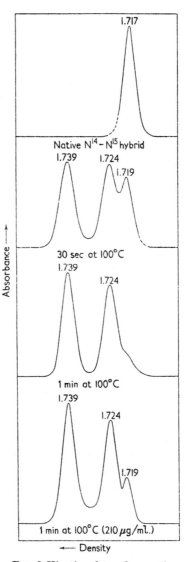

FIG. 2. Kinetics of strand separation. The results shown in the second and third tracings were obtained by heating at a concentration of 20 μg/ml. in SSC.

(d) *Dependence of the time of strand separation on viscosity*

In the theoretical considerations just mentioned in connection with the kinetics of strand separation, the variation of the rate of unwinding with the viscosity of the medium is also considered (Longuet-Higgins & Zimm, 1960). An equation is derived which predicts that strand separation will take longer as the viscosity of the medium is increased. In the tracing shown at the bottom of Fig. 2 the time necessary for separation into subunits is shown to have increased with increased DNA concentration. Similar results have also been obtained by using concentrated sucrose solutions to produce a medium of higher viscosity (Schildkraut, Wierzchowski & Doty, 1961).

(e) *DNA from bacteriophage ΦX174* .

The results of the studies with the biological hybrids indicate that the most convenient way to obtain single-stranded DNA is to heat for 10 min in SSC at a concentration of from 10 to 20 μg/ml. and quickly cool. Since one kind of DNA, that from bacteriophage ΦX174[†], is known to be single-stranded (Sinsheimer, 1959), it is expected that our thermal treatment would produce no density change in this material. This has, indeed, been found to be the case.

4. Renaturation

The general features of the specific recombination of single DNA strands to form the native helical structure have been described in the first publications (Marmur & Lane, 1960; Doty *et al.*, 1960), and methods of optimizing conditions to ensure the maximum renaturation were described in the preceding paper (Marmur & Doty, 1961). Here we have only one point to add to this: it has to do with the reduction or elimination of the remaining differences between renatured and native DNA. In the earlier work it was evident that renaturation was not complete. For example, the density had returned only about 75% of the way from the denatured to the native density. It was thought that this was due to the inequality in length of the recombined strands. DNA molecules are apt to be broken during isolation and the single strands suffer some hydrolysis during the thermal treatment. Using the measurements on rate of bond scission by Eigner, Boedtker & Michaels (1961), we would estimate that strands originally in molecules of 10,000,000 molecular weight would undergo about three scissions each as a result of a typical heat treatment: 10 min at 100°C and 100 min at 68°C. This would mean that two strands meeting in a complementary region may differ in length by as much as one quarter to one half, on the average. If they renatured completely, there would still remain 20 to 33% of the weight unrenatured and the incomplete recovery of the characteristics of the native DNA would be explained.

In the course of electron microscope studies with Professor C. E. Hall we frequently saw renatured molecules with circular regions at one or both ends that would be consistent with the protrusion of a single chain end beyond the re-formed helical region just described. One electron micrograph showing an unusually large concentration of these in a single photograph is shown in Plate II. While this interpretation lacks proof, the observation is nevertheless strikingly similar to what was expected from the point of view outlined.

From this evidence it appeared that in order to improve the extent of renaturation the unmatched single chain ends would have to be removed. The possibility of doing this suddenly became available with the discovery by Lehman (1960) of a phosphodiesterase from *E. coli* that selectively attacked single-stranded DNA. This was tested

† This material was kindly provided by Professor Robert L. Sinsheimer.

on renatured DNA from *B. subtilis* as shown in Fig. 3. The denatured DNA showed the characteristic increase in density. Upon renaturation at a concentration low enough to ensure that some denatured DNA remained, the result shown in the third frame was obtained. The density of the renatured DNA is, as expected, 0·004 g/cm³ higher than that of the native DNA. After treatment of this sample with the phosphodiesterase it is seen, in the bottom frame, that the denatured shoulder has been

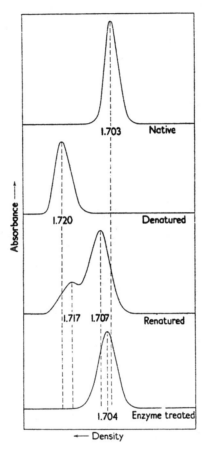

FIG. 3. Renaturation of *B. subtilis* DNA. When a solution of native DNA (band profile shown in top tracing) is heated for 10 min at 10 μg/ml. at 100°C in 1·9 X SSC and quickly cooled, the density increases 0·017 g/cm³ (second tracing). A portion of this solution was annealed and, as is evident from the third tracing, about 80% of the DNA renatures. This concentration (10 μg/ml.) was chosen so that some denatured material would still be present. Treatment with the *E. coli* phosphodiesterase causes the complete disappearance of the denatured band and a decrease in the buoyant density of the renatured band (bottom tracing).

completely removed and the density of the renatured band has been reduced to within 0·001 g/cm³ of the native DNA density. Thus the possibility of removing the denatured unmatched ends from renatured molecules has been demonstrated. As will be seen, this procedure greatly improves the resolution of bands encountered in the study of artificially produced hybrid molecules.

It may be of interest to mention at this point our failure thus far to produce renaturation of DNA from bacterial sources without thermal treatment of the type described here and in the preceding papers. We have employed urea and moderate temperatures, formamide, and low pH to produce strand separation; and then, by gradual withdrawal of the hydrogen-bond breaking agent attempted to renature the strands. Such attempts have thus far failed. While our experiments have not been exhaustive, it appears quite possible that renaturation cannot be achieved in this way. If so, a likely explanation would be that at room temperature the decreased Brownian motion of the chain segments is no longer sufficient to provide the mobility required for the exploration necessary to create nuclei, that is, for complementary regions to find each other rather than become frozen in mismatched pairings.

In connection with renaturation, we might return just briefly to the case of ΦX DNA and mention that the density does not decrease but remains exactly 1·723 g/cm³ after heating and annealing. This is as should be expected for single-stranded material whose complementary strands are not present during the annealing process.

5. Hybrid DNA Molecules produced by Renaturation

We now turn to the problem of demonstrating that the strands which unite in renaturation are not the same strands that were united in the native DNA but instead are complementary strands originating in different cells. This requires that renaturation be studied in a solution of two homologous DNA samples, one of which carries a distinctive label. With density gradient ultracentrifugation offering such good resolution it was natural to turn to the introduction of heavy atoms or to heavy isotope substitution. If the pairing of DNA strands is restrained only by the condition of complementarity one would expect that renaturation would lead to three bands, one heavy, one light and one intermediate, corresponding in density to a hybrid composed of one normal and one heavy strand. Moreover, the amount of the intermediate would be expected to be double that of either the heavy or light component provided that equal amounts of the two DNA samples had been mixed originally.

If this situation is found, then it will be of interest to attempt to form hybrid molecules from other pairs of DNA samples in which some differences exist. In order to have sufficient resolution to exploit this means of searching for such effects it is necessary to have a considerable density difference between the two samples. Practical considerations indicate that it should be at least 0·040 g/cm³. Thus the use of ¹⁵N label in one sample, such as first used by Meselson & Stahl (1958), is not sufficient since this gives a separation of only about 0·015 g/cm³.

(a) *Preliminary experiments*

Our first attempts to produce DNA with substantially higher density were with 5-bromouracil substitution for thymine inasmuch as this had been accomplished in both bacteria and bacteriophage (Dunn & Smith, 1954; Zamenhof & Griboff, 1954). A sample of such DNA from *B. subtilis* (Ephrati-Elizur & Zamenhof, 1959) was kindly supplied to us by Professor W. Szybalski. Unfortunately, when this sample was carried through the heating and annealing cycle it displayed a quite broad band with little indication of renaturation. The molecular weight had evidently been substantially reduced either by enzymatic attack during isolation or as a result of the greater heat sensitivity of this substituted DNA.

Nevertheless this material was mixed with high molecular weight, normal *B. subtilis* DNA, and put through the heating and annealing cycle. The banding of the mixture before and after heating is shown in Fig. 4. It is seen that the thermal treatment has produced a broad band without any resolution of the expected hybrid. Indeed the result is such as to indicate that the smaller heavy-labeled chains have been bound in a random manner to the larger, normal density chains producing a complete spectrum of densities which hides what little renaturation may have occurred.

Initial attempts to study 5-bromodeoxyuridine-labeled DNA from T4r⁺ bacteriophage led to similar disappointing results.

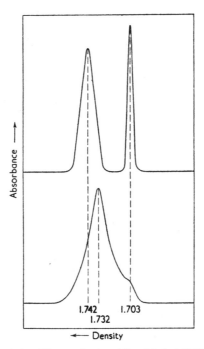

FIG. 4. Hybrid formation using 5-bromodeoxyuridine-labeled DNA. The upper tracing shows native normal and 5-bromodeoxyuridine-labeled DNA isolated from *B. subtilis*. The properties of the latter DNA have been described in detail elsewhere (Szybalski, Opara-Kubinska, Lorkiewicz, Ephrati-Elizur & Zamenhof, 1960). Portions of these two samples were heated together in 1·9 × SSC at a concentration of 20 μg/ml. each for 5 min at 100°C. The solution was then allowed to cool to room temperature over a period of about 5 hr and CsCl was added to obtain the proper density. The lower tracing suggests that the expected renatured species are present in solution, but the broadening of bands due to degradation and aggregation obscures the results.

(b) ¹⁵N-*deuterated DNA*

Under these circumstances we turned to a combination of ¹⁵N and deuterium to provide the density increase. After considerable manipulation this combined labeling has been shown to be successful in producing DNA with a density increase of about 0·040 g/cm³ and the details have now been published elsewhere (Marmur & Schildkraut, 1961a).

We were, therefore, in a position to proceed with the experiment that would demonstrate whether or not hybrid DNA molecules form from random mating of complementary strands. The steps are shown for *B. subtilis* DNA in Fig. 5. In the first two frames the density profiles of the normal and labeled native DNA samples are shown. These were then mixed so that the concentration of each was 5 μg/ml. in 1·9 × SSC, and the heating and annealing cycle was carried out. At this concentration only partial renaturation is expected. So low a concentration was chosen in order to minimize the association of renatured DNA molecules since this would lead to an un-necessary smearing of the pattern. Consequently, at this concentration, one would expect to find 5 species of different density: heavy denatured DNA, heavy renatured DNA, hybrid DNA, light denatured DNA and light renatured DNA. The result, seen in the third frame of Fig. 5, is of precisely this character.

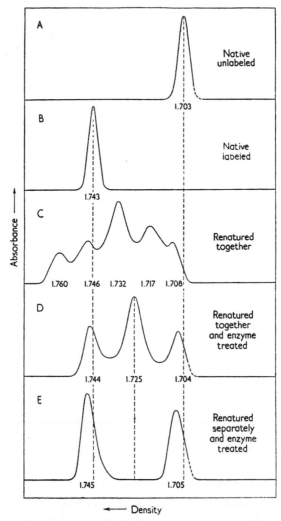

Fig. 5. The effect of *E. coli* phosphodiesterase on a heated and annealed mixture of heavy-labeled and normal *B. subtilis* DNA.

(A) and (B) Native samples.

(C) Mixed so that the final concentration of each was 5 μg/ml. and then heated and annealed.

(D) The sample shown in (C) was dialysed against 0·067 M-glycine buffer and incubated with the *E. coli* phosphodiesterase.

(E) Heated and annealed separately, treated with the phosphodiesterase and then mixed.

The application of the phosphodiesterase previously described would be expected to remove the first and fourth bands as well as shift the density of the remaining species back toward the native values. The band profile after this treatment is shown in the fourth frame and it is seen that the expected changes had been brought about.

As a control, the two native DNA samples were carried through the identical operations described above and then mixed and banded. The result is shown in the last frame where it is seen that the densities match within 0·001 g/cm³ that of the outer bands of the frame above and that no hybrid is evident.

Similar experiments were carried out on *E. coli* DNA: the band profiles corresponding to frames C and D of Fig. 5 are shown in Fig. 6. The result is seen to be the same.

It is to be noted that the area under the hybrid band in these two experiments is about equal in area to that of the two outer bands.† Thus the conclusion can be drawn that the earlier and much more primitive experiment along these lines (Doty *et al.*, 1960) has been substantiated and the formation of hybrid DNA molecules by random pairing of complementary strands can be taken as proved.

It is useful to note in passing that the phosphodiesterase treatment had one other effect not evident in the band profiles. Without enzyme treatment part of the hybrid band would begin to form much earlier than the other, indicating high molecular weight material suggestive of some aggregation despite our efforts to avoid it. However, after enzyme treatment all three bands appear at the same rate. Thus, the enzymatic treatment not only improves the resolution by removing denatured material but appears to break up aggregated renatured DNA molecules as well.

(c) *Further investigation of 5-bromodeoxyuridine-substituted DNA*

With the successful conclusion of the above work, we have returned again to the 5-bromodeoxyuridine-labeled bacteriophage DNA. With the guide that the above work provides it has been possible to overcome some of the problems that appear to be associated with the greater inherent instability of this material. As a consequence results nearly as satisfactory as those shown in Fig. 5 and 6 are being obtained at present. This system has been used to study DNA homologies among the T-even bacteriophages (Schildkraut, Marmur, Wierzchowski, Green & Doty, 1961).

(d) *Kinetics of hybrid formation*

As a further check on the above interpretation and in order to show further details of the process of hybrid formation, samples were withdrawn from the annealing mixtures, held at 68°C, during renaturation. These were quickly cooled, in order to "freeze" the distribution present and band profiles were obtained in the ultracentrifuge. A number of profiles are shown in Fig. 7. From these, the progress of the renaturation can be followed and it is seen that it reaches about 50% completion in the first hour.

6. Hybrid DNA Formation among Various Strains of *E. coli*

It is to be expected from their close taxonomic, physiological and genetic relationships that all strains identified as *E. coli* should yield DNA which will form the 5-band pattern upon the renaturation discussed in the previous section. Hybrid formation in a number of pairs has been examined. The results are summarized in Plate III where the ultraviolet absorption photographs are reproduced. It is seen that the 5-band pattern is produced in all but the one case where sequence homology did not exist.

---

† The linearity of response of our optical system is not yet such as to justify a precise measurement of the area under these curves.

PLATE I. Ultraviolet absorption photographs of ^{14}N-^{15}N-labeled DNA showing different stages of the thermally induced separation into subunits. The band at the far right has been used as a standard and is DNA isolated from *D. pneumoniae*. The other band in the top photograph is the biologically formed hybrid DNA. The second photograph shows the stability of the hybrid to a 20 min exposure at 93·8°C. At 100°C the number of molecules separating increases rapidly with time of exposure as shown in the next 3 photographs. The samples were heated in SSC at 20 μg/ml. for 30 sec, 1 min and 10 min, respectively.

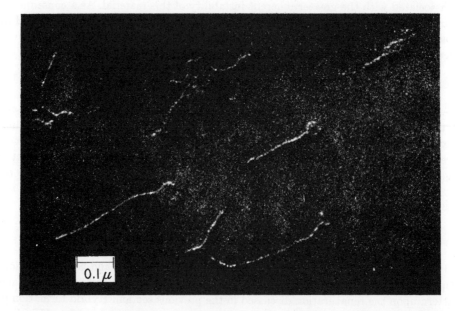

PLATE II. Electron micrograph of renatured *D. pneumoniae* DNA showing an unusually large concentration of renatured molecules with circular regions at one or both ends. Magnification × 100,000.

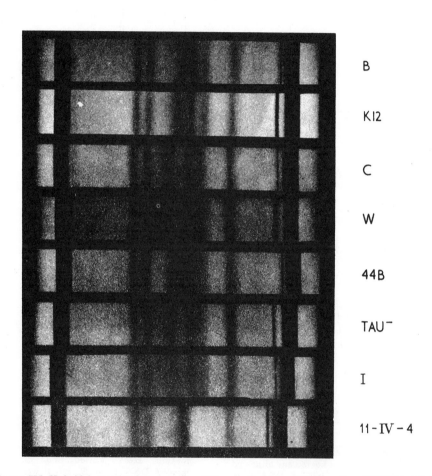

B

K12

C

W

44B

TAU⁻

I

11-IV-4

PLATE III. Hybrid formation between *E. coli* B and other *E. coli* strains. *E. coli* B DNA, labeled with ¹⁵N and deuterium, was mixed with DNA from each of the strains listed above and heated and annealed in separate experiments. The concentrations were 5 μg/ml. each, and the other conditions were as described in the section on methods. Each of 8 different ultracentrifuge runs is represented above by a typical ultraviolet absorption photograph. Six DNA bands appear in all but the last example. The photographs have been lined up according to the position of the standard band at the far right, which is DNA from *Cl. perfringens*.

This DNA isolated from *E. coli* 11-IV-4, an alkali-producing form, had a GC content of 67%, which is far from that characteristic of *E. coli* DNA. Belozersky (1957) has reported that these cells were produced by an alteration of the properties of *E. coli* CM caused by growing it together with Breslau bacteria No. 70 killed by heat.

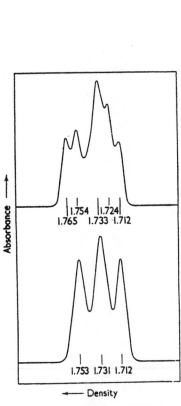

FIG. 6. The effect of *E. coli* phosphodiesterase on a heated and annealed mixture of heavy-labeled DNA from *E. coli* B and normal DNA from *E. coli* K12. The concentration of each sample was 5 µg/ml. each during the heating and annealing cycle.

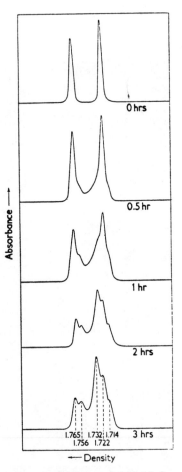

FIG. 7. Kinetics of hybrid formation. Heavy-labeled and unlabeled DNA from *E. coli* B were heated together at 100°C for 10 min under the usual conditions of 5 µg/ml. each and in 1·9 × SSC. The mixture was placed at 68°C and quickly cooled portions were centrifuged after the addition of CsCl. The tracings are shown above. These samples have, of course, not been treated with the phosphodiesterase.

The failure to obtain a hybrid band confirms the gross difference from *E. coli* DNA, but of course does not help to eliminate the argument that these special cells may have been a contaminant rather than a variant. As seen in the photograph the location of the peaks and their intensities is variable. For this reason three of the samples were treated with phosphodiesterase. When this was done the patterns gave three bands and showed greatly reduced differentiation. We are, therefore, now in a position to decide if there are any small reproducible differences in the renaturation of various *E. coli* pairs that may indicate small differences in sequence, but the effect is at most quite small. We can conclude that the homology is essentially complete.

7. The Aggregation of Renatured DNA

In exploring further the extent to which hybrids can form between somewhat different DNA strands we encountered more examples of the way in which renatured DNA molecules could aggregate so as to indicate hybrid formation even though it was indeed an artifact. While these difficulties were always removed by phosphodiesterase treatment it is of interest to record a few examples in order to indicate the nature of this process and to emphasize the need for phosphodiesterase treatment.

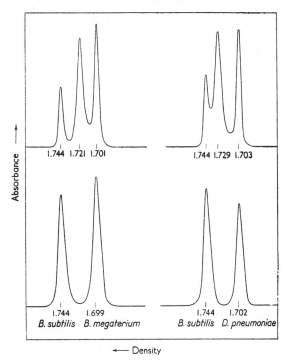

FIG. 8. Non-specific aggregation. ^{15}N-deuterated DNA from *B. subtilis* was heated with either DNA from *B. megaterium* or *D. pneumoniae* at a concentration of 50 μg/ml. each in 1·9 × SSC. The mixtures were annealed under the usual conditions. Portions were centrifuged in CsCl before (upper tracings) and after (lower tracings) treatment with the *E. coli* phosphodiesterase.

A number of experiments were carried out using the heavy-labeled *B. subtilis* DNA and various normal DNA samples at concentrations of about 50 μg/ml where no denatured strands remain after renaturation. (This is in contrast to the standard procedure which involves tenfold lower concentrations.) Results are shown in Fig. 8 for renatured mixtures of *B. subtilis* DNA with either *B. megaterium* or *D. pneumoniae* DNA. The band profiles before enzyme treatment are shown at the top. Here we see three well-resolved band patterns highly suggestive of hybrid formation. However, the enzyme treatment completely eliminates the central band and increases the amount of material in the outer bands. Thus the center band must have been the result of homogeneous renatured DNA molecules being held together by the association of unpaired, protruding chains. The rather remarkable feature is that the aggregation must consist mostly of two such renatured molecules, one of each type, since otherwise a whole range of densities would be displayed.

It was also possible to form aggregates between *B. subtilis* and *B. brevis* DNA by heating and annealing at a concentration of 40 μg/ml. each, and between *B. subtilis* and *B. macerans* DNA by heating and annealing at 50 μg/ml. each. The GC content of *B. subtilis* DNA is similar to that of *B. brevis* while that of *B. macerans* is somewhat higher. In this case the density of the aggregates was considerably higher than the average of the two renatured "parent" samples. This was true for practically every case of aggregation except two, one shown in Fig. 8 and, the other in Fig. 13. When *B. subtilis* and calf thymus DNA were heated together and annealed at a total concentration of 100 μg/ml., the heavy labeled *B. subtilis* DNA reformed completely while the calf thymus DNA maintained the denatured density. No intermediate band was visible.

As a final example we report the rather puzzling case of aggregation occurring upon mixing renatured solutions at room temperature. Heavy and light *B. subtilis* DNA at a concentration of 10 μg/ml. were heated and annealed separately; the band profiles are shown in the first 4 frames of Fig. 9. The two renatured solutions (C and D) were then mixed at room temperature and examined in the density gradient. The result is shown in Fig. 9 (E); it is seen that a central band has formed, apparently at the expense of the renatured molecules. Thus it would appear that unpaired chain ends which cannot be satisfied in the separated solutions can pair in the mixture. However, this artifact too was removed by phosphodiesterase treatment, as shown in the final frame of Fig. 9. It is important to note that this aggregation of renatured molecules at room temperature occurred between homologous molecules, not molecules from two different sources.

Thus the formation of bands of intermediate density that have the appearance of true hybrid DNA molecules has been clarified, and the importance of eliminating such false bands with phosphodiesterase treatment has been emphasized.

8. Interspecies Hybridization of DNA

We are now in a position to examine the possibility of forming DNA hybrids between strands of DNA coming from different bacterial species. By heating and annealing each of several DNA samples with the heavy-labeled *B. subtilis* DNA, it became evident, even without recourse to the enzyme treatment, that hybrid DNA molecules did not form. Such results are shown in Fig. 10. These and other results could be summarized by stating that hybrid DNA formation was not observed, even at double the usual DNA concentration, if the two DNA samples had significantly different base-compositions.

With the search for hybrid formation narrowed to DNA samples having essentially the same composition, it was natural to look first at DNA from two organisms of the same species which were known to be genetically related by virtue of their ability to undergo transformation with each other's DNA. This is true for *B. subtilis* and *B. natto* (Marmur, Seaman & Levine, 1961). The results for heating, annealing and enzyme treating this pair of DNA samples are shown in Fig. 11. It is seen that substantial

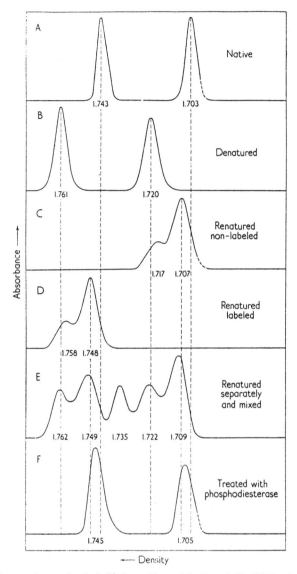

FIG. 9. Control experiment for hybrid formation with *B. subtilis* DNA. A stock solution in 1·9 × SSC of normal *B. subtilis* DNA was prepared so that the concentration was 10 µg/ml. The same was done for an ^{15}N-deuterated sample. After portions of these solutions received the following treatments, they were made up to the proper density with CsCl and centrifuged. The tracings are shown above.

(A) Mixed while still native.

(B) Heated separately for 10 min at 100°C, quickly cooled and mixed.

(C) and (D) Heated separately, annealed under identical conditions and banded separately.

(E) Mixed immediately after annealing while still in the 1·9 × SSC solvent. If the mixture was dialysed against a lower salt concentration (0·067 M-glycine buffer) before addition of CsCl or if each annealed sample was dialysed against 0·067 M-glycine buffer before mixing, 5 bands resulted.

(F) Mixed, dialysed against 0·067 M-glycine buffer and treated with the *E. coli* phosphodiesterase.

The same picture also resulted when portions of the annealed samples shown in B and C were separately treated with the enzyme and then mixed.

hybrid formation did occur although it may not have been as much as expected from random pairing of complementary strands since the three bands appear to have about equal areas.

For the next experiment we chose three members of the family Enterobacteriaceae which have similar base compositions (Lee, Wahl & Barbu, 1956) and which are genetically related. *E. coli* K12 shows a high degree of genetic exchange by conjugation and transduction with *E. coli* B and *Sh. dysenteriae* (Lennox, 1955; Luria & Burrous, 1957; Luria, Adams & Ting, 1960) whereas *Salm. typhimurium* mates well with K12 but is transduced only to a very limited extent, if at all (Zinder, 1960).

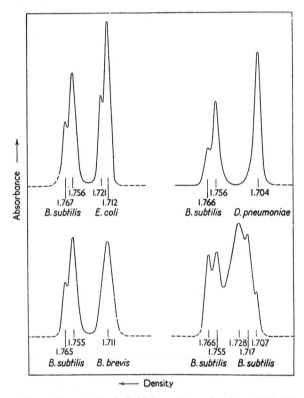

Fig. 10. The specific requirement for hybrid formation using heavy-labeled *B. subtilis* DNA. The three non-homologous samples were heated and annealed at a concentration of 10 μg/ml. each, under the same conditions already described. In the case where both samples were *B. subtilis* DNA, the concentrations were 5 μg/ml. each. Neither the labeled nor the unlabeled *B. subtilis* preparations were the same ones used to obtain the results shown in Fig. 5.

In this context, attempts were begun to prepare hybrid DNA molecules between ^{15}N-deuterated *E. coli* DNA and normal *Shigella* or *Salmonella* DNA. Unlabeled DNA isolated from *Sh. dysenteriae* was substituted for the unlabeled *E. coli* DNA in the procedure discussed in Section 5, and the heated and annealed mixture was treated with *E. coli* phosphodiesterase. Three bands were observed as shown in Fig. 12. The hybrid band does not contain twice as much DNA as either of the uniformly labeled renatured bands, as was observed in most previous examples. This indicates that not every *Sh. dysenteriae* DNA molecule is homologous to a corresponding *E. coli* DNA molecule. Genetic evidence also supports this partial degree of homology (Luria & Burrous, 1957).

Similar studies were carried out with DNA isolated from different strains and species of *Salmonella*. Heating and annealing was carried out with DNA isolated from each *Salmonella* organism listed in Table 1. The results for all samples were similar to those shown in Fig. 13 for the *Salm. typhimurium* and the heavy-labeled *E. coli* B DNA. No hybrid appears at 10 μg/ml. (top tracing) but an intermediate band does appear at 20 μg/ml. (middle tracing). It is, however, removed by enzyme treatment. The experiments were also repeated using heavy-labeled DNA from *E. coli* K12 since this strain has been used most successfully in the studies of genetic recombination. So far,

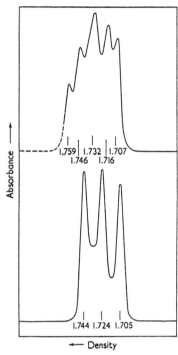

Fig. 11. Interspecies hybrid formation with *B. natto* DNA. Heavy-labeled *B. subtilis* DNA and unlabeled DNA from *B. natto* were heated and annealed at 5 μg/ml. each in 1·9 × SSC. The lower tracing shows the three bands produced when the mixture is treated, with the *E. coli* phosphodiesterase.

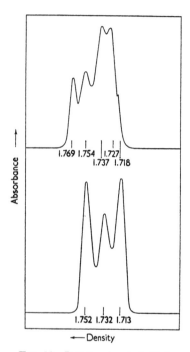

Fig. 12. Interspecies hybrid formation with *Sh. dysenteriae* DNA. Heavy-labeled *E. coli* DNA and unlabeled DNA from *Sh. dysenteriae* were heated and annealed at 5 μg/ml. each in 1·9 × SSC. When the mixture is treated with the phosphodiesterase and centrifuged in CsCl, only three bands appear.

the DNA of only one of the strains of *Salmonella* (*Salm. typhimurium*) has been tried but no hybrid was formed in this case either. We may then conclude that, in general, there seems to be no indication of sequence complementarity between the DNA of *Salmonella* and *E. coli* as measured by hybrid formation.

Thus in this case, where some homology was to be expected, none was found. It is likely that there is homology in some regions of the DNA molecules, but its failure to be displayed suggests that the homologous regions are dispersed or exist in only a few of the several hundred different molecules. Low concentrations of hybrid molecules would not be observed in the analytical ultracentrifuge, but could be isolated by using larger amounts of interacting DNA and working with the preparative swinging-bucket rotor. In this way we plan to search very carefully for small amounts of hybrid that may be formed between ^{15}N-deuterated DNA from *E. coli* K12 and DNA from any one of the *Salmonella* strains.

Hybrid formation among other members of the family Enterobacteriaceae is now being investigated. Preliminary results with heavy *E. coli* B DNA and normal DNA from a strain of *E. freundii* (5610–52) or of *Erwinia carotovora* (ATCC 8061) show no

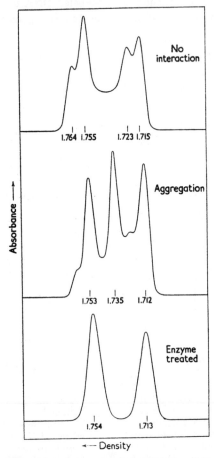

Fig. 13. Attempts at hybrid formation between *Salm. typhimurium* and *E. coli* DNA. The top tracing shows the results of heating and annealing DNA from *Salm. typhimurium* with ^{15}N-deuterated DNA from *E. coli*. Even at 10 μg/ml., which is double the usual concentration, no hybrids have been formed. In the central tracing an intermediate band appears as a result of heating at 20 μg/ml. each. This must be due to aggregates rather than double-stranded molecules since the band disappears after treatment with the phosphodiesterase (bottom tracing).

tendency toward hybrid formation. Unlabeled DNA isolated from another strain (17) of *E. freundii* did form hybrids with labeled *E. coli* B DNA. Since the classification of *E. freundii* is difficult, further studies with DNA of strains of this organism are necessary. Heavily-labeled DNA from *E. coli* K12, *Erwinia carotovora*, and *Salm. typhimurium* have also been prepared, allowing the study of other obvious possible relationships.

From this survey it can be concluded that equality of base composition and some genetic relation are necessary but not sufficient requirements for the formation of hybrid DNA molecules composed of strands from different organisms. The present work is only indicative in nature but it does suggest that, with the further development of this technique, the degree of sequential homology between two different DNA molecules can be quantitatively explored.

9. Discussion

Until now we have been talking strictly in terms of the concept of strand separation and recombination, in accordance with the evidence presented by Doty *et al.* (1960). It has been seen that some important new facts can now be added to this evidence. By using heavy-isotope-labeled and normal DNA for heating and annealing experiments, it has been demonstrated that the units that unite in renaturation could not be the same as those that were united in the native DNA. It has also been demonstrated that the hybrids formed are not simply physically entangled aggregates. The strong species specificity of hybrid formation indicates that the links that hold the subunits of the hybrid molecules together are determined by the base sequence in each subunit. These links break whenever the hydrogen bonded structure of the molecule is broken. In fact, the rate of separation into subunits corresponds very closely to what has been predicted for the rate of unwinding of the strands of a double helix. It is possible that certain unknown links are broken whenever the Watson-Crick structure is disrupted and that the base sequence determines the specific type of linkage. The most natural explanation, however, seems to be that the strands of the Watson-Crick double helix do separate completely and come back together again through the formation of hydrogen bonds between base-pairs of complementary strands.

The hypothesis that genetic information resides in the sequence of bases in DNA has created a demand for techniques that will allow the study of the linear order of nucleotides along the DNA strand. An approach to this problem has been made in the nearest neighbor studies of Josse, Kaiser & Kornberg (1961) which offers a statistical evaluation of sequences displayed in DNA of various sources. By inference it can be assumed that genetic compatibility is also a measure of similarity in sequences between the DNA of the two parental strains. The present report has outlined a method by which it is possible to measure the extent of renaturation between homologous and heterologous DNA strands and has attempted to add a new dimension to the study of sequences of DNA of microbial origin.

The methodology of molecular hybrid formation *in vitro* has been outlined and its use in the study of similarities of DNA of various groups of micro-organisms has been illustrated. The close correlation between genetic compatibility, taxonomy, and hybrid formation has been mentioned. These applications are discussed in much greater detail elsewhere (Marmur & Schildkraut, 1961*b*; Marmur, Schildkraut & Doty, 1961). Organisms whose taxonomic classification is in doubt might readily be classified by first determining their base composition and studying their interaction by heating and annealing. Such experiments are now being extended to other members of the family Enterobacteriaceae (*Aerobacter*, *Klebsiella*, and *Serratia*) as well as the Pseudomonadaceae (*Pseudomonas Xanthomonas* and *Acetobacter*). Moreover, it might also be possible, by collecting homogeneous fractions of DNA of animal or plant origin, to investigate relationships in a similar manner. Aside from its taxonomic importance the technique offers a rational approach to the study of genetic compatibility where genetic exchanges have not yet been demonstrated.

It is a pleasure to express our gratitude to Dr. L. Grossman for his generous gift of the *E. coli* phosphodiesterase and for his valuable advice. We are especially indebted to Mr. William Torrey for his expert technical assistance with many aspects of this work, and to Mrs. F. Seelig for aid in preparation of the manuscript. We would also like to thank Drs. L. Wierzchowski and Donald M. Green, and Mr. Robert Rownd for their aid. The authors are very grateful to Professor C. E. Hall of the Massachusetts Institute of Technology for the electron micrograph. This investigation was supported by a grant (C-2170) from the National Cancer Institute, United States Public Health Service.

Note added in proof

It can be argued that the DNA species of intermediate buoyant density observed in the CsCl density gradient could be formed by non-specific, end-to-end aggregation of heavy isotope labeled and unlabeled DNA. It is evident, however, that biological hybrids could not be used to form DNA species possessing buoyant densities characteristic of either renatured fully labeled or fully unlabeled DNA unless strand separation and subsequent recombination of similarly labeled strands occurs during the heating and annealing procedure. Preliminary experiments by R. Rownd and D. Green using biological hybrids labeled with both ^{15}N and deuterium in only one strand and isolated from either *Escherichia coli* or *Bacillus subtilis* have shown that the heating and annealing procedure does result in renatured labeled and fully unlabeled, as well as hybrid, DNA molecules. The proportions are the same as observed when a mixture of labeled and unlabeled DNA is used as the starting material.

REFERENCES

Belozersky, A. N. (1957). In *The Origin of Life on the Earth*. Reports on the International Symposium, ed. by A. Oparin, p. 194. U.S.S.R.: Academy of Sciences.

Belozersky, A. N. & Spirin, A. S. (1960). In *The Nucleic Acids*, ed. by E Chargaff & J. N. Davidson, Vol. III, p. 147. New York: Academic Press.

Cox, R., Marmur, J. & Doty, P. (1961). In preparation.

Davern, C. I. & Meselson, M. (1960). *J. Mol. Biol.* **2**, 153.

Doty, P., Marmur, J., Eigner, J. & Schildkraut, C. (1960). *Proc. Nat. Acad. Sci., Wash.* **46**, 461.

Dunn, D. B. & Smith, J. D. (1954). *Nature*, **174**, 305.

Eigner, J., Boedtker, H. & Michaels, G. (1961). *Biochim. biophys. Acta*, in the press.

Eigner, J. & Doty, P. (1961). In preparation.

Ephrati-Elizur, E. & Zamenhof, S. (1959). *Nature*, **184**, 472.

Ifft, J. B., Voet, D. H. & Vinograd, J. (1961). *J. Phys. Chem.*, in the press.

Josse, J., Kaiser, A. D. & Kornberg, A. (1961). *J. Biol. Chem.* **236**, 864.

Kuhn, W. (1957). *Experientia*, **13**, 301.

Lee, K. Y., Wahl, R. & Barbu, E. (1956). *Ann. Inst. Pasteur*, **91**, 212.

Lehman, I. R. (1960). *J. Biol. Chem.* **235**, 1479.

Lennox, E. S. (1955). *Virology*, **1**, 190.

Longuet-Higgins, H. C. & Zimm, B. H. (1960). *J. Mol. Biol.* **2**, 1.

Luria, S. E. & Burrous, J. W. (1957). *J. Bact.* **74**, 461.

Luria, S. E., Adams, J. N. & Ting, R. C. (1960). *Virology*, **12**, 348.

Marmur, J. (1961). *J. Mol. Biol.* **3**, 208.

Marmur, J. & Doty, P. (1959). *Nature*, **183**, 1427.

Marmur, J. & Doty, P. (1961). *J. Mol. Biol.* **3**, 585.

Marmur, J. & Lane, D. (1960). *Proc. Nat. Acad. Sci., Wash.* **46**, 453.

Marmur, J., Lane, D. & Doty, P. (1961). In preparation.

Marmur, J. & Schildkraut, C. L. (1961a). *Nature*, **189**, 636.

From the *Biophysical Journal*, Vol. 3, No. 1, 49–63, 1963.
Reproduced with permission of the authors and publisher.

PHYSICOCHEMICAL STUDIES ON

THE REACTION BETWEEN

FORMALDEHYDE AND DNA

DAVID FREIFELDER *and* PETER F. DAVISON

From the Department of Biology, Massachusetts Institute of Technology, Cambridge. Dr. Freifelder's present address is University Institute of Microbiology, Copenhagen, Denmark.

ABSTRACT   The reaction between formaldehyde and phage T7 DNA has been studied by optical absorbance and sedimentation measurements. Through the course of denaturation, OD_{260} and $s_{20,w}$ rise; after the attainment of full hyperchromicity the $s_{20,w}$ falls sharply, suggesting a decrease in molecular weight. Conditions in which formaldehyde causes cross-linking are defined. Some experimental applications of the denaturation technique are given. Evidence which suggests that preformed single-strand interruptions may exist in phage DNA is briefly discussed.

INTRODUCTION

When DNA solutions are heated through a critical temperature range, the secondary structure of the molecules is disrupted, and, as a result of this denaturation process, the optical absorbance of the solution at 260 mμ increases. It has recently been shown that through the temperature range in which denaturation occurs, segments within molecules become disordered (1), and at any temperature an equilibrium state exists in which a certain fraction of each molecule is denatured (2). As the temperature is raised the intramolecular denaturation increases until a step ensues which, in DNA from higher organisms at least, is irreversible. This irreversible step has been interpreted as the separation of the two strands of the double helix (3).

Until the onset of this irreversible process, cooling a solution of partially denatured DNA molecules results in the recovery of the original optical absorbance and viscosity (4); hence the physical properties of partly denatured molecules can be studied only at the ambient temperature. However, at elevated temperatures hydrolytic degradation proceeds at a detectable rate (5). As an alternative to studies at high temperatures, the reassociation of the separated strands may be blocked by a suitable reagent such as formaldehyde. It has been shown that in the presence of formaldehyde little or no decrease of optical absorbance ensues on cooling a partly or fully denatured DNA solution (6, 7), since the amino groups on the nucleotide bases react with the formaldehyde which thereby blocks the reformation of the hydrogen bonds dictating ordered reassociation of the strands (8).

In this paper we describe an investigation of the reaction between formaldehyde and a monodisperse DNA preparation from T7 bacteriophage (9). From these studies conditions which apparently eliminate strand hydrolysis have been defined. The conditions under which cross-linking occurs in the presence of formaldehyde have also been examined.

MATERIALS AND METHODS

1. *Growth of Phage and Preparation of DNA.* Bacteriophage T7 was grown as described elsewhere (9). Other phages were grown by similar techniques. Phage T4 was a shock-resistant mutant (10) which was purified by CsCl density gradient centrifugation. Phage DNA was prepared by a modification of the phenol technique (9).

Trout sperm and bacterial DNA were prepared by modified sodium dodecyl sulfate and phenol techniques (11, 12). Hybrid $N^{14}-N^{15}$ *E. coli* DNA was prepared as previously described (1).

2. *Formaldehyde Solutions.* Reagent grade 37 per cent HCHO (Merck) was used for all experiments. This material contains 12 per cent methanol added as a polymerization inhibitor. The solutions were neutralized with about 0.1 ml of M NaOH per 10 ml HCHO, and 0.1 ml of molar phosphate buffer, pH 7.8, was added. Neutralization of HCHO causes an immediate increase in optical density at 260 mμ (OD_{260}) which continues for a period of several hours, the rate being temperature-dependent. To avoid errors resulting from a varying optical blank and to reduce the concentration of polymeric forms of formaldehyde, the neutralized 37 per cent stock HCHO solution was always boiled for 10 minutes in a sealed ampoule before use. No further OD_{260} increase occurs after boiling. Fresh solutions were prepared daily.

3. *Thermal Denaturation Curves.* In the absence of HCHO, DNA melting-out curves were obtained by heating the solutions in 3 ml stoppered cuvettes in the thermostated compartment of the Beckman model DU spectrophotometer.

Since in the presence of HCHO the hyperchromicity is both time- and temperature-dependent, it is clear that the melting-out curve must refer to a controlled time of heating. We chose to heat each sample for 10 minutes. The denaturing solution was chilled in iced water before addition of the DNA. For each point on the denaturation profile 1 ml of the DNA in the denaturing solution was sealed in a 2 ml glass ampoule and rechilled. The ampoules were then completely immersed in a constant temperature bath and after 10 minutes' heating quenched in iced water; the OD_{260} was read shortly afterwards.

4. *Denaturation Kinetics.* The time course of DNA denaturation at a given temperature was followed by adding a small volume of a concentrated DNA solution to a preheated denaturing solution in a covered cuvette in the thermostated compartment of the spectrophotometer.

For kinetic measurements requiring several hours, solutions were stored in sealed ampoules in constant temperature baths until the absorbance was measured.

For studies relating time of denaturation to the sedimentation properties, the denaturing solution was placed in a screw-cap vial in a constant temperature water bath and heated for 15 minutes. The DNA solution—constituting less than 5 per cent of the volume of the solution—was then added. At various times samples were withdrawn with a pipette, quenched in small vials cooled in iced water, and subsequently studied in the ultracentrifuge.

5. *Alkaline Release of DNA from Phage.* (*a.*) *Denaturation by heating in formaldehyde.* To 0.67 ml of phage solution (ionic strength <0.002) was added 0.03 ml of M K_3PO_4. After 5 minutes at room temperature 0.05 ml M KH_2PO_4 was added, followed by 0.05 ml M phosphate buffer, pH 7.8, and 0.4 ml neutral 37 per cent HCHO. The solution was heated for 2 1/2 minutes at 70°C, cooled, and 0.4 ml 4 M NaCl was added before ultracentrifugation.

(*b*) *Denaturation by alkali.* To 0.05 ml of a phage solution (ionic strength <0.05) was added 0.05 ml M NaOH. After 1 minute at room temperature, 0.4 ml 37 per cent HCHO was added, followed by 0.1 ml M KH_2PO_4, 0.6 ml H_2O, and 0.4 ml M NaCl. The optical density of DNA treated in this way was unaffected by heating to 100°C for 10 minutes, a fact which indicates that the DNA had been fully denatured.

6. *Ultracentrifugation.* Sedimentation velocities were determined as described elsewhere (9). All solutions were diluted with one volume of 4 M NaCl to 3 volumes of the original solution. The final DNA concentration was always 20 ± 5 µg/ml. The sedimentation coefficients were corrected to 20°C and the viscosity of water ($s_{20,w}$), and are expressed in Svedbergs (S).

7. *Standard Denaturation Technique.* Of the several conditions that provided complete denaturation with minimal cross-linking and hydrolysis, the following standardized procedure was adopted for the final sedimentation studies. A solution containing 12 per cent HCHO and 0.1 M phosphate, pH 7.8, was preheated at 70°C for 5 minutes. The DNA (in a volume not greater than 1/20th of the denaturing solution) was added and, after heating for 2 minutes, the solution was quenched in iced water. If preheating was not feasible, heating was extended to 2 1/2 minutes.

RESULTS

1. *Optical Studies*

The relation between the OD_{260} of a DNA solution and temperature (the "melting-out" or denaturation curve) is a function of the composition of the DNA and of the pH, ionic strength, and composition of the solvent (13, 14).

Fig. 1 shows for three different HCHO concentrations the OD_{260} as a function of time after concentrated T7 DNA was added to preheated buffered HCHO solutions. The curves show that the rate of increase of OD_{260} rises with temperature and HCHO concentration.

The reaction at low temperatures is not shown in the graphs but does proceed at a measurable rate. For example in 4 per cent HCHO and 0.01 M phosphate at 25°C and 45°C, full hyperchromicity is reached in approximately 65 and 10 hours, respectively. These times are sufficiently long that at 25°C or lower, hyperchromicity can be considered as constant for the duration of most ultracentrifuge experiments.

Fig. 2 shows the DNA melting-out curves at four different concentrations of HCHO and at two salt concentrations. Two melting-out curves in the absence of HCHO, obtained by the usual procedures, are also given. In these cases a lower final hyperchromicity was attained (6). The maximum hyperchromicity is independent of ionic strength and HCHO concentrations within the range 1 to 12 per cent.

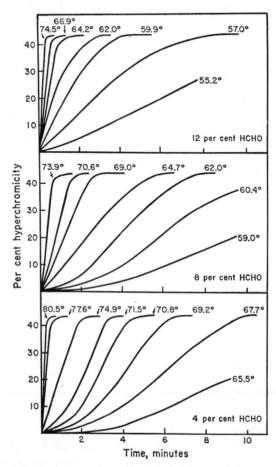

FIGURE 1   Percentage hyperchromicity as a function of time of incubation of T7 DNA in denaturing solutions at the temperatures indicated. The HCHO concentrations are given in the diagrams.

For constant heating time, denaturation occurs at lower temperatures with increasing HCHO concentration. Fig. 3a indicates for both salt concentrations employed the relationship between the HCHO concentration and the decrease in the temperature (T_m) at which the DNA is at 50 per cent of full hyperchromicity after 10 minutes' heating. The curves demonstrate that the decrease of T_m is not

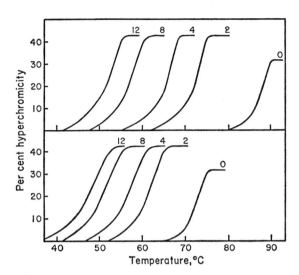

FIGURE 2   Melting curves for T7 DNA heated for 10 minutes in denaturing solutions containing HCHO in the percentages indicated. Upper curves, in 0.01 M phosphate, pH 7.8; lower, 0.1 M phosphate, pH 7.8.

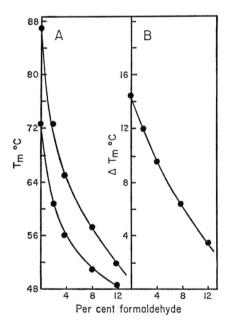

FIGURE 3A T_m obtained from the curves given in Fig. 2 as a function of HCHO concentration. Upper curve, 0.1 M phosphate, pH 7.8; lower, 0.01 M phosphate, pH 7.8.

FIGURE 3B   Influence of salt on the denaturation temperature of T7 DNA in 0, 2, 4, 8 and 12 per cent, HCHO denaturing solutions. Each point represents the difference between T_m measured in 0.1 and 0.01 M phosphate, pH 7.8.

FREIFELDER, D., AND DAVISON, P. F.   *Reaction between Formaldehyde and DNA*   **53** linear with respect to HCHO concentration. The curves for the two salt concentrations are not parallel, and a plot of ΔT_m (*i.e.*, $T_{m,0.1} - T_{m,0.01}$) *versus* HCHO concentration, as shown in Fig. 3*b*, indicates that with increasing HCHO concentration the effect of ionic strength on denaturation rate diminishes.

2. *Sedimentation Studies*

The transition from native to denatured DNA can be followed by sedimentation analysis of partially hyperchromic material. This is possible because the HCHO blocks the reformation of the hydrogen bonds disrupted during the heating period, yet the continued reaction of the native DNA with the HCHO is, after cooling, sufficiently slow to be ignored during the sedimentation experiment.

Native T7 DNA shows a sharp sedimenting boundary (see Fig. 1 in reference 9). The upper part of the trace shows a small curvature which is believed to be an artifact resulting from boundary disturbances and is not indicative of higher molecular weight material (9). This curvature appears on all the sedimentation diagrams (Figs. 4 and 8) and will be ignored in the following discussion. Partially or fully hyperchromatic T7 DNA shows a sharply defined leading component, followed by a fraction of more slowly sedimenting polydisperse material, the proportion of which increases with increased denaturation (Fig. 4). The sedimentation

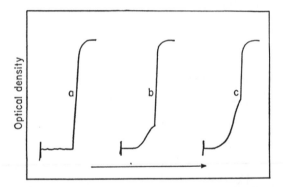

FIGURE 4 Photometric trace of UV absorption photograph of sedimenting T7 DNA denatured to different degrees.
a. 8 per cent hyperchromicity, median $s_{20,w} = 34.5$
b. 25 per cent hyperchromicity, median $s_{20,w} = 43.0$
c. 42 per cent hyperchromicity, median $s_{20,w} = 39.0$
The arrow represents the direction of sedimentation. The vertical bar at the left of each trace indicates the meniscus.

diagram is at all times biomodal, *i.e.*, the "tail" is not smoothly connected with the sharp boundary of the fast component. Since the slower material could have arisen from hydrolytic degradation under the denaturation conditions, attention was focused on the fast component, and the values of $s_{20,w}$ quoted in the following paragraphs refer to that fraction of the DNA. Consideration of the slower sedimenting material will be deferred to the section dealing with hydrolysis.

Fig. 5 shows $s_{20,w}$ as a function of temperature for several T7 DNA preparations which had been heated in sealed vessels for 10 minutes in 12 per cent HCHO, 0.1 M PO_4, pH 7.8, and then quenched in iced water. Comparison with the melting-out curve shows that $s_{20,w}$ rises with increasing hyperchromicity, continues to rise when no further increase in OD_{260} is detectable, and then drops sharply.

In a few DNA samples heating under conditions which normally gave the fully hyperchromic 39S boundary gave instead a double boundary consisting of varying ratios of 39 and 65S material. With longer heating times or higher temperatures the 65S component decreased and the 39S increased. This anomalous behavior was ascribed to the presence of divalent ions in the DNA preparation since, when the denaturation was performed in the presence of 10^{-3} M versene, no double boundary was observed, *i.e.*, the DNA solutions behaved as if they contained one molecular species, each molecule undergoing the same change simultaneously at a unique temperature under these denaturing conditions.

The $s_{20,w}$ of the fully denatured material is greater than that of the native DNA, and with further heating it does not decrease until hydrolytic degradation is detectable.

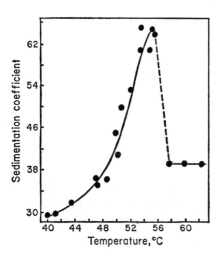

FIGURE 5 Sedimenation coefficient of the sharp boundary of T7 DNA fully or partly denatured by heating for 10 minutes at the indicated temperatures in 12 per cent HCHO, 0.1 M phosphate, pH 7.8.

The same data relating denaturation and $s_{20,w}$ is shown in Fig. 6 where the relationship between the percentage of full hyperchromicity and $s_{20,w}$ is plotted.

A similar relationship between $s_{20,w}$ and the progress of denaturation results when denaturation is carried out at constant temperature for various heating times. Fig. 7 shows the $s_{20,w}$ as a function of time after the addition of DNA to the denaturing solution preheated to 58.3°C. Again it is seen that $s_{20,w}$ increases with time and does not drop until past the onset of apparently full hyperchromicity.

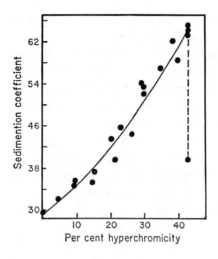

FIGURE 6 Sedimentation coefficient of the sharp boundary of T7 DNA, fully or partly denatured under several conditions, as a function of per cent hyperchromicity.

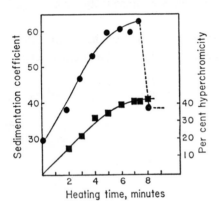

FIGURE 7 Sedimentation coefficient of the sharp boundary (●) and per cent hyperchromicity (■) of T7 DNA fully or partly denatured by heating at 58.3°C for the indicated times in 12 per cent HCHO, 0.1 M phosphate, pH 7.8.

3. *Hydrolysis*

As mentioned above, the percentage of T7 DNA in the slowly sedimenting "tail" of the sedimentation diagrams increases with hyperchromicity. In fully denatured DNA this tail includes material with $s_{20,w}$ ranging from 15 to 39 Svedbergs. That the percentage of DNA in the tail increased in old or degraded preparations suggested that some of this tail at least was an artifact; several experiments were therefore conducted to minimize the material in this tail. Careful control of the steps in the preparation and denaturation of the DNA brought the amount of material of lower $s_{20,w}$ down to 50 ± 5 per cent of the whole, and this percentage was observed consistently in 5 samples of DNA from 3 phage strains prepared by the phenol method.

To ensure that the 50 per cent tail was not a result of degradation brought about by the preparative or denaturation process, various modifications of these procedures were introduced. Two DNA preparations were made by alkali release; denaturation by the standard method again showed a 50 per cent tail. Two DNA samples, one prepared by phenol, the other by alkali release, were denatured in the following ways: (*a*) heating for 10 minutes at 60°C in 12 per cent HCHO, 0.1 M PO$_4$, pH 7.8; (*b*) heating for 2 minutes at 70°C in 12 per cent HCHO, 0.1 M PO$_4$, pH 7.8; (*c*) heating for 20 minutes at 40°C in 2 per cent HCHO, 0.2 M K$_3$PO$_4$, pH 11.6; (*d*) heating for 1 minute at 25°C in 0.5 M NaOH followed by addition of HCHO. For each of these samples the tail amounted to 50 per cent. After DNA was heated for 18 minutes at 70°C in 12 per cent HCHO, 0.1 M PO$_4$, pH 7.8, the tail on the sedimentation diagram was similar in shape and size to that of a sample heated for 2 minutes. However, when the DNA was heated for 30 minutes, the sedimentation diagram showed 80 per cent of the DNA in the slow tail; the tail also both lost its bimodality and gained very low $s_{20,w}$ material, presumptive evidence of hydrolysis.

These results suggest that the 50 per cent fraction of the DNA trailing the apparently homogeneous fast component is not an artifact. However, it may be deduced that any procedure which causes a loss of material from the fast component—with a concomitant increase in the tail—is producing breaks in the polymer strands. On the assumption that the fast component in the sedimentation diagram represents single polynucleotide strands half the molecular weight of the intact T7 DNA, the number of hydrolytic breaks in the single strands introduced by various conditions have been summarized in Table I. These numbers have been calculated on the basis of an empirical calibration of $s_{20,w}$ as a function of molecular weight for a series of formaldehyde-denatured DNA preparations; the derivation of this approximate relationship is presented below.

TABLE I

DEGREE OF DEGRADATON ACCOMPANYING VARIOUS CONDITIONS FOR DENATURATION

Treatment*	Median $s_{20,w}$	Per cent undegraded	Hydrolytic breaks/ 10^7 mol. wt. ($\pm 20\%$)
2 min. 70°C—12% HCHO	39	50	0
18 min. 70°C—12% HCHO	39	50	0
30 min. 70°C—12% HCHO	37	20	0.6
10 min. 75°C	34	10	0.8
10 min. 85°C	32	5	0.8
3 min. 100°C	34	10	0.8
3 min. 100°C	27	0	1.0
10 min. 100°C	26	0	1.0

* In each case in which HCHO was not present during heating, it was added to make 12% HCHO, and the solution was further heated for 2½ minutes at 70°C. For all experiments the solvent was 0.1 M phosphate, pH 7.8.

4. Cross-Linking

In certain circumstances the $s_{20,w}$ of the T7 DNA rises to the maximum value shown in Fig. 5, but with further heating or higher temperatures, $s_{20,w}$ does not decrease. If the drop in $s_{20,w}$ is interpreted as strand separation, then one can infer that the high $s_{20,w}$ material contains at least one heat-resistant interstrand cross-link. This cross-linking occurs when DNA is allowed to react with HCHO for long periods of time before heat denaturation. The degree of cross-linking can be determined by denaturing the DNA and measuring the amount of material sedimenting in the 65S peak. A sample with 25 per cent cross-linked material is shown in Fig. 8. Fig. 9

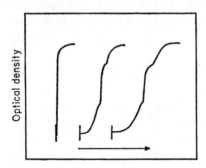

FIGURE 8 Photometric trace of UV absorption photographs of sedimenting T7 DNA taken at 0, 16, and 24 minutes afer reaching final speed of 33,450 RPM. The DNA was previously incubated in 12 percent HCHO, 0.1 M phosphate, pH 7.8, at 0°C for 11 1/2 hours, and then heated for 2 1/2 minutes at 70°C. Approximately 25 per cent of the strands are cross-linked. The arrow represents the direction of sedimentaion and the vertical bar at the left of each trace indicates the meniscus.

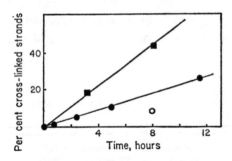

FIGURE 9 Percentage of cross-linked strands for T7 DNA pretreated (as described below) before heating in 12 per cent HCHO, 0.1 M phosphate, pH 7.8, for 2-1/2 minues.
● Incubated at 0°C in 12 per cent HCHO.
■ Incubated at 25°C in 12 per cent HCHO.
○ Incubated at 25°C in 3 per cent HCHO.

shows the percentage of cross-linked material as a function of time of incubation for various HCHO concentrations and temperatures. The cross-linking reaction is minimized at high pH (we are indebted to Dr. L. Grossman who pointed out this fact), e.g., a sample treated for 24 hours at 25°C in 4 per cent HCHO, 0.04 M borate, pH 9, shows no cross-linking. Cross-linking is not confined to T7 DNA nor to intact molecules, since it has been observed in T2 DNA, sheared T2 DNA ($s_{20,w} = 25$), E. coli DNA, and salmon sperm DNA. The cross-links are stable when briefly heated at 100°C. Cross-linking was also demonstrated and measured by storing $N^{14}-N^{15}$ "hybrid" DNA from E. coli in HCHO and subsequently denaturing and banding in a cesium chloride density gradient.

5. *Further Denaturation Studies*

The standardized HCHO denaturation was performed on DNA released by alkali from T1, T5, and λ (*hc*) phages, and on T4 and T5 phage DNA obtained by phenol extraction. With T1 and λ (*hc*) a bimodal boundary consisting of a sharp region and a tail resulted, as is the case for T7. The $s_{20,w}$ for the sharp region was 35 for T1 and 39 for λ (*hc*). The percentage of material in the tail was 55 per cent for T1 and 28 per cent for λ (*hc*). The DNA of both T4 and T5 gave on denaturation a broad boundary with a median $s_{20,w}$ of 54 and 41, respectively; no sharp leading boundary was seen in the photometric traces. For each DNA the boundary was identical after heating for 2 or 4 minutes.

Curves similar to those in Figs. 5 and 6 relating $s_{20,w}$ to hyperchromicity or temperature have been obtained for DNA isolated from stationary phase *E. coli* and *Pseudomonas fluorescens*. The changes are not so sharply defined or so striking as for the T7 DNA (probably because of polydispersity), but still an increase in $s_{20,w}$ followed by a fall was observed through the denaturation range.

For both bacterial DNA's the viscosity of each sample used in the sedimentation studies was also measured. As the OD_{260} increased the viscosity decreased continuously showing no inflection over the range where the $s_{20,w}$ fell.

The sedimentation coefficient was also determined for sheared and sonicated T7 DNA and for intact and sheared trout sperm DNA. From these values a curve relating the $s_{20,w}$ of HCHO-denatured material and molecular weight has been derived in the following way: for trout sperm DNA the molecular weight was assumed to be half that of the native material, whose molecular weight was estimated from the empirical calibration of Doty, McGill, and Rice (15). Two further points were provided by the fast component of fully denatured intact T7 DNA (39S) and the maximum $s_{20,w}$ reached before strand separation (65S); the molecular weight of T7 DNA has been reported in another paper (9). The curve is given in Fig. 10.

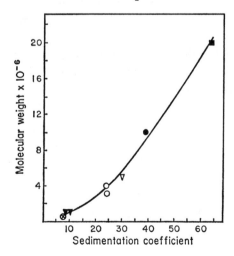

FIGURE 10 Relationship between $s_{20,w}$ and molecular weight for DNA denatured in HCHO.
⊗ Sonicated trout sperm DNA.
▼ Sonicated T7 DNA.
○ Sheared T7 DNA.
▽ Trout sperm DNA.
● Fully denatured T7 DNA.
■ Fully hyperchromic T7 DNA just prior to strand separation.

From the crudity of the derivation this curve is clearly suitable for only the most approximate calculations.

DISCUSSION

Some aspects of the denaturation of DNA in the presence of formaldehyde have been elucidated in the experiments reported above. The interpretation of the ultra-centrifuge diagrams has been greatly facilitated by the use of a monodisperse DNA. With such a homogeneous material the physical properties of the solution reflect the molecular changes in each molecule, whereas the interpretation of experiments on more conventional DNA preparations is obscured by the physical and chemical heterogeneity of the molecules.

Under the conditions of these experiments the optical absorbance of a DNA solution in the presence of formaldehyde increased at a rate determined by the temperature, solvent, and formaldehyde concentration until the DNA was fully denatured. Grossman, et al. (6) have shown that the equilibrium reaction between formaldehyde and the amino groups of the bases heavily favors the hydroxymethylated form of the bases. Therefore, if it is assumed that in any DNA molecule native and "denatured" segments are, in the absence of formaldehyde, in thermodynamic equilibrium with hydrogen bonds constantly being made and broken—a likely contingency since the free-energy changes involved in the separation of the base pairs are in the order of kT (16)—then it is to be expected that in the presence of formaldehyde the amino groups will be progressively titrated until each molecule of DNA has been fully denatured.

At high formaldehyde concentrations the denaturation rate increases, as anticipated. This rate change may not arise solely from the changes in the equilibrium conditions, since Haselkorn and Doty have shown that formaldehyde, in destabilizing polyinosinic acid (17), must exert a denaturing effect distinct from its reaction with the amino groups. The methanol present in the formaldehyde solution may also increase the denaturation rate.

The change in $s_{20,w}$ of the fast component with hyperchromicity indicates that the optical changes occurring in denaturation are accompanied by physical alterations in the molecule. The hydrodynamic behavior of a macromolecule is a function primarily of its molecular weight and its shape, i.e., frictional coefficient. The steady rise of $s_{20,w}$ is most simply interpreted as a progressive decrease in the friction coefficient of the molecules without any change in their mass. This would be the predicted behavior as the normally rigid DNA molecule assumed a more compact configuration as a result of the introduction of points of flexion at regions of hydrogen-bond rupture within the molecule, and the collapsing of the twin strands to a random coil configuration with the breakdown of hydrogen bonds at the ends of the molecules. Obviously with the progressive binding of the HCHO the weight of the DNA also increases, but this factor probably contributes negligibly to the increase of $s_{20,w}$.

It seems unlikely that the formaldehyde can penetrate the DNA structure and react with the bases without the base pairs—and hence each polynucleotide strand —locally separating. This separation presupposes a degree of unwinding of the double helix. On the completion of denaturation and the hydroxymethylation of all the amino groups on the bases, there arises a question: Do the paired strands of the DNA separate or remain entangled? These studies show that after the DNA has become fully denatured, the sedimentation coefficient suddenly falls. This behavior could reflect a sudden extension of a compact molecule or a decrease in molecular weight. The former explanation appears unlikely, since a progressive fall in viscosity of *E. coli* and *Pseudomonas* DNA was observed through the denaturation range. We therefore interpret the change as a halving of the molecular weight of the DNA resulting from the physical separation of the strands.

It is of course possible that the subunits separating at the completion of denaturation are paired double helices (18), but since the separation appears to be related to the dissociation of base pairs (see also (1)) we presently prefer the postulate of strand separation.

It may be emphasized that this relationship between $s_{20,w}$ and denaturation can be observed only in the presence of a reagent blocking renaturation; in the absence of such a reagent a partly denatured molecule would return to the native configuration on cooling (2), whereas after strand separation the solution would contain a mixture of single strands, aggregated single strands, and renatured molecules.

It is clear that detectable cross-linking does not occur under the conditions finally chosen for denaturation. Although the chemical link has not been identified, it is possible that it is a methylene bridge between adenine groups, forming the adenine dimer described by Alderson (19). Intrastrand dimers, perhaps of adjacent adenines, may also be formed after denaturation. Such structures would not be detected in the present experiments but could interfere with renaturation of HCHO-denatured DNA from which the HCHO has been removed by dialysis.

It should be noted that "full" hyperchromicity is not an adequate criterion of complete denaturation or strand separation, as we have pointed out elsewhere (1). The strands cannot separate, we believe, until the last and strongest hydrogen bonds are broken; but these strongest bonds are a minute fraction of the total and contribute negligibly to the total hyperchromicity.

The significance of the low $s_{20,w}$ tail in the sedimentation diagram of denatured T7 phage DNA is obscure. A homogeneous DNA would be expected to yield a physically monodisperse denatured product; for this reason the sharply defined fast sedimenting component is believed to represent separated intact strands of the native DNA. It therefore seems evident that the slow "tail" comprises fragments of the DNA—moreover, the fragments are polydisperse, a fact which would appear

FREIFELDER, D., AND DAVISON, P. F. *Reaction between Formaldehyde and DNA* 61 to rule out the possibility of a unique preformed break in one of the strands. Furthermore, from the bimodality of the sedimentation boundary, one can deduce that the loci for such breaks are not randomly distributed (20).

After full denaturation is achieved (2 minutes, 70°C), the sedimentation distribution is unchanged for heating times up to 20 minutes. Further heating results in both loss of material from the sharp boundary and the appearance of even slower material in the tail. Therefore, if this latter process represents phospho-ester bond hydrolysis, then the consistent appearance of a 50 per cent tail immediately upon denaturation—independent of the mode of DNA release or denaturation—demands the presence of preformed breaks or the rapid breakage of a very labile bond.

It is clear that the number of breaks is small, *i.e.*, since 50 per cent of the single strands are intact, either each phage DNA molecule contains one break or half of the molecules have no breaks and half have one in each strand. The absence of very slow material makes it unlikely that there are many molecules containing more than one or two breaks per strand.

Following the initial appearance of the 50 per cent tail, the kinetics of the loss of intact single strands with additional heatings shows a lag (20 minutes) and therefore represents a multi-event process. This process can be either the degradation of a multicomponent structure (18) by a single- or multi-hit mechanism or of a single component by a multi-hit mechanism, *e.g.*, if phospho-ester hydrolysis required preliminary depurination. These alternatives are not distinguishable in the present work. However, one can validly conclude that in the lag period there is hidden damage of some sort in the separated units.

If there exist preformed interruptions in the phage DNA strands, there should be some significance to the variation in the percentage of intact strands detected in the different phage species. Professor C. Levinthal has suggested that these interruptions might represent unrepaired chain breaks in molecules which have undergone pairwise exchange of double-strand material as proposed by Meselson and Weigle (21) for the mechanism of genetic recombination.

In earlier denaturation studies with formaldehyde, Berns and Thomas (7) concluded that there were no single-strand interruptions in T2 and T4 phage DNA. If their conclusion is true, then the broad sedimentation velocity distribution we observe for this DNA after denaturation must reflect a variety of molecular configurations rather than polydisperse fragments of the single strands. However, the mean $s_{20,w}$ of most DNA preparations we have studied increased on denaturation, suggesting that the separated strands have a $s_{20,w}$ greater than the parent DNA molecules. The fact that 25 per cent of the denatured T4 molecules have $s_{20,w} < 40$, *i.e.*, substantially lower than that of the undenatured molecules, strongly suggests that this material represents fragments of the single strands.

The authors are grateful to Professor Francis O. Schmitt, in whose laboratories these investigations were performed, for his interest and help; to Dr. L. Grossman of Brandeis University for several informative discussions on the chemical reactions between DNA and formaldehyde; and to Mrs. K. Y. Ho for her valuable technical assistance.

This investigation was supported by research grant E-1469 from the National Institute of Allergy and Infectious Diseases, National Institutes of Health, United States Public Health Service.

Dr. Freifelder was supported by a postdoctoral fellowship from the Division of General Medical Sciences, United States Public Health Service, 1961.

Received for publication, April 19, 1962.

REFERENCES

1. FREIFELDER, D., and DAVISON, P. F., 1962, *Biophysic. J.*, **2**, 249.
2. GEIDUSCHEK, E. P., 1962, *J. Mol. Biol.*, **4**, 467.
3. DOTY, P., MARMUR, J., EIGNER, J., and SCHILDKRAUT, C., 1960, *Proc. Nat. Acad. Sc.*, **46**, 453.
4. GEIDUSCHEK, E. P., and HERSKOVITS, T. T., 1961, *Arch. Biochem. and Biophysics*, **95**, 114.
5. EIGNER, J., BOEDTKER, H., and MICHAELS, G., 1961, *Biochim. et Biophysica Acta*, **51**, 165.
6. GROSSMAN, L., LEVINE, S. S., and ALLISON, W. S., 1961, *J. Mol. Biol.*, **3**, 47.
7. BERNS, K. I., and THOMAS, C. A., 1961, *J. Mol. Biol.*, **3**, 289.
8. STOLLAR, D., and GROSSMAN, L., 1962, *J. Mol. Biol.*, **4**, 31.
9. DAVISON, P. F., and FREIFELDER, D., 1962, *J. Mol. Biol.*, **4**, in press.
10. BRENNER, S., and BARNETT, L., 1959, *Brookhaven Symp. Biol.*, **12**, 86.
11. CHARGAFF, E., *in* The Nucleic Acids, (E. Chargaff, and J. N. Davidson, editors), New York, Academic Press, Inc., 1955, **1**, 307.
12. KIRBY, K. S., 1957, *Biochem. J.*, **66**, 495.
13. MARMUR, J., and DOTY, P., 1959, *Nature*, **183**, 1427.
14. HERSKOVITS, T. T., SINGER, S. J., and GEIDUSCHEK, E. P., 1961, *Arch. Biochem. and Biophysics*, **94**, 99.
15. DOTY, P., MCGILL, B. B., and RICE, S. A., 1958, *Proc. Nat. Acad. Sc.*, **44**, 432.
16. LONGUET-HIGGINS, H. C., and ZIMM, B. H., 1960, *J. Mol. Biol.*, **2**, 1.
17. HASELKORN, R., and DOTY, P., 1961, *J. Biol. Chem.*, **236**, 2738.
18. CAVALIERI, L. F., and ROSENBERG, B. H., 1961, *Biophysic. J.*, **1**, 301.
19. ALDERSON, T., 1960, *Nature*, **187**, 485.
20. FREIFELDER, D., and DAVISON, P. F., 1962, *Biophysic. J.*, **2**, 235.
21. MESELSON, M. and WEIGLE, J., 1961, *Proc. Nat. Acad. Sc.*, **47**, 857.

"REVERSIBLE" DNA

By E. Peter Geiduschek

COMMITTEE ON BIOPHYSICS, THE UNIVERSITY OF CHICAGO

Communicated by Raymond E. Zirkle, May 26, 1961

By chemical modification of DNA, we have produced materials whose helix-coil transitions are totally reversible. This communication concerns the properties of this "reversible" DNA, which we have made by two distinctly different processes: (1) Reaction with the chemical mutagen, sodium nitrite, at pH 4.2 (HNO_2-DNA), and (2) Reaction with the cytotoxic bifunctional alkylating agent, bis(β-chloroethyl)methylamine hydrochloride at pH 7.1 (HN2-DNA).

The spectrophotometric and density-gradient centrifugation experiments reported below provide evidence that the chemical reactions leading to these products need not substantially change the secondary structure of double-helical, native DNA. This suggests the hypothesis that the reversibility of denaturation in these chemically modified DNA's is controlled by covalent bonds which link complementary strands of the double helix.

Materials and Methods.—The DNA used in these experiments was isolated from salmon testes by the method of Simmons.[1] *Pseudomonas fluorescens* DNA, used as a density reference in CsCl density-gradient centrifugation, was prepared by a method to be described elsewhere.[2]

Ultraviolet absorbance measurements were made in the following manner. A DNA solution, in a stoppered cuvette, is heated to a given temperature in the thermostated cell compartment of a spectrophotometer and the absorbance (A) at 259 mμ is measured after equilibrium is attained. The cuvette is then removed, plunged into ice to insure rapid cooling (quenching), re-equilibrated in a second spectrophotometer maintained at 25°C, and the absorbance is measured again. This heating and quenching cycle is now repeated at successively higher temperatures. Although an entire denaturation curve can be performed on a single sample, a certain saving of time is achieved by using two cells simultaneously. Measurements at a series of temperatures yield the curves shown in Figures 1 and 3. Absorbances measured at the elevated temperatures yield the curves marked (d), while measurements, at 25°C, on quenched solutions yield the curves marked (i). Appropriate control measurements are made to verify that the properties of DNA at the highest temperatures are not substantially affected by previous cycles of heating.

Preparation of HNO_2-DNA: Salmon DNA (0.1%) in 0.01 M NaCl was mixed with 4 volumes of 1.25 M $NaNO_2$, 0.3 M Na acetate, pH 4.20, and equilibrated at 25°C. After a specified reaction time, the mixture was placed in ice and 0.5 volumes 2 M Na_2HPO_4 added to neutralize (to pH 6.5 approx.). There followed 24 to 48 hours' dialysis against several changes of 0.01 M NaCl at 1°C.

Preparation of HN2-DNA: HN2, DNA in 0.01 M NaCl and phosphate buffer were mixed to give: 4.3 μ moles DNA-P, 2.4 μ moles HN2, 0.07 M phosphate pH 7.1, total volume 7.7 ml. After 45 minutes' incubation at 25°C, the mixture was cooled in ice and 1 ml added to 9 ml 8.0 M $NaClO_4$, pH 7.05 for the thermal denaturation experiment, shown in Figure 3.

Results and Discussion.—The reversibility of the thermal denaturation of HNO_2-

DNA was demonstrated in three ways—spectrophotometrically, by viscosity measurements, and by equilibrium centrifugation in a CsCl density gradient. Results of the spectrophotometric experiments are shown in Figure 1. Two types of measurement of the temperature dependence of absorbance in low ionic strength, 51 vol. per cent methanol are represented there: those made at the ambient temperature and those made on solutions which have been heated to a given temperature, rapidly cooled (quenched) in ice, and re-equilibrated at 25°C. The two curves marked O show the results for unreacted salmon DNA: It will be noted that, in this solvent, irreversibly denatured DNA is almost completely hyperchromic, and that its absorbance increases only very slightly between 25 and 55°C. Conse-

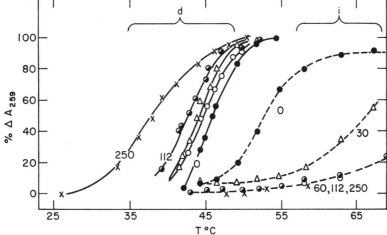

Fig. 1.—The reversible thermal denaturation of HNO_2-DNA (salmon testis) in 51% (v/v) methanol ($10^{-3}M$ NaCl, $10^{-3}M$ Tris-hydroxymethyl-aminomethane, pH 7.1).

$\%\Delta A_{259}$: Per cent of maximum absorbance increase (referred to measurement at the ambient temperature), uncorrected for thermal expansion of solvent.

$$\%\Delta A_{259} = \frac{100[A(T) - A(25°C)]_{d\ or\ i}}{[\Delta A_{max}]_d}$$

————, absorbance measured at temperature shown, (d).
--------, absorbance measured at 25°C on samples previously equilibrated at the temperature shown and quenched in ice, (i).

Time of reaction with HNO_2, pH 4.2, 25°C: zero time control, ●; 30 minutes, △; 60 minutes, ○; 112 minutes, ◑; and 250 minutes, ✕.

quently, reversible changes of absorbance upon heating and cooling can be unequivocally interpreted in terms of reversible changes of long-range order. The other curves of Figure 1 show the results of denaturation experiments on DNA which has been treated with 1 M nitrite at pH 4.2. As the amino groups of adenine, guanine, and cytosine react,[3] the thermal stability of the DNA is decreased. This is entirely in keeping with the fact that the amino groups are involved in interactions which stabilize the secondary structure of DNA. Maximum absorbance increases on thermal denaturation, $100\ \Delta A_{max}/A(25°C.)$, are 44, 43, 40, 39, and 39 (±2) per cent, respectively, for DNA's that have been reacted for 0, 30, 60, 112, and 250 minutes. With regard to the irreversible component of denaturation, on the other hand, one soon observes a drastic change. DNA which has been reacted with nitrous acid for more than one hour cannot be irreversibly denatured at 60°C which, in this medium, suffices to denature untreated DNA completely. Similar results are obtained when the denaturation is followed in aqueous 7.2 *M* NaClO$_4$, or 0.01 *M* NaCl.

When 70-minute-reacted HNO$_z$-DNA is heated in 0.01 *M* NaCl at 100°C for 15 minutes and quenched in ice, the specific viscosity (zero shear, measured

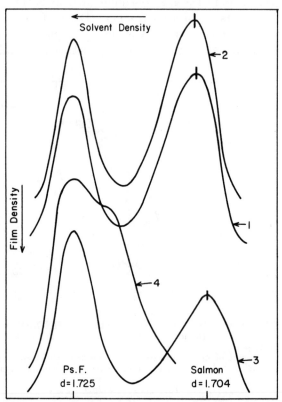

Fig. 2.—Equilibrium centrifugation of normal and HNO$_2$-treated DNA (salmon) in a CsCl density gradient (7.7 molal CsCl, 0.01 *M* Tris, pH 8; 40 hours centrifugation at 44,770 rpm, 25°C). DNA from *Pseudomonas fluorescens* is used as a density marker in all four experiments. Data are presented in terms of densitometric traces of absorption camera pictures through the region of the DNA "bands." Plate density baselines are displaced for ease of comparison. The buoyant densities of native salmon and Pseudomonas DNA are taken from the data of Rolfe and Meselson,[5] Sueoka, Marmur, and Doty.[4]

Curve 1, HNO$_2$-DNA (70-minutes reaction); *curve 2*, HNO$_2$-DNA heated for 15 minutes at 100°C in 0.01 *M* NaCl and quenched; *curve 3*, unreacted DNA; and curve 4, same, but heated for 15 minutes at 100°C in 0.01 *M* NaCl, and quenched.

at 25°C) decreases only 13 per cent. Unreacted DNA subjected to the same heating and quenching cycle decreases its specific viscosity by a factor of ten.

CsCl density-gradient centrifugation of heated and unheated, reacted and control DNA, demonstrates this reversibility in yet a third way (Fig. 2). The effective densities of heated and unheated HNO$_2$-DNA differ by less than 0.001 gm/cm^3 and the bandwidths are almost the same (curves 1 and 2). The latter result is in keeping with the slight viscosity change upon heating, and the fact that for salmon DNA, which is relatively heterogeneous with respect to the base composition of individual molecules, only a part of the observed bandwidth is due to Brownian motion. Curves 3 and 4 show the familiar behavior of untreated DNA in which heating to 100°C results in irreversible denaturation and its concomitant increased effective density.

Preliminary results on the kinetics of the reaction of nitrite with DNA amino groups have been reported.[6] On the basis of these results one would estimate that in one hour at 25°C, pH 4.2, less than 2 per cent of the amino groups of DNA have reacted. The complete modification of such a fundamental property of DNA as its irreversibility of denaturation at such a low extent of reaction is remarkable. In conjunction with current work on the factors controlling the irreversibility of denaturation, it strongly suggests the possibility that covalent crosslinks between

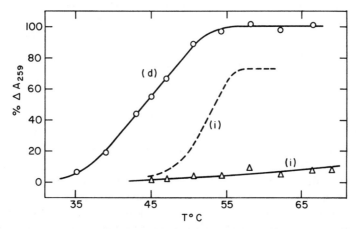

FIG. 3.—The reversible thermal denaturation of HN2-DNA in 7.2 M NaClO$_4$, pH 7. %ΔA_{259}, percent of maximum absorbance increase (referred to measurement at the ambient temperature) uncorrected for thermal expansion of solvent. O, absorbance measured at temperature shown, (d). △, absorbance measured at 25°C on samples previously equilibrated at the temperature shown and quenched in ice, (i). --------, (i) curve of untreated DNA, in this medium.

Comparison of properties of HN2-DNA and untreated DNA in this medium: (a) midpoint denaturation temperature, measured at ambient temperature ($T_{1/2,d}$), 44.1 and 44.2°C, respectively; (b) maximum absorbance increase upon thermal denaturation, (uncorrected for solvent expansion) 34 and 36%, respectively.

polynucleotide chains are being formed. We have, therefore, turned to a reagent whose crosslinking ability was suggested a number of years ago, namely the bifunctional nitrogen mustard, HN2.[7] In fact, we have found HN2 to be capable of converting DNA into a reversibly denaturable form (Fig. 3).

Our original interest in "reversible" DNA stemmed from the light that it casts on the mechanism of denaturation. The following interpretation, consistent with the reversibility of partial denaturation in unmodified DNA,[8] may be offered for the properties of "reversible" DNA: Wherever covalent bonds are formed between complementary chains, the contiguous base pairs are constrained to remain in a small-volume element, even when the secondary structure is completely disordered. Let the environmental conditions once more become thermodynamically favorable to the re-establishment of the ordered structure, and those nucleotides, which have been constrained by a covalent crosslink, re-associate in the "correct" conjugate sequence. By so doing, they establish the registration of the rest of the two polynucleotide strands, and also provide a nucleus, from the ends of which the reformation of the specifically paired, double helix may be propagated. Similar considerations appear to apply to the control of denaturation reversibility in globular[9] and fibrous[10] proteins. Undoubtedly, as the folding of the internal structure becomes simpler, the ability of crosslinks to control the reversal of denaturation becomes greater, and from this point of view, the "efficiency" of DNA crosslinks should be very great indeed.

It should be made clear that the reversible denaturation of HNO_2- and HN2-"reversible" DNA on the one hand, and the "re-naturation" described by Marmur, Lane, Doty, Eigner, and Schildkraut,[11, 12] on the other, are distinctly different phenomena: (1) "Reversible" DNA can be made from heterogeneous animal sources, and not merely from bacterial or viral DNA; (2) the reversibility can be displayed in low-ionic-strength media, such as that of Figure 1, in which completely denatured bacteriophage T2 DNA "re-natures" extremely slowly, if at all,[8] as well as in high-ionic-strength media, where both processes occur; and (3) "reversible" DNA cannot be kinetically blocked (quenched) near 0°C.

Finally, it is necessary to comment on the fact that the ability of nitrous acid to produce reversible DNA, presumably by crosslinking, has not been noted before. It is probable that the intramolecular condensation products formed in this reaction are unstable under the rigorous conditions previously used for isolating and identifying RNA and DNA deamination products[6] and have escaped detection for that reason. It should also be pointed out that if the interpretation of the structural basis of DNA reversibility is correct, a very small number of crosslinks should suffice to endow a DNA molecule with reversibility, so that the acquisition of this property becomes a most sensitive assay of a variety of, as yet unspecified, reaction products.

The question of the extent to which chemical mutagenic experiments on TMV, DNA, and bacteriophage, utilizing nitrous acid,[13-16] need re-evaluation depends on a determination of the relative frequencies of crosslinking and other reactions. It seems reasonable to anticipate that crosslinking would produce entirely different modifications of genetic material from those ascribed to base transformations, perhaps leading to inactivation of sections of the phage genome rather than point mutations. The nature of these inactivations probably differs in single-stranded and double-stranded polynucleotides. In the single-stranded polynucleotides, the compact, partly ordered configuration existing at the high ionic strength of the nitrous acid reaction, would permit covalent bonds to join distant elements of the nucleotide *sequence*, whereas in native DNA, crosslinks would form between complementary strands. Perhaps, the demonstration of the chemical and physical effects of nitrous acid described in this communication accounts, in part, for the high lethality of this *in vitro* mutagen.

I should like to express my thanks to B. S. Strauss for most helpful discussions, and to W. Bloom, R. B. Uretz, and R. E. Zirkle for their comments on the manuscript. This research was supported by U. S. Public Health Service Grant C5007.

Note added in proof: I should like to direct attention to the paper of Marmur and Grossman [these Proceedings, **47**, 778 (1961)] which presents convincing evidence that a similar type of "reversible" DNA formation occurs as the result of irradiation with ultraviolet light.

[1] Simmons, N., as quoted by V. L. Stevens, and E. L. Duggan, *J. Am. Chem. Soc.*, **79**, 5703 (1957).

[2] Hamaguchi, K., and E. P. Geiduschek (manuscript in preparation).

[3] Levene, P. A., and W. A. Jacobs, *Ber.*, **43**, 3150 (1910).

[4] Sueoka, N., J. Marmur, and P. Doty, *Nature*, **183**, 1429 (1959).

[5] Rolfe, R., and M. Meselson, these PROCEEDINGS, **45**, 1039 (1959).

[6] Vielmetter, W., and H. Schuster, *Biochem. Biophys. Res. Comm.* **2**, 324 (1960).

[7] Goldacre, R. J., A. Loveless, and W. C. Ross, *Nature*, **163**, 667 (1949).

[8] Geiduschek, E. P., *Federation Proc.*, **20**, 353 (1961).

[9] White, F. H., Jr., *J. Biol. Chem.*, **235**, 383 (1960).

[10] Veis, A., and J. Cohen, *Nature*, **186**, 720 (1960).

[11] Marmur, J., and D. Lane, these PROCEEDINGS, **46**, 453 (1960).

[12] Doty, P., J. Marmur, J. Eigner, and C. Schildkraut, these PROCEEDINGS, **46**, 461 (1960).

[13] Schuster, H., and G. Schramm, *Zeits. Naturforschung*, **13b**, 697 (1958).

[14] Mundry, K. W., and A. Gierer, *Z. Vererbungslehre*, **89**, 614 (1958).

[15] Litman, R., and H. Ephrussi-Taylor, *Co. Re. Acad. Sci.* (*Paris*) **249**, 838 (1959).

[16] Bautz-Freese, E., and E. Freese, *Virology*, **13**, 19 (1961).

From *Nature*, Vol. 194, No. 4835, p. 1274, 1962. Reproduced with permission of the author and publisher.

Proof that the Replication of DNA involves Separation of the Strands

IN their description of the specific pairing of the nucleotides of deoxyribonucleic acid (DNA), Watson and Crick[1] proposed that this pairing was the basis for a mechanism of duplication: if unwinding occurred at the time of replication, each strand of the double helix could act as a template for the creation of a new molecule so that each daughter molecule would be a 'hybrid' of a parental strand and a new strand. Since then there have been several demonstrations that DNA replication does indeed result in an equal division of the parent molecule between the two daughters[2,3]. This, however, has not put the hypothesis beyond the realm of cavil. The DNA which divides between the two daughters might, the argument goes[4,5], consist of a pair of double helices: if so, no unwinding of the strands would be necessary to produce the observed result, and the mechanism of replication could be radically different from that proposed by Watson and Crick.

Decision could be reached if some DNA were found which (a) becomes hybrid on replication and (b) can be isolated intact in its entirety and, in this state, shown to be two-stranded. Bacterial DNA is unsuitable; it becomes hybrid[2], but it has not yet been isolated intact and so there is no unambiguous estimate of its native length-to-mass ratio (that is, strandedness). Conversely, $T2$ bacteriophage DNA is known to be largely or totally two-stranded[6], but the transfer of parental $T2$ DNA to the progeny is accompanied by so much fragmentation that only small sections of the recipient molecules are known to be hybrid[7]; these small hybrid sections could conceivably be four-stranded without markedly disturbing the ratio of total length to total mass. However, the case of λ-bacteriophage DNA appears to be straightforward. Parental λ-DNA is transferred to progeny particles which, though typical in other respects[8], contain DNA which is half parental and half new (that is, is manifestly hybrid)[9]. Further, the entire DNA of λ resides in a single molecule which can be extracted intact[10,11]. If this molecule can be shown to have the ratio of total length to total mass of two-stranded DNA, then strand separation must occur.

To prepare labelled λ DNA for autoradiography, lysogenic *E. coli CR*34 (λ) thy⁻ was induced by irradia-

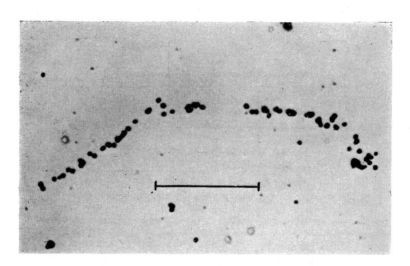

Fig. 1. Autoradiograph showing two molecules of λ-bacteriophage DNA, labelled with tritiated-thymine and exposed under Kodak 'AR 10' stripping film for 56 days. The scale represents 10μ

tion with ultra-violet light and then incubated in A medium[9] supplemented with 2 μgm./ml. tritiated-thymine (11·2 c./mM). After 100 min., the bacteria were lysed with chloroform. The resulting phage was sedimented with 10,000-fold excess of unlabelled λ. resuspended and extracted with phenol[12]. This preparation of λ-DNA was diluted to a total concentration of 5 μgm./ml. in M/100 phosphate buffer pH 7, heated at 75° C. for 20 min.[11], and collected on glass microscope slides for autoradiography. Though most labelled molecules were seen to be folded or tangled, some could be measured. These ranged up to 23μ in length (Fig. 1). This estimate of length, when combined with the reported value of 46 × 10[6] for the molecular weight[11], gives a ratio of about 2 × 10[6] mol. wt./μ. This is the expected ratio for a DNA double helix in the B configuration[13]. Further support for the conclusion that λ-DNA is two-stranded came from the finding that the density of grains per unit length along these molecules was approximately the same as that along T2 DNA, labelled with the same tritiated-thymine and exposed under film for the same time; if T2 DNA is largely two-stranded so must be λ-DNA.

In short, λ-DNA is known to form hybrids on replication. It is shown here to have the ratio of total length to total mass of two-stranded DNA. Therefore the replication of DNA must involve separation of the polynucleotide strands and the Watson–Crick model for DNA replication seems to be correct.

JOHN CAIRNS
Department of Microbiology,
Australian National University, Canberra.

[1] Watson, J. D., and Crick, F. H. C., *Cold Spring Harb. Symp. Quant. Biol.*, **18**, 123 (1953).

[2] Meselson, M., and Stahl, F. W., *Proc. U.S. Nat. Acad. Sci.*, **40**, 783 (1958).

[3] Simon, E. H., *J. Mol. Biol.*, **3**, 101 (1961).

[4] Bloch, D. P., *Proc. U.S. Nat. Acad. Sci.*, **41**, 1058 (1955).

[5] Cavalieri, L. F., and Rosenberg, B. H., *Biophys. J.*, **1**, 317, 323, 337 (1961).

[6] Cairns, J., *J. Mol. Biol.*, **3**, 756 (1961).

[7] Kozinski, A. W., *Virology*, **13**, 124 (1961).

[8] Kellenberger, G., Zichichi, M. L., and Weigle, J. J., *Proc. U.S. Nat. Acad. Sci.*, **47**, 869 (1961).

[9] Meselson, M., and Weigle, J. J., *Proc. U.S. Nat. Acad. Sci.*, **47**, 857 (1961).

[10] Kaiser, A. D., and Hogness, D. S., *J. Mol. Biol.*, **2**, 392 (1960).

[11] Hershey, A. D., *et al.*, *Yearbook Carnegie Inst.*, 455 (1961).

[12] Mandell, J. D., and Hershey, A. D., *Anal. Biochem.*, **1**, 66 (1960).

[13] Langridge, R., *et al.*, *J. Mol. Biol.*, **2**, 19 (1960).

Questions

1. The parental strands must unwind and this unwinding requires energy. Propose a source of energy.
2. If there is a polymerizing enzyme, as Watson and Crick suggest, how many binding sites must the enzyme have, and what chemical reaction must it catalyze?
3. The model is designed for linear DNA. If the DNA were circular, how would the model have to be changed?
4. *E. coli* DNA has a molecular weight of 2.6×10^9 and is replicated in 40 minutes. What is the unwinding speed in revolutions per minute? *E. coli* is a cylindrical bacterium about $1\,\mu$ in diameter and $4\,\mu$ long. Discuss some of the replication problems due to geometric constraints.

ARTICLE 16   Meselson and Stahl

1. In their interpretation of this experiment, Meselson and Stahl assumed that the DNA had been fragmented during isolation. What would have been the result if replication had been semi-conservative, if the bacterium had contained a single DNA molecule, and if the DNA had not been fragmented during isolation? Suppose each bacterium had contained two DNA molecules that replicated simultaneously? That replicated sequentially?
2. Are the data consistent with the possibility that replication of a DNA molecule sometimes starts a second round of replication before the first round is completed?
3. What is the evidence that each of the two subunits of DNA is a single strand?
4. Calculate the density difference between ^{14}N and ^{15}N DNA. Between ^{12}C and ^{13}C DNA.

ARTICLE 17   Marmur and Lane
ARTICLE 18   Doty, Marmur, Eigner, and Schildkraut
ARTICLE 19   Schildkraut, Marmur, and Doty
ARTICLE 20   Freifelder and Davison
ARTICLE 21   Geiduschek

1. What assumptions have been made in articles 17, 18, and 19 in concluding that renatured DNA is a normal double helix? Could it have single-stranded ends? Could it be four-stranded?
2. In Figure 1 of Article 19 by Doty et al., the melting curve for quickly cooled DNA and slowly cooled DNA are different from one another. Why? Why are they both so different from that for native DNA?
3. What result would Freifelder and Davison (article 20) have observed if strand separation did not occur? What would they have observed if strand separation had occurred, but the single strands were not continuous and consisted instead of 10 to 20 broken phosphodiester bonds each?
4. Consider ^{14}N-^{15}N DNA as prepared by Meselson and Stahl, (article 16). Suppose this DNA were heated in formaldehyde as in article 20 by Freifelder and Davison and then centrifuged in CsCl. Draw a graph showing the fraction of DNA having hybrid density as a function of temperature.
5. Draw a melting curve that would result if two DNA samples each having a different melting curve were mixed.

•6. If the nitrite treatment used in article 21 were very short, one can imagine a situation in which on the average there would be one cross-link per molecule. According to the Poisson distribution $1/e$ will have no cross-links. Draw the results of a CsCl experiment, as in Geiduschek's paper, for DNA so treated. Also draw a pair of curves as in Figure 3 of this paper.

7. In Geiduschek's work (Figures V-5 and V-6) could an i-curve ever have a lower T_m than a d-curve for the same DNA?

8. Optical density is always lower in an i-curve than in a d-curve. Why?

9. Does the d- or i-curve indicate when strand separation occurs?

10. How would d- and i- curves be related if the DNA was heated in formaldehyde?

11. How would you expect DNA concentration to affect either a d-curve or an i-curve?

•12. Why should molecular weight affect an i-curve? Should this be less than or greater than the effect on a d-curve? Why? Should there be an effect on reversible DNA?

13. In the presence of formaldehyde the denaturation of normal DNA is always irreversible since the formaldehyde reacts with the amino groups of the bases. How would reversible DNA behave if heated in formaldehyde?

ARTICLE 22 **Cairns**

1. If Cairns knew the absolute efficiency of detection of ^3H by autoradiography, he could have determined M/l from the grain density. How could he have determined this efficiency and what other information would he have needed?

References

Baldwin, R. L., and E. M. Shooter. 1963. "The Alkaline Transition of BU-containing DNA and Its Bearing on the Replication of DNA." *J. Mol. Biol.* 7, 511–526.

Cavalieri, L. F., J. F. Deutsch, and B. H. Rosenberg. 1961. "The Molecular Weight and Aggregation of DNA." *Biophys. J.* 1, 301–316.

Cavalieri, L. F., and B. H. Rosenberg. 1961. "The Replication of DNA. I. Two Molecular Classes of DNA." *Biophys. J.* 1, 317–322; "II. The Number of Polynucleotide Strands in the Conserved Unit of DNA." *Biophys. J.* 1, 323–336. "III. Changes in the Number of Strands in *E. coli* DNA During Its Replication Cycle." *Biophys. J.* 1, 337–351.

Geiduschek, E. P. 1962. "On the Factors Controlling the Reversibility of DNA Denaturation." *J. Mol. Biol.* 4, 467–487.

Grossman, L., S. S. Levine, and W. S. Allison. 1961. "The Reaction of Formaldehyde with Nucleotides and T2 Bacteriophage DNA." *J. Mol. Biol.* 3, 47–60.

Inman, R. B., and R. L. Baldwin. 1962. "Helix-Random Coil Transition in Synthetic DNA's of Alternating Sequence." *J. Mol. Biol.* 5, 172–184. The lower stability of BU-containing DNA in alkali was first reported here.

Meselson, M., and J. J. Weigle. 1961. "Chromosome Breakage Accompanying Genetic Recombination in Bacteriophage." *Proc. Nat. Acad. Sci.* 47, 857–868. In this paper it is shown that phage λ DNA replicates semi-conservatively.

Additional Readings

Josse, J. and J. Eigner. 1966. "Physical Properties of Deoxyribonucleic Acid." *Ann. Rev. Biochem.* 35, 789–834. This is a careful and critical review of the state of knowledge of DNA in 1966.

Zimm, B. H., and N. R. Kallenbach. 1962. "Selected Aspects of the Physical Chemistry of Polynucleotides and Nucleic Acids." *Ann. Rev. Phys. Chem.* 13, 171–194.

FACTORS DETERMINING
THE STABILITY OF DNA

Hydrophobic Forces in the DNA Molecule

For years it seemed obvious from the Watson-Crick structure that the force holding the two strands together was hydrogen bonding. Therefore it was assumed that if the hydrogen bonds were broken, the strands would separate. Since a GC pair has three hydrogen bonds and an AT pair has two (see Appendix B), it was expected that the stability of DNA would increase with increasing GC content. This was investigated by studying the melting curves (see Fig. V-2) for DNA samples having different GC contents. Most of this work was carried out in the laboratory of Paul Doty, and was reported by J. Marmur and him in Article 23.

Hydrogen Bonds Are Not
the Only Stabilizing Factor

After the work of Marmur and Doty on heat denaturation, other potential denaturants were sought and found. The kinds of reagents that

destabilize (denature) DNA are substances that interact with the amino, keto, or hydroxyl groups engaged in hydrogen bonding—e.g., urea, formaldehyde, formamide, dimethysulfoxide (and others shown in Fig. VI-1)—and the degree of denaturation increases with increasing concentration of the reagent. If hydrogen bonding is the major stabilizing force, a substituent in the denaturing reagent that does not affect hydrogen bonding—e.g., the addition of a methyl group in urea—would not be expected to affect the denaturing power of the reagent. The fact that it does was one of the early surprises. Also, increasing ionic strength should not affect the stability once the negative charge on the phosphate is completely neutralized (see question 1 for Article 23). However, Kozo Hamaguchi and Peter Geiduschek (1961) found the contrary. They discovered that for certain anions—principally trichloracetate (Cl_3CCOO^-), trifluoracetate (F_3CCOO^-), thiocyanate (CNS^-), and perchlorate (ClO_4)—at very high molarity the

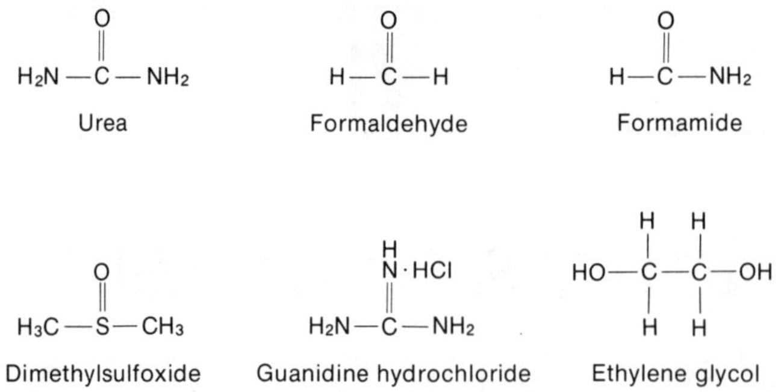

Figure VI-1
Chemical formulas for several common denaturants. Note that many contain groups that can form hydrogen bonds. [From *Physical Biochemistry* by D. Freifelder. W. H. Freeman and Company. Copyright © 1976.]

melting temperature T_m began to decrease with increasing molarity, as shown in Fig. VI-2 (see also Article 23). This phenomenon was observed whether the temperature dependence of the relative viscosity (a measure of the rigidity of the DNA molecule), optical rotation (a measure of the helicity of the DNA molecule), or optical absorbance (Fig. VI-3) was measured. Since none of these anions were known to be agents that break hydrogen bonds, and for a given anion concentration, T_m still increased substantially with increasing guanine-cytosine content (Fig. VI-4), Hamaguchi and Geiduschek stated,

"That the difference in the contributions of AT and GC nucleotide pairs can be varied so greatly . . . by substances which are not, in the usual sense, hydrogen bond breaking agents strongly suggests that the difference in the stabilities of AT and GC-rich DNA helices does not arise exclusively from hydrogen bond contributions."

What, then, is the cause of the denaturing power of these concentrated electrolytes? The clue came from a study of the effect these salts have on the solubility in water of reasonably insoluble or immiscible organic compounds. Hamaguchi and Geiduschek found that those anions which decreased T_m the most were the most effective in dissolving such substances. However, this was not the complete answer since methylammonium chloride had the same effect on solubility, but was not a denaturant.

A study by nuclear magnetic resonance spectroscopy of the effects these compounds have on the hydrogen bonding between water molecules showed that the denaturants were water-structure breakers whereas methylammonium chloride was a water-structure former. Therefore they drew the following conclusion:

Insofar as the denaturing anions exert their effect on DNA through their modification of the structure of water they can be classed as *hydrophobic bond breaking agents*. The structural details of these effects are not, at present, established. The interpretation which has been offered above, is only qualitative. In fact, until recently there has been little information about the structure of water in concentrated electrolyte solutions, on the basis of which to give these qualitative notions precise structural definition. It is undoubtedly for this reason that the specific details of previous speculations about the function of water in stabilizing the secondary structure of DNA and proteins have proved to be more provocative than accurate. However, current structural studies may ultimately provide the basis for more fruitful model building.

Further Evidence for the Role of Hydrophobic Forces

Another tool for studying denaturation was made available at the time by Lawrence Levine, who had found that it was possible to

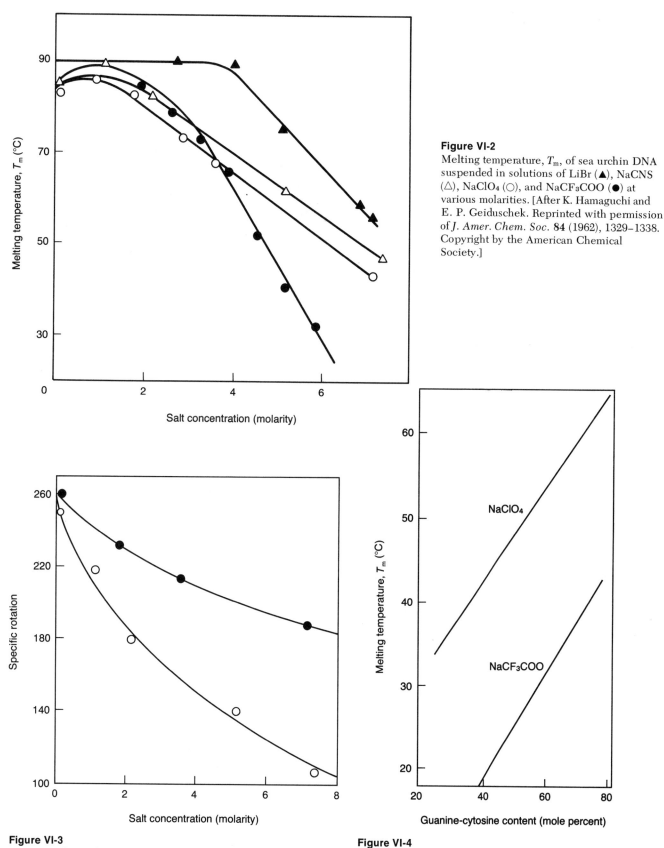

Figure VI-2
Melting temperature, T_m, of sea urchin DNA suspended in solutions of LiBr (▲), NaCNS (△), NaClO₄ (○), and NaCF₃COO (●) at various molarities. [After K. Hamaguchi and E. P. Geiduschek. Reprinted with permission of *J. Amer. Chem. Soc.* **84** (1962), 1329–1338. Copyright by the American Chemical Society.]

Figure VI-3
Variation of specific rotation of sea urchin DNA with salt concentration for NaClO₄ (solid circles) and KCNS (open circles). [After K. Hamaguchi and E. P. Geiduschek. Reprinted with permission of *J. Amer. Chem. Soc.* **84** (1962), 1329–1338. Copyright by the American Chemical Society.]

Figure VI-4
Variation of T_m with base composition of several DNA samples suspended in either 7.2 *M* NaClO₄ or 6.5 *M* NaCF₃COO. DNA was obtained from different organisms in order to vary base composition. [After K. Hamaguchi and E. P. Geiduschek. Reprinted with permission of *J. Amer. Chem. Soc.* **84** (1962), 1329–1338. Copyright by the American Chemical Society.]

TABLE VI-1
Reagents and their effects on DNA stability

Reagent	Molarity giving 50% denaturation[1]	Adenine solubility in 1 M reagent	Sodium pyrophosphate solubility in 3 M reagent
Methanol	3.5	0.0159 M	0.129 M
Ethanol	1.2	0.0177 M	0.062 M
n-propanol	0.54	0.0226 M	0.054 M
Formamide	1.9	0.0154 M	0.166 M
N,N-dimethyl-formamide	0.60	0.0277 M	0.043 M
Acetamide	1.1	0.0206 M	0.177 M
Propionamide	0.62	0.0225 M	0.105 M
Urea	1.0	0.0177 M	0.215 M
Ethylurea	0.60	0.0225 M	0.052 M

Source: Data taken from L. Levine, J. Gordon, and W. P. Jencks, 1963, *Biochemistry* **2** (1963), 168–173, with permission of the American Chemical Society.

[1]Denaturation determined by complement fixation using antibody to single-stranded DNA of *E. coli* phage T4. The DNA was in aqueous solution at an ionic strength of 0.043. The reagent was added and the sample was heated to 73°C.

prepare an antibody that could be directed against single-stranded DNA but that would not interact with normal double-stranded DNA. Therefore, denaturation could be detected by following the interaction of this antibody with DNA. Using complement fixation, an immunological technique having the advantage that a large number of potential denaturants could be studied in a relatively short time, Levine, Julius Gordon, and William Jencks examined the stability of DNA in the presence of a wide variety of reagents (Table VI-1). The pattern that emerged was that the destabilizing effect of the reagents is unrelated to their ability to break hydrogen bonds but that the stability of DNA is determined by the solubility of adenine* in the solution of the reagent, stability decreasing as solubility increases. In addition they observed that the solubility of sodium pyrophosphate, a compound whose negative charge density is like that of the sugar phosphate backbone of DNA, decreases as the solubility of adenine increases. Once again evidence was accumulating that hydrophobic forces contributed to the stabilizing of DNA. However, allowing the possibility that this is not the only stabilizing factor (e.g., the increase in stability with high

*Unfortunately, only adenine was studied. Useful information about the effect of GC content on T_m might have come from studies of other bases.

GC content always had to be considered), they cautiously stated:

It should be emphasized that the forces involved in the denaturation of DNA are not necessarily the same as those which are important in maintaining the structure of native DNA. Although it is quite likely that hydrophobic forces contribute to the stability of native DNA, the effectiveness of hydrophobic denaturing agents does not prove this hypothesis and in particular the conclusion that hydrogen bonding of DNA does not contribute to the denaturing activity of the compounds examined does not imply that hydrogen bonding is not important in maintaining the structure of native DNA.

Base Stacking

The idea that a force other than hydrogen bonding might stabilize DNA and that hydrogen bonds might serve mainly to keep the strands in register (i.e., the bases properly aligned for pairing) was also evolving in another lab from a rather different line of experiments. Ignacio Tinoco, Jr., a molecular spectroscopist, had been studying the interaction between free nucleotides in solution in order to understand why single-stranded DNA had a lesser absorbance than the free bases. The groundwork for his work had been laid more than 50 years earlier in the German dye

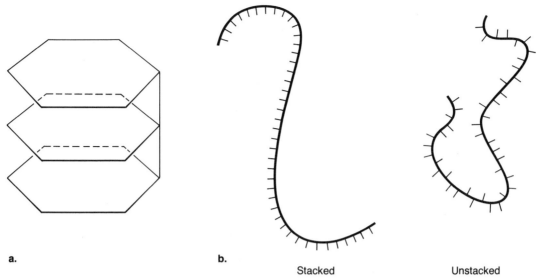

a.   b.   Stacked   Unstacked

Figure VI-5
(a) Drawing of a trinucleotide with stacked bases.
(b) Structures of stacked and unstacked polynucleotides. The stacked polynucleotide is more extended because the stacking tends to decrease the flexibility of the molecule. [From *Physical Biochemistry* by D. Freifelder. W. H. Freeman and Company. Copyright © 1976.]

industry when it was observed that dyes containing conjugated ring systems decreased in relative absorbance and often changed color as the concentration was increased; it had been proposed that this was an effect of stacking of the dye molecules in solution. In 1962 Robert Devoe and Tinoco carried out a theoretical analysis of some of the optical properties of nucleotides and predicted that stacking would occur in polynucleotides (Fig. VI-5). This meant that one of the widespread ideas about polynucleotide structure—i.e., that in the absence of hydrogen bonding a single strand would have no secondary structure and would be a random coil—might be incorrect.

The idea of stacking interactions was pursued in various laboratories using the related techniques of optical rotatory dispersion (ORD) and circular dichroism (CD), two methods of measuring optical activity as a function of wavelength.* The most important

---

*These methods measure the wavelength dependence of the ability of an optically active group to rotate plane-polarized light (ORD) and the differential absorption of right and left circularly polarized light (CD) The physical basis of ORD and CD is the same, and they yield the same information.

observation was that, whereas the nucleotides themselves are optically inactive, polynucleotides, DNA, and even dinucleotides show optical activity. Since optical activity is a result of a certain type of asymmetry, the bases must be arranged in space so that asymmetry is produced. Tinoco and his students showed that one could actually calculate the shape of the ORD and CD spectra by assuming that the bases are stacked. They pointed out that for polynucleotides, the existence of an ORD or CD spectrum means that the single strand is helical. This observation suggested that it might be interactions between adjacent ring systems rather than hydrogen bonding that gives DNA its extended rigid structure, and many papers in support of this idea soon appeared. The basic idea in each of these papers was that a single-stranded polynucleotide incapable of forming any hydrogen bonds seems to have a rigid, helical structure under certain conditions of pH and ionic strength and that this structure is disrupted by known denaturants of double-stranded DNA. An example of this is a study of poly-N^6-hydroxyethyladenylic acid (pHEA). The pHEA contains no groups that can hydrogen bond, yet it

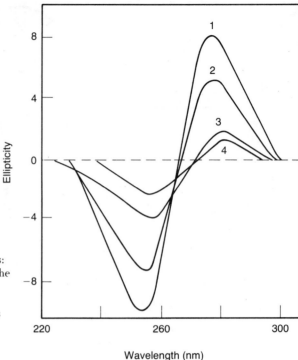

Figure VI-6
Circular dichroic spectra of pHEA at various temperatures: (1) 4°C, (2) 22°C, (3) 35°C, (4) 50°C. Note that the size of the peaks decreases with increasing temperature, indicating a loss of helicity. (Ellipticity is the term used in circular dichroism analysis for the difference in the refractive indices of left and right circularly polarized light at each wavelength.) [After K. E. Van Holde, J. Brahms, and A. M. Michelson, *J. Mol. Biol.* **12** (1965), 726–739.]

shows a strong CD spectrum whose intensity is markedly reduced by heating (Fig. VI-6).

In another study, in which yeast transfer RNA (tRNA) was used, helicity was demonstrated by the criterion of an intense ORD band (Figure VI-7). This helicity was found to be unaffected by formaldehyde, a reagent known to break hydrogen bonds, yet eliminated by ethylene glycol, one of the denaturants described by Levine, Gordon, and Jencks, which increases the solubility of the nucleotides. Hence we have two complementary situations: (1) with pHEA helicity can be present even though there are no hydrogen bonds; (2) with tRNA helicity is not disrupted by an agent known to eliminate hydrogen bonds but rather by a substance that does not affect hydrogen bonding. The conclusion of both studies is that a helical structure can be maintained in the absence of hydrogen bonds.

Current View of the Forces Stabilizing the DNA Structure

From these and many related experiments the following information on DNA stabilization has emerged. DNA consists of two components, the bases and the sugar-phosphate backbone. The bases have quite low solubility in water, owing to the aromatic ring, whereas the sugar-phosphate chain has the very high solubility of sugars and the phosphate group. High solubility in water indicates the ability to interact with water molecules; a substance of low solubility cannot interact significantly with water and tends to interact with itself (a so-called *hydrophobic force*). Hence the stacking tendency of a single-stranded polynucleotide may be thought of as the consequence of the attempt of the bases to minimize contact with water. Therefore we can understand why DNA is double-stranded with internal, stacked bases—this arrangement allows the sugar-phosphate chain to be highly solvated by water and essentially removes the bases from the aqueous environment. The hydrogen bonding then maintains the proper register between the two strands. Hence base stacking is a major cause of DNA stability, and reagents that either increase the solubility of the bases or decrease the solubility of the sugar-phosphate chain reduce the stacking tendency and will be denaturants.

The greater stability of high GC DNA raises the question whether hydrogen bonding too is a stability factor. Note however that in most solvents the GC effect induces only about a 20°

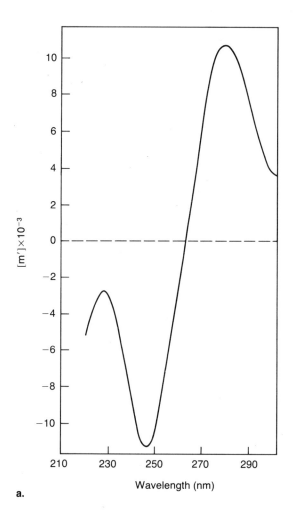

a.

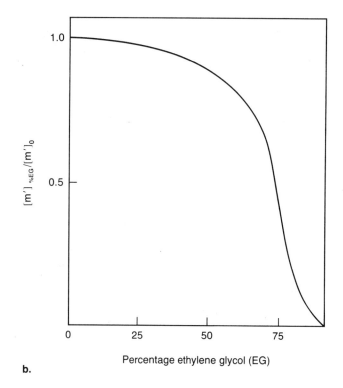

b.

Figure VI-7
(a) ORD curve of yeast tRNA. The symbol $[m']$ is the reduced mean residue rotation. The high value of the peak at 279 nm is indicative of extensive helicity.
(b) The decrease of $[m']$ at the peak value is a function of the amount of added ethylene glycol. $[m']_{\%EG}$ is the value when ethylene glycol (indicated by percent) is present and $[m']_0$ is the value when none is present. [After G. D. Fasman, C. Lindblow, and E. Seaman, *J. Mol. Biol.* **12** (1962) 630–640.]

change in T_m, whereas with reagents such as trichloracetate, which should not affect hydrogen bonds, for a 50% GC DNA the decrease in T_m with trichloracetate concentration is roughly 80°. One must conclude that, whereas hydrogen bonds probably *contribute* to the stability of DNA, they are not of major importance and the effect of base composition probably lies mainly in the strength of the stacking interactions of the various bases with one another.* Thus we may make the important suggestion that what strand separation is necessary during replication (see page 177) might be

accomplished by local environmental changes that would increase solubility of the bases, for example, by binding of a molecule (e.g., a protein) that would increase base solubility. Indeed it is known that certain proteins, one of which is necessary for DNA replication, cause denaturation of DNA at room temperature.

From the information presented here on the factors that stabilize DNA, how would you answer the following questions?

1. At pH 12 the strands of DNA separate. A possible explanation is that the amino group is protonated and therefore cannot hydrogen bond. Give another explanation that does not involve hydrogen bonding.

---

*Note that none of these considerations explain why the two strands of DNA actually unwind. Presumably when there is nothing to hold the strands together, they will unwind owing to the disordering effect of thermal motion.

2. In distilled water the two strands of DNA separate at room temperature. The resulting single strands are fairly rigid. Explain both phenomena.

3. If DNA is put into 100% methanol, its optical density increases by 37%. If the methanol concentration is reduced 20-fold by dilution of the methanolic solution into water, the normal optical density and double-stranded structure are restored. Explain.

•4. It has been stated that a single-stranded polynucleotide is helical if there is stacking and this stacking tendency has been used to explain the stability of DNA. Since stacking in a single-strand polynucleotide is between adjacent bases, how can stacking explain the tendency of the two strands of the double helix to stay together? In order to answer this question, you must remember that DNA does not have the two-dimensional ladder structure usually drawn but that the two strands are coiled around one another.

From *Journal of Molecular Biology*, Vol. 5, 109–118, 1962. Reproduced with permission of the authors and publisher.

Determination of the Base Composition of Deoxyribonucleic Acid from its Thermal Denaturation Temperature

J. MARMUR† AND P. DOTY

Department of Chemistry, Harvard University, Cambridge, Massachusetts, U.S.A.

(*Received 2 November 1961, and in revised form 5 April 1962*)

The previously discovered linear relation between the base composition of DNA, expressed in terms of percentage of guanine plus cytosine bases, and the denaturation temperature, T_m, has been further investigated. By means of measurements on 41 samples of known base composition the previously observed relation has been confirmed. It can be summarized thus: for a solvent containing $0 \cdot 2$ M-Na$^+$, $T_m = 69 \cdot 3 + 0 \cdot 41$ (G–C) where T_m is in degrees Centigrade and G–C refers to the mole percentage of guanine plus cytosine. The deviations of experimental points from this relation are no more than that expected from the uncertainties of base analysis and the variations of a half degree in the reproducibility of determining the T_m. Consequently it appears that the measurement of the T_m is a satisfactory means of determining base composition in DNA. The T_m values are most simply measured by following the absorbance at 260 mμ as a function of temperature of the DNA solution and noting the midpoint of the hyperchromic rise. Only 10 to 50 μg of DNA are required.

A number of other DNA samples of unknown base composition have been examined in this manner and their base compositions recorded.

1. Introduction

The extensive chemical base analyses of DNA by Chargaff (1955) and his associates in which they have shown the equivalence of the base pairs adenine and thymine and guanine and cytosine, and the general acceptance of the Watson–Crick structure for DNA, have made it possible to determine the average base composition of DNA by correlating it with some physical property of the macromolecule. Such a relationship has been demonstrated between the mole percent guanine plus cytosine (G–C) and the buoyant density of DNA in a CsCl density gradient (Sueoka, Marmur & Doty, 1959; Rolfe & Meselson, 1959; Schildkraut, Marmur & Doty, 1962).

It has also been shown that the base composition of DNA is related to its thermal denaturation temperature. When DNA is heated in solution, a sharp increase in its extinction coefficient occurs at the temperature where the transition takes place from the native, double-stranded structure to the denatured state. The temperature corresponding to the midpoint of the absorbance rise, the T_m, is linearly related to the average DNA base composition; a higher G–C content confers a higher thermal stability (Marmur & Doty, 1959).

Because of its ease of determination and the reproducibility of the results, the thermal transition profile is useful in determining the mole percent G–C of polymerized DNA. Other properties of the DNA preparation, such as its heterogeneity

† Present address: Graduate Department of Biochemistry, Brandeis University, Waltham, Massachusetts.

with respect to base composition as well as the presence of denatured, contaminating regions can be detected and estimated.

2. Materials and Methods

DNA was isolated by the method of Marmur (1961) from micro-organisms harvested in the logarithmic phase of growth. The samples of DNA from salmon sperm and calf thymus were prepared by the method of Simmons (Litt, 1958). The bacteriophage DNA was a gift from Dr. H. Van Vunakis of Brandeis University. Samples of poly d-AT and d-GC were generously supplied by Drs. J. Adler, J. Josse and A. Kornberg of Stanford University. It is important that the DNA be in the native configuration. The protein content of the DNA samples was estimated to be approximately 0·2 to 0·5%. Small variations in the amount of protein contaminating the DNA preparations did not significantly alter the denaturation temperature.

The base compositions of the DNA, determined by chemical analyses, were gathered from published reports of other workers.

Determination of the T_m: Unless otherwise stated, the solvent used for the DNA was 0·15 M-NaCl plus 0·015 M-sodium citrate adjusted to pH 7·0 ± 0·3 and referred to hereafter as saline–citrate. The DNA was at a concentration of approximately 20 μg/ml. and contained in a 3 ml. glass-stoppered quartz cuvette having a 1 cm light path. Small (1 ml.) cuvettes were also used when a limited supply of the sample was available. The samples were placed in the Beckman DU spectrophotometer chamber whose temperature could be raised by circulating hot water through two thermal spacers on either side of it. To protect the photocell, spacers with slowly circulating tap water at room temperature are placed adjacent to the thermal spacers circulating the hot water. A special cover for the cuvette chamber can be easily fashioned with a small hole for a thermometer to fit snugly into the chamber with its bulb immersed in a cuvette occupying either end of the chamber. This arrangement leaves enough space for two samples and a blank. The temperature drop between the thermostat and the cuvette varies from 0 to 6°C over the operating temperature range 25 to 100°C.

In determining the T_m of the DNA, the temperature of the chamber can be raised quickly to about 5°C below the estimated onset of the melting region after first measuring the optical density at 25°C. When temperature equilibrium has been attained, the temperature is then raised about 1°C at a time, allowing about 10 min for equilibration at each temperature. The optical density is read at 260 mμ and corrected for thermal expansion of the solution. A sharp increase in the absorbance occurs in the transition range during which the DNA denatures. When no further increase occurs on raising the temperature, the denaturation can be considered to be complete. When the T_m is greater than 90°C, ethylene glycol is added to the thermostatted water. Temperatures of 104°C in the chamber can easily be attained in this manner. If one uses glass-stoppered cuvettes that fit well, the loss of water due to evaporation during the heating cycle can be kept to less than 2%.

The optical density at each temperature, corrected for thermal expansion, is divided by the value at 25°C and the ratio (relative absorbance) plotted *versus* the temperature of the solution. The temperature corresponding to half the increase in the relative absorbance is designated as the T_m.

When the thermometer used to record the cuvette temperature was standardized, its readings, in °C, compared to the corrected values (in parentheses) were as follows: 65 (65·5); 70 (70·5); 80 (80·5); 85 (85·8); 90 (91); 100 (101). The temperatures recorded in Figs. 1 to 3 of this and previous communications are uncorrected; those of the Tables 1 to 5 have been corrected, as have been the values of Fig. 4.

3. Experimental Results

Dependence of the T_m *on the ionic strength.* The thermal transition of DNA is greatly influenced by the ionic strength of the solution. Figures 1 and 2 clearly show this dependence for DNA isolated from *E. coli* (K12) and *D. pneumoniae* (R-36A). It is also apparent that, because of its higher G+C content, the *E. coli* (G+C = 50%) has a higher T_m than the *D. pneumoniae* DNA (G+C = 39%) at equivalent KCl concentrations.

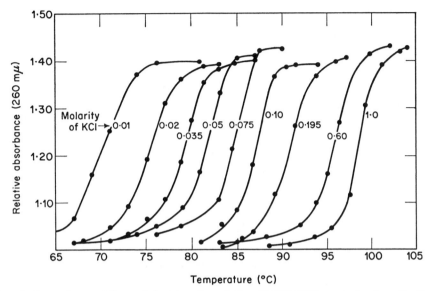

FIG. 1. Dependence of thermal denaturation of *E. coli* K12 DNA on ionic strength. *E. coli* DNA, suspended in various concentrations of KCl in glass-stoppered quartz cuvettes, was heated in the Beckman model DU spectrophotometer chamber and the relative absorbance (corrected for thermal expansion) measured at the elevated temperatures. The temperature readings are uncorrected.

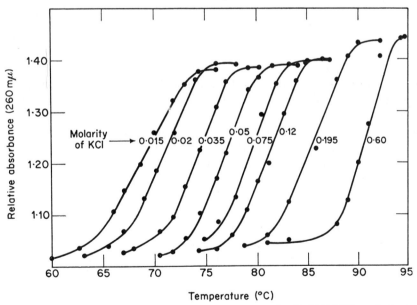

FIG. 2. Dependence of thermal denaturation of *D. pneumoniae* (R-36A) DNA on ionic strength. The experiments were carried out in an identical manner to that described in Fig. 1.

Effect of molecular weight on the T_m. Calf thymus DNA subjected to sonic dis-integration (Doty, McGill & Rice, 1958) to a molecular weight of 620,000 was com-pared to native DNA of a higher molecular weight (8×10^6) with respect to their T_m values in saline–citrate. It can be seen from Fig. 3 that the T_m is essentially unaltered by degrading the double-stranded molecule.

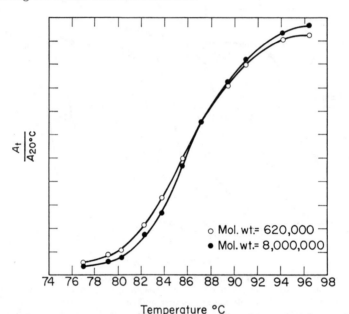

FIG. 3. Effect of molecular weight on the thermal denaturation of calf thymus DNA in 0·15 M-NaCl plus 0·015 M-sodium citrate. Sonically treated and untreated samples of DNA were thermally denatured and the absorbance at the elevated temperatures relative to that of native material at 20°C ($A_t/A_{20°C}$) was plotted as a function of the temperature to which the DNA solutions were exposed. The temperature readings are uncorrected.

Reproducibility of the T_m *within strains.* In order to evaluate the reproducibility of the T_m values, a number of determinations was carried out on portions of the same sample and on different preparations from the same source. Eight determinations of a single sample of *E. coli* DNA and measurements of twelve different *D. pneumoniae* preparations showed the same degree of reproducibility of the T_m values. The standard deviation was $\pm 0·4$°C. From the slope of the curve (Fig. 4) relating mole percent G–C to T_m, the maximum error in the estimation of the base composition from the T_m is seen to be of the order of ± 1 mole per cent G–C. Thus, by determining the T_m value several times on the same preparation, the uncertainty arising from the measurement itself can be kept within the equivalent of 1 mole per cent.

When seven different strains of *E. coli* (B, C, K12, TAU⁻, I, 44B, and Crooks) were used as sources of DNA, it was found that the variability of their denaturation temperatures was approximately the same as the T_m values of different DNA prepara-tions isolated from the same strain as well as of repeated determinations carried out on a single sample of DNA.

Relationship between T_m *and G–C contents.* In Fig. 4 we have plotted the values of T_m for samples isolated from a variety of sources as a function of the G–C content of the sample for two ionic strengths. The line at the lower ionic strength

(0·01 M-phosphate plus 0·001 M-EDTA) is parallel to the line at the higher ionic strength (0·15 M-NaCl plus 0·015 M-sodium citrate). The lines have a slope of 0·41°C per 1% rise in the G–C content and are separated by 20°C. The relation in saline–citrate

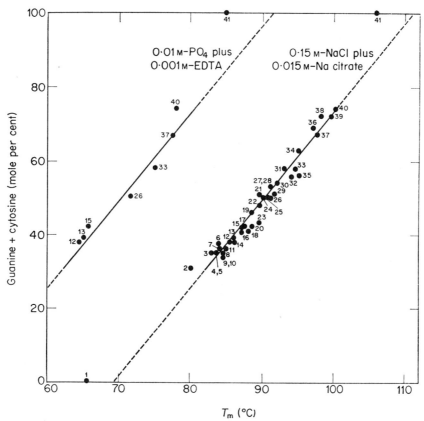

Fig. 4. Dependence of the denaturation temperature, T_m, on the guanine plus cytosine (G–C) content of various samples of DNA. The DNA samples were dissolved in either of the two solvents shown in the Fig. The T_m represents the midpoint of the hyperchromic increase of individual DNA absorbance–temperature profiles, carried out and plotted as shown in Figs. 1 to 3, and has been plotted as a function of the G–C content. The numbers next to each T_m value in the Fig. refer to the DNA extracted from the following organisms: 1, poly d-AT; 2, *Cl. perfringens*; 3, *Past. tularensis*; 4, *M. pyogenes* var. *aureus*; 5, *B. cereus*; 6, *Pr. vulgaris*; 7, *B. thuringiensis*; 8, T2r+; 9, T4r+; 10,T6r+; 11, *S. cerevisiae*; 12, *B. megaterium*; 13, *D. pneumoniae*; 14, *H. influenzae*; 15, Calf thymus; 16, *N. catarrhalis*; 17, Chicken liver; 18, Salmon sperm; 19, Wheat germ; 20, *B. subtilis*; 21, *B. licheniformis*; 22, T7; 23, *V. cholerae*; 24, T3; 25, *N. flavescens*; 26, *E. coli*; 27, *Salm. typhosa*; 28, *Sh. dysenteriae*; 29, *N. meningitidis*; 30, *Salm. typhimurium*; 31, *Br. abortus*; 32, *A. aerogenes*; 33, *S. marcescens*; 34, *Ps. fluorescens*; 35, *Azot. vinelandii*; 36, *Myco. phlei*; 37, *Ps. aeruginosa*; 38, *Sar. lutea*; 39, *M. lysodeikticus*; 40, *Strep. viridochromogenes*; 41, poly d-GC. The first and last samples, the enzymatically prepared polymers, were obtained through the generosity of Drs. A. Kornberg, J. Josse and J. Adler.

The T_m values have been corrected and are those marked with an asterisk in Tables 1 to 3. The line shown to the right was fitted to the points shown by the method of least squares.

can be represented by the equation $T_m = 69\cdot3 + 0\cdot41$ (G–C). The T_m values for enzymatically synthesized poly d-GC and *S. viridochromogenes* could only be determined at the lower ionic strength. The expected values of T_m for these samples

TABLE 1

Mole per cent G–C and T_m *of DNA from micro-organisms*

Organism	Strain or source	T_m (°C)	Mole per cent G–C From T_m	Mole per cent G–C Literature
Enterobacteriaceae†				
*Proteus vulgaris	ATCC 9484	85	37	36·5‡
Pr. mirabilis	35	85·3	38	
Pr. rettgeri	3478	86	39·5	
*Shigella dysenteriae	15	90·5	50	53·4‡
*Salmonella typhosa	643, ETS9	90·5	50	53·3‡
Salm. arizona	PC 145	90·5	50	
Escherichia freundii	5619–52	90·5	50	
*E. coli	B, C, W, K12	90·5	50	50·1‡
Salm. ballerup	Walter Reed	91	51·5	
"Aerobacter"	1041	91	51·5	
Pr. morganii	ATCC 8019	91	51·5	
*Salm. typhimurium	LT-2	91	51·5	50·2‡
Erwinia carotovora	ATCC 8061	91	51·5	
Paracolobactrum aerogenoides	McK	92·5	55	
Klebsiella pneumoniae	23	92·5	55	
*Aerobacter aerogenes	1088	93·5	57·5	57·1‡
*Serratia marcescens	Harvard Med.	93·5	57·5	58‡
Lactobacillaceae				
*Diplococcus pneumoniae	R-36A	85·5	39	38·5‡
Lactobacillus acidophilus	Blechman	85·5	39	
Streptococcus salivarius	I-R14Sm^r	85·5	39	
Leuconostoc mesenteroides	ATCC 12291	85·5	39	
Bacillaceae				
*Clostridium perfringens	876	80·5	26·5	31‡
Cl. tetani	Mandel	80·5	26·5	
Cl. chauvei	Mandel	80·5	26·5	
Cl. madisonii	Seeley	80·5	26·5	
Bacillus alvei	ATCC 6344	83	33	
*B. cereus	MB 19	83	33	35‡
*B. thuringiensis	ATCC 10792	83·5	34	35·9‡
B. megaterium-cereus	ATCC 14B22	83·5	34	
B. circulans	ATCC 4513	84	35	
*B. megaterium	U. of Penn.	85	37	37·6‡
B. lentus	ATCC 10840	85	37	
B. sphaericus	ATCC 4525	85	37	
B. pumilus	NRS 236	85·5	39	
B. laterosporus	ATCC 64	86	40	
B. firmus		86·5	41	
B. subtilis var. atterimus	ATCC 6460	87·5	43	
B. brevis	ATCC 9999	87·5	43	
*B. subtilis	168	87·5	43	42·4‡
B. natto	MB 275	87·5	43	
B. niger	ATCC 6454	87·5	43	
B. subtilis var. niger	ATCC 6455	87·5	43	
B. stearothermophilus	194	88	44	
B. polymyxa	ATCC 842	88	44	
*B. licheniformis	NRS 243	88·5	46	50·7§
B. macerans	ATCC 7069	90·5	50	

TABLE 1 (*continued*)

Organism	Strain or source	T_m (°C)	Mole per cent G–C	
			From T_m	Literature
Brucellaceae				
*_Pasteurella tularensis_	Detrick	83	33	34·7‡
Hemophilus suis	3090	85·5	39	
*_H. influenzae_	RD	85·5	39	
H. parainfluenzae	G. Leidy	85·5	39	
H. aegypti	G. Leidy	86	40	
Moraxella bovis	M. Mandel	86·5	41	
P. pestis	AVO₂, EV6	88·5	46	
*_Brucella abortus_	19	92·5	55	57·9‡
Spirillaceae				
*_Vibrio cholerae_	20A10	89	47	43·3‡
Pseudomonadaceae				
Acetobacter gluconicum	2G	92	54	
Acetobacter ascendens	ATCC 9323	92·5	55	
*_Pseudomonas fluorescens_	ATCC 949	94·5	60	63·0‡
*_Ps. aeruginosa_	NRRL B-23	97	66	67‡
Corynebacteriaceae				
Listeria monocytogenes	Detrick	85·3	38	
Corynebacterium xerosis	ATCC 9016	93·5	57·5	
Azotobacteriaceae				
*_Azotobacter vinelandii_	ATCC 9104	94·5	60	56·3‡
Mycobacteriaceae				
*_Mycobacterium phlei_	A. Brodie	97	66	66·4‡
Micrococcaceae				
*_M. pyogenes_ var. _aureus_	NRRL B-313	83·5	34	34‡
*_Sarcina lutea_	26C (Mandel)	98	68	72‡
*_M. lysodeikticus_	NRRL B-287	99·5	72	71·9‡
Athiorhodaceae				
Rhodospirillum rubrum	S-1	94·5	60	
Neisseriaceae				
*_Neisseria catarrhalis_	Ne 11 (Catlin)	86·5	41	40·7¶
N. perflava	Ne 20 (Catlin)	90	49	50·4¶
N. sicca	Ne 12 (Catlin)	90	49	51·5¶
*_N. flavescens_	13120 (Catlin)	90	49	49·2¶
*_N. meningitidis_	Ne 15 (Catlin)	91	51·5	51·3¶
Rhizobiaceae				
Rhizobium japonicum	555	95	61	
Streptomycetaceae				
Streptomyces albus	G	100·5	74	
*_S. viridochromogenes_	93	100·5	74	73·8‡
Rickettsiaceae				
Coxiella burnetii	Paretsky	87	42	
Mycoplasmataceae				
Mycoplasma gallisepticum	PPLO 5969	84	35	

† According to the classification in Bergey's *Manual of Determinative Bacteriology*, Seventh Edition.

* Values used to plot Fig. 4. ‡ Belozersky & Spirin (1960).

§ Envreinova, Bunina & Kusnetzova (1959). ¶ Catlin & Cunningham (1961).

in saline–citrate are obtained by adding the above difference. Both sets of values are included in Fig. 4.

It is interesting to note that the presence of glucose and hydroxymethylcytosine in the DNA (bacteriophages T2, T4 and T6) does not displace their T_m values very much from the line. Also, DNA isolated from a thermophile grown at an elevated temperature (*B. stearothermophilus*) exhibits a T_m which is compatible with its G–C content (Marmur, 1960). T_m values are greatly influenced by the presence of certain polyamines (Mahler, Mehrotra & Sharp, 1961), denaturing agents and extremes in pH.

It has also been found that T_m values are unchanged when the denaturation is carried out in D_2O as the solvent (Kucera, unpublished results). The presence of deuterium in the non-exchangeable positions of DNA also has no influence on its T_m (Marmur & Schildkraut, 1961).

T_m values that fall off the line may arise from the use of strains different from those whose mole per cent G–C are listed in the literature, inaccuracies in the chemical determinations of their G–C content or some yet unknown or unevaluated variables. The values listed for the chemical analyses of the DNA base compositions from the same tissue vary to a greater extent than that obtained from the T_m. Thus, the reported guanine plus cytosine content of calf thymus DNA, compiled by Chargaff (1955), varies from the extremes of 41·3 to 45·0 mole per cent.

Determination of the G + C content from the T_m. DNA was isolated from a number of organisms whose G–C content have not yet been recorded or whose published values show considerable variance. The mole per cent G–C was then evaluated from Fig. 4 using their T_m values determined in 0·15 m-NaCl plus 0·015 m-sodium citrate. The base compositions thus determined are listed in Tables 1 to 3. A careful examination of the G–C values will show that, in general, those organisms which are genetically and/or taxonomically related have similar base compositions (Lee, Wahl & Barbu, 1956).

4. Discussion

Since the results of this further investigation of the dependence of T_m on the composition of DNA are quite straightforward, only two further comments are necessary: one regarding the applicability of the method to the determination of the base composition in DNA and another concerning the relation of these results to the related study of the dependence of buoyant density on the composition of DNA (Schildkraut *et al.*, 1962).

The extension of the correlation between T_m and G–C to more than forty samples indicates the high reliability of the method for independent base composition determination and the absence of unexpected interfering situations. Moreover, the generality of the relation clearly supports the original interpretation of the molecular melting phenomenon being of a co-operative type that averaged out the fluctuations of base composition along the chain and responded to the mean composition of the chain (Doty, 1956; Marmur & Doty, 1959).

The advantages of the use of T_m determinations are clear. The result can be obtained easily on a very small amount of DNA, ordinarily 50 μg although 15 μg suffices with special cuvettes. Of course, care must be taken to insure that the DNA was not denatured in preparation, that the cation concentration is maintained at 0·2 m-Na^+ without divalent metal ion contamination and that the heating is not carried out too rapidly.

TABLE 2

Mole per cent G–C and T$_m$ of DNA from bacteriophage

Bacteriophage	Host	T$_m$ (°C)	Mole per cent G–C From T$_m$	Literature†
*T2r+	E. coli	83	33	35
*T4r+	E. coli	84	35	34·4
*T6r+	E. coli	83	33	34·2
α	"B. megaterium"	86·5	41	
λ^c	E. coli	89	47	50·0
*T7	E. coli	89·5	48	48·0
*T3	E. coli	90	49	49·6
S20 v	S. marcescens	92·5	55	

* Values used to plot Fig. 4.
† Sinsheimer (1960).

TABLE 3

Mole per cent G–C and T$_m$ of DNA from plants and animals

Organism	Tissue	Method of preparation and/or source	T$_m$ (°C)	Mole per cent G–C† From T$_m$	Literature
		Plants			
Tobacco	Leaf	Marmur	85·5	39	
*Wheat	Germ	Josse	88·5	46	46·6†
		Animals			
Human	Spleen	Kirby	86·5	41	41·4§
Mouse	Spleen	Kirby	86·5	41	41·9‖
Rat	Liver	Kirby	86·5	41	41·4‖
Drosophila melanogaster	Whole animal	Kirby	86·5	41	
*Calf	Thymus	Simmons	87	42	41·9§
*Chicken	Liver	Kirby	87·5	43	41·7‖
Chicken	Embryo liver	Kirby	87·5	43	41·7‖
*Salmon	Sperm	Simmons	87·5	43	41·2§
		Miscellaneous			
*Saccharomyces cerevisiae			85	37	35·7¶
Euglena gracilis			90	49	

* Values used to plot Fig. 4.
† Includes the substitution of cytosine by methylcytosine.
‡ Josse (personal communication).
§ Chargaff (1955).
‖ Kirby (1959).
¶ Belozersky & Spirin (1960).

The limitations of the method are its obvious insensitivity to other than the four normally occurring bases and its restriction to the range of composition for which it has been established, that is from 25 to 75 mole per cent G–C. This latter point is emphasized by the fact that the helical complexes made from synthetically prepared homologous polynucleotides do not fit the extrapolation of the linear relation.

Indeed, the divergences for poly GC and poly AT correspond to 5°C or 12 mole per cent G–C. This is quite outside probable error and suggests that a slightly different structure is taken up when only one kind of base pair needs to be accommodated in the helix. Conversely, when two base pairs, which have slightly different dimensions, occur in the helical framework, it is quite likely that a different adjustment of the parameters is required for optimal fit from those that best satisfy the case where only one base pair is involved.

The linear relation between T_m and G–C obtained here and that derived between buoyant density and G–C (Schildkraut et al., 1962) are, of course, expected to be mutually consistent. A direct examination of compositions deduced from the measurement of T_m and buoyant density on identical samples shows that the base compositions deduced from the linear relations are generally within the expected probable error. A few deviations occur in the very low G–C region (25 to 33%). However, in no case was the difference greater from that corresponding to either 2·5° or 0·006 g/c.cm. Thus, although some further work remains to be done when samples of very low G–C content become available, the agreement can be taken as satisfactory over the range of 30 to 75% and compositions estimated from either T_m or buoyant density taken as interchangeable.

The authors are indebted to Drs. C. L. Schildkraut and N. Sueoka for discussions during the course of this work and to Dr. J. Josse for a sample of wheat germ DNA, to Dr. B. W. Catlin for DNA from *Neisseria* and to Dr. K. S. Kirby for DNA samples from animal sources prepared by the phenol method. Dr. D. Paretsky supplied the DNA from *Coxiella burnetii*. The *Mycoplasma* was grown for us by Dr. R. Cleverdon. Several samples of DNA and strains of bacteria were supplied by Drs. M. Mandel and S. Falkow. They also contributed greatly in rewarding discussions concerning the taxonomic classification of several bacterial groups.

This work was supported by grants from the United States Public Health Service (C–2170) and the National Science Foundation (G–13990). Part of the work was carried out at Brandeis University.

REFERENCES

Belozersky, A. N. & Spirin, A. S. (1960). In *The Nucleic Acids*, ed. by E. Chargaff & J. N. Davidson, vol. 3, p. 147. New York: Academic Press.

Catlin, B. W. & Cunningham, L. S. (1961). *J. Gen. Microbiol.* **26**, 303.

Chargaff, E. (1955). In *The Nucleic Acids*, ed. by E. Chargaff & J. N. Davidson, vol. 1, p. 308. New York: Academic Press.

Doty, P. (1956). *Proc. Nat. Acad. Sci., Wash.* **42**, 791.

Doty, P., McGill, B. B. & Rice, S. A. (1958). *Proc. Nat. Acad. Sci., Wash.* **44**, 432.

Envreinova, T. N., Bunina, N. V. & Kusnetzova, N. Y. (1959). *Biokhimiya,* **24**, 912.

Kirby, K. S. (1959). *Biochim. biophys. Acta,* **36**, 117.

Lee, K. Y., Wahl, R. & Barbu, E. (1956). *Ann. Inst. Pasteur,* **91**, 212.

Litt, M. (1958). Ph.D. dissertation, Harvard University, Cambridge, Massachusetts.

Mahler, H. R., Mehrotra, B. D. & Sharp, C. W. (1961). *Biochem. Biophys. Res. Comm.* **4**, 79.

Marmur, J. (1960). *Biochim. biophys. Acta,* **38**, 342.

Marmur, J. (1961). *J. Mol. Biol.* **3**, 208.

Marmur, J. & Doty, P. (1959). *Nature,* **183**, 1427.

Marmur, J. & Schildkraut, C. L. (1961). *Nature,* **189**, 636.

Rolfe, R. & Meselson, M. (1959). *Proc. Nat. Acad. Sci., Wash.* **45**, 1039.

Schildkraut, C. L., Marmur, J. & Doty, P. (1962). *J. Mol. Biol.* **4**, 430.

Sinsheimer, R. L. (1960). In *The Nucleic Acids*, ed. by E. Chargaff & J. N. Davidson, vol. 3, p. 187. New York: Academic Press.

Sueoka, N., Marmur, J. & Doty, P. (1959). *Nature,* **183**, 1429.

Questions

ARTICLE 23 Marmur and Doty

1. Why is there an effect of salt concentration on T_m?

2. Why should heat break hydrogen bonds?

3. Would you expect these curves to be affected by the molecular weight of the DNA?

4. Suppose you obtained a melting curve, but the total increase in optical density was only 20%, the transition being normally sharp. How might you explain this? Would T_m from such a measurement be believable?

•5. How should T_m be related to strand separation? To the temperature at which strand separation occurs? Does your answer depend upon whether the DNA is homogeneous (a phage DNA) or heterogeneous (fragmented bacterial DNA)?

References

Fasman, G. D., C. Lindblow, and E. Seaman. 1965. "Optical Rotatory Dispersion Studies on the Conformational Stabilization Forces of Yeast Soluble Ribonucleic Acid." *J. Mol. Biol.* **12**, 630–640.

Hamaguchi, K., and E. P. Geiduschek. 1961. "The Effect of Electrolytes on the Stability of the DNA Helix." *J. Amer. Chem. Soc.* **84**, 1329–1338.

Levine, L., J. A. Gordon, and W. P. Jencks. 1963. "The Relationship of Structures to the Effectiveness of Denaturing Agents for Deoxyribonucleic Acid." *Biochem.* **2**, 168–175.

Van Holde, K. E., J. Brahms, and A. M. Michelson. 1965. "Base Interactions of Nucleotide Polymers in Aqueous Solution." *J. Mol. Biol.* **12**, 726–739.

Additional Readings

Bloomfield, V. A., D. M. Crothers, and I. Tinoco. 1974. *Physical Chemistry of Nucleic Acids.* Harper and Row.

Crothers, D. M., and B. H. Zimm. 1964. "Theory of the Melting Transition of Synthetic Polynu-cleotides: Evaluation of the Stacking Free Energy." *J. Mol. Biol.* **9**, 1–9. This is a thermodynamic treatment of the denaturation process leading to the conclusion that stacking is the major factor of the forces stabilizing the double helix.

Helmkamp, G., and P. O. P. Ts'o. 1961. "The Secondary Structures of Nucleic Acids in Organic Solvents." *J. Amer. Chem. Soc.* **83**, 138–142.

Herskovits, T. T., S. J. Singer, and E. P. Geiduschek. 1961. "Nonaqueous Solutions of DNA. Denaturation in Methanol and Ethanol." *Arch. Biochem. Biophys.* **94**, 99–114. The idea of hydrophobic bonds as a stabilizing agent is presented here.

Marmur, J., R. Rownd, and C. L. Schildkraut. 1963. "Denaturation and Renaturation of Deoxyribonucleic Acid." *Prog. Nucleic Acid Res.* **1**, 232–300.

Ts'o, P. O. P. 1974. *Basic Principles in Nucleic Acid Chemistry.* Academic Press.

Zimm, B. H., and N. R. Kallenbach. 1962. Selected Aspects of the Physical Chemistry of Polynucleotides and Nucleic Acids." *Ann. Rev. of Phys. Chem.* **13**, 171–194. The physical basis of hypochromicity is explained here.

SUBSTRUCTURE OF THE DOUBLE HELIX

Variants of the Basic DNA Model:
Fine Points of Structure

In the simplest version of the Watson-Crick model it is assumed that naturally occurring DNA is a linear molecule consisting of two continuous polynucleotide strands, both of same length (i.e., having the identical number of nucleotides) and terminating simultaneously (i.e., no single-stranded ends), and held together by adenine-thymine and guanine-cytosine base pairs (see Fig. A-4 of Appendix A). Indeed, although that description was probably applicable to the samples used in the X-ray analysis, it is not generally accurate.

Other Bases in DNA

The first departure from this model came when G. R. Wyatt and Seymour Cohen (1952) showed that the DNA of the *E. coli* phages T2, T4, and T6 contains the base hydroxymethyl-cytosine (HMC) instead of cytosine. Their work was followed shortly by Robert Sinsheimer's observation (1954) that the HMC is glucosylated (Fig. VII-I).

Since HMC pairs with guanine in the same way as cytosine does, the finding of a new DNA base was not difficult for the scientific community to accept, and it was guessed that other "unusual bases" would be found. In the next few years, methylated derivatives of adenine, guanine, and cytosine were also found in the DNA of many organisms. It was, however, surprising when uracil, a base formerly found only in RNA, was found in thymine-free DNA in some of the phages of the bacterium *Bacillus subtilis*.

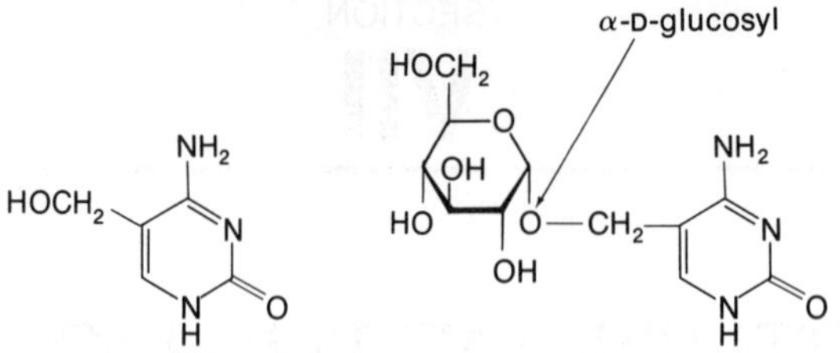

Figure VII-1
Chemical structure of 5-hydroxymethylcytosine (left) and a glucosylated derivative found in *E. coli* phage T4 DNA.

Interruptions in the Single Strands

Another unforseen detail of DNA structure came from Peter Davison's laboratory. He and I had noticed for a variety of phage DNA molecules that the single strands obtained by heat denaturation did not have a single sedimentation rate and hence were probably not all of the same size; we proposed that single-strand breaks would have to be present in many double-stranded phage DNA molecules, probably even when the DNA was contained within the phage (Fig. VII-2). This work is described in Article 24. The response to this paper was one of extreme skepticism, and it was thought that our findings were due to sedimentation artifacts. That the differences in *s* we observed were results of real differences in *M* was not conclusively proved until Albrecht Kleinschmidt and I showed by electron microscopy that the rapidly moving material consisted of long molecules and the slowly moving material consisted of short ones.*

At the same time Alfred Hershey and his coworkers were reaching the same conclusion. Hershey had shown, after Davison's study of hydrodynamic shear degradation, that DNA could be broken successively in two—into half molecules, quarter molecules, and so on—by stirring with a high speed mixer. He was attempting to obtain a calibration curve relating sedimentation coefficient *s* and molecular weight *M* with such a series of molecules. The advantage of this procedure (as described on page 494) was that, since *s* was apparently related to *M* by an equation of the form $s = kM^a$ where *k* and *a* are constants, one could evaluate *a* from such a series, *without knowing either* s *or* M, *merely by measuring the ratio of* s-*values for two samples 1 and 2 for which* M_1/M_2 *was known.* Hershey completed this work in collaboration with Betty Burgi using T2 DNA. The DNA of phage T5 was then examined by Burgi, Hershey, and Laura Ingraham, who learned—much to their surprise—that the first breakage products were not of equal size. Furthermore when either of these products was sheared, each again yielded fragments of unequal size. They also suggested that T5 DNA might contain single-strand breaks, one being in one strand 60% of the distance from one end (Fig. VII-3). Their work is described in Article 25. It lent support to the idea proposed by us in Article 24. T5 DNA was examined more closely during the next seven years by several groups using gel electrophoresis, sedimentation, and electron microscopy. It was ultimately shown that each T5 DNA molecule contains several single-strand breaks at unique positions; also, in some, but not all, of the molecules one or a small number of additional breaks occur at apparently random positions.

All there is to know about single-strand breaks has not yet been uncovered. What is clear though are the following points: (1) single-strand breaks are present in most natural phage DNA molecules; (2) their biological significance is unknown; (3) most are simply broken phosphodiester bonds although occa-

---

*This was a necessary element in the proof since *s* is a function of both the size and shape of a molecule, and it is possible though unlikely that some single strands in a population of phage DNA have a different shape—perhaps owing to something binding to the DNA.

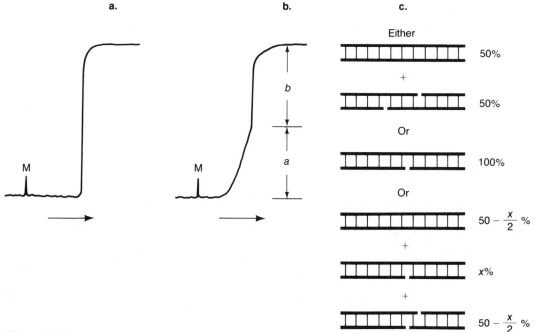

Figure VII-2
Sedimentation analysis of *E. coli* phage T7 DNA; M is the meniscus, and the arrow indicates the direction of sedimentation. (a) Native DNA has a single sharp boundary, indicating that no molecules are broken. (b) Sedimentation in alkali to separate the strands, which shows that some single strands are intact (*b*) and some are broken (*a*). Because *a* = *b*, the concentration of intact molecules equals the concentration of broken molecules. The number of broken molecules is, of course, greater than the number intact. (c) Three possible interpretations of the data. [From *Physical Biochemistry* by D. Freifelder. W. H. Freeman and Company. Copyright © 1976.]

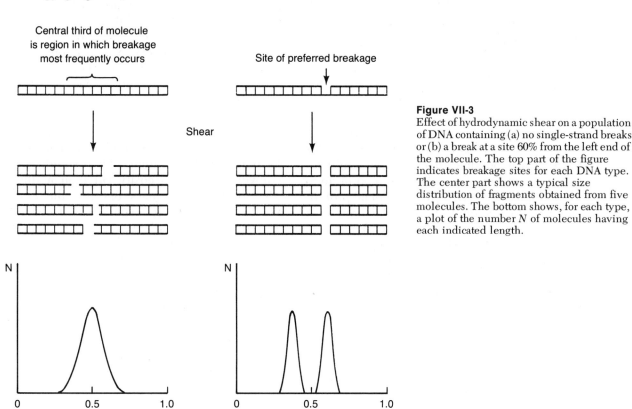

Figure VII-3
Effect of hydrodynamic shear on a population of DNA containing (a) no single-strand breaks or (b) a break at a site 60% from the left end of the molecule. The top part of the figure indicates breakage sites for each DNA type. The center part shows a typical size distribution of fragments obtained from five molecules. The bottom shows, for each type, a plot of the number *N* of molecules having each indicated length.

a. DNA with no single-strand breaks

b. DNA containing one single-strand break 60% from left end sionally a deoxyribose is also missing; (4) with the exception of *E. coli* phage T5 and possibly *Pseudomonas aeroginosa* phage B3, the breaks seem to be randomly distributed along the DNA chains; (5) how they are made is unknown.

Single-Stranded Termini

Another surprise—this also from Hershey's laboratory—was revealed by methylated albumin-kieselguhr (MAK) chromatography, a technique that fractionates DNA according to molecular weight (see Appendix D). Hershey found that the DNA from *E. coli* phage λ behaves anomalously in MAK chromatography—that is, the DNA acts as if its molecular weight were far greater than the known value, and therefore it is extraordinarily difficult to extract the DNA from the column. He knew from other work that single-stranded DNA binds very tightly to a MAK column and guessed that λ DNA might have single-stranded regions. At the same time he, Burgi, and Ingraham had also observed that the *s* of λ DNA changed with time of storage of the DNA, yielding several forms of λ DNA, each with a unique *s* value. They reported their observations and this new idea in Article 26.

The structural transitions observed in Article 26 and the behavior of the DNA on the MAK column suggested to Hershey and his colleagues that the cohesive sites were single strands with complementary base pairs since base pairing was the only intramolecular interaction that had been demonstrated for DNA. It was possible that cohesion could result from binding of a contaminating protein to unique sites on the DNA or to a special protein that specifically binds to DNA at any, or a unique, site. But this idea seemed unlikely because (1) the DNA had been deproteinized by shaking with phenol* and (2) the cohesion is both facilitated and strengthened in concentrated salt solutions whereas DNA-protein binding usually is inhibited at high ionic

strength. If the cohesive ends were single strands, the base pairing rule would require that the two ends be different. This was proved again by Hershey and Burgi in a simple but elegant paper (Article 27). Then in Article 28 H. B. Strack and A. D. Kaiser give the chemical evidence measuring the size and proving the terminal location for the single-stranded regions.

Terminal Redundancy and Cyclic Permutation

As we will see in Section IX, the work on λ just described ultimately led to the discovery that λ DNA molecules could circularize. The idea of circular DNA also arose from a parallel development in the study of the genetics of *E. coli* phage T4, as described in this section. In 1964 a striking finding was made in the genetics laboratories of George Streisinger, Franklin Stahl, and Robert Edgar. They had been doing genetic mapping* studies with *E. coli* phage T4 for some time in an effort to understand genetic recombination. As the number of mapped genes increased, it became apparent that the genetic map of T4 is circular. The basic argument was that if one maps a series of genes A, B, C . . . Z and it is found that A is near B, D near E, Y near Z, etc. and Z *is near A*, then the map is circular.

The question of whether the T4 DNA molecule is also circular immediately arose. Davison and I confronted this problem with viscometry, a technique that provides information about the length of the DNA molecule.† We reasoned that a circular DNA should have a lower intrinsic viscosity [η] than a linear DNA of the same mass because the circle would have a smaller axial ratio. However, since the empirical relation between [η] and M had not yet been extended beyond $M = 8 \times 10^6$ and the M of T4 DNA was greater than 100

---

*Shaking with phenol causes most proteins to precipitate. Since phenol and water do not mix, when a DNA solution containing protein is shaken with phenol and then centrifuged, the phenol and water form two separate phases with precipitated protein at the interface.

*Genetic mapping refers to determining the positions of genes with respect to one another. It is usually but not always true that the relative genetic distances between genetic markers are the same as the relative physical distances on the DNA molecule.

†Actually the information yielded is the ratio of length to width or the axial ratio. Since in general all DNA molecules have the same width, viscometry measures length. However when considering the possibility of circularity, one must be more precise.

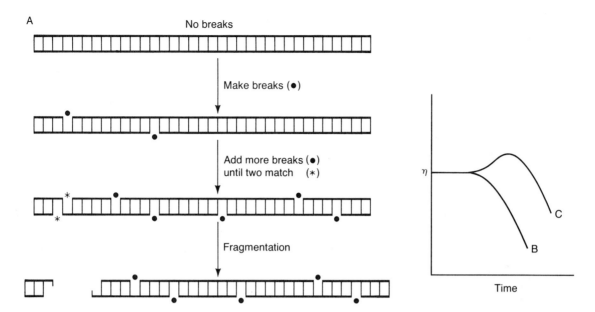

Figure VII-4

(a) Process of the matching of single-strand breaks in DNA to make double-strand breaks. The dots indicate the positions of the breaks. The asterisks indicate a pair of breaks that result in a double-strand break; they are separated by only one base pair, which is insufficient to maintain the integrity of the molecule.

(b) Viscosity of T2 DNA as a function of time of digestion with pancreatic DNase. The DNase produces single-strand breaks, which ultimately match to produce double-strand breaks, resulting in a decrease in the viscosity. The curve is flat at first because single-strand breaks do not affect the viscosity.

(c) Expected curve (as in curve b) if the DNA were circular. At the time that the first double-strand breaks occurred, the circle would become linear and the viscosity would increase. Viscosity would not decrease until a second double-strand break had formed. [From *Physical Biochemistry* by D. Freifelder. W. H. Freeman and Company. Copyright © 1976.]

$\times 10^6$, a direct measurement of $[\eta]$ and M would not have been worthwhile. Therefore, we measured viscosity η as a function of time of exposure to the enzyme DNase. Since DNase makes single-strand breaks, we reasoned that, when enough single-strand breaks had been made, a double-strand break would result, a circle (if it existed) would become linear, and η would go up. Subsequent double-strand breaks would cause η to decrease because then M would decrease (Fig. VII-4). The data showed no increase in η demonstrating that T4 DNA was not circular but linear. Even before this result was communicated, Streisinger, Edgar, and Georgetta Denhardt had already rejected the idea of a circular DNA and, with extraordinary insight, had proposed that the circular map is a result of the fact that T4 DNA is "cyclically permuted" and "terminally redundant." This work is described in Articles 29 and 30.

The idea that the phage DNA does not have defined ends plus the fact demonstrated by Davison and by Hershey that T4 DNA molecules are all the same size (Articles 4 and 8) immediately provoked the question of how the phage controls the length of its DNA. The simple proposal was made that the phage packages a "headful" of DNA (see Appendix B), and is described in Article 31.

As shown in Articles 29–31, two important ideas to come out of this work were those of terminal redundancy and cyclic permutation. Physical proof of these properties came from the laboratory of Charles Thomas, who realized that such a population of molecules could be converted in the laboratory to DNA circles by special techniques. This work is described in Article 32 by Thomas and L. A. MacHattie, and Article 33 by MacHattie and others.*

---

*In subsequent papers from Thomas' laboratory it was shown that not all phage DNA molecules are cyclically permuted and terminally redundant. For example, phage T5 is neither, and phage T7 is terminally redundant but not permuted.

Molecules of Increased Length Generated by Replication

The physical demonstration of cyclic permutation and terminal redundancy in T4 (Articles 32 and 33) and the genetic analysis described in Articles 29–31 reinforced the headful model. However, simple replication of linear DNA to give two daughter molecules by the Watson-Crick scheme could not account for the generation of a permuted population from a single phage. Streisinger and Stahl pointed out that if the intracellular, recently replicated, phage DNA were in the form of a long polymer consisting of many DNA units and if a nonredundant molecule is less than one headful, then packaging headful units successively would generate cyclically permuted, terminally redundant molecules (Appendix B). They suggested that such long polymers could be the result of recombination, although some years later another scheme (called the Rolling Circle model), by which the polymers could be generated by replication, was proposed. Their idea was supported by the discovery that Fred Frankel made while working in Hershey's laboratory: that is, of the various types of DNA found in *E. coli* that has been infected with phage T4, one of them is more easily broken by stirring and sediments more rapidly than DNA extracted from the phage. This observation was explored in a series of papers. As he relates in Article 34, Frankel proved that cells infected with T4 contained molecules several times longer than the DNA of the intact phage.*

---

*For some time these long molecules were known in laboratory jargon as "spaghetti." Later Charles Thomas proposed the currently accepted term, *concatemer*.

From *Journal of Molecular Biology*, Vol. 8, 1–10, 1964.
Reproduced with permission of the authors and publisher.

Interruptions in the Polynucleotide Strands in Bacteriophage DNA

P. F. DAVISON, D. FREIFELDER† AND B. W. HOLLOWAY‡

Biology Department, Massachusetts Institute of Technology, Cambridge Massachusetts. U.S.A.

(*Received 15 July 1963*)

Denatured phage DNA preparations are heterogeneous in the analytical ultracentrifuge. Preparative zone sedimentation confirmed that the sedimentation diagrams reflect a true polydispersity of sedimentation coefficients among the polynucleotide strands. It is concluded that many phage particles—but not all—have one or more interruptions in the polynucleotide sequences, and the number of interruptions is consistent in a population of any one phage.

1. Introduction

We have previously reported studies on the physical properties of DNA preparations extracted from bacterial viruses (Davison, 1959; Davison & Freifelder, 1962a; Freifelder & Davison, 1963). From these studies and those of other workers, it was anticipated that DNA from viral sources could be obtained homogeneous with respect to molecular weight (or homogeneous to a degree limited only by the natural processes controlling the size of the viral genome). Since DNA has a very low diffusion coefficient, a homogeneous DNA preparation should sediment with a sharp boundary in the ultracentrifuge, and the integral distribution diagram of the DNA in the cell throughout a sedimentation run should approximate a step function. Within experimental limitations (Davison & Freifelder, 1962a) such diagrams have been obtained by ourselves and others (Meselson, 1959; Watson & Littlefield, 1960) from a variety of viral DNAs (see Fig. 1). We have also shown (Davison, 1959; Davison & Freifelder, 1962a) that at a DNA concentration of 20 μg/ml. these boundaries are not self-sharpened and a mixture of degraded and intact DNA can be readily identified; we may therefore accept these sharp sedimentation boundaries as evidence of homogeneity of the DNA.

It is widely believed that when DNA is denatured the individual polynucleotide strands of the DNA double helix separate from one another (Marmur, Rownd & Schildkraut, 1963). If these single polynucleotide strands from a viral DNA are uninterrupted, then the integral distribution diagram of denatured DNA should qualitatively resemble that of the native DNA, with perhaps a slightly broadened boundary due to the higher diffusion coefficient of the collapsed single strands. In order to investigate this prediction the DNA was denatured and centrifuged in formaldehyde solutions which, by titrating the amino group on the purine and pyrimidine bases, obviate intramolecular hydrogen-bond formation and inter-molecular

† Present address, Donner Laboratory, University of California, Berkeley, California, U.S.A.
‡ Present address, Department of Bacteriology, University of Melbourne, Australia.

1

aggregation. As previously reported, this expectation was not realized (Freifelder & Davison, 1963). In the case of T7 bacteriophage DNA it was found that the sedimentation diagram of the denatured DNA differed markedly from that of the native material (Fig. 2); about 50% of the molecules sediment in a sharp "front" while the remainder have lower sedimentation coefficients and form a broad boundary behind the front. We interpreted this sedimentation diagram on qualitative grounds to indicate that in such preparations 50% of the polynucleotide strands are intact—the "front"—while 50% of the strands contain one or more interruptions (or very labile bonds which rupture during isolation or denaturation of the DNA)—the "tail" of the diagram. For brevity throughout this paper we refer only to single-strand interruptions although the possibility of highly labile bonds is not dismissed.

In the previous study we looked briefly at denatured DNA from other phages and found the sedimentation diagrams to be qualitatively similar. In the present investigation we have looked systematically at a collection of phages grown on various hosts and have confirmed our earlier observations.

Since we have not determined the molecular weights of our denatured DNA preparations, our conclusions that single-strand interruptions exist might be invalid if the observed sedimentation heterogeneity could be attributed to one or more of the following: (1) a variation in the frictional coefficient of the molecules among a population of uniform molecular weight, (2) an artificial addition to the molecular weight, for example, by non-uniform binding of formaldehyde to the polynucleotide strands, or (3) a boundary anomaly in the centrifuge. In this paper we shall describe experiments which eliminate these alternatives, and we conclude that there are single-strand interruptions in phage DNA.

2. Experimental

Table 1 lists the phages, bacterial hosts, methods of growth, techniques for releasing DNA and denaturation methods we have used. The significance of the numbers in the table is as follows.

(a) Method of growth

The liquid media used were: (1) Fraser medium (Fraser & Jerrel, 1953); (2) a solution containing 0·8% Difco nutrient broth, 1% Difco Casamino acids, 2% glycerol, 0·003 M-CaCl$_2$, 0·0025 M-tris pH 7·8. The methods of growth have been previously described (Davison & Freifelder, 1962b).

For phage production on solid media, bacteria were grown in an appropriate liquid medium to approximately 2×10^9/ml. 5 ml. of such a suspension was mixed with 10^7 phage particles in 100 ml. soft agar, and the solution was poured over a 0·1 square meter surface of suitable 1·5% agar in a large dish. After confluent lysis the phage was extracted in broth or a buffered salt solution from the macerated soft agar layer. The following media were used.

(1) λ agar. 1% Bacto-tryptone, 0·5% NaCl, 0·01 M-MgSO$_4$, 1 mg/ml., thiamine hydrochloride, 1·5% agar.
(2) Difco nutrient agar, 0·1% NaCl.
(3) TY agar (Romig & Brodetsky, 1961).
(4) 1% Bacto-peptone, 1% Difco nutrient broth, 0·5% NaCl, 1·5% agar.

(b) Methods of release of native DNA from phage

(1) Shaking with phenol (Davison & Freifelder, 1962a).
(2) Titration to high pH by addition of 0·02 M-NaOH to give a final NaOH molarity 0·002.
(3) Heating for 5 min at 65°C in 0·1 M-NaCl, 0·001 M-EDTA pH 7·8.
(4) Osmotic shock (Anderson, 1950).

TABLE 1

Preparation of phage, and isolation and denaturation of DNA

Phage	Host	Method of growth	Methods of DNA release	Methods of denaturation
T2, T4, T4-05 (Brenner)	*Escherichia coli* B	Liquid 1	1, 2	1, 2, 3
T5	*E. coli* B	Liquid 2	1, 2	1, 2, 3, 5
T7L†	*E. coli* B	Liquid 1	1, 2	1, 2, 3, 4, 5
T7M†	*E. coli* B	Liquid 1	1, 2, 3	1, 2, 3, 4, 5
λ	*E. coli* C600 *E. coli* W3110	Solid 1	1, 2	1, 2, 3, 4, 5
B³‡	*Pseudomonas aeruginosa* 1C	Solid 2	1, 2	1, 2, 3, 4, 5
D3‡	*Ps. aeruginosa* 1C	Solid 2	1, 3	2, 3
E79‡	*Ps. aeruginosa* 1C	Solid 2	1, 2	1, 2, 3, 5
F116‡	*Ps. aeruginosa* 1C	Solid 2	1, 2	1, 2, 3, 5
SP-8§	*Bacillus subtilis*	Solid 3	1, 2, 3	1, 2, 3, 5
Osaka 1 ‖	*B. subtilis*	—	1, 2, 3	1, 2, 3, 5
α¶	*B. megatherium*	—	1, 2, 3	1, 2, 3, 5
c††	*B. cereus* G1	Solid 4	1, 2	1, 2, 3, 5

† Davison & Freifelder (1962*b*).
‡ Holloway, Egan & Monk (1960).
§ Romig & Brodetsky (1961).
‖ Generously provided by S. Okubo and M. Stodolsky, University of Chicago.
¶ Generously provided by F. Graziosi, University of Rome, and G. Tenecchini-Valenti, University of Chicago.
†† Norris (1961).

(c) Methods of denaturation

The DNA was denatured by heat or alkali treatment. For samples that were to be maintained at neutral pH, formaldehyde was added to prevent renaturation (Freifelder & Davison, 1963). The following procedures were employed.

(1) 0·1 M-Na$_3$PO$_4$ was added to a phage solution to make the solution 0·043 M. After 5 min at room temperature, 0·5 vol. neutral, boiled 37% HCHO and 0·04 vol. M-KH$_2$PO$_4$ were added. The solution was heated 5 min at 70°C and quickly cooled.

(2) To a DNA solution of ionic strength between 0·05 and 0·2 were added 0·5 vol. 37% HCHO, pH 7 to 9, and 0·01 vol. 0·1 M-EDTA. The solution was heated 5 min at 70°C.

(3) Same as in 2, but ionic strength lower than 0·05. Heated 3 min at 70°C.

(4) Addition of M-NaOH to a phage to make 0·1 to 0·5 M final concentration.

(5) One vol. phage or DNA solution, 1 vol. 0·2 M-NaOH mixed for 30 sec, 1 vol. 37% HCHO added and mixed thoroughly, then 0·125 vol. M-KH$_2$PO$_4$.

Method 5 was convenient and applicable to all phages except D3.

Preparation of solvents

In order to prevent spurious and irreproducible measurements it was found necessary to autoclave all solvents. When this precaution was not taken, nucleases, presumably from contaminating micro-organisms, could frequently be detected.

Storage of phage stocks

The suspending buffered saline is described elsewhere (Davison & Freifelder, 1962*b*). For T5 and E79 0·002 M-CaCl$_2$ was added.

γ-Irradiation

Glass ampoules of deoxygenated samples of phage in buffered saline were irradiated using a ^{60}Co source.

Sucrose gradient centrifugation

One ml. of denatured phage DNA (O.D.† > 20) in 12% HCHO, 0·01 M-PO$_4^{3-}$ pH 7·8, 0·001 M-EDTA, 0·25 M-NaCl was layered on 20 ml. of a 5 to 25% by weight linear sucrose gradient containing 6% HCHO, 0·1 M-PO$_4^{3-}$ pH 7·8, 1 M-NaCl. All solutions were at 4°C. Centrifugation was carried out at 4°C for 16 hr at 24,000 rev./min in the SW25 rotor of the Spinco model L ultracentrifuge. Fifty to sixty samples per tube were collected dropwise.

Ultracentrifugation

Methods of centrifugation and scanning the u.v. photographs have been described (Davison & Freifelder, 1962a). In the case of the denatured DNA from the large phages for which no front could be seen at the usual DNA concentration, centrifugation was carried out at O.D. 0·7 to 1·2. It was then necessary to expose the u.v. photograph for such a length of time that the film response was no longer linear. However, at this concentration it was possible to visualize a front containing less than 10% of the total DNA. Because of the non-linearity of the film, the percentage of material in the front was not determined. The $S'_{20,w}$, however, could be measured since the concentration dependence of $S'_{20,w}$ is small. In some negatives the densitometer did not respond to the low contrast across the front; the $S'_{20,w}$ was then evaluated by direct measurement on high-contrast photographic enlargements. The median $S'_{20,w}$ of the tail of these preparations was determined as usual at 20 μg/ml.

When fractions from a sucrose gradient were centrifuged, they were pooled or diluted as necessary to give an O.D. 0·4 to 0·5 in a solvent composition of 2% HCHO, 0·5 M-NaCl. The sucrose concentration of these samples was uncontrolled. A reference DNA was always included in order that the $S'_{20,w}$ could be calculated.

3. Results

(i) *The occurrence of the tail*

Native DNAs of each of the bacteriophages listed in Table 1 have sharp ultracentrifugal boundaries similar to those described previously (Davison & Freifelder, 1962a) (Fig. 1). However, the boundary for any of the denatured DNAs (Figs. 2, 3 and 4)

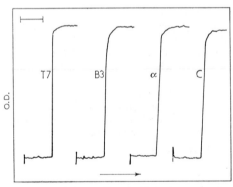

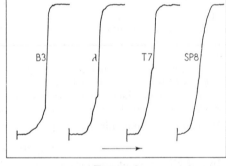

FIGURE 1   FIGURE 2

FIG. 1. Photometric traces from u.v. absorption photographs of native bacteriophage DNA sedimenting in the ultracentrifuge. Speed 33,450 rev./min. Concentration 20 μg/ml. Solvent M-NaCl, 0·01 M-phosphate pH 7·8. The vertical bar at the left of each trace is the meniscus. The arrow shows the direction of sedimentation. The scale indicates the radial distance in the centrifuge cell.

FIG. 2. Photometric traces of u.v. absorption photographs of denatured bacteriophage DNA sedimenting in the ultracentrifuge. Speed 33,450 rev./min. Concentration 15 to 20 μg/ml. Solvent 9% HCHO, 0·005 M-phosphate pH 7·8, 0·002 M-EDTA, M-NaCl.

† Abbreviations used in this paper are O.D. = optical density at 260 mμ; $S'_{20,w}$ = sedimentation coefficient corrected to *solvent* viscosity and density at 20°C, with $\bar{v}$ assumed to be 0·55.

is not that of a homogeneous material; in each case a broad boundary indicative of material polydisperse with respect to sedimentation coefficient is obtained. In the diagrams of all but the largest phages there is also present a characteristic "front" of faster material. Table 2 gives for each phage the $S'_{20,w}$ of the native DNA and of the front

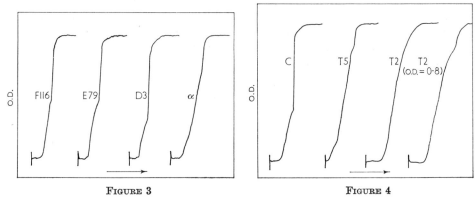

FIGURE 3   FIGURE 4

FIG. 3. Photometric traces of u.v. absorption photographs of denatured bacteriophage DNA sedimenting in the ultracentrifuge. Conditions as for Fig. 2.

FIG. 4. Photometric traces of u.v. absorption photographs of denatured bacteriophage DNA sedimenting in the ultracentrifuge. Conditions as for Fig. 2.

TABLE 2

Sedimentation characteristics of various phage DNAs

Phage	Native DNA	Front of denatured DNA	Tail of denatured DNA	% tail
T2, T4	52 to 56 s	64 s	41 s	> 90
T5	43·5 s	50 to 54 s	35 s	> 90
T7	29 s	39 s	34 s	50
λ	31 s	40 s	33·5 s	28 to 30
B3	28·5 s	38 s	30·5 s	15 to 17
D3	39·5 s	48·5 s	39·5 s	35 to 40
E79	38 s	47·5 s	32 s	40 to 42
F116	36·5 s	45·5 s	40 s	30 to 40
SP-8	52 to 56 s	—	37 s	100
Osaka 1	52 to 56 s	61 s	41 s	> 90
α	31·5 s	41·5 s	35 s	55 to 60
C	35·5 s	44 s	33 s	38 to 40

All samples were centrifuged in a 30 mm Kel-F cell at 33,450 rev./min at a concentration of 20 μg/ml. Each S value is a median $S'_{20,w}$. The reproducibility of these measurements is ±2% except where indicated.

and tail of the denatured DNA. The percentage of material in the tail of the sedimentation diagram has been found to be reproducible from one preparation to the next for a given phage when the phage is propagated under a variety of culture conditions and on a number of bacterial hosts (Freifelder, Holloway & Davison, unpublished results). For instance, from more than fifteen preparations of four different mutants of coliphage T7, DNA was extracted by three methods, and denatured by various combinations of time, temperature, pH, ionic strength, and formaldehyde concentration; in every case, 50% of the polynucleotide strands appeared in the tail of the diagram. This same reproducibility has been found for the other phages.

For phages T2, T4, T5, SP8 and Osaka 1 no front is visible at 20 μg/ml. However, as described in the experimental section, at higher concentrations the front is detectable in all cases but that of SP8. In order to justify the validity of the measurement of $S'_{20,w}$ at this high concentration, the effect of concentration on the $S'_{20,w}$ of the front of other phages was determined. For B3 the $S'_{20,w}$ of the front (and the tail percentage) was independent of concentration from 8 to 20 μg/ml. For T7 there was similarly no concentration dependence in the range 15 to 30 μg/ml.

The boundary shape of sedimenting denatured DNA is independent of centrifuge speed from 27,690 to 42,040 rev./min.

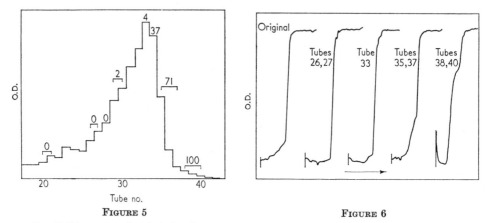

FIGURE 5   FIGURE 6

FIG. 5. Histogram of optical absorbance of fractions collected dropwise after zone centrifugation of denatured B3 DNA through a sucrose gradient. The bracketed fractions were sedimented in the analytical centrifuge and the figures indicate the percentage of tail measured in the diagrams.

FIG. 6. Photometric traces of u.v. absorption photographs of denatured B3 DNA recovered from zone centrifugation experiment (Fig. 5).

(ii) *Sucrose gradient fractionation of denatured phage DNA*

Since certain anomalies have been found in the sedimentation diagrams of native DNA (Davison & Freifelder, 1962a) it is necessary to demonstrate that the sedimentation diagram of denatured phage DNA reflects a true polydispersity of sedimentation coefficient among the sedimenting molecular species. Evidence that there are species with different sedimentation characteristics is provided by their separation by means of zone centrifugation of the denatured DNA through a sucrose gradient. Figure 5 shows the sedimentation distribution of polynucleotides recovered from zone sedimentation of B3 DNA (released by the Na_3PO_4 method and denatured by heating for 5 minutes at 68°C in 12% HCHO). Fractions from various parts of the zone were recentrifuged in the analytical ultracentrifuge and the percentage of material in the tail for each fraction tested is indicated in the figure above the particular fraction. Figure 6 shows the actual sedimentation profile for several of these fractions. More fast-moving material is found as one proceeds to the faster fractions in the gradient, an observation consistent with the assumption of sedimentation heterogeneity. Of particular significance is the fact that material completely free of any tail is obtainable from the fast side of the band. We interpret this result to show that a population of homogeneous DNA molecules containing no interruptions in the single strands could, as anticipated, give rise to denatured strands which sediment in a step-function boundary. Similarly, fractions from the slow side of the band contain only tail material.

The validity of the fractionation is confirmed by summing the amounts of tail and front material for each fraction and comparing to the proportions present in the original preparations. The agreement is excellent, the slow material representing 17% of the total sedimenting optical density in the gradient, compared to the 16% tail measured in the unfractionated material. Qualitatively similar results have been obtained by fractionating the DNA from phage T7 and λ.

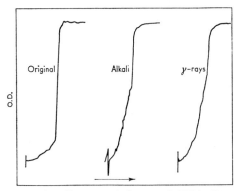

FIG. 7. Change of sedimentation diagram of denatured DNA (B3) with strand scission by (a) alkaline hydrolysis, (b) γ-irradiation. Conditions as for Fig. 2.

(iii) Sedimentation without formaldehyde

Since solutions of HCHO at room temperature contain polymeric forms, the sedimentation heterogeneity of denatured DNA could have been a result of selective binding of such polymers. This is clearly not the case because identical boundaries are obtained when DNA is centrifuged in 0·1 M-NaOH, 0·2 M-NaCl, or 0·5 M-NaOH, 0·5M-NaCl, either with or without formaldehyde. Moreover, the boundary shape is the same in alkali as that obtained in the usual 1 M-NaCl, 9% HCHO, 0·01 M-PO$_4^{3-}$ pH 7·8 solvent.

(iv) Sedimentation in low ionic strength solutions

If the denatured DNA in formaldehyde is as expected in the random-coil configuration, the $S'_{20,w}$ should decrease with decreasing ionic strength because the mutual repulsion of the charged phosphate groups will, in the absence of sufficient neutralizing counterions, cause an extension of the molecule. If the sedimentation heterogeneity of denatured DNA is a result of dispersity in molecular weight, the fraction of material in the tail should be independent of ionic strength. The $S'_{20,w}$ of the denatured DNA of phage B3 over a range of ionic strength from 0·02 to 0·5 varied from 10 to 39 s; although at low ionic strength the tail is not as clearly resolved from the front as at high ionic strength, the relative proportions of front and tail are unchanged. Similar results obtain for DNA from phages λ and T7.

(v) Effect of irradiation and hydrolysis on tail size

If we attribute the tail of the sedimentation diagram to molecular weight heterogeneity, then it should be possible to increase the proportion of the tail at the expense of the front material by any process producing single-strand breaks, e.g. γ-ray irradiation, or thermal or alkaline hydrolysis. Typical sedimentation diagrams for denatured DNA degraded by these techniques are shown in Fig. 7. As expected, the percentage of material in the tail of the diagram is increased by such treatments.

(vi) *Kinetics of hydrolysis*

Since hydrolysis can increase the number of low molecular weight strands—i.e. increase the size of the tail in the sedimentation diagram—it is important to show that the tail normally observed is itself not a consequence of the breakage of phosphoester bonds during the denaturation process (a somewhat unlikely possibility in view of the reproducibility of the tail size under various conditions of denaturation). We have previously reported that the sedimentation boundary of phage T7 is unchanged for at least 20 minutes at 70°C in 12% HCHO, 0·01 M-PO_4^{3-} pH 7·8 (Freifelder & Davison, 1963).

In a series of experiments on the effect of alkali we found that B3 and λ DNA could both survive 4 hours in 0·1 M-NaOH at 25°C with no detectable strand scission. This slow rate of hydrolysis justifies the use of the alkaline denaturation method and the centrifugation in alkaline solutions. Hydrolysis in 0·5 M-NaOH was faster, although no increase in tail size was seen after 1 hour. The kinetics of hydrolysis were clearly non-linear: after 4 hours in 0·5 M-NaOH no intact strands, e.g. no front, remained.

4. Discussion

We have shown that centrifugally homogeneous viral DNAs after denaturation give non-homogeneous sedimentation diagrams. When precautions are taken to avoid strand hydrolysis the sedimentation diagram of denatured DNA from each virus studied is characteristic of that virus, and reproducible despite variations in the methods employed for DNA isolation and denaturation.

That the inhomogeneous sedimentation distribution is not a boundary anomaly but is indicative of a real polydispersity of sedimentation coefficient was shown by zone centrifugation. DNA fractions collected through the sedimenting zone showed the expected progressive change in heterogeneity. It was possible to obtain from the zone an ultracentrifugally homogeneous sample of denatured DNA.

Since the ultracentrifugation diagram of denatured viral DNAs must be accepted as indicative of sedimentation polydispersity, it remains to be decided whether the range of sedimentation coefficients observed reflects a range of molecular weights within the population of polynucleotide strands or whether the molecules are of uniform length but solvated or folded to differing degrees.

Although it is not *a priori* unlikely that one polynucleotide strand may be solvated to a different extent from its complementary strand or that the partial specific volume for the two strands may differ, it does not seem possible to explain the complexity of the sedimentation diagrams by these factors. The distribution of material in the sedimenting boundary is the same in formaldehyde as in alkali solutions; moreover, the appearance of tails containing consistently less than 50% of the polynucleotide strands argues against any substantial differentiation between the strands.

The consistency of the sedimentation profiles for any single phage DNA in a series of solvents covering a range of pH and ionic strength is also difficult to explain on the model of uniformly long molecules adopting a variety of configurations. Since the polynucleotide strands in solutions of high ionic strength probably adopt a random coil configuration, the front of the sedimentation diagram may be attributed to such a population of molecules. Molecules moving slower than this front would therefore be more extended, and one would need to explain how for each phage consistent fractions of the molecules appear to be extended through a range of configurations corresponding to the observed range of sedimentation coefficients.

From these arguments it appears improbable that the sedimentation polydispersity is due to anything but a variation in molecular weight; it is reasonably concluded that there exists within the viral DNAs a consistent number of strand interruptions or weak points at which the strands are broken in the course of isolation and denaturation. Support for this interpretation of the sedimentation diagrams is given by the changes observed in the diagrams when the DNAs are subjected to treatments which introduce single-strand scissions. As shown, these degradative procedures cause the appearance of DNA with very low sedimentation coefficients and increase the size of the tail in the diagrams.

We therefore conclude that each bacteriophage population contains a number of particles in whose DNA there are present single-strand interruptions. The sizes of the tails of the sedimentation diagrams show that in the case of B3 phage more than 70% of the viruses have DNA molecules with both strands intact, whereas in the case of T2 coliphage fewer than 5% of the viruses, and perhaps none, contain uninterrupted polynucleotide strands. Further experiments to study the effects of growth conditions, bacterial host and other variables on the number of single-strand interruptions will be published separately. Thus far, none of our observations has indicated the biological role of these single strand interruptions. It may be seen from Table 2 that the size of the tail is not proportional to the length of the DNA (cf. T7 and λ), but in general the larger phages have more strand interruptions.

The distribution of material in the tail of the diagram of the undegraded polynucleotide chains is non-random (see Freifelder & Davison, 1962), and from the shape of the diagram it appears that the breaks or weak links are located preferentially at sites near the middle of the strands. This point was particularly noticeable in the case of T5 DNA where a sharp false front was detected (see Fig. 8); this material we now believe is attributable to strands half the length of the intact strands. The presence of single strand *breaks*, as opposed to labile bonds, close to the center of the strands should make such molecules particularly labile to scission by hydrodynamic shear forces. Such a vulnerable population has been detected by controlled shearing experiments and these findings will be reported separately.

It is apparent that the presence of these strand interruptions does not destroy the infectivity of the phage since the larger T-even phages have an infective efficiency close to 100% yet few of their strands are intact. Whether the interruptions are *necessary* for infection we have not yet determined.

The examination of the separated strands of the T-even phages provided the most exacting test of the conditions used to release and denature the DNA. Heat and alkali treatments which produced no detectable damage in the DNA of smaller phages had to be modified or abandoned for the large viruses, but techniques were developed (method 5 in Experimental section) which gave reproducible sedimentation diagrams, and these results confirmed previous preliminary findings (Freifelder & Davison, 1963) that few of the polynucleotide strands are uninterrupted.

However, the demonstration of a small front for the large T-even DNAs removes the objection that the broad distribution of the denatured DNA might merely be an anomalous spreading of a boundary of mono-disperse, very high molecular weight, polynucleotide strands. The detection of single-strand interruptions cannot be reconciled with the finding by Berns & Thomas (1961) that the denatured strands of T2 and T4 are intact.

A by-product of this investigation is a graph relating the sedimentation coefficient of native DNA and the unbroken polynucleotide strands obtained from them (Fig. 8). When molecular weights can be assigned to the separated strands, such a calibration series may be useful for determining the molecular weights of other viral DNAs. Such measurements should complement those on the native DNA, for the latter measurements are complicated by speed-dependence, aggregation and the difficulties of extrapolation to zero concentration.

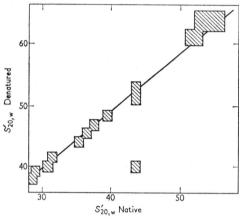

FIG. 8. Relationship between the sedimentation coefficients of native bacteriophage DNA (measured at 20 μg/ml. in M-NaCl, 33,450 rev./min) and the *front* of the denatured DNA diagram i.e. the intact single strands (measured under the conditions specified in Fig. 2). For phage T5 an apparent front with an anomalously low $S'_{20,w}$ was observed at 20 μg/ml. When the concentration of DNA was increased a small front with an appropriate $S'_{20,w}$ was detected. The false front is an extreme example of the non-random distribution of the DNA in the sedimentation diagram and probably reflects a preferential break point close to the center of one of the strands.

The authors are grateful to Professor Francis O. Schmitt and Professor S. Luria for their encouragement and interest in this work and for the facilities they provided. They also thank Mrs. K. Y. Ho for her valuable technical assistance. This research was supported by grant E 1469 from the National Institute of Allergy and Infectious Diseases, National Institutes of Health; United States Public Health Service, grant G 8808 from the National Science Foundation, and by grants from the Trustees under the Wills of Charles A. and Marjorie King.

One of us (David Freifelder) was supported by post-doctoral fellowship (no. 7617) from the Division of General Medical Sciences, National Institutes of Health.

REFERENCES

Anderson, T. F. (1950). *J. Appl. Phys.* **21**, 70.
Berns, K. I. & Thomas, C. A. (1961). *J. Mol. Biol.* **3**, 289.
Davison, P. F. (1959). *Proc. Nat. Acad. Sci., Wash.* **45**, 1560.
Davison, P. F. & Freifelder, D. (1962a). *J. Mol. Biol.* **5**, 643.
Davison, P. F. & Freifelder, D. (1962b). *J. Mol. Biol.* **5**, 635.
Fraser, D. & Jerrel, E. A. (1953). *J. Biol. Chem.* **205**, 291.
Freifelder, D. & Davison, P. F. (1962). *Biophys. J.* **2**, 235.
Freifelder, D. & Davison, P. F. (1963). *Biophys. J.* **3**, 49.
Holloway, B. W., Egan, J. B. & Monk, M. (1960). *Austr. J. Exp. Biol.* **38**, 321.
Marmur, J., Rownd, R. & Schildkraut, C. L. (1963). In *Progress in Nucleic Acid Research*, vol. 1, ed. by J. N. Davidson & W. E. Cohn. New York: Academic Press.
Meselson, M. (1959). *The Cell Nucleus*, p. 240. London: Butterworth & Co.
Norris, J. R. (1961). *J. Gen. Microbiol.* **26**, 167.
Romig, W. R. & Brodetsky, A. M. (1961). *J. Bact.* **82**, 135.
Watson, J. D. & Littlefield, J. W. (1960). *J. Mol. Biol.* **2**, 161.

From *Virology*, Vol. 28, No. 1, 1966. Reproduced with permission of the authors and publisher.

Preferred Breakage Points in T5 DNA Molecules Subjected to Shear

ELIZABETH BURGI, A. D. HERSHEY, AND LAURA INGRAHAM

Carnegie Institution of Washington, Genetics Research Unit, Cold Spring Harbor, New York

Accepted August 30, 1965

T5 DNA molecules subjected to critical rates of shear break into two pieces measuring 0.6 and 0.4 of the original molecular length. The fragments of length 0.6, isolated and subjected to a higher rate of shear, break into two pieces measuring 0.4 and 0.2 of the original molecular length. The molecules must contain two to four sites of preferred breakage at prescribed locations. Preferred breakage points have not been detected and may be absent in DNA's of some other species.

INTRODUCTION

DNA molecules from phages T2 and lambda, under critical rates of shear, break near their centers of length, according to theoretical expectation (Hershey and Burgi, 1960; Levinthal and Davison, 1961; Burgi and Hershey, 1961, 1963; Kaiser, 1962). DNA from phage T5 behaves differently, in a manner suggesting that the molecules contain acentric points of preferred breakage (Hershey *et al.*, 1963b). The results presented below show that there are at least two such points per molecule and that they are similarly situated in all or most of the molecules.

MATERIALS AND METHODS

T5 DNA, usually labeled with tritium or radiophosphorus, was prepared by the phenol method from a heat-stable (*st*) mutant of the phage (Hershey *et al.*, 1963b; Burgi, 1963). For the preparation of H^3-labeled stocks, labeled uracil was substituted for the uridine formerly used. Breakage under shear was accomplished by stirring DNA solutions in buffered 0.6 M NaCl for 30 minutes at 5°C (Hershey *et al.*, 1962). Breakage products were fractionated by chromatography (Mandell and Hershey, 1960) and analyzed by sedimentation through density gradients of sucrose (Burgi and Hershey, 1963).

RESULTS

Primary Breakage Products

When T5 DNA is stirred at speeds just sufficient to break part or all of the DNA, breakage products of two sedimentation-rate classes are invariably seen (Fig. 1). The relative amounts of the two products, and their sedimentation rates, do not depend significantly on the fraction of the molecules broken, indicating that the first and the last molecules to break yield the same products, and that both products are derived from the same molecules.

If the original DNA and its breakage products are linear structures whose only hydrodynamically important difference is length, the lengths can be calculated from the sedimentation rates (Burgi and Hershey, 1963). On this basis, the data of Fig. 1 assign to the breakage products the lengths 0.6 and 0.4, measured in relation to the unbroken DNA of unit length (molecular weight 75–80 million; Burgi and Hershey, 1963). Since the areas under the two sedimenting peaks also correspond very nearly to a 60:40 ratio, the interpretation of the sedimentation rates as measures of a length and mass difference is correct. Molecules of T5 DNA therefore contain sites of preferred breakage at points distant about 40% of the molecular length from one or both ends.

Not all molecules break at the prescribed

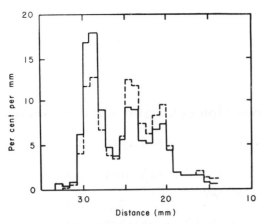

FIG. 1. Sedimentation pattern of T5 DNA partly broken under shear. Solid line: P³²-labeled DNA stirred at 10 μg/ml and 1050 rpm. Broken line: H³-labeled marker DNA stirred separately. Left to right, unbroken DNA, 0.6 and 0.4 unit fragments. Centrifugation, 4.3×10^9 rpm² hr. Distance measured from the meniscus.

positions, because chromatographic analysis reveals small numbers of fragments of diverse lengths, including half lengths. Whether the exceptional fragments arise from exceptional molecules, or result from accidents of breakage, we have not yet determined.

Secondary Breakage Products

When a sample of partly broken T5 DNA is restirred at a higher speed sufficient to produce additional breaks, the fragments 0.6 unit long disappear, and most of the DNA now sediments as a single band at the rate characteristic of fragments 0.4 unit long. This behavior confirms the postulated length difference of the initial breakage products, and suggests that the 0.6 unit fragments themselves break acentrically. Breakage of the 0.6 unit fragments is best seen after isolating them.

On passage through a column of methylated serum albumin, a partly broken sample of T5 DNA separates into two bands corresponding to broken and unbroken fractions and, at times, the band corresponding to broken DNA is visibly double. At all times, sedimentation analysis of individual fractions shows that the 0.4 and 0.6 unit fragments are resolved, the shorter ones

eluting at lower salt concentrations than the longer ones. Most of the recovered fragments belong to one class or the other, though neither class is perfectly homogeneous, and the exact composition of each class remains unknown. Figure 2 shows the sedimentation character of a fraction consisting mainly of 0.6 unit fragments.

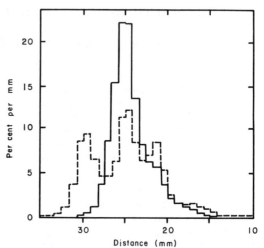

FIG. 2. Sedimentation pattern of isolated 0.6 unit fragments. Solid line: a single chromatographic fraction, 21% of the total DNA, isolated from the P³²-labeled breakage products of Fig. 1. Broken line: unfractionated H³-labeled marker DNA also shown in Fig. 1. Centrifugation 4.3×10^9 rpm² hr.

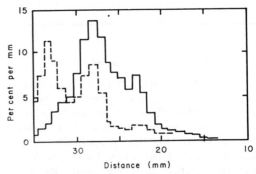

FIG. 3. Sedimentation pattern of isolated 0.6 unit fragments after further breakage under shear. Solid line: P³²-labeled material shown in Fig. 2 restirred at 2 μg/ml. Broken line: same marker DNA as in Figs. 1 and 2. Centrifugation 5.6×10^9 rpm² hr. The band containing unbroken marker DNA, having sedimented nearly to the bottom of the tube, is not shown in the figure.

When a solution containing 0.6 unit fragments, isolated as described, is stirred at a series of increasing speeds, the starting material gives way to two simultaneously generated products sedimenting at rates corresponding to 0.4 and 0.2 unit lengths, respectively. Figure 3 shows the result after complete breakage of the starting material. The masses of the two breakage products, measured as areas in the sedimentation pattern, do not exactly correspond to the lengths estimated from rates. The discrepancy can be attributed to the inhomogeneity of the starting material (Fig. 2), because the length assignment derived from sedimentation rates is supported by the chromatographic properties of the breakage products. Specifically, the 0.4 unit fragments generated on initial breakage of intact molecules, and on breakage of 0.6 unit fragments, are chromatographically indistinguishable, whereas the fragments that sediment more slowly are also eluted from the columns at lower salt concentrations. Therefore the 0.6 unit fragments contain sites of preferred breakage at points distant about 40% of the original molecular length from one or both ends.

DISCUSSION

Our results show that T5 DNA molecules contain at least two preferred breakage points at the same or nearly the same location in all or most of the molecules. The acentric breakage we observe implies that the molecules contain weak spots, which could be distributed in any of at least three ways: (1) two spots at positions 0.4 and 0.6 of the total molecular length measured from either end; (2) two spots at positions 0.4 and 0.8 measured from a specified end; or (3) four spots dividing the molecule into 5 equal segments. The individual spots must be short compared with 20% of the molecular length, and there cannot be more than four unless they are clustered, or unless others lie very close to the original molecular ends. (A weak spot lying too far from the center of a molecule or fragment subjected to shear would not in general reveal itself as a site of preferred breakage.)

It is likely, though not logically necessary, that weak spots in prescribed locations signify prescribed nucleotide sequences at those locations. If so, our results suggest that the molecules are not circularly permuted in the sense postulated for T4 DNA (Streisinger et al., 1964). In fact, Thomas and Rubenstein (1964) have shown that nucleotide sequences in T5 DNA are not circularly permuted. Our own evidence in this connection is somewhat trivial since it leaves open the possibility of a segmental permutation compatible with hypothesis (3) of the preceding paragraph.

In earlier work, the DNA's of T2 and T5 were found to exhibit equal fragility per unit length under shear (Hershey et al., 1962). This fact and results reported here might suggest that both DNA's contain weak spots, which do not reveal themselves as preferential breakage points in T2 DNA precisely because the molecules are circularly permuted (Thomas and Rubenstein, 1964). That interpretation seems unlikely because (1) lambda DNA breaks under shear at centrally clustered points (Burgi and Hershey, 1963); (2) lambda DNA molecules are not circularly permuted (Kaiser, 1962; Hershey et al., 1963a); and (3) lambda DNA also matches T2 DNA in fragility per unit length (Burgi and Hershey, 1963). Possibly all three DNA's contain weak spots, which in lambda happen to be centrally clustered. More likely only T5 DNA contains singular weak spots, which weaken the molecules so little that in fact they sometimes break centrally. A small overall effect could have been missed in the comparison with T2 DNA.

Concerning the structure of the weak spots one can only list hypotheses. (1) They may be linkages of unknown type, nearly as strong as the known covalent bonds. (2) They may be points at which one of the polynucleotide chains of the helix is interrupted (Davison et al., 1964). (3) They may be regions bounded by interruptions in both strands, regions within which only the secondary forces responsible for helical structure hold the molecules together. (4) As suggested to us by C. A. Thomas, Jr., they may be rare nucleotide sequences within which secondary structure is characteristically weak if, as seems likely, secondary structure contributes appreciably to the overall strength of the molecule. It may be noted that T5 DNA does contain interrupted polynucleotide chains (Davison *et al.*, 1964; also our preparations). Whether or not the gaps are responsible for the observed breakage characteristics remains to be ascertained.

REFERENCES

BURGI, E. (1963). Changes in molecular weight of DNA accompanying mutations in phage. *Proc. Natl. Acad. Sci. U.S.* **49**, 151–155.

BURGI, E., and HERSHEY, A. D. (1961). A relative molecular weight series derived from the nucleic acid of bacteriophage T2. *J. Mol. Biol.* **3**, 458–472.

BURGI, E., and HERSHEY, A. D. (1963). Sedimentation rate as a measure of molecular weight of DNA. *Biophys. J.* **3**, 309–321.

DAVISON, P. F., FREIFELDER, D., and HOLLOWAY, B. W. (1964). Interruptions in the polynucleotide strands in bacteriophage DNA. *J. Mol. Biol.* **8**, 1–10.

HERSHEY, A. D., and BURGI, E. (1960). Molecular homogeneity of the deoxyribonucleic acid of phage T2. *J. Mol. Biol.* **2**, 143–152.

HERSHEY, A. D., BURGI, E., and INGRAHAM, L. (1962). Sedimentation coefficient and fragility under hydrodynamic shear as measures of molecular weight of the DNA of phage T5. *Biophys. J.* **2**, 423–431.

HERSHEY, A. D., BURGI, E., and INGRAHAM, L. (1963a). Cohesion of DNA molecules isolated from phage lambda. *Proc. Natl. Acad. Sci. U.S.* **49**, 748–755.

HERSHEY, A. D., GOLDBERG, E., BURGI, E., and INGRAHAM, L. (1963b). Local denaturation of DNA by shearing forces and by heat. *J. Mol. Biol.* **6**, 230–243.

KAISER, A. D. (1962). The production of phage chromosome fragments and their capacity for genetic transfer. *J. Mol. Biol.* **4**, 275–287.

LEVINTHAL, C., and DAVISON, P. F. (1961). Degradation of deoxyribonucleic acid under hydrodynamic shearing forces. *J. Mol. Biol.* **3**, 674–683.

MANDELL, J. D., and HERSHEY, A. D. (1960). A fractionating column for analysis of nucleic acids. *Anal. Biochem.* **1**, 66–77.

STREISINGER, G., EDGAR, R. S., and DENHARDT, G. H. (1964). Chromosome structure in phage T4. I. Circularity of the linkage map. *Proc. Natl. Acad. Sci. U.S.* **51**, 775–779.

THOMAS, C. A., JR., and RUBENSTEIN, I. (1964). The arrangements of nucleotide sequences in T2 and T5 bacteriophage DNA molecules. *Biophys. J.* **4**, 93–106.

From the *Proceedings of the National Academy of Sciences*, Vol. 49, No. 4, 748–755, 1963. Reproduced with permission of the authors.

COHESION OF DNA MOLECULES ISOLATED FROM PHAGE LAMBDA

By A. D. Hershey, Elizabeth Burgi, and Laura Ingraham

GENETICS RESEARCH UNIT, CARNEGIE INSTITUTION OF WASHINGTON, COLD SPRING HARBOR, LONG ISLAND, NEW YORK

Communicated March 25, 1963

Aggregation of DNA is often suspected but seldom studied. In phage lambda we found a DNA that can form characteristic and stable complexes. A first account of them is given here.

Materials and Methods.—DNA was extracted from a clear-plaque mutant (genotype cb^+) of phage lambda[1] by rotation[2] or shaking[3] with phenol. Sodium dodecylsulfate, ethylenediaminetetraacetate, citrate, or trichloroacetate was sometimes included in the extraction mixture without effect on the properties of the DNA. Phenol was removed by dialysis, with or without preliminary extraction with ether, against 0.1 or 0.6 M NaCl.

Sedimentation coefficients were measured[4] at 10 μg DNA/ml in 0.1 and 0.6 M NaCl in aluminum cells at 35,600 rpm with consistent results, and are reported as $S_{20,w}$.

Zone sedimentation[5] of labeled DNA's[6] was observed in 0.1 M NaCl immobilized by a density gradient of sucrose. A sample, usually containing less than 0.5 μg of DNA in 0.15 ml of 0.1 M NaCl, was placed on 4.8 ml of sucrose solution, and the tube was spun for 5 or 6 hr at 28,000 rpm in an SW39L rotor of a Spinco Model L centrifuge at 10°C.

Solutions containing 5–40 μg DNA/ml in 0.1 or 0.6 M NaCl were stirred on occasion for 30 min at 5°C with a thin steel blade turning in a horizontal plane.[7] Since we used two stirrers of different capacities, stirring speeds given in this paper are comparable only within a context.

Salt solutions were buffered at pH 6.7 with 0.05 M phosphate.

Results.—Disaggregation and breakage: Solutions containing 0.5 mg/ml of lambda DNA in 0.1 M NaCl acquire an almost gel-like character on standing for some hours in a refrigerator. Diluted to 10 μg/ml, the DNA exhibits in the optical centrifuge an exceedingly diffuse boundary sedimenting at 40–60 s (Fig. 1A). If the diluted solution is aged for several days, the sedimentation rate may fall somewhat (not below 40 s), but the boundary remains diffuse and often appears double.

Stirring the diluted solution at 1,300 to 1,700 rpm yields a single component sedimenting at 32 s (Fig. 1B). The product so obtained is stable for a week or more in the cold in 0.1 M NaCl. We call this process disaggregation by stirring.

If samples of the diluted solution are stirred at increasing speeds between 1,800 rpm and 2,100 rpm, one sees a stepwise transition from 32 s to 25.2 s components, each by itself exhibiting a sharply sedimenting boundary (Figs. 1C and 1D). We call this phenomenon breakage. Broken DNA can form aggregates, but the characteristic 32 s species cannot be regained.

Aggregation: Disaggregation, in contrast to breakage, is reversible, as shown by the following experiment. Lambda DNA at 40 μg/ml in 0.6 M NaCl was

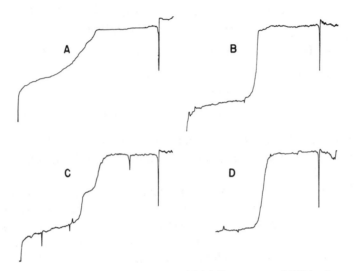

FIG. 1.—Sedimentation pattern of initially aggregated DNA after stirring at several speeds. *A*, unstirred; *B*, 1,600 rpm; *C*, 1,900 rpm; *D*, 2,000 rpm. The meniscus shows at the right.

disaggregated by stirring at 1,700 rpm, and samples at either 40 μg/ml or 10 μg/ml in the same solvent were warmed to 45°C. After measured time intervals, the tubes were chilled and their contents diluted to 10 μg/ml, if necessary, with cold 0.6 *M* NaCl. Sedimentation coefficients were measured over the course of some hours. Unheated samples showed the same sedimentation rate at the beginning and end of the series of measurements. The heated samples were analyzed in random order, so that the results reflect mainly the duration of heating, not the duration of subsequent storage.

The results, presented in Figure 2, show that the sedimentation rate of the DNA increases rapidly on heating at 40 μg/ml, and less rapidly on heating at 10 μg/ml. The reversibility of disaggregation, and the dependence of rate of aggregation on concentration of DNA, justify our choice of language.

Similar experiments showed that heating in 0.1 *M* NaCl under the same conditions does not cause appreciable aggregation. Aggregation occurs in that solvent at higher DNA concentrations, however. Thus, aggregation is accelerated by high DNA concentrations, high temperatures, and high salt concentrations.

Linear molecules: According to the description given above, aggregated lambda DNA can be reduced under shear to a uniform 32 s product, which is evidently the structure subject to breakage at higher rates of shear. The maximum stirring speed withstood by 32 s lambda DNA is 1,800 rpm at 10 μg/ml. When

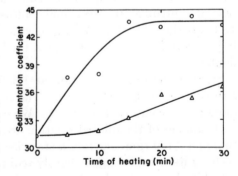

FIG. 2.—Aggregation at 45°C. Circles, 40 μg DNA/ml in 0.6 *M* NaCl; triangles, 10 μg/ml. The scale on the ordinate refers to observed sedimentation coefficients.

T2 DNA is stirred under the same conditions, it is reduced to fragments sedimenting at 31 s. Thus, lambda DNA exhibits a fragility under shear that is appropriate
to linear molecules[7] sedimenting at about 32 s. We therefore conclude that 32 s
lambda DNA consists of linear molecules. These and other DNA structures are best
identified by zone centrifugation, as illustrated below.

Linear molecules can also be prepared (irrespective of the initial state of aggregation of the DNA) by heating a solution in 0.1 or 0.6 M NaCl to 75°C for 10 min
and cooling the tube in ice water (Fig. 3A, solid line). This procedure is effective
at concentrations up to 10 μg/ml at least.

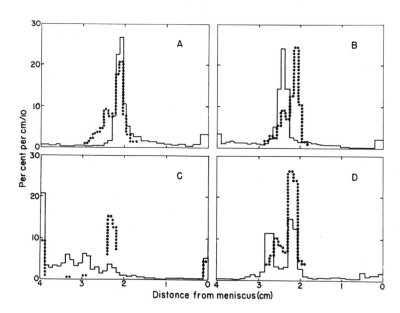

FIG. 3.—Zone sedimentation of several molecular forms. Solid lines: P[32]-
labeled linear molecules (A), folded molecules (B), aggregates (C), dimers and
linear monomers (D). Broken lines, a preparation of H[3]-labeled marker DNA
containing linear and some folded molecules.

Linear molecules are obtained directly by extracting the DNA (by rotating, not
shaking, the tubes)[6] at 2 μg/ml into 0.1 M NaCl (Fig. 3A, dotted line). Control
experiments showed that the mechanical operations involved in the extraction do
not destroy previously formed complexes in solutions diluted to 2 μg/ml.

We conclude that the 32 s form of lambda DNA is analogous to more conventional phage DNA's and is a typical double-helical molecule.

Folded molecules: Another form of lambda DNA we usually prepare by heating
a dilute solution (5 μg/ml or less) in 0.6 M NaCl to 75°C for 1 min, and allowing
the container to cool slowly (0.4° per min at 65°) in the heating bath with the
heater disconnected. The resulting product sediments as a narrow band moving
1.13 times faster than linear molecules in zone centrifugation (Fig. 3B). The
expected sedimentation coefficient is $32 \times 1.13 = 36.2$ s. Material prepared as
described and then concentrated by dialysis against dry sucrose followed by 0.6
M NaCl shows in the optical centrifuge a sharp boundary at 37 s.

The formation of 37 s material is equally efficient at several DNA concentrations between 5 μg/ml and 0.1 μg/ml (at higher concentrations it is obscured by simultaneous aggregation). The 37 s product, therefore, is composed of monomers that we shall call folded molecules.

When a dilute solution containing either linear or folded molecules in 0.6 *M* NaCl is heated to 75°C, one gets only linear molecules by rapid cooling and only folded molecules by slow cooling. Partial conversion of linear to folded molecules occurs on heating to 45°C for 30 min followed by rapid cooling, and nearly complete conversion at 60°C. Thus, at 75°C linear molecules are the stable form of lambda DNA. At low temperatures, folded molecules are more stable but the conversion is slow. The slow cooling from 75°C serves to find a temperature near 60°C at which the conversion to folded molecules is rapid and the product is stable.

Folded molecules are formed on heating and slow cooling in 0.1 *M* NaCl as well as in 0.6 *M* NaCl, but the conversion is not complete at the lower salt concentration. Some molecular folding also occurs when linear molecules are stored at low concentration and low temperature for a few weeks in 0.1 *M* NaCl or a few days in 0.6 *M* NaCl. This is the origin of the faster-sedimenting component of the tritium-labeled marker DNA whose sedimentation pattern appears in Figure 3.

Folded molecules can be converted back into linear molecules by stirring as well as by heating, though the margin between the stirring speed required to accomplish this and the speed sufficient to break linear molecules is rather narrow.

It should be added that heating DNA at 10 μg/ml and 45°C in 0.6 *M* NaCl produces many folded molecules whose formation competes with the simultaneous aggregation. For this reason the dependence of rate of aggregation on DNA concentration is not truly represented in Figure 2.

Folded molecules themselves do not aggregate. Solutions concentrated for analytical centrifugation continue to yield sharp boundaries after aging in 0.6 *M* NaCl. Neither do folded molecules form complexes with linear molecules. This was shown by mixing P³²-labeled folded molecules with unlabeled linear molecules (20 μg/ml) and aging the mixture in 0.6 *M* NaCl for 4 days at 5°C. A similar mixture containing labeled linear molecules served as control. Zone centrifugation of each mixture with added H³-labeled marker DNA showed that the labeled linear molecules but not the folded molecules had formed complexes with the unlabeled DNA.

The similarity between the conditions, other than DNA concentration, controlling formation and destruction of folded molecules, and formation and destruction of aggregates, suggests that similar cohesive forces are involved in both phenomena. The folding implies that each molecule carries at least two mutually interacting cohesive sites, which join to form a closed structure. The uniformity of structure of folded molecules, indicated by the narrow zone in which they sediment, suggests that there are not more than two cohesive sites, and that these are identically situated on each molecule.

Dimers and trimers: Aggregated DNA often shows multiple boundaries in the optical centrifuge and always shows multiple components in zone centrifugation. An example, prepared by heating linear molecules for 30 min at 45°C and 40 μg/ml in 0.6 *M* NaCl, is shown in Figure 3C. Since the characteristic folding seen in monomers is incompatible with aggregation, as already described, it is likely that some of the differently sedimenting products of aggregation are polymers differing in mass rather than configuration.

One form of aggregate can be obtained in moderately pure state by allowing aggregation to occur during a day or so in the cold at 100 μg/ml in 0.1 M NaCl (Fig. 3D). Such material contains a fraction of the molecules in linear form, and presumably contains in addition mainly the smaller and more stable aggregates. One of these, as shown in the figure, always predominates, and we assume that it is a dimer. It sediments 1.25 times faster than linear molecules.

In a study of zone centrifugation to be reported separately, we found a relation

$$\frac{D_2}{D_1} = \left(\frac{L_2}{L_1}\right)^{0.35} \tag{1}$$

between molecular lengths (L) and distances sedimented (D) of two DNA's, which is valid for linear molecules. According to this relation, dimers are about twice as long as linear molecules of lambda DNA. The only alternative compatible with the sedimentation rate is a second form of folded monomer, which is ruled out by the requirement for high DNA concentrations during formation. Therefore, dimers are tandem or otherwise open structures. (For definitions of "open" and "closed," see hereafter.)

In more completely aggregated material (Fig. 3C) one sees few or no linear molecules, a very few folded molecules (fewer the more concentrated the solution in which aggregation occurred), a considerable fraction of dimers sedimenting 1.25 times faster than linear molecules, and another characteristic component sedimenting 1.43 times faster than linear molecules. According to its sedimentation rate, the last component could be a tandem trimer or a folded or side-by-side dimer. We believe that it is an open trimer for the following reasons.

A folded dimeric structure is ruled out because material sedimenting at rate 1.43 does not form when a dilute solution containing dimers (similar to that shown in Fig. 3D) is aged for two weeks in the cold in 0.1 or 0.6 M NaCl, or is heated in 0.6 or 1.0 M NaCl at 45°C. At high DNA concentrations, trimers do form under these conditions. At low DNA concentrations, dimers and trimers are stable and one sees only the conversion of linear to folded monomers.

A side-by-side dimeric structure can be ruled out on the basis of susceptibility to hydrodynamic shear. Figure 4 shows the result when samples of a mixture of trimers, dimers, and folded and linear monomers are stirred at increasing speeds. Trimers disappear first, being converted to dimers or linear molecules or both. Next to go are dimers. Folded monomers are much more resistant, but can be reduced to linear monomers at stirring speeds just insufficient to break the molecules. Thus, trimers, as expected if they are open structures, are more fragile than open dimers, whereas closed dimers should be more stable. We note, however, that a small amount of the material sedimenting at the rate of trimers is relatively resistant to stirring and could signify a minority of closed dimers.

We note also that destruction of dimers and trimers does not liberate any folded monomers, a result consistent with the evidence from sedimentation rates for an open polymeric structure, and with our finding that folded monomers do not form complexes. The fact that aggregation and folding are mutually exclusive processes implies that both utilize the same limited number of cohesive sites, which must be small in size to account for the open polymeric structure. As already suggested by the unique configuration of folded monomers, there may be only two sites per molecule.

Specificity of aggregation: If tracer amounts of P³²-labeled lambda DNA are mixed with unlabeled lambda DNA at 25 μg/ml in 0.6 *M* NaCl, and the mixture is brought to 75°C for 1 min and allowed to cool slowly, subsequent zone sedimentation with added H³-labeled marker shows that most of the P³²-labeled DNA has been converted to aggregates and a small remainder to folded molecules. When the same procedure is followed with H³-labeled or unlabeled T5 DNA substituted for the unlabeled lambda DNA, the T5 DNA sediments (at its normal rate) 1.20 times faster than the P³²-labeled lambda DNA, which now consists entirely of folded molecules. Thus, lambda DNA shows no tendency to form complexes with T5 DNA, T5 DNA itself does not form stable aggregates, and T5 DNA does not inhibit molecular folding in lambda DNA. The cohesive sites in lambda DNA are therefore mutually specific, as our model requires.

Role of divalent cations: Divalent cations probably do not play any specific role in the phenomena described in this paper. In NaCl solutions, molecular folding and aggregation are not inhibited by added citrate or ethylenediaminetetraacetate. Neither are these processes appreciably accelerated, in the presence of NaCl, by added calcium or magnesium ions. In a solution of 0.01 *M* MgCl₂, 0.01 *M* CaCl₂, and 0.01 *M* tris (hydroxymethyl) aminomethane, pH 7.2,[8] linear monomers at 10 μg/ml are about as stable as they are in NaCl solutions.

Interpretation of sedimentation rates: Equation (1) shows that if two identical DNA molecules were joined end to end their sedimentation rate would increase by the factor 1.27, evidently owing to the loss of independent mobility. Perhaps the result would be about the same whether they were joined end to end or to form a V, a T, or an X. Thus, we are led to the definition of an *open dimeric structure* as one formed by the joining of two linear molecules at a single point, recognizable by a 1.27-fold increase in sedimentation rate. The principle of independent mobility of parts suggests that, as the structure departed from the tandem arrangement, its sedimentation rate

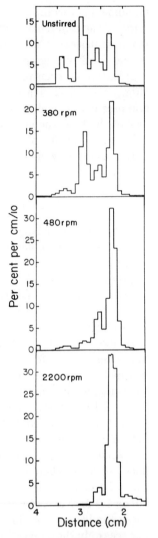

Fig. 4.—Successive destruction by stirring of trimers, dimers, and folded molecules, as seen by zone centrifugation. Linear molecules began to break at 2,400 rpm in this series. The starting material was prepared by heating a sample of DNA, already containing some spontaneously formed folded molecules and dimers, for 30 min at 45°C and 40 μg/ml in 1.0 *M* NaCl, and diluting to 5 μg/ml in buffered water.

could only increase, not decrease, and in the order V, T, X.

Our results also show that molecules of lambda DNA undergo some sort of folding, apparently as the result of bonding between two cohesive sites lying at some distance from each other on each molecule. If that interpretation is correct, it would appear that when the molecule (regarded as two halves joined end to end) forms an additional point of attachment between its parts, the sedimentation rate increases by an additional factor of 1.13, evidently owing to a further loss of independent mobility of parts. Thus, we are led to the definition of a *closed dimeric structure* as one formed by joining two linear molecules at two points. Such a structure ought to sediment 1.27 × 1.13 or 1.43 times faster than the linear monomer. We have not found closed dimers, but the question remains how the factor 1.13 would depend on the point of closure of a threadlike molecule. The principle of independent mobility of parts suggests that the sedimentation rate would approach or pass through a maximum as the fraction of the molecular length contained in the loop increased. In some measure it may be possible to answer such questions empirically by determining the locations of cohesive sites on the molecules.

Discussion.—Lambda DNA can exist in at least four characteristic forms that we call linear monomers, folded monomers, open dimers, and open trimers, which sediment respectively at the rates 1.0, 1.13, 1.25, and 1.43, expressed in arbitrary units. These structures are interconvertible with certain restrictions according to the scheme

open polymers ⇌ linear monomers ⇌ folded monomers

As the scheme indicates, linear monomers are subject to two distinct processes: aggregation, seen at DNA concentrations exceeding 10 μg/ml, and folding, seen at any concentration but forced to compete with aggregation at high DNA concentrations. Both processes are accelerated as the temperature is raised to about 60°C, beyond which only linear monomers are stable, and as the salt concentration is raised from 0.1 to 1.0 M. Both processes are rapidly reversed at 75°C or by hydrodynamic shear. All four structures are stable at low temperatures, low DNA concentrations, and low salt concentrations, except for a slow conversion of linear to folded molecules.

Since folded molecules exist in only one stable configuration, and since molecular folding and aggregation are mutually exclusive processes, we postulate that each molecule carries two cohesive sites in prescribed locations, and that these are responsible for both processes. To account for the considerable effect of molecular folding on sedimentation rate, the sites must lie rather far apart along the molecular length. To account for the moderate effect of dimerization on sedimentation rate, the cohesive sites must be small compared to the total molecular length.

According to the proposed model, one might anticipate two dimeric forms, open (that is, joined by one pair of cohesive sites) and closed (joined by two). We find only open polymers, though a minority with closed structures is not excluded. Failure to detect closed polymers may be explained, at least in part, by the fact that the rate of folding must decrease as the length of the linear structure increases.[9]

Whether all details of our model are correct or not, it is clear that lambda DNA forms a limited number of characteristic complexes, not the continuously variable series that might be expected if the molecules could cohere at random. The limited number of mutually specific cohesive sites implied thereby suggests a specialized biological function, one that remains to be identified.

Summary.—The DNA of phage lambda undergoes reversible transitions from linear to characteristically folded molecules, and from linear monomers to open polymers. Some conditions favoring one state or another have been defined. It may be surmised that each molecule carries two specifically interacting cohesive sites.

This work was aided by grant CA-02158 from the National Cancer Institute, National Institutes of Health, U.S. Public Health Service. Its direction was determined in part in conversation with Dr. M. Demerec about heterochromatin, synapsis, deletions, and speculations to be pursued. Professor Bruno Zimm contributed useful suggestions about the manuscript.

[1] Kellenberger, G., M. L. Zichichi, and J. Weigle, these PROCEEDINGS, **47**, 869 (1961).

[2] Frankel, F. R., these PROCEEDINGS, **49**, 366 (1963).

[3] Mandell, J. D., and A. D. Hershey, *Anal. Biochem.*, **1**, 66 (1960).

[4] Burgi, E., and A. D. Hershey, *J. Mol. Biol.*, **3**, 458 (1961).

[5] Hershey, A. D., E. Goldberg, E. Burgi, and L. Ingraham, *J. Mol. Biol.*, **6**, 230 (1963).

[6] Burgi, E., these PROCEEDINGS, **49**, 151 (1963).

[7] Hershey, A. D., E. Burgi, and L. Ingraham, *Biophys. J.*, **2**, 423 (1962).

[8] Kaiser, A. D., *J. Mol. Biol.*, **4**, 275 (1962).

[9] Jacobson, H., and Stockmayer, W. H., *J. Chem. Phys.*, **18**, 1600 (1950).

From the *Proceedings of the National Academy of Sciences*, Vol. 53, No. 2, 325–328, 1965. Reproduced with permission of the authors.

COMPLEMENTARY STRUCTURE OF INTERACTING SITES AT THE ENDS OF LAMBDA DNA MOLECULES*

BY A. D. HERSHEY AND ELIZABETH BURGI

CARNEGIE INSTITUTION OF WASHINGTON, GENETICS RESEARCH UNIT,
COLD SPRING HARBOR, NEW YORK

Communicated December 21, 1964

Molecules of DNA from phage lambda are characterized by terminal cohesive sites that permit end-to-end joining to form rings and chains.[1-3] The joining is reversible, depending on temperature and salt concentration in a manner that suggests denaturation and renaturation of DNA.[1] Joining and disjoining occur at temperatures well below those causing denaturation of the DNA as a whole, which occurs at temperatures appropriate to the over-all base composition. These facts suggest a typical, helical, double-stranded structure modified by short, unpaired ends that are complementary to each other in base sequence. The model predicts specific joining of left to right ends. That prediction is verified by the results of an experiment described below.

Methods.—Lambda DNA labeled with P[32] was prepared according to Burgi[4] and sheared to half-length fragments by stirring an ice-cold solution containing 10 μg/ml in 0.1 M NaCl at 2150 rpm for 30 min in a laboratory stirrer.[5] Molecular weights were measured from the sedimentation rate in sucrose.[6] Equilibrium density-gradient centrifugation in CsCl was carried out in a Spinco model E centrifuge in the usual way except that a rotor temperature of 5°C was maintained to prevent the rejoining of lambda DNA fragments that occurs in CsCl solutions at

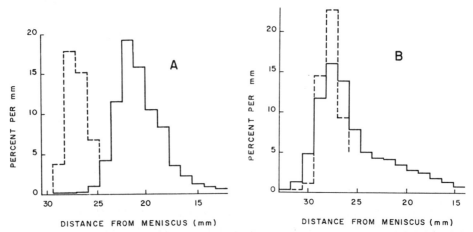

Fig. 1.—(*A*) Sedimentation in a density gradient of sucrose[6] of sheared, P[32]-labeled lambda DNA (solid line) and unsheared, H[3]-labeled marker DNA (broken line).   (*B*) Same, after thermal treatment of the P[32]-labeled DNA to join molecular ends.

higher temperatures.   Samples were spun for about 30 hr at 44,770 rpm and the bands formed were analyzed by densitometry of ultraviolet photographs.   To ensure that a DNA sample contained only disjoined molecules or fragments, the solution in 0.1 *M* NaCl was heated to 75°C for 1 min and quickly cooled shortly before use, or examined immediately after stirring.   To bring about rejoining of fragments, the solution containing 10 μg DNA/ml was made 0.6 *M* in NaCl, heated to 75°C, and allowed to cool slowly (0.11°/min at 65°) in the heating bath with the heater disconnected.[1]   Neither of these treatments affects the buoyant density or sedimentation rate of the DNA, except for discrete, reversible changes that reflect joining or disjoining of molecular ends.

Results.—Molecular-weight measurements: Figure 1*A* shows the sedimentation in sucrose of a sample of P[32]-labeled lambda DNA previously broken by stirring. It sediments as a single band at the rate expected for half-length fragments.

Figure 1*B* shows the sedimentation of an otherwise identical sample heated and slowly cooled after stirring.   Most of the DNA sediments at the rate characteristic of unbroken linear molecules.   There is some material that sediments more slowly, but none that sediments much faster, than the main component.

The behavior is that expected according to the following argument.   Lambda DNA molecules cohere only by their two ends.[1]   Half-length fragments cohere by their single remaining natural end and can rejoin only in pairs.   A few fragments produced by more than one break per molecule and lacking a natural end do not rejoin at all.   (An alternative interpretation of the incomplete rejoining of halves would be that a minority of the original molecules, though possessing complementary ends and able to form rings, differ from the majority of molecules and cannot interact with them.)

Density measurements: Density measurements yield additional information about rejoining because the two halves of the lambda DNA molecule differ in density.   For DNA of the wild-type phage used here, the difference is about 0.009 gm/ml,[7] greater than that found in a lambda *dg* DNA.[8]   Following Hogness and Simmons,[8] the denser half may be called the left half of the molecule.

Figure 2A shows the resolution of right and left halves into two bands when a DNA sample stirred in the manner described is centrifuged in CsCl. The two bands contain roughly equal amounts of DNA. The band corresponding to right halves is narrower than the other because the boundary between denser and less dense DNA lies somewhat to the left of the molecular center, so that variations in length about the mean half length affect mainly the density of left halves.[7]

Figure 2B shows the result when the stirred DNA is heated and slowly cooled before centrifugation. One sees a single band lying at the same position on the density scale as that formed by natural molecules of lambda DNA (Fig. 2C).

If the two halves could rejoin in random pairs, one would expect only half of the DNA in the band of Figure 2B to lie at the position of native molecules. Instead, most of it does, and only small amounts of material, presumably fragments that failed to rejoin (Fig. 1B), exhibit the densities of the individual halves. Rejoined pairs therefore consist mainly or entirely of right and left halves.

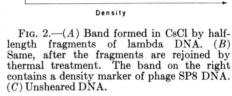

FIG. 2.—(A) Band formed in CsCl by half-length fragments of lambda DNA. (B) Same, after the fragments are rejoined by thermal treatment. The band on the right contains a density marker of phage SP8 DNA. (C) Unsheared DNA.

Discussion.—The conditions under which lambda DNA molecules undergo end-to-end joining and disjoining[1] suggested that the interacting structures might be complementary, unpaired polynucleotide chains. The finding reported here, that the joining occurs between right and left ends, is consistent with that hypothesis. Additional support for the same hypothesis is coming from more specific, but still incomplete, enzymic analysis of structure (Burgi, see ref. 7; A. D. Kaiser, personal communication).

The structure may be thought of as a means for achieving the initial step in formation of the stable molecular rings found in bacteria infected with phage lambda.[9] More generally, structures of the postulated type may play a role in the synaptic phase of molecular-genetic recombination.[10]

Summary.—Half-length fragments of DNA from phage lambda rejoin in pairs when subjected to thermal annealing. The pairs consist mainly or exclusively of right and left halves. The reactive sites at the two ends of the original molecule are therefore different from each other and, at least in a formal sense, complementary in structure. It also follows that the majority of molecular ends of each class are more or less identical, or belong to one of very few types.

* This work was aided by gifts of phage SP8 DNA from Dr. Julius Marmur, and by Public Health Service research grant HD 01228 from the National Institute of Child Health and Human Development.

[1] Hershey, A. D., E. Burgi, and L. Ingraham, these Proceedings, **49**, 748 (1963).

[2] Ris, H., and B. L. Chandler, in *Synthesis and Structure of Macromolecules,* Cold Spring Harbor Symposia on Quantitative Biology, vol. 28 (1963), p. 1.

[3] MacHattie, L., and C. A. Thomas, Jr., *Science,* **144**, 1142 (1964).

[4] Burgi, E., these Proceedings, **49**, 151 (1963).

[5] Hershey, A. D., E. Burgi, and L. Ingraham, *Biophys. J.,* **2**, 423 (1962).

[6] Burgi, E., and A. D. Hershey, *Biophys. J.,* **3**, 309 (1963).

[7] Hershey, A. D., *Carnegie Inst. Wash. Yr. Bk.,* **63**, 580 (1964).

[8] Hogness, D. S., and J. R. Simmons, *J. Mol. Biol.,* **9**, 411 (1964).

[9] Young, E. T., II, and R. L. Sinsheimer, *J. Mol. Biol.,* in press.

[10] Meselson, M., *J. Mol. Biol.,* **9**, 734 (1964).

From *Journal of Molecular Biology*, Vol. 12, 36–49, 1965. Reproduced with permission of the authors and publisher.

On the Structure of the Ends of Lambda DNA

H. B. Strack† and A. D. Kaiser‡

Department of Biochemistry, Stanford University School of Medicine
Palo Alto, California, U.S.A.

(*Received 16 December 1964*)

The structure of the ends of a molecule of λ DNA has been deduced from the effect of DNA polymerase and exonuclease III on the infectivity of λ DNA and from the ability of the two ends of the molecule to cohere.

The infectivity of λ DNA is destroyed when it primes DNA synthesis catalyzed by DNA polymerase. Inactivation depends on the presence of deoxynucleoside triphosphates and on the amount of enzyme added. The inactivating synthesis behaves like a repair of single-stranded regions, that is, their conversion to double strands, because it occurs at 15°C. Richardson, Inman & Kornberg (1964) found that DNA polymerase catalyzes repair but not net synthesis at temperatures lower than 20°C.

Infectivity returns if the polymerase product is subsequently exposed to exonuclease III. The specificity of exonuclease III is such that it would be expected to remove nucleotides added by polymerase. This result shows that inactivation by polymerase is not due to degradation. In a complementary fashion the infectivity of native λ DNA is destroyed by exonuclease III and restored by subsequent synthesis catalyzed by DNA polymerase.

The two ends of a molecule of λ DNA can cohere, specifically, to each other (Hershey, Burgi & Ingraham, 1963; Ris & Chandler, 1963). To account for cohesiveness and for inactivation by DNA polymerase, it is proposed that 5′-hydroxyl (or 5′-phosphate)-terminated single strands protrude from both ends of the otherwise double-stranded molecule and that the two protruding single strands have complementary base sequences. Cohesion would result from base pairing between the protruding single strands.

Synthesis primed by native λ DNA and catalyzed by DNA polymerase at 15°C would be expected to repair partially single-stranded ends and to destroy their cohesiveness. As the infectivity of λ DNA for helper-infected bacteria requires cohesiveness, polymerase would thus inactivate λ DNA. Treating the polymerase product with exonuclease III restores infectivity, according to this model, because removal of the added nucleotides restores cohesiveness.

1. Introduction

The infectivity of bacteriophage λ resides in a single molecule of DNA which can be isolated intact and biologically active from mature phage particles (Kaiser & Hogness, 1960; Meyer, Mackal, Tao & Evans, 1961). Molecules of λ DNA have specific cohesive sites. They form closed or circular molecules when two sites on the same molecule cohere, and dimers, trimers, etc., when sites on different molecules cohere (Hershey, Burgi & Ingraham, 1963; Ris & Chandler, 1963; MacHattie & Thomas, 1964). It is likely that the cohesive sites are located at the ends of the molecule, since closed

† Present address: Max Planck Institut für Biologie, Tübingen, Germany.
‡ Present address: Medical Research Council Laboratory of Molecular Biology, Hills Road, Cambridge, England.

molecules seen in the electron microscope are unbranched and since the mean contour length of closed molecules is the same as that of open molecules (MacHattie & Thomas, 1964; Inman, unpublished observations).

In the work to be reported here the structure of the ends of λ DNA has been investigated by the effect of specific enzymes on its biological activity. The two enzymes studied, *Escherichia coli* exonuclease III and *E. coli* DNA polymerase, were selected because they alter the structure of the ends of DNA molecules in known ways and because they have been purified away from endonucleases. *E. coli* exonuclease III (Richardson, Lehman & Kornberg, 1964) removes 5'-mononucleotides stepwise from the two 3'-hydroxyl ends of a double-stranded DNA molecule. The enzyme has little or no activity on single-stranded DNA, since it hydrolyzes heat-denatured DNA to a limit of 3% at 45°C and since it stops degrading native DNA when 40% has been digested.

The DNA polymerase of *E. coli* incorporates 5'-mononucleotides into DNA in 3'-5' phosphodiester linkages (Bessman, Lehman, Simms & Kornberg, 1958). Polymerase can catalyze the initiation and continuation of new chains, a process which has been called "net synthesis". Polymerase can also complete chains terminated by a 3'-hydroxyl group which lie apposed to longer chains terminated by a 5'-hydroxyl group (Richardson, Inman & Kornberg, 1964). DNA molecules of this type can be produced by the action of exonuclease III on native DNA. Polymerase catalyzes the addition of 5'-mononucleotides to terminal 3'-hydroxyl groups until the ends of the two apposite chains are in register. This process has been called "repair". At 37°C both repair and net synthesis take place, but at 20°C only repair (Richardson, Inman & Kornberg, 1964).

The biological activity of λ DNA is a measure of its integrity and, because different genes can be assayed independently, the integrity of different parts of the molecule can be examined separately. The linkage map of λ has been shown to be colinear or congruent with the order of genes on the molecule of λ DNA by recombination experiments with labeled phage (Meselson & Weigle, 1961; Kellenberger, Zichichi & Weigle, 1961) and by mechanical breakage experiments with λ DNA (Radding & Kaiser, 1963; Hogness & Simmons, 1964; Kaiser & Inman, manuscript in preparation). Therefore the biological activity of the ends or the middle of the molecule can be examined by assaying markers located at the ends or middle of the linkage map.

2. Materials and Methods

Three different genotypes of λ, λ wild type, λh and λhlco₁ (Kaiser, 1962) were used as sources of DNA. Wild type λ and λh were grown by ultraviolet-light induction of lysogenic strains of *E. coli* K12, and λhlco₁ was grown by infection of sensitive bacteria on agar plates. The phage stocks were purified by differential centrifugation and equilibrium sedimentation in a density-gradient of CsCl. The DNA was isolated by extraction with water-saturated phenol at 5°C. Details of the procedures for purification and extraction will be found in Kaiser & Hogness (1960).

Six different DNA preparations were used in the course of this investigation. Preparation D2 was made from λhlco₁. Examination of its sedimentation in alkali (0·1 M-NaOH, 0·9 M-NaCl, 0·001 M-sodium EDTA) (Studier, 1965) showed that approximately two-thirds of the single chains were without breaks; it resembles in this respect the preparations studied by Davison, Freifelder & Holloway (1964). Preparation D3 was made from λ wild type and preparation D4 from λhlco₁. Preparations DF1 and DF2 were both made from wild type λ. Their preparation differed from the others in that the phenol was buffered with 0·5 M-tris at pH 7·5. Both DF1 and DF2 were purified further by sedimentation in sucrose gradients. Two hundred and seventy μg of DNA in 3 ml. were layered on to 24 ml. of 5 to 25% sucrose in 0·01 M-tris–HCl buffer (pH 7·1), 0·001 M-sodium EDTA and sedimented for 27 hr at 25,000 rev./min and 5°C in the SW25 rotor of the Spinco model L ultracentrifuge. Fractions corresponding to a peak in absorbance at 260 mμ, approximately two-thirds of the way down the tube, were pooled.

Preparation D^{32}P was made from λh grown in ^{32}PO$_4^{3-}$ medium as described by Kaiser (1962). This DNA when sedimented in alkali failed to show any major component of homogeneously sedimenting material, indicating that most of the strands had one or more breaks.

Substrates

The deoxynucleoside triphosphates of adenine, guanine, cytosine and thymine were used as purchased from the California Corporation for Biochemical Research. ^{32}P-labeled deoxyribonucleotides were prepared as described by Lehman, Bessman, Simms & Kornberg (1958) and were the generous gifts of A. Kornberg, J. Jackson, and L. Bertsch.

Enzymes

E. coli DNA polymerase, hydroxylapatite fraction, prepared as described by Richardson, Schildkraut, Aposhian & Kornberg (1964), was the generous gift of C. C. Richardson and A. Kornberg. This preparation had an activity of 20,000 units/ml.; one unit of enzyme would incorporate 10 mμmoles of labeled nucleotide into an acid-insoluble form in 30 min at 37°C.

Two preparations of *E. coli* exonuclease III were employed. Both were prepared according to the procedure of Richardson & Kornberg (1964). The first, a gift of C. C. Richardson, had an activity of 6000 units/ml. (Richardson, Lehman & Kornberg, 1964). This preparation was used only for the experiment reported in Table 2. All other experiments were performed with a second enzyme preparation, prepared according to Richardson & Kornberg (1964), but subsequently concentrated by dialysis against 30% w/w Carbowax in 70% 0·02 M-potassium phosphate buffer at pH 6·5, 0·01 M-2-mercaptoethanol at 0°C, then dialyzed against 0·02 M-potassium phosphate at pH 6·5, 0·01 M-2-mercaptoethanol overnight at 5°C. The concentrate had an activity of 3200 units/ml. and a specific activity of 55,000 units/mg protein.

Enzyme-catalyzed reactions

The detailed conditions are reported with each experiment separately. The following reaction conditions were found to be suitable for both exonuclease III and polymerase: 67 μmoles/ml. tris–HCl buffer (pH 7·1), 6·7 μmoles/ml. MgCl$_2$, 10 μmoles/ml. 2-mercaptoethanol. Exonuclease III was found to be more stable in 0·01 M-2-mercaptoethanol at pH 7·1 than in 0·001 M-2-mercaptoethanol at pH 8. Unless otherwise stated, enzyme was the last component to be added. Zero time samples were taken before the addition of enzyme.

λ DNA infectivity assays

Samples from enzyme reaction mixtures were diluted 100-fold or more into 0°C TCM (0·01 M-tris–HCl buffer, pH 7·1, 0·01 M-CaCl$_2$, 0·01 M-MgCl$_2$) before adding them to helper-infected bacteria. Three different systems of recipient and indicator bacteria were used. The first (assay A) employed K12 gal_1^- (λ i^{434}) as the recipient, λ i^{434} as the helper, and C600 (λ i^{434} mi) as the indicator. The details of this assay are given by Radding & Kaiser (1963). The second (assay B) employed K12 gal_1^- (λ i^{434}) as both the recipient and the indicator and λ i^{434} as the helper. The third (assay C) employed K12 gal_4^- pm^+ as the recipient, λ $sus_A sus_J i^{434} c_I$ (Radding & Kaiser, 1963) as the helper phage and K12 pm^-, K12 pm^+ (λ i^{434}) and K12 pm^- (λ $sus_A sus_J i^{434}$) as the indicator strains.

The preparation of helper-infected bacteria in assays B and C was modified from assay A in several ways, thereby achieving specific infectivities of 0·2 to 1$\times$10^8 plaques/μg DNA. The recipient bacteria were grown in Hl glucose medium (Kaiser & Hogness, 1960) and harvested in the exponential phase of growth at a density of 1$\times$10^9 cells/ml. The bacteria were chilled to 0°C, sedimented and suspended at a density of 2$\times$10^9 cells/ml. in medium I†, incubated 10 min at 37°C, 5 min at 0°C and finally mixed with an equal volume of helper phage in medium I at 0°C. The concentration of helper phage was adjusted so as to be 10^{10}/ml. after dilution into the bacteria. Adsorption occurred during 10 min at 0°C; the complexes were then incubated 5 min at 37°C, chilled to 0°C for 5 min and sedimented at 5°C. The infected bacteria were suspended at a density of 2×10^9/ml. at 0°C in TCM and held at 0°C for 60 to 240 min before use.

DNA diluted in TCM (0·1 ml.) was mixed with helper-infected bacteria (0·2 ml.) and the mixture incubated 25 min at 37°C. Pancreatic DNase was added at a final concentration of 15 μg/ml. and incubation continued for 5 min. The assay mixtures were then plated in suitable dilution on the appropriate indicator strains. In assay C, dilution had to be at least tenfold.

Under these assay conditions, the number of plaques obtained is proportional to the concentration of DNA in the assay mixture, for concentrations below 0·15 μg/ml. The activities are reported as the number of plaques obtained divided by the DNA concentration in the assay tube, the concentration being measured by the absorbance at 260 mμ (A_{260}) of the solution. For λ DNA, $A_{260} = 1$ corresponds to 50 μg DNA/ml.

3. Results

(1) Inactivation of λ DNA by DNA polymerase

Incubation of native λ DNA with DNA polymerase and the four deoxyribonucleoside triphosphates led to a rapid loss of infectivity. The results in Table 1 show that the extent of inactivation increases with increasing amounts of enzyme and that

TABLE 1

Inactivation of λ DNA by DNA polymerase; dependence on the triphosphates and on the amount of enzyme

	Fraction of initial activity
Complete system 1	0·002
minus dATP	0·05
minus dTTP	0·1
minus dGTP	0·8
minus dCTP	0·8
minus dATP, dTTP, dGTP, dCTP	0·9
minus enzyme	1·0
Complete system 2	
with 0·082 unit DNA polymerase	0·8
0·25	0·4
0·74	0·12
2·2	0·024
6·7	0·004

Complete system 1 consisted of 4 mμmoles λ DNA, preparation DF1, 8 units of DNA polymerase, 0·05 mμmole each dATP, dGTP, dCTP, dTTP, 33 μmoles tris–HCl buffer (pH 7·1), 3·3 μmoles MgCl$_2$ and 5 μmoles 2-mercaptoethanol in a total volume of 0·5 ml. The mixtures were incubated for 8 min at 15°C. The initial activity of i^λ was $1·9 \times 10^9$ and the assay was C.

Complete system 2 consisted of 5·7 mμmoles λ DNA, preparation DF2, the amounts of DNA polymerase indicated in the Table, 0·04 mμmole each dATP, dGTP, dCTP, dTTP, 26 μmoles tris–HCl buffer (pH 7·1), 2·6 μmoles MgCl$_2$, 4 μmoles 2-mercaptoethanol in a total volume of 0·4 ml. The mixtures were incubated for 10 min at 15°C. The activity of i^λ was measured by assay C. Initially it was $1·3 \times 10^9$.

† Medium I contains 10^{-2} M-tris–HCl buffer (pH 7·1), 6×10^{-5} M-MgCl$_2$, 0·6 mg/ml. glucose, 6×10^{-4} M-potassium phosphate buffer, 5×10^{-4} M-(NH$_4$)$_2$SO$_4$, 4×10^{-10} M-FeSO$_4$.

inactivation depends on the presence of the triphosphates. Although some inactivation did occur in the absence of the deoxynucleoside triphosphates of adenine or thymine, it was less than in their presence. Virtually no inactivation occurred in the absence of the deoxynucleoside triphosphates of guanine or cytosine. Thus inactivation takes place under conditions of DNA synthesis.

Richardson, Inman & Kornberg (1964) have shown that molecules of T7 DNA, rendered partially single-stranded by exposure to *E. coli* exonuclease III, prime two sorts of synthetic reactions catalyzed by DNA polymerase. One sort, called "repair", restores the nucleotide units removed by exonuclease III. The other sort, called "net synthesis", increases the amount of DNA over that present in the native, completely double-stranded molecules of T7 DNA. At 37°C, partially single-stranded T7 DNA primed both repair and net synthesis. At 20°C, however, only the repair reaction took place.

The inactivation of native λ DNA by DNA polymerase occurred at 15°C. It even occurred at 0°C at about one-third the rate observed at 15°C. Thus by analogy with T7 experiments, the inactivation of λ behaves like a repair of single-stranded ends.

Kinetics of inactivation of λ DNA by DNA polymerase are depicted in Fig. 1. The marker sus_A is located at the left end of the linkage map, sus_J about one-fourth of the way in from the left end and i^λ about one-third of the way in from the right end. Fragments carrying either $sus_A sus_J$ or i^λ but not both have been prepared and found to be infective (Radding & Kaiser, 1963), so enzymic destruction of one marker but not the others could in principle have been detected as inactivation of one marker but not the others. This was not observed; rather, as is shown in Fig. 1, all three markers were lost at the same rate as if the sites of polymerase-catalyzed synthesis are essential for the activity of the whole molecule.

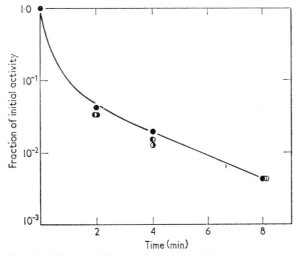

Fig. 1. Inactivation of λ DNA by DNA polymerase at 15°C. A mixture containing 37 mμmoles λ DNA, preparation D3, 1·2 units DNA polymerase, 0·1 mμmole each dATP, dGTP, dCTP, dTTP, 67 μmoles tris–HCl buffer (pH 7·1), 10 μmoles 2-mercaptoethanol and 6·7 μmoles MgCl₂ in an initial volume of 1 ml. was incubated at 15°C. At the times indicated, portions were removed, diluted 100-fold or more in TCM at 0°C and assayed using the procedure of assay C. Portions of the same assay mixture were plated on K12 pm^+ (λi^{434}) to measure total i^λ activity, on K12 pm^- to measure total $sus_A^+ sus_J^+$ activity and on K12 pm^- ($\lambda\, sus_A sus_J i^{434}$) to measure the linked $sus_A^+ sus_J^+ i^\lambda$ activity. The activity of i^λ (◑) was initially 6·2 × 10⁹; the activity of $sus_A^+ sus_J^+$ (◐) was initially 5·0 × 10⁹; the linked activity $sus_A^+ sus_J^+ i^\lambda$ (●) was initially 4·1 × 10⁹.

(2) *Reversal of polymerase-catalyzed inactivation by exonuclease III*

One possible explanation of the inactivation of λ DNA by polymerase is that nuclease activity was present in the enzyme preparation. In fact, DNA polymerase is known to have an associated exonuclease activity (Lehman & Richardson, 1964) which is optimum at alkaline pH. Small amounts of endonuclease activity, although previously undetected, might also have been present.

These possibilities have been ruled out by exposing DNA inactivated by polymerase to subsequent hydrolysis with exonuclease III. A two-stage experiment is illustrated in Fig. 2. During the first stage, lasting 32 minutes, λ DNA was inactivated by DNA polymerase and the four triphosphates to 2×10^{-5} of its initial infectivity. Then the reaction mixture was heated to 75°C for five minutes to denature the polymerase activity. In the second stage, exonuclease III was added to the heated first-stage mixture and the incubation continued. Within five minutes the infectivity increased by more than 10^3 times up to 10% of its initial value; it then decreased again. The restoration of activity was shown by a control experiment to be due to the enzyme and not to the diluent in which it was added.

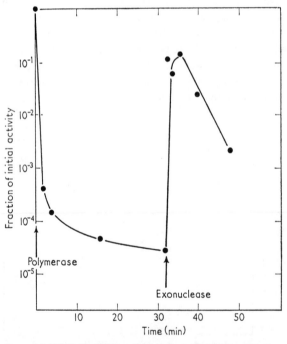

Fig. 2. Inactivation of λ DNA by DNA polymerase and reactivation by exonuclease III. A mixture containing 29·5 mμmoles λ DNA, preparation D4, 48 units of DNA polymerase, 40 mμ-moles each dATP, dGTP, dCTP, dTTP, 57·5 μmoles tris–HCl buffer (pH 7·1) and 0·58 μmole MgCl₂ in an initial volume of 0·8 ml. was incubated at 15°C. The triphosphates were the last component to be added to the mixture.

At 32 min 0·7 ml. of the mixture was heated to 75°C for 5 min, then cooled rapidly to 0°C; 6 μmoles 2-mercaptoethanol in 0·06 ml. and 40 μmoles of tris–HCl buffer at pH 8 in 0·04 ml. were added. To 0·4 ml. of this mixture 0·64 unit of exonuclease III was added (in a very small volume); to the remaining 0·3 ml. enzyme diluent was added. Both were incubated at 37°C and portions removed at the times indicated, diluted into TCM at 0°C and assayed for total i^λ activity by the procedure of assay B. The initial activity of i^λ was $3·1 \times 10^9$.

A two-stage experiment employing the markers sus_A^+, sus_J^+ and i^λ (Fig. 3), shows that just as polymerase inactivates the pair of markers $sus_A^+ sus_J^+$ at the same rate as i^λ, so exonuclease III restores activity to the pair $sus_A^+ sus_J^+$ and to i^λ, selected separately or together, at the same rate.

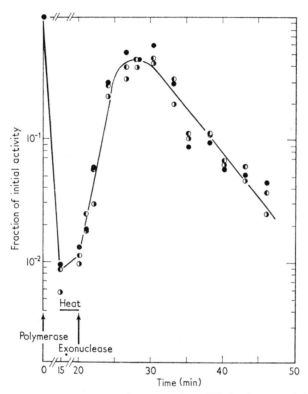

FIG. 3. Reactivation of $sus_A^+ sus_J^+$ and of i^λ by exonuclease III. A mixture containing 40 mμmoles λ DNA, preparation D3, 1·2 units of DNA polymerase, 0·1 mμmole each dATP, dGTP, dCTP, dTTP, 67 μmoles tris–HCl buffer (pH 7·1), 6·7 μmoles MgCl$_2$, and 10 μmoles 2-mercaptoethanol in an initial volume of 1 ml. was incubated at 15°C.

At 15 min the mixture was heated to 75°C for 5 min, then cooled to 0°C. To 0·95 ml. of the heated mixture 3·4 units of exonuclease III were added and incubation continued at 37°C. At the times indicated portions were removed, diluted and assayed by the procedure of assay C. Portions of the same assay mixture were plated on K12 pm^+ (λi^{434}) to measure total i^λ activity, on K12 pm^- to measure total $sus_A^+ sus_J^+$ activity and on K12 pm^- ($\lambda\, sus_A sus_J\, i^{434}$) to measure the linked $sus_A^+ sus_J^+$ i^λ activity. The activity of i^λ (◗) was initially $1·9 \times 10^9$; the activity of $sus_A^+ sus_J^+$ (◖) was initially $1·9 \times 10^9$; the linked activity $sus_A^+ sus_J^+$ i^λ (●) was initially $1·4 \times 10^9$.

If polymerase inactivates because it adds nucleotides to native λ DNA, then exonuclease III ought to restore infectivity by removing those added nucleotides, restoring the *status quo ante*. This indeed is what seems to have happened.

Prolonged treatment with exonuclease III may be seen in the experiments of Figs 2 and 3 to destroy activity. This point was explored with native DNA (see the next section).

(3) *Inactivation of λ DNA by exonuclease III*

Exonuclease III acting on native λ DNA rapidly destroyed its infectivity, as exemplified by the experiment illustrated by Fig. 4. Several observations, summarized in Table 2, indicate that inactivation depends on the exonuclease activity of the enzyme. Richardson & Kornberg (1964) showed that a divalent cation is required for enzymic activity and that the enzyme loses catalytic activity in the absence of a sulfhydryl compound. The results in Table 2 show that incubation of the enzyme at 37°C in the absence of a sulfhydryl compound destroys the capacity of exonuclease III to inactivate λ DNA and that inactivation fails to occur if magnesium is left out. The rate of inactivation is proportional to enzyme concentration: 8 units of enzyme catalyzed inactivation at the rate of 0·3 lethal hit per minute, and 17 units, 0·6 hit per minute.

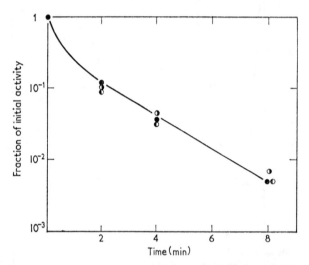

FIG. 4. Inactivation of $sus_A^+ sus_J^+$ and of i^λ by exonuclease III. A mixture containing 36 mμmoles λ DNA, preparation D3, 12·8 units of exonuclease III, 67 μmoles tris–HCl buffer (pH 7·1), 6·7 μmoles MgCl$_2$ and 10 μmoles 2-mercaptoethanol in an initial volume of 0·8 ml. was incubated at 37°C. At the times indicated, portions were removed, diluted and assayed according to the procedure of assay C. Portions of the same assay mixture were plated on K12 pm^+ (λi^{434}) to measure the total i^λ activity, on K12 pm^- to measure the total $sus_A^+ sus_J^+$ activity and on K12 pm^- (λ $sus_A sus_J$ i^{434}) to measure the linked $sus_A^+ sus_J^+ i^\lambda$ activity. The activity of i^λ (◗) was initially 6·2 × 10⁹; the activity of $sus_A^+ sus_J^+$ (◖) was initially 5·0 × 10⁹; the linked activity of $sus_A^+ sus_J^+ i^\lambda$ (●) was initially 4·1 × 10⁹.

Figure 4 also shows that the activity of the pair of markers $sus_A^+ sus_J^+$ is lost at the same rate and concurrently with the activity of i^λ, as if the activity of the whole molecule were being destroyed by action at the ends.

Using ³²P-labeled λ DNA, the amount of inactivation was compared with the amount of DNA digested. The inactivation of ³²P λ DNA was biphasic; a rapid initial rate was followed by a slower final rate equal to that observed under identical conditions with λ DNA not containing ³²P. However, radioactive material was made acid-soluble at the same rate during both periods. The high initial rate of inactivation might be due to a sensitizing effect of the many single-strand breaks present in ³²P λ DNA. (Please consult Materials and Methods for the properties of ³²P λ DNA). By comparing the final rate of inactivation with the rate at which ³²P-containing material is made acid-soluble, an upper limit to the number of nucleotides released for a given amount of inactivation is obtained. Interpolation of the results in Table 3 gives a release of

0·4% of the total ^{32}P, or about 400 nucleotides per molecule of λ DNA, for one lethal hit during the second phase of inactivation. Therefore the average molecule loses activity when fewer than 400 nucleotides have been removed.

TABLE 2

Inactivation of λ DNA by exonuclease III; requirements and dependence on the amount of enzyme

	Fraction of initial activity
Complete system 1	
Enzyme incubated without SH	0·003
Minus Mg^{2+} plus 5×10^{-4} M-EDTA	0·5
Minus enzyme	1·0
	1·0
Complete system 2	
With 8 units of enzyme at 2 min	0·56
at 4 min	0·29
at 8 min	0·078
With 17 units of enzyme at 1 min	0·61
at 2 min	0·27
at 4 min	0·08

Complete system 1 consisted of 24 mμmoles λ DNA, preparation D2, 30 units of exonuclease III, 150 μmoles tris–HCl buffer (pH 8), 1·5 μmoles MgCl$_2$ and 2·25 μmoles 2-mercaptoethanol in a total volume of 3 ml. The mixtures were incubated at 37°C for 32 min and assayed by assay A. Enzyme was incubated without an SH compound in 0·01 M-tris–HCl buffer (pH 8), for 30 min at 37°C.

Complete system 2 consisted of 96 mμmoles λ DNA, preparation D4, the amount of exonuclease III indicated, 120 μmoles tris–HCl buffer (pH 7·1), 12 μmoles MgCl$_2$ and 18 μmoles 2-mercapto-ethanol in a total volume of 1·7 ml. Activities were measured by assay B.

TABLE 3

Hydrolysis of ^{32}P λ DNA by exonuclease III

Time (min)	Percentage ^{32}P acid-soluble	Fraction of initial infectivity
0	< 0·1	1·0
5	0·7	0·01
10	1·4	0·002

60 mμmoles ^{32}P λ DNA with 2×10^6 cts/min/μmole, preparation D^{32}P, 1·9 units exonuclease III, 9 μmoles 2-mercaptoethanol, 60 μmoles tris–HCl buffer (pH 8·0), and 0·6 μmole MgCl$_2$ in a total volume of 0·84 ml. were incubated at 37°C. At the times specified, 0·3-ml. portions were removed, 0·2 ml. calf thymus DNA was added as carrier and the mixture was precipitated by the addition of 0·5 ml. 7% perchloric acid at 0°C. After centrifugation, 0·5 ml. of the supernatant fluid was plated for the measurement of radioactivity. The value for 0 time acid-soluble ^{32}P was obtained from a second, equivalent mixture with enzyme diluent rather than enzyme added. This mixture was incubated for 30 min at 37°C before precipitation. The infectivities were measured by assay B.

(4) *Reversal of exonuclease III–catalyzed inactivation by polymerase*

Restoration of λ DNA infectivity catalyzed by polymerase following inactivation catalyzed by exonuclease III depends on limiting the extent of synthesis, because of the inactivating effect of synthesis described earlier. Restoration was demonstrated by using low concentrations of the triphosphates and also by using a third-stage treatment with exonuclease III.

Figure 5 displays the infectivity changes during treatment with exonuclease III, polymerase and finally exonuclease III again. Reactivation catalyzed by polymerase can be seen in two ways. First, after two minutes of polymerase treatment, the infectivity was six times higher than the infectivity at the end of the first exonuclease treatment. Second, after one minute of the second exonuclease treatment, the infectivity was nine times higher than the infectivity at the end of the first exonuclease

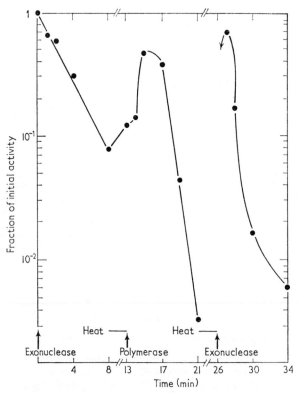

FIG. 5. Reactivation by DNA polymerase. A mixture containing 96 mμmoles λ DNA, preparation D4, 8 units of exonuclease III, 120 μmoles tris–HCl buffer (pH 7·1), 12 μmoles MgCl$_2$ and 18 μmoles 2-mercaptoethanol in an initial volume of 1·7 ml. was incubated at 37°C.

At 8 min the mixture was heated to 75°C for 5 min, then cooled to 0°C. To 1·3 ml. of the heated mixture 40 units of polymerase and 0·28 mμmole each of dATP, dGTP, dCTP, dTTP were added, giving a final volume of 1·58 ml. This mixture was incubated at 15°C.

At 21 min, the mixture was again heated to 75°C for 5 min, then cooled to 0°C. To a 0·32-ml. sample of this mixture 3·2 units of exonuclease III were added in a very small volume and this mixture was incubated at 37°C. A second sample without exonuclease III was incubated at 37°C as a control. This sample (results are not given in the Figure) showed no change in activity whereas the sample with enzyme showed an increase as indicated.

At the times indicated on the Figure, portions were withdrawn, diluted at least 100-fold in TCM at 0°C and assayed for $i^λ$ activity according to the procedure of assay B. The initial activity was 5·5 × 10⁹.

treatment. The infectivity during the second exonuclease treatment rose to 69% of the initial infectivity. Evidently part of the reactivation by polymerase had been obscured by an overshooting in the synthetic phase.

The capacity of polymerase-catalyzed synthesis to restore infectivity destroyed by exonuclease III shows that exonuclease III destroyed infectivity by removing nucleotides from the 3'-hydroxyl end of a double-stranded portion of the molecule. Had the molecule not been double-stranded at the point where exonuclease III acted, there would not have been a complementary template strand for polymerase to copy.

4. Discussion

The inactivation and reactivation phenomena produced by treatment of λ DNA with *E. coli* DNA polymerase and *E. coli* exonuclease III can be accounted for on the basis of three assumptions.

(1) That the structure of native λ DNA is

$$5'\underline{3'\underline{}}_{3'}5' \qquad \text{(I)}$$

(2) That the two protruding single strands have complementary base sequences, permitting them to cohere and to form thereby closed molecules, dimers, trimers, etc. The structure of a closed molecule, for example, would be

$$\text{(II)}$$

(3) That a molecule must have a *free* cohesive end to be infective.

Granting these assumptions, we would first like to show how the results of each of the enzyme treatments can be explained and why structures alternative to (I) are less likely. Finally, arguments specifically supporting the second and third assumptions will be given.

Inactivation of λ DNA by polymerase can be explained, since polymerase would be expected to catalyse the "repair" of a structure such as (I) with protruding 5'-terminated single strands to produce a molecule with its ends in register:

$$\begin{array}{l}3'\text{-----}\underline{}5'\\5'\underline{}\text{-----}3'\end{array} \qquad \text{(III)}$$

--- = newly synthesized by polymerase.

This molecule would not be infective because the ends would not be cohesive. Treating structure III (the polymerase product) with exonuclease III would lead first to the removal of the newly added nucleotides, restoring the original structure (I), which would be again cohesive and therefore infective. Thus exonuclease III would reactivate λ DNA inactivated by polymerase.

Exonuclease III acting on native λ DNA would remove 5'-nucleotides from the 3'-hydroxyl termini. On the basis of assumptions (1) and (2), it would be expected to increase the total length of the protruding single strands without increasing the length of their mutually complementary parts. It is reasonable that the treated molecules would lose their capacity to cohere, either because the complementary regions would be less likely to come together or because the base-paired region would be less stable with single-strand regions on either side of it.

Exonuclease III action on λ DNA "repaired" by DNA polymerase would eventually pass the length of the 3'-strands characteristic of native λ DNA. Consequently exonuclease III would first reactivate DNA exposed to polymerase, then inactivate it.

In an analogous way, using DNA inactivated by exonuclease III as a primer for DNA polymerase would lead first to a return to structure (I) and high infectivity. Continued polymerase-catalyzed synthesis would overshoot the native length of the 3'-terminated strand until a molecule with structure III was produced. Therefore, polymerase would first reactivate λ DNA inactivated by exonuclease III, then inactivate it.

One structure to be considered as an alternative to (I) has protruding 3'- (rather than 5'-) hydroxyl-terminated single strands. Its ends could be cohesive. Whether exonuclease III would be active on a molecule with such a structure is not known. If, on one hand, exonuclease cannot catalyze the removal of nucleotides from a protruding 3'-hydroxyl-terminated single strand, then exonuclease III treatment should not inactivate λ DNA. If, on the other hand, exonuclease can catalyze the hydrolysis of protruding 3'-hydroxyl-terminated single strands, then polymerase would find no template strand to copy and so could not regenerate the original structure and restore activity destroyed by exonuclease. Thus, depending on the specificity of exonuclease III, this structure fails to explain either inactivation by exonuclease or restoration by polymerase of DNA inactivated by exonuclease.

Another pair of alternative structures has no protruding single strands. Molecules which are entirely doubly helical might be cohesive if a sequence found at one end of the molecule were duplicated at the other in the same or in the inverted order. Structure (I) explains more readily than either of these alternative structures why polymerase catalyzes the inactivation of λ DNA at 15°C and even at 0°C. Richardson, Inman & Kornberg (1964) found that although the priming activity of native double-stranded DNA fell to a very low level when the temperature was lowered from 37 to 20°C, the priming activity of DNA with protruding 5'-hydroxyl-terminated single strands remained high. Moreover, both of these alternative structures imply that a "left" end could cohere to another left end, or a "right" end to another right end. But Hershey & Burgi (1965) have demonstrated that the ends of λ DNA do not join at random, but instead join specifically in pairs of one left and one right end.

Various structures which are combinations of the four discussed above might also be considered. In general, these structures share the disadvantages of the pure structures of which they partake. One of them however, is not ruled out by the data under consideration: a structure with a protruding 5'-terminated single strand at one end and a protruding 3'-terminated single strand at the other.

Next we shall summarize the experimental support for assumption 2, that the two protruding single strands have complementary base sequences. The experiment of Hershey & Burgi (1965), referred to above, shows that the two ends of λ DNA are complementary to each other, whatever the details of their structure may be.

Although it is not excluded that molecular groups other than nucleotides occur at the ends of λ DNA and are responsible for cohesion, none of the experiments reported here or elsewhere, to our knowledge, indicates their presence. Rather, the chemical basis of cohesion of the ends of λ DNA is likely to be found in the formation of base pairs, since the conditions described by Hershey *et al.* (1963) favoring the bonding together of the cohesive sites, namely, raising the salt concentration to 0·6 M and slow cooling starting from 75°C, are conditions which would promote the formation of base pairs.

The transition between closed and open forms appears to be co-operative and it occurs below 65°C in 2 M-ammonium acetate (Kaiser & Inman, manuscript in preparation), a temperature which might be considered too low for a DNA transition in molar salt at neutral pH (Marmur & Doty, 1959, 1962). However, the protruding single strands seem to be very short, since preliminary experiments indicate that the point of maximum reactivation is reached when no more than 20 nucleotides (measured as ^{32}P) per λ molecule incorporated by the action of DNA polymerase on native λ DNA are made acid-soluble by exonuclease III. A short double helical structure would be expected to denature at a lower temperature than a long structure of the same base composition.

The third assumption, that a molecule of λ DNA must have a free cohesive end to be infective in helper-infected bacteria, is supported by experiments of Kaiser & Inman (1965, manuscript in preparation). They found that when whole, half or sixth molecules lose their free cohesive sites by cohesion to themselves or to other molecules, they also lose part and perhaps all of their infectivity.

The requirement of a free cohesive end for infectivity helps to explain why the various inactivation and reactivation phenomena produced by DNA polymerase and by exonuclease III affect the marker pair $sus_A^+ sus_J^+$ and the marker i^λ at the same rate, despite the broad distribution of these markers along the molecule. It is because the infectivity of the whole molecule depends on the integrity of the cohesive ends. A parallel loss of activity of different markers is also seen when molecules or fragments cohere (Kaiser & Inman, manuscript in preparation).

Finally, it might be noted that if structure (I) is correct, there could be terminal phosphate residues on the 5′-ends of λ DNA but not on the 3′-ends, since a 3′-phosphate is known to inhibit the addition of nucleotides by *E. coli* DNA polymerase (Richardson *et al.*, 1963).

The advice and supply of materials by members of the Department of Biochemistry, Stanford University, made this work possible; in these regards the contributions of Arthur Kornberg, I. Robert Lehman, and Charles C. Richardson are especially to be mentioned. We would also like to thank Charles Richardson for his guidance in the purification of exonuclease III.

This work was supported in part by grants from the National Institutes of Health of the United States Public Health Service.

REFERENCES

Bessman, M. J., Lehman, I. R., Simms, E. S. & Kornberg, A. (1958). *J. Biol. Chem.* **233**, 171.
Davison, P. F., Freifelder, D. & Holloway, B. (1964). *J. Mol. Biol.* **8**, 1.
Hershey, A. D. & Burgi, E. (1965). *Proc. Nat. Acad. Sci., Wash.* in the press.
Hershey, A. D., Burgi, E. & Ingraham, L. (1963). *Proc. Nat. Acad. Sci., Wash.* **49**, 748.
Hogness, D. S. & Simmons, J. R. (1964). *J. Mol. Biol.* **9**, 411.

Kaiser, A. D. (1962). *J. Mol. Biol.* **4**, 275.

Kaiser, A. D. & Hogness, D. S. (1960). *J. Mol. Biol.* **2**, 392.

Kellenberger, G., Zichichi, M. L. & Weigle, J. J. (1961). *Proc. Nat. Acad. Sci., Wash.* **47**, 869.

Lehman, I. R., Bessman, M. J., Simms, E. S. & Kornberg, A. (1958). *J. Biol. Chem.* **233**, 163.

Lehman, I. R. & Richardson, C. C. (1964). *J. Biol. Chem.* **239**, 233.

MacHattie, L. A. & Thomas, C. A. Jr. (1964). *Science*, **144**, 1142.

Marmur, J. & Doty, P. (1959). *Nature*, **183**, 1427.

Marmur, J. & Doty, P. (1962). *J. Mol. Biol.* **5**, 109.

Meselson, M. & Weigle, J. J. (1961). *Proc. Nat. Acad. Sci., Wash.* **47**, 857.

Meyer, F., Mackal, R., Tao, M. & Evans, E. (1961). *J. Biol. Chem.* **236**, 1141.

Radding, C. M. & Kaiser, A. D. (1963). *J. Mol. Biol.* **7**, 225.

Richardson, C. C., Inman, R. B. & Kornberg, A. (1964). *J. Mol. Biol.* **9**, 46.

Richardson, C. C. & Kornberg, A. (1964). *J. Biol. Chem.* **239**, 242.

Richardson, C. C., Lehman, I. R. & Kornberg, A. (1964). *J. Biol. Chem.* **239**, 251.

Richardson, C. C., Schildkraut, C. L., Aposhian, H. V. & Kornberg, A. (1964). *J. Biol. Chem.* **239**, 222.

Richardson, C. C., Schildkraut, C. L., Aposhian, H. V., Kornberg, A., Bodmer, W. & Lederberg, J. (1963). In *Symposium on Informational Macromolecules*, ed. by H. Vogel, V. Bryson and J. Lampen, p. 13. New York: Academic Press, Inc.

Ris, H. & Chandler, B. L. (1963). *Cold Spr. Harb. Symp. Quant. Biol.* **28**, 1.

Studier, F. W. (1965). *J. Mol. Biol.* **11**, 373.

From the *Proceedings of the National Academy of Sciences*, Vol. 51, No. 5, 775–779, 1964. Reproduced with permission of the authors.

CHROMOSOME STRUCTURE IN PHAGE T4, I. CIRCULARITY OF THE LINKAGE MAP*

By George Streisinger, R. S. Edgar, and Georgetta Harrar Denhardt

INSTITUTE OF MOLECULAR BIOLOGY, UNIVERSITY OF OREGON, AND BIOLOGY DIVISION, CALIFORNIA INSTITUTE OF TECHNOLOGY

Communicated by A. D. Hershey, March 26, 1964

On extraction, particles of phage T4 each yield a single DNA molecule;[1-3] and in genetic crosses, all the markers prove to be linked to one another.[4] Crossing also generates short heterozygous regions that are distributed randomly over the genome.[5]

Doermann and Boehner[6] have suggested that at least some of the heterozygous regions involve interruptions of some sort in the DNA chain. Berns and Thomas[7] and Cummings,[8] on the other hand, interpret results of their experiments on the physical properties of T4 DNA to mean that the single strands are continuous.[9]

The paradox could be resolved by making two assumptions about the chromosome of phage T4. First, that it is a linear molecule with a terminal redundancy:

$$a\ b\ c\ d\ e\ f\ldots\ldots\ w\ x\ y\ z\ a\ b\ c$$

Second, that after several rounds of replication the chromosome becomes circularly permuted as a result of genetic recombination within the region of redundancy.[10] The ends of a population of chromosomes would thus be randomly distributed over the genome:

$$g\ h\ i\ j\ k\ l\ldots\ldots\ c\ d\ e\ f\ g\ h\ i$$

$$m\ n\ o\ p\ q\ r\ldots\ldots\ i\ j\ k\ l\ m\ n\ o,\ \text{etc.}$$

One of several predictions of the above model is that the linkage map of phage T4 would be circular. A test of this prediction forms the substance of the present communication.

Materials and Methods.—The phage strains used were derivatives of T4D containing the markers r73;[4] tu41, tu42b, tu44, tu45a, and r48;[11] h42;[12] ac41;[13] rEDb-48;[14] am85 and am54, obtained from Dr. R. H. Epstein; and r67, obtained from Dr. A. H. Doermann. The bacterial strains were *Escherichia coli* B, S/6, K12(λ), CR63, and K12(λ)/4.

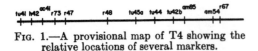

FIG. 1.—A provisional map of T4 showing the relative locations of several markers.

Media: Broth—1 liter H₂O, 10 gm bacto-tryptone, 5 gm NaCl. Tryptone plates—bottom layer, broth with 1.1 per cent bacto-agar; top layer, broth with 0.7 per cent bacto-agar. EHA plates—bottom layer contained 1 liter H_2O, 13 gm bacto-tryptone, 2 gm sodium citrate · 2 H_2O, 1.3 gm glucose, 14 gm bacto-agar, 8 gm NaCl; top layer contained 1 liter H_2O, 13 gm bacto-tryptone, 2 gm sodium citrate · 2 H_2O, 3 gm glucose, 7 gm bacto-agar, 8 gm NaCl. Salt-poor EHA plates—bottom layer was identical to EHA but contained only 2.5 gm NaCl per liter; top layer contained no added NaCl.

Acriflavin-neutral (Nutritional Biochemicals Corp.) was added, when required, to the bottom layer only, at a concentration of 0.25 µg per ml.

Results.—A linkage map of phage T4 is shown in Figure 1. With the exception of *am*85, *am*54, and *r*67, the relative order of the markers indicated in the map was established previously by two- and three-factor crosses.[4,11-13] We now find that *am*54 is closely linked to *r*67 (1.5% recombinants), and *am*85 is closely linked to *tu*42b (3.5% recombinants). The order *tu*44—*am*85—*am*54 is established by means of the three-factor cross *tu*44 *am*54 × *am*85 (cross 14, Table 1). The order *tu*44—*tu*42b—*r*67 is confirmed by cross 15, Table 1.

The information cited above is compatible with either of two alternatives. The linkage map may have ends, in which case *r*67 lies closer to *ac*41 than to *h*42, as represented in Figure 1. Or the linkage map could be a circle formed by joining the ends shown in Figure 1, in which case *r*67 would be expected to lie closer to *h*42 than to *ac*41.

The cross *r*67 *h*42 *ac*41⁺ × *r*67⁺ *h*42⁺ *ac*41 (cross 1, Table 1) distinguishes between these alternatives. The progeny of the cross were plated under conditions that permitted the recognition of the *h*42 *ac*41 recombinants. About 65 per cent of these were *r*, and 35 per cent *r*⁺, showing that the *r* marker is more closely linked to *h* than to *ac*. In other words, the results of cross 1 are inconsistent with previous data summarized in Figure 1 unless the linkages are represented on a circle.

The same linkage was tested in cross 2, Table 1, with a different arrangement of the parental markers. The results are consistent with those of cross 1, showing that the marker frequencies depend on linkages between loci only, and do not reflect a bias associated with particular markers.

Additional three-factor and four-factor crosses are listed in Table 1. They are similar in principle to crosses 1 and 2 and place all the markers in a unique order on a circular map. In some crosses, unequal multiplicities of infection were used to increase the sensitivity of the linkage tests.[4] In all crosses except 3, 14, and 15, the frequency of recombinants was measured among the early phage progeny (1–10 per bacterium), and usually the drift toward genetic equilibrium of unselected alleles was demonstrated by additional sampling at later times during phage growth. Arguments concerning the validity of these types of linkage test have been presented by Streisinger and Bruce.[4] Scoring procedures are described in Table 2. Figure 2 identifies linkages tested in the individual crosses and summarizes the circular map with which all results are consistent.

Discussion.—Our results demonstrate genetic circularity and are thus compatible with the model presented in the introduction to this paper. The circularity of the

TABLE 1

No.	Cross — Parent 1	Cross — Parent 2	Multiplicity Parent 1	Multiplicity Parent 2	Recombinant ratio measured*	Progeny Phage — Value of ratio	Progeny Phage — Control ratio	Progeny Phage — Value of control ratio
1	r67 / h42	+ / ac41	9.8	10.3	112/·12	0.65	1··/···	0.53
2	h42 / +	r67 / ac41	7.5	7.4	112/·12	0.66	1··/···	0.54
3	ac41 / rEDb48	h42 / +	4.2	4.7	112/·1·	0.09		
4	r67 / +	+ / r47	5.0	0.5	211/·1·	0.11		
5	+ / r47	r73 / +	3.1	0.7	212/·1·	0.02		
6	r67 / +	r73 / r47	5.8	0.6	221/·21	0.58		
7	+ / r47	r73 / +	5.1	0.7	212/·12	0.33		
8	r48 / +	r47 / +	8.6	0.9	221/·21	0.53		
9	+ / r48	r73 / +	7.2	0.8	212/21·	0.29		
10	r73 / r47	+ / tu45	8.6	0.8	122/12·	0.43		
11	+ / r47	r73 / tu45	8.4	0.9	212/21·	0.27		
12	tu44 / +	r48 / am85	8.8	9.6	122/12·	0.40	1··/···	0.60
13	+ / tu44	r48 / +	7.5	7.4	112/·12	0.74	1··/···	0.55
14	tu44 / am54	+ / am85	6	6	112/·12	0.72		
15	+ / tu42	tu44 / +	6.3	7.8	112/·12	0.80		
16	r48 / am54 / tu41	+ / am85 / +	7.1	7.6	2121/·12·	0.74	2··/····	0.53
					212·/·12·	0.12		
					·121/·12·	0.46		
17	r48 / +	+ / am54	9.7	9.4	1212/·21·	0.36	2··/····	0.54
	am85 / tu41	+			121·/·21·	0.13		
					·212/·21·	0.46		
					·212/·21·	0.37		

* 1 or 2 indicates a marker derived from parent 1 or 2, and a dot (·) indicates a marker derived from either parent. The ratio 112/·12 in cross 1, for instance, represents r67 h42 ac41/(r67 h42 ac41 plus r+ h42 ac41).

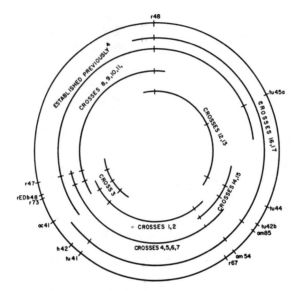

FIG. 2.—The circular map of T4. The locations of markers used for any one
cross are connected by an arc.

genetic map of T4 has been confirmed by two-factor crosses made with a large set
of amber mutants[15] and temperature-sensitive mutants.[16] Foss[17] has been able to
demonstrate genetic circularity in T4 by means of a single, ingeniously devised four-
factor cross.

TABLE 2

PROCEDURES FOR SCORING THE PROGENY OF CROSSES

Crosses	Plates	Bacteria	Scoring
1, 2	Salt-poor EHA with acriflavin	Phage preadsorbed to S/6, plated on a mixture of K/4 and S/6	Only $h42$ $ac41$ forms clear plaques, classified as r or r^+ by inspection.
3	EHA with acriflavin	Mixed K/4 and S/6	Only $ac41$ forms plaques, classified as r or r^+ and h or h^+ by inspection.
4–11	Tryptone	K12(λ)	Only $r73^+$ $r47^+$ forms plaques, classified as r or r^+ by inspection.
12, 13	EHA	S/6	$tu45^+$ $tu44^+$ plaques selected and classified as r or r^+ by inspection.
15	EHA	S/6	Genotypes tu^+ r^+ and tu^+ r classified by inspection.
14, 16–17	EHA	S/6	Only am^+ forms plaques, classified as r or r^+ and tu or tu^+ by inspection.

It should be emphasized that our results, while demonstrating genetic circularity,
are by no means a critical test of the model we present; a number of other models
would account for the results equally well. For instance, Stahl[18] has pointed out
that a circular genetic map would be obtained if the chromosome were a nonpermuted rod and if genetic exchanges frequently occurred in pairs. A decision with
respect to the molecular basis of circularity calls for other, more critical tests.

Summary.—The linkage map of phage T4 is circular.

The authors wish to acknowledge their great indebtedness to F. W. Stahl, without whose encouragement they would not have taken their rather bizarre notions seriously. Many of the fea-

tures of the proposed model were developed in the course of conversations and correspondence with F. W. Stahl, M. Meselson, and M. Fox.

* This investigation was supported by research grants from The National Foundation (to R.S.E.), the National Science Foundation (G-14055 to G.S.), and the National Institute of Allergy and Infectious Diseases (E-3892 to G.S.).

[1] Rubenstein, I., C. A. Thomas, Jr., and A. D. Hershey, these PROCEEDINGS, **47**, 1113 (1961).

[2] Davison, P. F., D. Freifelder, R. Hede, and C. Levinthal, these PROCEEDINGS, **47**, 1123 (1961).

[3] Cairns, J., *J. Mol. Biol.*, **3**, 756 (1961).

[4] Streisinger, G., and V. Bruce, *Genetics*, **45**, 1289 (1960).

[5] Hershey, A. D., and M. Chase, in *Genes and Mutations*, Cold Spring Harbor Symposia on Quantitative Biology, vol. 16 (1951), p. 471.

[6] Doermann, A. H., and L. Boehner, *Virology*, **21**, 551 (1963).

[7] Berns, K. I., and C. A. Thomas, Jr., *J. Mol. Biol.*, **3**, 289 (1961).

[8] Cummings, D. J., *Biochim. Biophys. Acta*, **72**, 475 (1963).

[9] P. F. Davison, D. Freifelder, and B. W. Holloway [*J. Mol. Biol.*, **8**, 1 (1964)] obtained physical evidence suggesting that the strands are not continuous. Thus, the question of strand interruptions is not yet resolved. Although it is true that our experiments were motivated by Berns and Thomas' conclusions and are compatible with them, they do not in fact have any direct bearing on the question of whether strand interruptions do exist.

[10] Particular models for the generation of a population of circularly permuted chromosomes will be discussed in detail in a subsequent communication. The following two alternatives may help to clarify the present discussion: (1) the ends of a particular chromosome, after its injection into a bacterium, could be imagined to pair and to join, forming a circular structure. This circle might then be opened by cuts in the two chains of the DNA molecule, the cuts being staggered in relation to each other. (2) The chromosome could be imagined to replicate and the progeny chromosomes to join end-to-end at the region of terminal redundancy. The polymers thus formed would have to be cut into phage-sized pieces before maturation

[11] Doermann, A. H., and M. Hill, *Genetics*, **38**, 79 (1953).

[12] Edgar, R. S., *Virology*, **6**, 215 (1958).

[13] Edgar, R. S., and R. H. Epstein, *Science*, **134**, 327 (1961).

[14] Edgar, R. S., R. P. Feynman, S. Klein, I. Lielausis, and C. M. Steinberg, *Genetics*, **47**, 179 (1962).

[15] Epstein, R. H., personal communication.

[16] Edgar, R. S., and I. Lielausis, *Genetics*, in press.

[17] Foss, H. M., and F. W. Stahl, *Genetics*, **48**, 1659 (1963).

[18] Stahl, F. W., personal communication.

ARTICLE 30

From the *Proceedings of the National Academy of Sciences*, Vol. 54, No. 5, 1333–1345, 1965. Reproduced with permission of the authors.

CHROMOSOME STRUCTURE IN PHAGE T4, II. TERMINAL REDUNDANCY AND HETEROZYGOSIS*

By Janine Séchaud,† George Streisinger, Joyce Emrich, Judy Newton, Henry Lanford, Helen Reinhold, and Mary Morgan Stahl

INSTITUTE OF MOLECULAR BIOLOGY, UNIVERSITY OF OREGON, EUGENE

Communicated by A. D. Hershey, September 15, 1965

In a previous communication[1] it was suggested that the chromosome of each particle of phage T4 is terminally redundant, and that after several rounds of replication (following infection), progeny chromosomes arise which have circularly permuted genetic sequences. Among the members of a population of phage particles, the ends of the chromosomes would be randomly distributed over the genome, and a phage particle carrying different alleles at one of its redundant loci would be heterozygous.

Nomura and Benzer[2] have reported that heterozygotes for rII deletion mutants (from mixed infections with r^+ phage) occur with about a third of the frequency of those for rII point mutants. It thus seemed likely that heterozygosity in phage T4 is of at least two kinds. One kind may be the result of heterozygosity within the molecule, as was first suggested by Levinthal,[3] whereas another kind may be due to the terminal redundancy described above. Deletion mutants may be unable to form heterozygotes of the internal kind since these would involve violations of duplex complementarity (regardless of their precise molecular structure), whereas point mutants may form heterozygotes of both kinds.

The chromosome of a phage particle will be terminally redundant and heterozygous for marker a provided that (1) the chromosome terminates near marker a and includes marker a, and (2) the last recombinational event, occurring somewhere within the chromosome, has been such that the other end of the chromosome contains marker a^+. *Terminal-redundancy heterozygotes are formed and lost by the process of recombination; their frequency (as a function of time after infection) will depend on the rate of recombination but not on the rate of replication.* In a pool of vegetative chromosomes, the frequency of terminally redundant heterozygotes would be expected to change in time as does the frequency of recombination for very distantly linked markers; it would be expected to increase until an equilibrium value was reached and to remain constant after that. Because of the high frequency of recombination in phage T4, one would expect that the equilibrium value would be reached very early during the latent period.

Internal heterozygotes are most probably formed by an event that results in the recombination of markers on either side of the heterozygous region;[3] thus, they may well represent the primary products of recombination. They would be expected to disappear upon the semiconservative replication of the chromosome bearing them. The chromosome of a phage particle will be internally heterozygous for marker a provided that (1) a recombinational event leading to internal heterozygosis will have occurred at the site of marker a, and (2) the chromosome is incorporated into a mature particle before it undergoes replication. *Internal heterozygotes are formed by recombination and are lost by replication: their frequency (as a function of*

1333

time after infection) will depend both on the rate of recombination and on the rate of replication.

The frequency of heterozygotes, as a function of time after infection, has been observed to remain constant.[4] In a standard cross, the formation and disappearance of heterozygotes is at equilibrium throughout the period during which phage are withdrawn from the vegetative pool. If the rate of replication could be substantially decreased during this period, the frequency of internal heterozygotes would be expected to rise.

The thymidine analogue fluorodeoxyuridine (FUDR) inhibits DNA synthesis in phage-infected bacteria[5] but does not interfere with recombination.[6] It would be expected that in phage-infected, FUDR-treated bacteria, the frequency of internal heterozygotes would increase as a function of time, whereas the frequency of terminal-redundancy heterozygotes would remain constant. A test of this prediction forms the substance of this communication.

Materials and Methods.—The phage strains used were T4B standard type and mutants *r*607, *r*287, *r*205, *r*1364, *r*W8-33, *r*H23, *r*168 (all obtained from Dr. S. Benzer); T4D standard type and mutants *ts*N30 and *r*EDb42 (obtained from Dr. R. S. Edgar); and T4 h^{2+} (obtained from Dr. G. Stent). Stent's strain was derived by him by crossing T4 and T2, crossing a progeny strain with the T2 host range to T4, and repeating the process a number of times. The h^{2+} locus of Stent's strain was incorporated by us into the appropriate T4B and T4D strains by exposing T4 h^{2+} phage to a dose of ultraviolet light sufficient to produce about 20 phage lethal hits, crossing the UV-treated T4 h^{2+} with the appropriate phage strain and selecting a (rare) progeny strain with the h^{2+} host range. The standard-type T4 strain is designated h^{4+}.

The bacterial strains used were *Escherichia coli* B, B/2, B/4, S/6, and K12 112-12(λh), hereafter called K.

Media: Broth: H_2O, 1 liter; bacto-tryptone, 10 gm; NaCl, 5 gm. Tryptone bottom agar: broth with 1.1% bacto-agar. Tryptone top agar: broth with 0.7% bacto-agar. EHA bottom agar: tryptone bottom agar, 1 liter; sodium citrate·$2H_2O$, 1.1 gm; glucose, 1.3 gm. EHA top agar: tryptone top agar, 1 liter; sodium citrate·$2H_2O$, 1.1 gm; glucose, 3.0 gm. M-9: H_2O, 1 liter; Na_2HPO_4, 7 gm; KH_2PO_4, 3 gm; NH_4Cl, 1 gm; supplemented after autoclaving with glucose, 4 gm; $MgSO_4$, 10^{-3} M.

Fluorodeoxyuridine was a gift from Dr. R. Duschinsky of Hoffman-LaRoche, Inc.

Phage stocks were prepared by stabbing a 4- to 5-hr plaque with a needle and rinsing the needle into 20 ml of B cells grown to a concentration of about 10^8 cells per ml in aerated M-9 at 37°. The stocks were lysed by the addition of chloroform after about 5 hr of aeration. Stocks of *ts* mutants were prepared at 30°. DNase was added to the lysates, and the lysates were centrifuged at low speed to remove bacterial debris. Stocks of h^{2+} phage were incubated at 65° for $2^1/_2$ min immediately after lysis and centrifugation.[7]

Assay of phage: For platings on K or on B, the bacteria were grown to a concentration of 2×10^8 cells per ml in aerated broth at 37°, and 0.5-ml amounts were used per 2.5 ml of tryptone top agar to overlay plates containing 36 ml of tryptone bottom agar.

For platings on B/2 + B/4, appropriate amounts of phage were added to B grown to a concentration of 2×10^8 cells per ml in aerated broth at 37°. After 5 min of incubation at 37°, one part of the phage-B mixture was added to two parts of an equal mixture of B/2 and B/4, each grown to a concentration of 2×10^8 per ml in aerated broth at 37°. One part of the final mixture was added to five parts of melted tryptone top agar at 45°, and 3.0-ml samples were immediately distributed to as many plates containing 36 ml of bottom tryptone agar as were necessary.

For platings on S/6, the bacteria were grown to a concentration of 2×10^8 cells per ml in aerated broth at 37°, chilled, and concentrated tenfold by centrifugation. Two drops of the concentrated bacteria were added to 2 ml of EHA top agar, 0.05 ml of the appropriate dilution of phage was added, and the mixtures were poured on plates containing 36 ml EHA bottom agar. In some cases phage were first adsorbed to concentrated B and were then diluted and plated on S/6.

Results.—*The frequency of heterozygotes as a function of time:* The frequency of heterozygotes for point and deletion *r*II mutants was measured at various times after infection in the presence of FUDR.

Since little DNA synthesis takes place in the presence of FUDR, the removal, through maturation, of the few DNA molecules in the vegetative pool needs to be prevented. This can be accomplished by the addition of chloramphenicol (CAM) to the infected bacteria at about 9 min after infection. The chloramphenicol is removed at various times after infection, and the infected bacteria remain in the presence of FUDR for (usually) another 60 min and are then lysed by the addition of chloroform. The infected bacteria are thus always in the presence of FUDR.

The frequency of heterozygotes as a function of time is illustrated in Figure 1. As expected, the frequency for point *r*II mutants increases, that for deletion *r*II mutants stays constant.

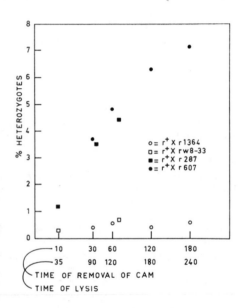

FIG. 1.—The frequency of heterozygotes after various periods of incubation in FUDR. Bacteria of strain B were grown to a concentration of 10^8 per ml in aerated M-9 supplemented with 0.5% bacto-casamino-acids, at 37°. To these cultures 20 μg of tryptophane per ml, $4 \times 10^{-5} M$ FUDR, and $2 \times 10^{-4} M$ uracil were added just before infection with an average of five phage particles of each type per bacterium. Nine min after infection, 250 μg of chloramphenicol per ml were added to the cultures, and at various times after that, aliquots of the cultures were chilled and washed twice by centrifugation, both the first and the second resuspension being with M-9 containing bacto-casamino-acids, uracil, and FUDR, but no chloramphenicol. The *time of removal of chloramphenicol* refers to the time, in minutes, after infection at which the cultures were chilled, before centrifugation. The cultures were then further incubated with aeration at 37°, and were subsequently lysed by the addition of chloroform, the *time of lysis* indicating the time in minutes after infection (omitting the time required for centrifugation). The lysates were plated on S/6, and the frequency of mottled plaques was determined after overnight incubation.

The frequency of recombinants as a function of time: The frequency of recombinants is expected to increase in the presence of FUDR and chloramphenicol in the same way as did the frequency of heterozygotes. That this is so is shown by the results presented in Figure 2.

The frequency of heterozygotes in the presence and absence of FUDR: The frequency of heterozygotes formed by a number of different deletion *r*II mutants and by point *r*II mutants was compared. In each case FUDR was added near the time of infection, chloramphenicol was added 9 min after infection and removed at about 120 min after infection, and the infected

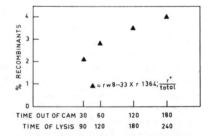

FIG. 2.—The frequency of recombinants after various periods of incubation in FUDR. The procedure used was exactly as described for Fig. 1, except that platings were on B and on K. Plaques appearing on plates containing K represent r^+; those appearing on B represent all types.

bacteria were lysed about 60 min after the removal of chloramphenicol. Control cultures were prepared in the absence of FUDR. The frequency of heterozygotes increased in the presence of FUDR for each of the point mutants examined, but remained the same as the control frequency for the deletion mutants (Table 1).

TABLE 1

FREQUENCY OF r/r^+ HETEROZYGOTES IN THE PRESENCE AND ABSENCE OF FUDR

Nature of r mutant	Infecting phages	Mottled Plaques in the Absence of FUDR		Mottled Plaques in the Presence of FUDR	
		Per cent	Number counted	Per cent	Number counted
Point	$r607 \times r^+$	1.0	45	7.1	142
	$r287 \times r^+$	0.7	42	5.9	29
	$r205 \times r^+$	0.9	45	3.0	91
Deletion	$r1364 \times r^+$	0.5	24	0.6	44
	$rW8\text{-}33 \times r^+$	0.4	34	0.4	31
	$rH23 \times r^+$	0.5	42	0.4	34

Bacteria of strain B were grown to a concentration of 10^8 per ml, in aerated M-9 supplemented with 0.5% bacto-casamino-acids, at 37°; 20 μg of tryptophane per ml were added to the bacteria just before they were to be infected. For mixed infections in the absence of FUDR, the bacteria were then infected with an average of about five phage particles of each type per cell, incubated with aeration at 37°, and lysed by the addition of chloroform at about 25 min after infection. For mixed infections in the presence of FUDR, 4×10^{-5} M FUDR and 2×10^{-4} M uracil were added just before infection, the bacteria were infected with an average of about five phage particles of each type per cell, and 250 μg chloramphenicol per ml were added 9 min later. After 120 min of incubation, with aeration, at 37°, the bacteria were washed twice by centrifugation, both the first and second resuspension being with M-9 containing bacto-casamino-acids, uracil, and FUDR, but no chloramphenicol. The infected bacteria were lysed by the addition of chloroform after a further 60 min of incubation with aeration. The lysates were in all cases treated with DNase and then centrifuged at low speed to remove bacterial debris. The lysates were plated on S/6, and the frequency of mottled plaques was determined after overnight incubation.

The behavior of h^{2+}/h^{4+} heterozygotes: Genetic markers (other than rII deletions) that would not form internal heterozygotes seemed very desirable for a number of experiments described below and elsewhere. The host-range difference between phages T2 and T4 is under control of allelic markers; no recombinants possessing a combination of host ranges have been observed.[8] That the host-range difference might be due to "complex" genetic changes was suggested by the observations that a change from T2-like to T4-like host range (or vice versa) could not be induced by treatment with mutagens,[9] and that the frequency of h^{2+}/h^{4+} heterozygotes was low[8] and of the order observed by Nomura and Benzer[2] for deletion mutants.

It seemed important to determine whether the h^{2+}/h^{4+} heterozygotes were similar to those formed by deletion mutants. The frequency of heterozygotes among the progeny phage produced in bacteria mixedly infected with h^{2+} and h^{4+} phage and incubated with and without FUDR was therefore measured. As indicated in Table 2, the frequency was similar in the presence and absence of FUDR, indicating that h^{2+}/h^{4+} heterozygotes behave in the manner expected for terminal-redundancy heterozygotes.[10]

Double heterozygosis: Plaques that contain two types of phage particles could be formed either by a heterozygous phage particle or else by a clump of several parti-

TABLE 2

FREQUENCY OF h^{2+}/h^{4+} HETEROZYGOTES IN THE PRESENCE AND ABSENCE OF FUDR

Infecting phages	Clear Plaques in the Absence of FUDR		Clear Plaques in the Presence of FUDR	
	Per cent	Number counted	Per cent	Number counted
$rH23\ h^{2+} \times rH23\ h^{4+}$	0.85	75	0.85	52

The procedure was exactly as described for Table 1, except that the lysates were plated on B/2 + B/4 and the frequency of clear plaques was determined after overnight incubation. Each plate contained fewer than 75 plaques.

cles.   In order to determine whether clumps of particles are responsible for the formation of plaques containing both h^{2+} and h^{4+} particles (h^{2+}/h^{4+} plaques), a cross of $h^{2+}r^+ \times h^{4+} r168$ was performed.   The marker $r168$ is a small deletion that is about 20 units distant from h^{2+}.   Among the progeny of this cross, about 30 per cent of h^{2+}/h^{4+} plaques were found to contain both r and r^+ phage (Table 3)

TABLE 3

ANALYSIS OF h^{2+}/h^{4+} PROGENY PLAQUES AFTER FRACTIONATION THROUGH SEDIMENTATION IN A DEUTERIUM GRADIENT

Fraction	Total number of plaques examined	Number of h^{2+}/h^{4+} plaques found	Frequency of h^{2+}/h^{4+} plaques	Fraction of h^{2+}/h^{4+} plaques that contained both r and r^+ phage
Peak drop	4.6×10^4	90	2.0×10^{-3}	0.07
1	2.3×10^3	3	—	0
2	2.9×10^3	10	3.4×10^{-3}	0.10
3	2.1×10^3	15	6.7×10^{-3}	0.32
4	2.5×10^3	50	2.0×10^{-2}	0.72
Unfractionated lysate	2.0×10^4	53	2.7×10^{-3}	0.30

Fractions from the deuterium-gradient sedimentation of the progeny of a cross of $h^{2+} r^+ \times h^{4+} r168$ described in Fig. 3 were plated on $B/2 + B/4$; each plate contained fewer than 40 plaques.   Clear plaques were picked and replated on $B/2 + B/4$ to test for the presence of h^{2+} and h^{4+} phage (and to exclude the rare plaques containing, instead, h^{2+} and h^2 phage).   The clear plaques were also plated on B and were scored with respect to whether they contained r, r^+, or both r and r^+ phage.

and could thus have been formed by clumps.   As will be shown, the actual fraction of spurious heterozygotes is somewhat less than 30 per cent, too low to invalidate the experiment with FUDR presented above.   For other purposes, however, it was necessary to isolate the true heterozygotes.

Since it was expected that clumps would sediment more rapidly than single phage particles, the progeny of the cross were centrifuged in a deuterium oxide gradient[11] and fractions were collected from the bottom of the centrifuge tube (Fig. 3).   The frequency of h^{2+}/h^{4+} plaques and the proportion of these that contained both r and r^+ phage were measured for the peak drop and for the more rapidly moving fractions.   As shown in Table 3, the frequency of h^{2+}/h^{4+} plaques was con-

FIG. 3.—The fractionation of progeny phage of a cross through sedimentation in a deuterium gradient.   Bacteria of strain B were grown to a concentration of 10^8 per ml in aerated broth at $37°$, chilled, and concentrated by centrifugation.   The bacteria were infected at a concentration of 10^9 per ml with an average of six phage particles each of $h^{2+} r^+$ and $h^{4+} r168$.   After 2 min incubation at $37°$, they were diluted in broth at $37°$ and were incubated at that temperature;   30 min after infection, the bacteria were lysed by the addition of chloroform.

A continuous gradient of D_2O was produced in a centrifuge tube by mixing broth prepared with D_2O and broth prepared with H_2O.   An aliquot of broth containing the progeny of the cross was delivered to the surface of the gradient and the tube was centrifuged for 15 min at 20,000 rpm in the SW39 head of a Spinco centrifuge.   After centrifugation, a hole was punched in the bottom of the tube, and individual drops were collected in tubes containing 0.5 ml of broth.   The contents of tubes were pooled as indicated in the figure to yield fractions 1–4.

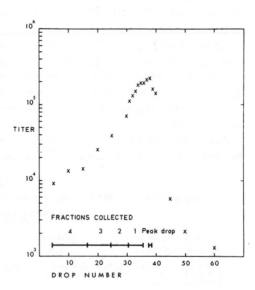

siderably greater, and a greater proportion of them contained both r and r^+ phage in the more rapidly moving fractions than in the peak drop. In the most rapidly moving fraction examined, 72 per cent of the h^{2+}/h^{4+} plaques contained both r and r^+ particles, whereas in the peak drop, only 7 per cent of the h^{2+}/h^{4+} plaques contained both r and r^+ particles.

The fast-moving fraction was heated for a length of time sufficient to inactivate 90 per cent of the phage particles. This treatment drastically reduced the frequency of apparent heterozygotes, an indication that at least a large proportion of them were clumps. Thus the rapidly sedimenting particles forming mixed plaques are mainly clumps, justifying the assumption that those sedimenting normally are authentic heterozygotes. If so, the data of Table 3 show that about 7 per cent of the authentic h^{2+}/h^{4+} heterozygotes are also heterozygous for the r marker, and that only about 25 per cent of the particles forming mixed plaques are clumps.

Recombination for neighboring markers among heterozygotes: Terminally redundant heterozygotes ought to be recombinant with respect to neighboring markers. In a cross of $a\ b\ c \times a^+b^+c^+$, for instance, terminal-redundancy heterozygotes that are b/b^+ would be expected to have either of two possible arrangements of markers:

$$\underline{b\ c \qquad a^+\ b^+} \quad \text{or} \quad \underline{b^+\ c^+ \qquad a\ b},$$

both arrangements being recombinant with respect to the markers a and c.

In order to determine whether particles that are terminal-redundancy heterozygotes for a particular marker are recombinant for neighboring markers, a cross of rEDb42 h^{4+} tsN30 $\times r^+h^{2+}$ ts^+ was performed. The r and ts markers are on opposite sides of the host-range marker and yield about 20 and 17 per cent recombinants, respectively, in crosses with the host-range marker. The progeny from the cross were centrifuged in a deuterium gradient and the contents of the peak drop were plated at 43° on B/2 + B/4. Only ts^+ phage (or phage heterozygous for ts/ts^+) produce plaques under these conditions. Among the plaques appearing at 43°, h^{2+}/h^{4+} heterozygotes were scored with respect to the r marker. A majority of the h^{2+}/h^{4+} heterozygotes that were ts^+ were r and thus recombinant, in contrast to the phage in the total population (Table 4). This cross was repeated with various other combinations of these same markers in order to show that the result did not depend on the particular alleles used (Table 4). We may thus conclude that h^{2+}/h^{4+} heterozygotes are recombinant with respect to neighboring markers

TABLE 4

Assortment of Outside Markers Among h^{2+}/h^{4+} Heterozygotes

Cross	Type of plaque examined	Number of plaques examined	Fraction of Plaques That Contained		
			r	r^+	r and r^+
$r\ h^{4+}\ ts \times r^+\ h^{2+}\ ts^+$	Clear ts^+	85	0.59	0.18	0.23
	All ts^+	50	0.20	0.80	0
$r^+\ h^{4+}\ ts \times r\ h^{2+}\ ts^+$	Clear ts^+	145	0.25	0.49	0.26
	All ts^+	50	0.80	0.20	0
$r^+\ h^{4+}\ ts^+ \times r\ h^{2+}\ ts$	Clear ts^+	28	0.68	0.14	0.18
	All ts^+	49	0.37	0.63	0

Bacteria of strain B were grown to a concentration of 10^8 per m lin aerated broth at 37°, chilled, and concentrated by centrifugation. The bacteria were infected at a concentration of 10^9 per ml with an average of six phage particles of each parental type. After 2 min at 30°, they were diluted into broth at 30° and incubated at that temperature; 55 min after infection, the bacteria were lysed by the addition of chloroform. An aliquot of the progeny of each cross was centrifuged in a deuterium gradient as described in Fig. 3. The peak drop was plated on B/2 + B/4 and the plates were incubated at 43°. At this temperature, only plaques containing ts^+ phage are formed. Each plate contained fewer than 40 plaques. Clear plaques or else a random sample of all the plaques present were picked and analyzed as described in Table 3.

as would be expected on the basis of the terminal-redundancy model. A similar result had been obtained previously by Doermann and Boehner,[12] who examined heterozygotes formed by an *r*-deletion mutant. In our crosses an appreciable fraction of the h^{2+}/h^{4+} heterozygotes were simultaneously heterozygous for r/r^+ even though clumps had presumably been eliminated by the centrifugal fractionation.

Discussion.—The experiments reported here demonstrate the existence of two types of heterozygotes and thus confirm observations of others.[2] Heterozygotes of one type do not accumulate under conditions of limited DNA synthesis, and the behavior of this type of heterozygote confirms an important prediction of the terminal-redundancy model for the chromosome of phage T4.

An alternate model for this type of heterozygote may be considered: phage particles could consist of a whole chromosome and an additional small piece of a chromosome. This model is excluded by the observation presented here that h^{2+}/h^{4+} heterozygotes are recombinant for neighboring markers, and by the results of others[12] that deletion-mutant heterozygotes are recombinant for neighboring markers.

Summary.—Point-mutant heterozygotes increase in frequency during phage growth under conditions of limited DNA synthesis but deletion-mutant heterozygotes do not. Heterozygotes for the h^{2+}/h^{4+} marker behave as deletion-mutant heterozygotes do and are recombinant with respect to neighboring markers. These observations support the terminal-redundancy model for the chromosome of phage T4.

The authors are grateful to Dr. R. Duschinsky of Hoffmann-LaRoche, Inc., for a generous gift of fluorodeoxyuridine.

* This investigation was supported by research grants to G. Streisinger from the National Science Foundation (G-14055) and the National Institute of Allergy and Infectious Diseases (E-3892). Part of this investigation was carried out at the Biological Laboratory in Cold Spring Harbor, New York, during the summer of 1962. The authors are grateful for the hospitality exhibited by the staff of the Laboratory.

† Present address: Laboratoire de Biophysique, Université de Genève, Geneva, Switzerland.

[1] Streisinger, G., R. S. Edgar, and G. H. Denhardt, these Proceedings, **51**, 775 (1964).

[2] Nomura, M., and S. Benzer, *J. Mol. Biol.*, **3**, 684 (1961).

[3] Levinthal, C., *Genetics*, **39**, 169 (1954).

[4] Hershey, A. D., and M. Chase, in *Cold Spring Harbor Symposia on Quantitative Biology*, vol. 16 (1951), p. 471.

[5] Cohen, S. S., J. G. Flaks, H. D. Barner, M. R. Loeb, and J. Lichtenstein, these Proceedings, **44**, 1004 (1958).

[6] Simon, E., personal communication; and Frey, Sister Celeste, personal communication.

[7] Sagik, B. P., *J. Bacteriol.*, **68**, 430 (1954).

[8] Streisinger, G., *Virology*, **2**, 377 (1956).

[9] Streisinger, G., unpublished observations.

[10] The frequency of h^{2+}/h^{4+} heterozygotes in this experiment is high owing to the presence of the long deletion *r*H23 in both parental stocks of the cross. The influence of deletions on the frequency of terminal-redundancy heterozygotes will be described in a subsequent communication.

[11] We are grateful to Professor Sidney Brenner, who suggested deuterium oxide as a convenient medium for this type of centrifugation.

[12] Doermann, A. H., and L. Boehner, *Virology*, **21**, 551 (1963).

From the *Proceedings of the National Academy of Sciences*, Vol. 57, No. 2, 292–295, 1967. Reproduced with permission of the authors.

CHROMOSOME STRUCTURE IN PHAGE T4, III.
TERMINAL REDUNDANCY AND LENGTH DETERMINATION*

By George Streisinger, Joyce Emrich, and Mary Morgan Stahl

INSTITUTE OF MOLECULAR BIOLOGY, UNIVERSITY OF OREGON, EUGENE

Communicated by A. D. Hershey, December 14, 1966

In previous communications it was suggested that chromosomes of phage T4 are circularly permuted and terminally redundant.[1,2] Heterozygotes formed by deletion r_{II} mutants and r^+, and by the host range alleles h^{2+} and h^{4+}, were shown to behave as if they were due to this terminal redundancy.[2]

After infection and replication the chromosome of any one phage particle is assumed to become circularly permuted so that the genetic location of the beginning of any one progeny chromosome is randomly distributed over the genome. This could take place through the formation of recombinants by whole chromosomes or by fragments of chromosomes:

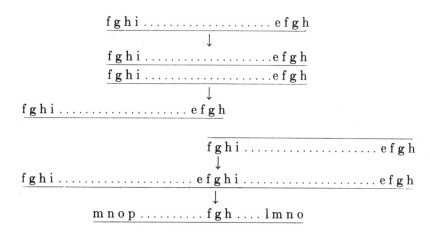

It is difficult to imagine, under this model, what factor *intrinsic to the chromosome itself* would determine its total length. We assume, therefore, that the mean length of the chromosome of a mature phage particle is determined by some extrinsic factor, such as, for instance, the amount of DNA that can be contained in a phage head. The mean length of the chromosome would therefore correspond to a "headful" of DNA.

If this notion is correct, we can expect that the length of the chromosome will not be affected by deletion of a part of it. Deletions should be compensated for by a lengthening of the region of terminal redundancy by an amount equal to the length of the deletion. A test of this prediction forms the substance of this communication.

The materials and the methods have been described previously,[2] except for the T4B mutant strains $r196b$, $rH88$, and $r1272$, which were obtained from Dr. S. Benzer.

Results.—The frequency of h^{2+}/h^{4+} heterozygotes among progeny phage carrying deletions: The frequency of heterozygosis for a given marker will depend on the mean length of the heterozygous region in which that marker can be included.   In order to compare the length of the region of terminal redundancy in chromosomes carrying deletions with those carrying point mutations, the crosses $rH23\ h^{2+}$ × $rH23\ h^{4+}$ and $r607\ h^{2+}$ × $r607\ h^{4+}$ were performed.   The mutant $r607$ is an r_{II} point mutant, whereas $rH23$ is a deletion covering the entire r_{II} region.   The markers h^{2+} and h^{4+} are allelic and can form heterozygotes only within the region of terminal redundancy.[2]   As shown by the results presented in Table 1, the frequency of plaques containing both h^{2+} and h^{4+} phage (h^{2+}/h^{4+} plaques) was higher among the progeny of crosses involving a long deletion than among the progeny of crosses involving a point mutation.   Clumps which may have contributed to the formation of h^{2+}/h^{4+} plaques[2] were not eliminated in this experiment.   The results nevertheless suggest that the deletion of a long section of the genome increases the length of the region of terminal redundancy.

TABLE 1

The Frequency of h^{2+}/h^{4+} Plaques in the Presence of a Deletion

Cross	Frequency of h^{2+}/h^{4+} progeny	Number of h^{2+}/h^{4+} plaques counted
$rH23\ h^{2+}$ × $rH23\ h^{4+}$	10×10^{-3}	75
$r607\ h^{2+}$ × $r607\ h^{4+}$	5×10^{-3}	146

Bacteria of strain B were grown to a concentration of 10^8 cells per milliliter in aerated broth at 37°C, and were chilled and concentrated by centrifugation.   The bacteria were infected at a concentration of 5×10^8 per milliliter with an average of about five phage particles of each parental type.   After 5 min at 37°C they were diluted into broth at 37°C and incubated at that temperature; chloroform was added to the cultures at 35 min after infection.   The lysates were plated on B/2 + B/4, and the frequency of clear plaques was determined after overnight incubation. The average number of plaques per plate was less than 70.

It was expected that the frequency of h^{2+}/h^{4+} heterozygotes would depend on the length of the deletion carried by the parental stocks.   Three crosses using deletions of various lengths were therefore performed: $rH23\ h^{2+}$ × $rH23\ h^{4+}$, $rH88\ h^{2+}$ × $rH88\ h^{4+}$, and $r168\ h^{2+}$ × $r168\ h^{4+}$, where $rH23$ is the longest deletion, $rH88$ is shorter, and $r168$ is shorter still.   In order to eliminate clumps from among the progeny of crosses the progeny phage of each cross were partially sedimented in a deuterium gradient and the phage from the peak drop of the gradient were examined.   This procedure eliminates most of the clumps.[2]

Since the frequency of heterozygotes may vary from cross to cross, and since the scoring of h^{2+}/h^{4+} heterozygotes is sensitive to the conditions of plating, two independent sets of the three crosses were performed.   The progeny of the three

TABLE 2

The Frequency of h^{2+}/h^{4+} Plaques in the Presence of Deletions of Various Lengths

Set	Cross	Frequency of h^{2+}/h^{4+} progeny	Number of h^{2+}/h^{4+} plaques counted
	$rH23\ h^{2+}$ × $rH23\ h^{4+}$	9.0×10^{-3}	99
1	$rH88\ h^{2+}$ × $rH88\ h^{4+}$	4.0×10^{-3}	67
	$r168\ h^{2+}$ × $r168\ h^{4+}$	2.6×10^{-3}	60
	$rH23\ h^{2+}$ × $rH23\ h^{4+}$	4.1×10^{-3}	26
2	$rH88\ h^{2+}$ × $rH88\ h^{4+}$	2.7×10^{-3}	27
	$r168\ h^{2+}$ × $r168\ h^{4+}$	2.0×10^{-3}	51

Bacteria of strain B were grown to a concentration of 10^8 cells per milliliter in aerated broth at 37°C and were chilled and concentrated by centrifugation.   The bacteria were infected at a concentration of 10^9 per milliliter with an average of about five phage particles of each parental type.   After 5 min at 37°C they were diluted into broth at 37°C and incubated at that temperature; chloroform was added to the culture 30 min after infection.   An aliquot of the progeny of each cross was partially sedimented in a deuterium gradient.[2]   The peak drop was plated on B/2 + B/4 and the frequency of clear plaques was determined after overnight incubation.   The average number of plaques per plate was less than 30.

crosses of any one set were plated and scored at the same time. As shown in Table 2, the frequency of h^{2+}/h^{4+} heterozygotes was greatest among the progeny of the cross with the longest deletion and decreased as the length of the deletion decreased.

The frequency of r and r^+ markers among h^{2+}/h^{4+} heterozygous progeny of $r^+h^{2+} \times$ rh^{4+} crosses: Progeny of crosses of $r^+h^{2+} \times rh^{4+}$ will contain h^{2+}/h^{4+} heterozygotes. Some of these will be r, others will be r^+, and some may be r/r^+ heterozygotes. One would expect the frequencies of the r and r^+ types to be equal among the h^{2+}/h^{4+} heterozygotes when the r genotype is a point mutation or a short deletion. When it is a long deletion, more of the h^{2+}/h^{4+} heterozygotes should be r than r^+, owing to the longer terminal redundancy of the r chromosomes. The results presented in Table 3 indicate that progeny of crosses involving long deletions did indeed contain more rh^{2+}/h^{4+} heterozygotes than r^+h^{2+}/h^{4+} heterozygotes.

TABLE 3

The Effect of Deletions on the Frequency of r Types among h^{2+}/h^{4+} Progeny

Cross	Frequency of h^{2+}/h^{4+} plaques	Number of h^{2+}/h^{4+} plaques tested	Fraction r	Fraction r^+	Fraction mixed	Ratio $r:r^+$
		(a) Unfractionated Progeny				
$rH23\ h^{4+} \times r^+h^{2+}$	11.0×10^{-3}	173	0.48	0.17	0.35	2.8
$rH23\ h^{2+} \times r^+h^{4+}$	7.5×10^{-3}	70	0.72	0.09	0.20	8.3
$r1272\ h^{4+} \times r^+h^{2+}$	13.0×10^{-3}	119	0.45	0.14	0.40	3.2
$r196b\ h^{4+} \times r^+h^{2+}$	7.0×10^{-3}	82	0.33	0.18	0.49	1.8
$r607\ h^{4+} \times r^+h^{2+}$	6.0×10^{-3}	87	0.30	0.25	0.45	1.2
$r607\ h^{2+} \times r^+h^{4+}$	3.4×10^{-3}	96	0.31	0.42	0.27	0.8
		(b) Peak Fraction of Partially Sedimented Progeny				
$r1272\ h^{4+} \times r^+h^{2+}$	5.1×10^{-3}	206	0.81	0.16	0.03	4.9
$r168\ h^{4+} \times r^+h^{2+}$	2.2×10^{-3}	76	0.39	0.52	0.09	0.8
$r168\ h^{4+} \times r^+h^{2+}$	2.0×10^{-3}	90	0.48	0.45	0.07	1.1

The crosses and platings were performed as described for Tables 1 and 2; the average number of plaques per plate was less than 35.

The progeny of the crosses in part (b) were partially sedimented in sucrose gradient[3] or a deuterium gradient[2] and the peak drop was plated.

$rH23$ and $r1272$ are long deletions covering all of the r_{II} region; $r196b$ and $r168$ are very short deletions and $r607$ is a point mutation.

The h^{2+}/h^{4+} plaques that contained only r or only r^+ phage were probably formed by heterozygotes rather than by clumps, since most h^{2+}/h^{4+} plaques formed by clumps of phage would be expected to contain both r and r^+ phage. A fraction of the h^{2+}/h^{4+} plaques (Table 3A) *did* contain both r and r^+ phage and this fraction probably originated from clumps.

In order to eliminate clumps, the phage from the progeny of three crosses were partially sedimented in a deuterium (or else a sucrose) gradient, and phage from the peak drop of the gradient were plated. Since only a small fraction of the h^{2+}/h^{4+} plaques from the peak drop contained both r and r^+ phage, most of the clumps were in fact eliminated. As shown in Table 3B, the progeny of crosses involving long deletions again contained more rh^{2+}/h^{4+} heterozygotes than r^+h^{2+}/h^{4+} heterozygotes, whereas control crosses involving point r mutants contained approximately equal frequencies of the two types.

In the crosses described in Table 3 it is noteworthy that the total frequency of h^{2+}/h^{4+} heterozygotes was greater in crosses involving long deletions than in crosses involving point mutants, confirming the results presented in Tables 1 and 2.

Discussion.—The results presented here confirm a rather bizarre prediction of the terminal redundancy model of the chromosome of phage T4. We have been unable to devise any other model that would incorporate this feature and the others that have been presented in previous papers.[1, 2]

The essential features of the model suggested by our experiments are: (1) The chromosome of mature phage particles is linear, with a terminal redundancy.   (2) The chromosomes of progeny phage represent circular permutations of the chromosome of the parent.   (3) A mature phage particle contains, on the average, a "headful" of DNA.

The model of the T4 chromosome that has been presented here has a number of implications concerning the mechanism of recombination in phage T4.   These implications will be discussed in a subsequent communication.

Summary.—Crosses of phage carrying long deletions yield a greater frequency of terminal redundancy heterozygotes than crosses of phage carrying short deletions or point mutations.   This suggests that each particle of phage T4 contains, on the average, a "headful" of DNA.

The experiment reported in this communication was conceived during the course of a conversation with Dr. M. Fox.   We are grateful to Dr. F. W. Stahl for many helpful discussions.

* This investigation was supported by research grants from the National Science Foundation (G14055 and 466-GB-2261) and from the National Institute of Allergy and Infectious Diseases (E3892).

[1] Streisinger, G., R. S. Edgar, and G. H. Denhardt, these PROCEEDINGS, 51, 775 (1964).

[2] Séchaud, J., G. Streisinger, J. Emrich, J. Newton, H. Lanford, H. Reinhold, and M. M. Stahl, these PROCEEDINGS, 54, 1333 (1965).

[3] A continuous gradient of sucrose was produced in a centrifuge tube by mixing continuously increasing amounts of broth containing 0.5% sucrose with continuously decreasing amounts of broth containing 20% sucrose.   Partial sedimentation in a sucrose gradient could not be performed reproducibly in subsequent experiments and further sedimentations were therefore performed with a deuterium gradient.

From the *Proceedings of the National Academy of Sciences*, Vol. 52, No. 5, 1297–1301, 1964. Reproduced with permission of the authors.

CIRCULAR T2 DNA MOLECULES

By C. A. Thomas, Jr., and L. A. MacHattie

DEPARTMENT OF BIOPHYSICS, JOHNS HOPKINS UNIVERSITY

Communicated by A. D. Hershey, September 30, 1964

Each T2 bacteriophage particle yields a single *linear* duplex DNA molecule, an example of which may be seen in the electron micrograph (Fig. 2B). Although each molecule contains the same genetic message, the order of the nucleotides appears to be different from molecule to molecule. Renaturation experiments indicate that those sequences that are found near the ends of some molecules are found near the middles of others.[1] This is precisely what would be expected if each linear molecule had a nucleotide sequence which was a different circular permutation of a common basic sequence. (A collection of linear molecules with circularly permuted sequences can be generated by making a single random break in each of a collection of identical circular molecules.) Other bacteriophage DNA molecules such as T5 (ref. 1) and λ (ref. 2) are known to be "unique" in that most of them have the same nucleotide sequence. The unusual situation in regard to T2 or the related T4 is probably the physical basis for the circular genetic map observed in this phage.[3]

If these molecules are circular permutations of each other, and if they consist mainly of two continuous polynucleotide chains as shown previously,[4] then denaturation followed by reannealing as depicted in Figure 1 should lead to the formation of artificial circular molecules.

Experiments.—DNA molecules from T2 bacteriophage were extracted with phenol and purified by chromatography[5] as described previously.[4] These molecules were diluted tenfold into 0.20 M NaOH at a final concentration of 1.25 γ/ml. After one min, $^1/_{10}$ vol of 3.0 M NaCl, 0.30 M Na citrate, was added and the solution dialyzed for 10 hr against 0.30 M NaCl, 0.03 M Na citrate. This solution was then heated at 65°C for 40 min and cooled to 4°C. Visualization in the electron microscope was accomplished by the method of Kleinschmidt[6] as previously described.[7] The present procedure renders duplex DNA clearly visible, but single polynucleotide chains are not seen. An aliquot of the same solution receiving all treatments except that of denaturation by NaOH served as a control. In addition to the experiments done on T2 whole molecules, T2 half molecules (41% relative length or 54 million molecular weight), and T5 whole molecules were given the same experimental and control treatments.

Results.—Grids prepared from the solutions of unbroken T2 molecules that had been treated with NaOH followed by reannealing at 65° showed many closed circles,

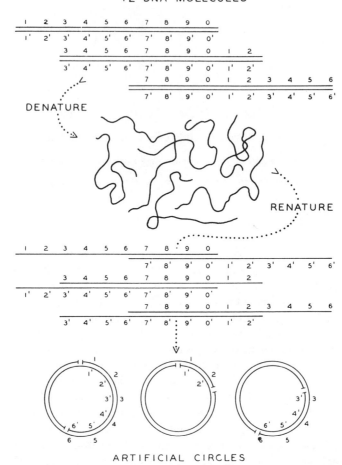

FIG. 1.—The diagram shows how a collection of linear duplex DNA molecules with sequences that are circular permutations of each other can produce circular molecules by chain separation followed by random association and reformation of the duplex structure. In principle nearly every final molecule could be circular.

while the grids made from the control solutions showed no circles, only linear molecules. Examples of the two types are shown in Figure 2. The measured contour lengths of many circular and linear molecules are compiled in Figure 3. All circles had nearly the same contour length ($55 \pm 3 \mu$) as the unbroken linear control molecules ($56 \pm 2 \mu$).[8] The linear molecules found in the circularized preparations had a variety of lengths as might be expected if these were overlapped structures (Fig. 1) which had not yet circularized.

The circles were abundant. In order to estimate their frequency, the entire areas of two (200 mesh) grid squares were photographed and inventoried. Of the 168 molecules seen, 25 ran into grid wires or were otherwise obscured. Of the remaining 143, 29 were circles, 4 of which must be classified as doubtful since it was impossible to verify strand continuity in these cases. The remaining 114 molecules were of various lengths with about half of them longer than 35μ.

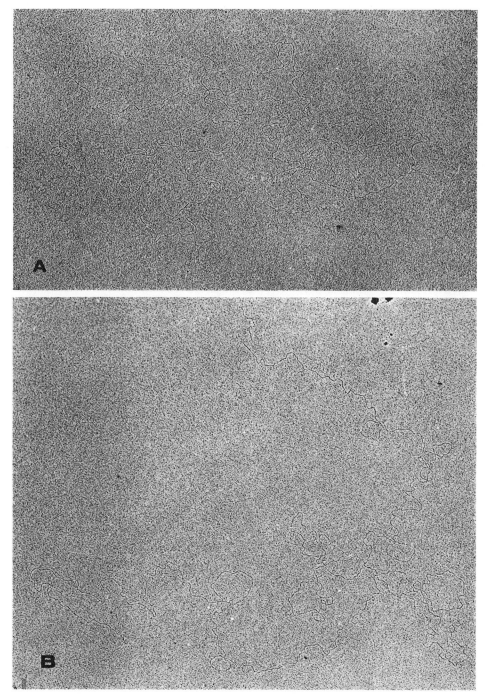

FIG. 2.—(A)   A circular T2 DNA molecule from a solution that had been exposed to denaturation and renaturation conditions.   (B)   T2 DNA molecule from a solution that had been exposed to renaturation conditions only.

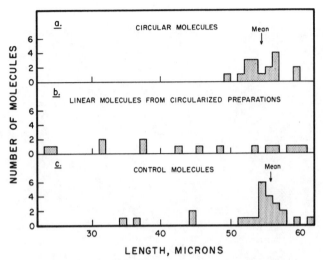

Fig. 3.—Histograms of measured contour lengths of duplex T2 DNA: (*a*) and (*b*), after denaturation and renaturation; (*c*) after renaturation treatment only.

We were interested to learn whether polynucleotide chains which were interrupted could form circles under these conditions. It is theoretically possible for four polynucleotide chains of less than full length to cooperate to form a circle of full contour length, albeit possibly with single-chain regions. The basic experiment with undenatured controls was performed with column-fractionated fragments of 41 per cent relative length, repeated again with a fourfold increased concentration of these fragments, and unbroken T2 molecules as a control. As expected, circles were found in the solutions which contained whole molecules, but no circles were found in the solutions containing 41 per cent fragments. We take this to mean that the polynucleotide chains must be continuous over the entire polynucleotide map in order for circle formation to be likely under these conditions.

When T5 whole molecules are examined in the same way, no circles are found in spite of an exhaustive search. None are expected because T5 DNA molecules are thought to have identical sequences.[1]

Discussion.—The finding that T2 DNA molecules can be caused to assume the form of a closed loop or circle by exposure to conditions which cause the separation and subsequent reunion of polynucleotide chains is in exact accord with expectation, provided that the original linear T2 DNA molecules have nucleotide sequences which are different circular permutations of each other. The fact that the contour length of the circular molecules is almost the same as that of the undenatured linear molecules means that the cycle of permutation extends over the majority of the molecule. This point is further supported by the inability of fragments of half-size to form circles. This question was left unresolved by earlier studies.[1]

On the basis of genetic experiments[3] Streisinger has proposed that each T4 molecule, in addition to being a different circular permutation of a common sequence, posesses a terminal reduplication of part of the genetic message, and that this terminal redundancy is responsible for a certain group of heterozygous phage particles. Thus, the nucleotide sequences found at the left end of each molecule would be repeated at the right end of the same molecule. These duplicated sequences would be different in different molecules depending on the permutation they represent. If circle formation comes about by the scheme shown in Figure 1, these repeated sequences would not be formed into the circle, but be left as unpaired single polynucleotide chains attached to the circle at two points. Since the present procedure does not reveal single polynucleotide chains, one would not expect to see them. The contour length of the circle would correspond to only one complete genetic map; thus, the fact that the observed mean contour length of the circles (55 μ) is shorter than that of the linear control molecules (56 μ) may be significant.

It might be supposed that the circle formation could be the result of complementary pairing of the end redundancies to produce a molecule of the following form:

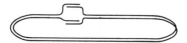

This appears unlikely to us for the following reasons: (1) it presumes that the original chains never completely separate before renaturation; (2) it supposes that these terminal polynucleotide chains find their complement at the opposite end of the molecule rather than reuniting with their original partners which are close by and already in register; and (3) this scheme predicts the formation of loops as shown above about 1 μ in length. No convincing example of such a loop exists in our collection of circular molecules.

Summary.—T2 DNA molecules can be caused to assume a circular form by denaturation with alkali followed by renaturation at neutral pH. This result confirms the hypothesis that native linear T2 DNA molecules have nucleotide sequences which are various circular permutations of a common sequence. When broken T2 DNA molecules or whole T5 DNA molecules are treated in the same way, no circles are found.

We are grateful to Messrs. John Abelson and Marcus Rhoades for chromatographed DNA samples. During this period one of us (L. A. M.) held a Damon Runyon Fellowship (#327). This work was supported by the National Science Foundation (G-10726) and the Atomic Energy Commission (AT(30-1)2119).

[1] Thomas, C. A., and I. Rubenstein, *Biophys. J.*, **4**, 93 (1964).

[2] Kaiser, A. D., and D. S. Hogness, *J. Mol. Biol.*, **2**, 392 (1960); Kaiser, A. D., *J. Mol. Biol.*, **4**, 275 (1962); Hogness, D. S., and J. R. Simmons, *J. Mol. Biol.*, in press.

[3] Streisinger, G., R. S. Edgar, and G. H. Denhardt, these PROCEEDINGS, **51**, 775 (1964).

[4] Thomas, C. A., T. C. Pinkerton, and I. Rubenstein, *Informational Macromolecules* (Academic Press, Inc., 1963), p. 89; Thomas, C. A., and T. C. Pinkerton, *J. Mol. Biol.*, **5**, 356 (1962); Berns, K. I., and C. A. Thomas, *J. Mol. Biol.*, **3**, 289 (1961).

[5] Mandell, J., and A. D. Hershey, *Anal. Biochem.*, **1**, 66 (1960).

[6] Kleinschmidt, A. K., and R. K. Zahn, *Z. Naturforsch.*, **14b**, 770 (1959); Kleinschmidt, A. K., D. Lang, D. Jacherts, and R. K. Zahn, *Biochim. Biophys. Acta*, **61**, 857 (1962).

[7] MacHattie, L. A., and C. A. Thomas, *Science*, **144**, 1142 (1964).

[8] While this is longer than the previously published value of 47 μ, [6] it is still somewhat shorter than 63 μ, the length predicted for a duplex in the B-form with a molecular weight of 133 million.[9]

[9] Rubenstein, I., C. A. Thomas, and A. D. Hershey, these PROCEEDINGS, **47**, 113 (1961).

From *Journal of Molecular Biology*, Vol. 23, 355–363,
1967. Reproduced with permission of the authors and
publisher.

Terminal Repetition in Permuted T2
Bacteriophage DNA Molecules

L. A. MacHattie, D. A. Ritchie,† C. A. Thomas, Jr.

Department of Biophysics, The Johns Hopkins University
Baltimore, Maryland 21218, U.S.A.

AND C. C. Richardson

Department of Biological Chemistry, Harvard Medical School
Boston, Massachusetts 02115, U.S.A.

(*Received 30 August 1966*)

Partial degradation (1 to 3·5%) of T*2‡ bacteriophage DNA molecules by
E. coli exonuclease III predisposes the intact linear molecules (but not fragments) to form circular molecules upon annealing. This indicates that the
nucleotide sequences at both ends of the molecule are identical (or nearly so).
This physical evidence agrees quantitatively with the model of the T-even
chromosome devised by Streisinger and collaborators to interpret the genetics
of these bacteriophages.

Continued studies on the circular molecules formed by the annealing of
denatured T2 DNA show that the number of different permutations is large
and that there is no preferred permutation in the collection.

1. Introduction

In this communication we report the results of experiments designed to test the
hypothesis that the linear DNA molecule liberated from a T2 bacteriophage particle
has the same sequence of nucleotides at the beginning and end of the molecule.
The hypothetical structure to be tested is illustrated in Fig. 1. This model has been
proposed by Streisinger, Edgar & Denhardt (1964) to account for one class of
heterozygotes in T-even phage.

Each molecule is shown having terminally recurring nucleotide sequences: regions
1 and 2 in the first molecule, 3 and 4 in the next, 5 and 6 in the last. That T2 DNA
molecules contain different circular permutations of a common over-all sequence,
as depicted, has been established previously (Thomas & Rubenstein, 1964; Thomas
& MacHattie, 1964).

To test this hypothesis, ^{32}P-labeled T*2 DNA molecules (see footnote ‡) were
digested under controlled conditions with exonuclease III, the specific enzyme that
removes nucleotides stepwise from the 3′ ends of polynucleotide chains in duplex
molecules (Richardson & Kornberg, 1964; Richardson, Lehman & Kornberg, 1964;
Richardson, Inman & Kornberg, 1964). This process exposes the 5′-ended polynucleotide chains which would have sequences complementary to each other—if

---

†Present address: Institute of Virology, University of Glasgow, Glasgow, W.1, Scotland.
‡ T*2 is T2 phage grown on the non-glucosylating host *E. coli* B/4_0.

355

the hypothesis of terminal repetition were correct. Under conditions that promote annealing, these 5'-ended chains should be capable of uniting to form a duplex and in the process convert the linear molecules into circular ones. When this experiment is performed, electron microscopic examination reveals that circles can indeed be formed when 1 to 3·5% of the nucleotides have been released by exonuclease III. Control experiments using fragments of T2 DNA molecules show that circles are formed only from unbroken molecules, and then only when the proper amount of digestion has occurred. From these facts and others described below, we conclude that T2 DNA molecules do contain terminally repeating sequences. The length of this terminal repetition can be estimated at 1 to 3% of the intact molecule.

2. Materials and Methods

(a) *Growth of ^{32}P-labeled non-glucosylated T*2 phage*

A saturated broth culture of *Escherichia coli* strain $B/4_0$ (the progeny viral DNA grown in this strain is not glucosylated, Hattman & Fukasawa, 1963) was diluted 100-fold into 25 to 40 ml. of TCG medium (Kozinski & Szybalski, 1959) modified to contain, per liter: glucose 4·0 g, Difco Casamino acids 1·0 g and KH_2PO_4 0·082 g. The culture was grown with aeration at 37°C. At a bacterial titer of 5×10^7 cells/ml., carrier-free [^{32}P] ortho-phosphate (Phosphotope Injection, Squibb) was added to give a specific activity of 0·2 $\mu c/\mu g$ phosphorus. When the concentration was 4×10^8 cells/ml., the bacteria were infected with T2H wild type phage at a multiplicity of 5 phage/cell. One hour later, spontaneous lysis was completed by bubbling with chloroform. These lysates contained 3 to 5×10^{10} phage/ml. plating on the permissive host, *Shigella dysenteriae* strain SH16. Such lysates contained 6×10^8 phage/ml. plating on the restricting host *E. coli* BB. Most of these are glucosylated parental phage that failed to adsorb (unpublished results).

(b) *Purification of phage*

The crude lysates were usually incubated for 30 min at 37°C with 10 $\mu g/ml.$ each of DNase and RNase (Worthington Biochemical Corp.). The phage were purified by 2 cycles of low-speed (5000 g for 8 min) and high-speed (35,000 g for 25 min) centrifugations and the pellets resuspended in the adsorption medium of Hershey & Chase (1952) by gentle agitation for several hours at 4°C. Where measured, the A_{260}/titer ratio was 1 to $1·15 \times 10^{-11}$ and the ultraviolet absorption spectrum of the phage suspension was normal.

(c) *Isolation of phage DNA*

The purified phage suspension was treated at room temperature with freshly distilled, water-saturated phenol. A 2-ml. sample of the purified lysate containing about 5×10^{11} phage/ml. plus an equal volume of phenol was rolled at 60 rev./min about the long axis of the tube for 30 min. After chilling in ice, the 2 phases were separated by brief centrifugation and the lower phenol phase discarded. This extraction was then repeated twice with briefer periods of rolling. The resulting DNA solution was transferred gently and dialysed 5 to 6 times against a large volume of 0·05 M-NaCl, 0·01 M-Tris–HCl buffer, pH 8. Prior to dialysis the tubing was washed by boiling for 10 min, first in 10% $NaHCO_3$ solution and then in water followed by thorough rinsing in demineralized water.

The resulting preparations of unbroken T*2 DNA molecules displayed a single sharp zone upon neutral or alkaline sucrose sedimentation, a single peak on methylated albumin-keiselguhr chromatography and contour lengths of $55 \pm 2 \mu$ as seen by electron microscopy.

(d) *Shear breakage of DNA*

A 50-ml. solution of 6 $\mu g/ml.$ unbroken T*2 DNA molecules in 0·05 M-NaCl, 0·01 M-Tris–HCl buffer, pH 8, was stirred at 800 rev./min for 40 to 60 min in a Virtis flask cooled in an ice–water bath. This produced about equal weights of whole and half-size fragments as determined by sucrose-gradient zone sedimentation at neutral pH.

(e) *Hydrolysis of T*2 DNA by exonuclease III*

Exonuclease III extracted from *E. coli* was the phosphocellulose fraction purified and assayed as previously described (Richardson & Kornberg, 1964). This fraction had a specific activity of 11,000 units/mg protein.

Partially degraded DNA was obtained by incubation with exonuclease III at 37°C. The reaction mixtures (4·5 ml.) contained 330 μmoles of Tris–HCl buffer, pH 8·0, 3·3 μmoles of $MgCl_2$, 33 μmoles of 2-mercaptoethanol and 100 to 150 mμmoles of ^{32}P-labeled T*2 DNA. After chilling in ice, 10 units of enzyme were added and the reaction started by transfer to 37°C. In some cases 10 additional units of enzyme were added during the course of the reaction to increase the rate of degradation.

The extent of hydrolysis was determined by measuring the trichloracetic acid-soluble radioactivity. A 0·1- or 0·2-ml. portion of the reaction mixture was added to 500 μg of carrier salmon sperm DNA (Calbiochem) to give a final volume of 0·5 ml. The DNA was precipitated by the addition of 0·5 ml. of a cold 10% solution of trichloroacetic acid. After 10 min at 0°C the precipitate was sedimented by centrifugation at 12,100 g for 10 min at 0°C. The radioactivity of a 0·5-ml. portion of the supernatant was determined either: (a) by drying onto a planchet and counting with a Nuclear Chicago low-background gas-flow Geiger counter or, (b) by addition of the sample to 20 ml. of an aqueous dioxane fluor (Richardson *et al.*, 1964) and counting by scintillation. The total radioactivity was determined by counting an unprecipitated sample of the reaction mixture (counting technique (a) only).

For annealing and electron microscopy, 0·2- to 0·3-ml. samples were removed from the reaction mixture and added to one-tenth the volume of 20 × SSC (3·0 M-NaCl, 0·30 M-sodium citrate) to stop the reaction, and stored at 4°C.

(f) *Annealing of DNA*

In order to avoid the thermal chain scission that occurs at elevated temperatures, samples of the reaction mixture (stored in 2 × SSC) were made 7·2 M in $NaClO_4$ by the addition of 3 vol. of 9·6 M-$NaClO_4$, buffered to pH 7·2 with 0·1 M-Tris–HCl. Annealing was achieved by incubation at 25°C for 3·5 hr (Geiduschek, 1962). Annealing can also be achieved in 2 or 5 × SSC at 65°C (Ritchie, Thomas, MacHattie & Wensink, to be published.

(g) *Electron microscopy*

Visualization in the electron microscope was accomplished by the method of Kleinschmidt, Lang & Zahn (1960) as simplified in this laboratory (MacHattie & Thomas, 1964). An equal volume of 0·04% cytochrome *c* dissolved in 4·0 M-NaCl or 7·2 M-$NaClO_4$ was mixed with the annealed DNA. The resulting solution (containing either 3·6 or 7·2 M-$NaClO_4$) was spread on a clean air–water surface and grids prepared in the usual manner. The electron micrographs were enlarged by projection, the images traced, and their contour lengths measured.

In order to estimate the frequency of circles or other structures, a grid square was selected at random, then photographed in a systematic manner and every DNA molecule or other structure seen was traced, measured and catalogued. In certain cases "flowers" and linear fragments were neglected. Occasionally linear molecules were identified and catalogued, but not photographed.

3. Results

(a) *Circles*

Limited exonuclease III degradation of intact T*2 DNA molecules allows many of them to join their ends upon annealing to form circular structures. Electron micrographs of undegraded and degraded molecules, both of which have received the same annealing treatment, are shown in Plate I.

No unambiguous circular structures that are continuous over their entire length have ever been found *without* exonuclease III treatment (Plate I(a)). In contrast,

(a)

(b)

PLATE I. T*2 DNA molecules.

(a) Undegraded, annealed, linear molecule. Contour length 54·4 μ.
(b) Degraded (3·5%), annealed, circular molecule. Contour length 54·9 μ. The NaClO₄ causes the molecules to assume a more compact configuration than usual and appears to cause about 5% shortening of the mean molecular lengths.

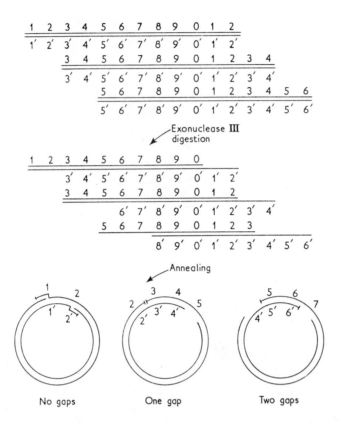

FIG. 1. Experimental scheme.

A collection of T2 DNA molecules is depicted by two parallel lines corresponding to the complementary chains of the duplex. Each molecule is shown having a different circular permutation of a common sequence and a terminal repetition of its first sequence. Exonuclease III exposes the complementary 5'-ended chains at the ends and circle formation takes place upon annealing. If the degradation proceeds beyond the limits of the terminal repetition, then two single-chain "gaps" will bracket a duplex segment, the length of which is the length of the terminal repetition.

more than ten perfectly continuous circles have been found after limited exonuclease III degradation in four separate experiments (Plate I(b)).

Many more of the circular molecules included in this analysis contain discontinuities bridged by what is apparently a single polynucleotide chain, as would be expected by Fig. 1. The conditions employed fail to render single-stranded regions clearly visible. However, in those cases where the ends of a molecule are so connected, it is classified as a circle for this analysis. In addition to these two classes, other molecules are found with their ends so close together that they appear to be circular even though no single-stranded region can be seen. Rather than rely on the psychological impression of circularity, the straight line end-to-end distances of all molecules in experiment III were measured and plotted (Fig. 2(a)). As can be seen, a large number of linear molecules can be found with their end-to-end distances distributed in a fashion expected for randomly coiled linear structures (random walk). Near the origin, a sharp peak is seen resulting from the circular

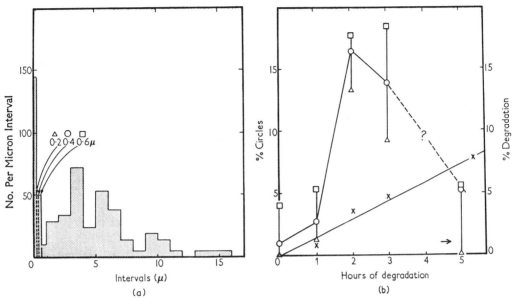

Fig. 2. The frequency of circles and their definition.

(a) Histogram of end-to-end separation of all molecules seen in experiment III. This includes the undegraded control molecules. Since most of the molecules that were visually classified as circles actually contained short gaps, the longest such gap in each molecule has been treated as an end-to-end separation. This distribution is seen to depart from the theoretical expectation for linear molecules in that the frequency of near-zero values (0 to 0·6 μ) is disproportionately high. Clearly, we have the overlapping of two distributions: truly linear and truly circular molecules. For purposes of this analysis, molecules with apparent ends no more than 0·2, 0·4 or 0·6 μ apart (see arrows) are classed as circles. The consequence of these limits on the frequency of circles is shown in (b). By the 0·6 μ criterion, more than 50 circles have been identified and measured.

(b) Frequency of circular molecules as related to extent of exonuclease III degradation prior to annealing. ($\times$) Fraction of nucleotides rendered trichloroacetic acid-soluble by exonuclease III. ($\triangle$, $\bigcirc$ and $\square$) The frequency of circles defined as molecules having apparent end separations of less than 0·2, 0·4 and 0·6 μ, respectively (see (a)). The low "%circles" value at 5 hr is probably spurious (see text).

molecules. The fraction of circular molecules depends to some extent upon the length of the gap considered acceptable. The arrows in Fig. 2(a) mark three different gap distances. The effect of these limits upon the frequency of circle assignment is shown in Fig. 2(b) by the length of the vertical bars. Note that for the highest degradation value (7·4%), even the most generous of these gap criteria (0·6 μ) probably ruled out many molecules that were in fact circles, since in this case each "gap" region would be expected to contain 2 μ or more of single chain.

Here it can be seen that efficient circle formation is promoted by 1 to 3·5% exonuclease III degradation. The presence of circles implies that the initial and terminal sequences (in either chain) were identical, or nearly so.

(b) Specificity of the ends

It might be thought that the exposed sequences need not be complementary to entangle and form effectively circular structures. To test this possibility one needs a collection of DNA molecules that are known to be *not* terminally repetitious. Since no intact phage DNA molecules were known to have this property, we selected shear-broken half-molecules of T*2. These fragments were considered not likely to be terminally repetitious.

A mixture containing approximately equal numbers of intact and half molecules was degraded with exonuclease III and samples were removed for annealing and examination by electron microscopy. In two such experiments, analysis by the 0·6 μ criterion of Fig. 2(a) yielded a total of five circles among 29 whole molecules, while three out of 117 half-length fragments showed end-to-end separations that could not be distinguished from circles. These data are statistically significant and support our conclusion that whole molecules can form circles, while fragments cannot. More convincing evidence to this point will be supplied in the following paper on T3 and T7 DNA (Ritchie et al., 1967). (These molecules are shorter and easier to analyse, and being non-glucosylated they are uniformly susceptible to exonuclease III.)

Thus we conclude that circle formation is the consequence of the specific interaction of complementary sequences found at the ends of the intact molecules, but unlikely to exist in other regions of the molecule.

(c) Length of the terminally repetitious region

As shown in Fig. 1, circular DNA molecules that have been degraded beyond the limits of the terminal repetition would be expected to possess single-stranded regions. A pair of single-stranded regions in these circular molecules would, in principle, demark the extent of the terminal complementary sequences. At present there are five circular molecules in our collection with pairs of gaps or faintly visible single chains separated by duplex segments 0·3, 0·4, 0·5, 1·0 and 1·7 μ long, covering a range of 0·5 to 3% of the total contour length. An example is shown in Plate II. These values agree favorably with the finding that 1 to 3·5% of the nucleotides must be solubilized before circle formation becomes frequent (Fig. 2(b)).

(d) Equivalence of T-even and T*-even DNA molecules

The glucose on T2 DNA renders it a poor substrate for exonuclease III (Richardson, 1966). Therefore, these experiments were conducted with DNA molecules isolated from phage grown on the non-glucosylating host E. coli B/4c. These phage, called T*2, are virtually non-glucosylated (Hattman & Fukasawa, 1963). This raises the question of whether the T2 and T*2 DNA molecules are genetically and physically equivalent in all respects other than glucose. The following evidence indicates that this is so.

(1) The contour length distribution of T2 and T*2 DNA molecules is indistinguishable, being 56 ±2 and 55 ±2 μ respectively (reported by Thomas, 1966).

(2) The buoyant densities of T2 and T*2 DNA molecules in CsCl are almost identical, in agreement with the results of Boyle, Ritchie & Symonds (1965).

(3) When a pair of T4B rII point mutants (r147 and r271) are crossed, the frequency of wild-type recombinants is found to be the same whether the cross is performed in a glucosylating (E. coli strain R2) or a non-glucosylating (E. coli strain W4597) host bacterium.

(4) For the same pair of rII point mutants in T4B, the frequency of mottled plaques (heterozygous particles) is found to be the same among the progeny from crosses in glucosylating and non-glucosylating hosts.

(5) Crosses between T4B wild type and an rII deletion mutant (r196) are found

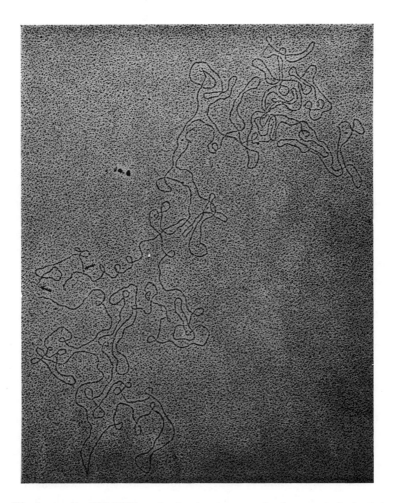

PLATE II. A circular T*2 DNA molecule showing two closely spaced single-chain regions separated by a duplex segment 1·7 μ long (see arrows). If the identification is correct, this duplex segment should be the terminally repetitious sequence for this molecule.

to produce the same frequency of mottled plaques among the progeny from normal and non-glucosylating host cells. This indicates that the frequency of terminal redundancy heterozygotes (Streisinger *et al.*, 1964; Séchaud *et al.*, 1965) is the same irrespective of the level of glucosylation.

4. Discussion

From the above results we conclude that there are sequences at the beginning of the T*2 DNA molecules that are effectively identical to the sequences at the end. On the basis of the physical and genetic equivalence of T*2 DNA molecules with those from T2 and T4, it is likely that this conclusion applies to these phage DNA molecules as well. Taken together with the previous work, which establishes that T2 DNA molecules are circular permutations of a common nucleotide sequence, it emerges that these DNA molecules are physically both *circularly permuted and terminally repetitious.*

Thus it appears that the physical structure of the T2 DNA molecule is in exact accord with the model proposed by Streisinger and his collaborators to explain the circular genetic map and the unusual kind of heterozygosis that is found in bacteriophage T4 (Streisinger *et al.*, 1964; Séchaud *et al.*, 1965). The length of the terminal repetition as determined by the frequency of terminal-redundancy heterozygotes is of the order of 1%. This agrees with our estimate of 1 to 3%.

5. The Variety of Permutations Present

The conclusion that T2 DNA molecules are terminally repetitious *and* are different permutations of each other, leads one to the question: How many different permutations are present in a T2 lysate? There must be at least two in order to satisfy our earlier experiments (Thomas & Rubenstein, 1964; Thomas & MacHattie, 1964); at the other extreme there could be nearly 2×10^5 different permutations if the beginning of the genetic text could occur at any nucleotide pair in the molecule. Our evidence on this point comes from further repetition of experiments that are already published (Thomas & MacHattie, 1964). The intact T2 molecules are treated with $0 \cdot 30$ M-Na_3PO_4 for a time sufficient to allow the separation of the intact chains. Upon reneutralization and annealing, the circularly permuted single chains assemble themselves into circular duplex molecules as shown in Fig. 3.

As can be seen, the terminal repetitions cannot find unengaged complementary partners and therefore will be left as single chains, connected at *two* locations on the circle. The protein film procedure does not render these single chains highly visible, yet they can frequently be seen as "bushes" connected to the circular duplex.

The evidence that these distinctive "bush" structures, seen in Plate III, do represent single polynucleotide chains is as follows: they are never found on grids of DNA molecules that have not been treated with exonuclease III or denatured. Molecules that have been partially degraded with exonuclease III show bushes almost exclusively at the ends of the molecule, where they are expected from the known action of this nuclease. The denatured and re-annealed T2 DNA preparation reported here showed "bushes" on 120 of 200 ends of linear duplex molecules (without regard to length), where single chains would be expected due to the mismatch of randomly selected partner chains. The same DNA sample prior to denaturation showed no bushes on a random sample of 200 ends. The only reservations in the interpretation

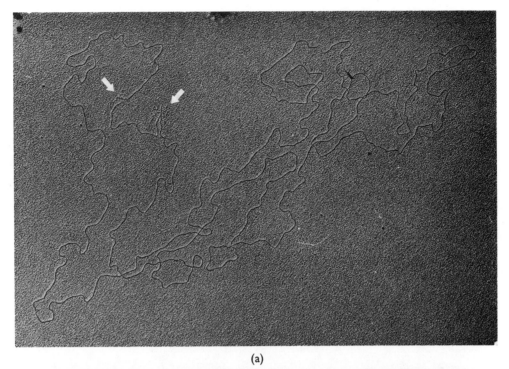

(a)

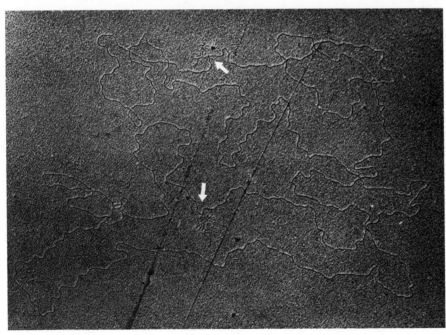

(b)

PLATE III (a) and (b). Examples of denatured and annealed artificial circular T2 duplexes bearing two single-chain "bushes".

These bushes (indicated by arrows) should be the repetitious terminal segments of the two component chains. Notice that the separation of the two bushes is significantly different in (a) and (b).

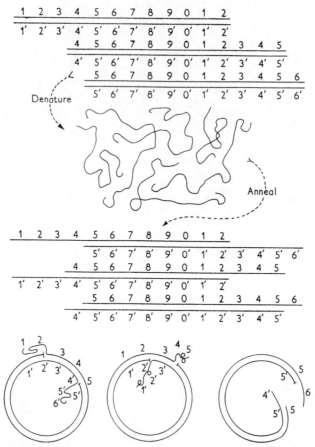

Fɪɢ. 3. Circle formation by denaturing and annealing a permuted collection of duplexes.

Notice that each permutation is also terminally repetitious. These repetitious terminals cannot find complementary partners and are left out of the circular duplex. Their separation depends on the relative permutation of the partner chains.

of "bushes" are these: (a) single chains do not invariably show up as "bushes"; they may remain invisible, or when subjected to slight tension, they may stretch out to form a faint line (this was the usual case in the single-chain regions of T*2 circles prepared in the presence of NaClO$_4$), and (b) after such treatments as melting and annealing, a few short pieces of single chains may be found scattered around on the grid, producing a certain background of "bushes" that are not necessarily connected to any duplex DNA.

Analysis of the "bushes" seen on a randomly selected sample of 26 annealed circular T2 DNA molecules in the experiment reported here showed that although some circles showed 0, 1 or 3 bushes, the majority showed two (see insert Fig. 4).

As can be seen from Fig. 3, the separation of these bushes depends on the relative permutation of the component single chains in the circular duplex. Therefore by measuring the length of duplex connecting the two bushes, one may determine approximately the relative permutation. Plate III shows two circular T2 molecules each bearing two bushes that are separated by substantially different lengths of duplex. The "interbush" distances in twelve two-bush circular molecules where they could

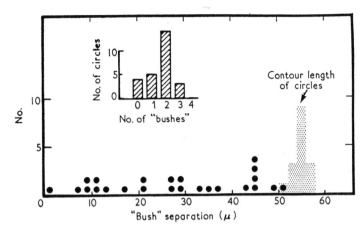

FIG. 4. "Bush" separation, a measure of relative permutation of the chains in denatured and annealed circular T2 DNA.

Each two-bushed circular molecule presents two interbush distances, their sum being the total contour length. The uniform scatter suggests that no preferred permutations exist.

Insert: number of bushes per circle showing the predominance of two-bushed circles.

be unambiguously measured are plotted in Fig. 4. This histogram shows that many different permutations exist and that no special permutation is preferred.

It is a pleasure to acknowledge the invaluable technical assistance of Mr Pieter Wensink. Dr Salil K. Niyogi provided some very useful suggestions. We thank Dr John N. Abelson and Mr Marcus M. Rhoades for many enlightening discussions. We thank Dr A. D. Hershey for his timely encouragement. This work was supported by the National Institutes of Health (E-3233), the Atomic Energy Commission (AT-30-1-2119), and National Science Foundation (GB-3316). The research of one of us (L. A. MacH.) is supported by a National Institutes of Health Career Development Award (5-K3-GM-25, 342).

REFERENCES

Boyle, J. M., Ritchie, D. A. & Symonds, N. (1965). *J. Mol. Biol.* **14**, 48.
Geiduschek, E. P. (1962). *J. Mol. Biol.* **4**, 467.
Hattman, S. & Fukasawa, T. (1963). *Proc. Nat. Acad. Sci., Wash.* **50**, 297.
Hershey, A. D. & Chase, M. (1952). *J. Gen. Physiol.* **36**, 39.
Kleinschmidt, A., Lang, D. & Zahn, R. K. (1960). *Z. Naturf.* **16**, 730.
Kozinski, A. W. & Szybalski, W. (1959). *Virology*, **9**, 260.
MacHattie, L. A. & Thomas, C. A., Jr. (1964). *Science*, **144**, 1142.
Richardson, C. C. (1966). *J. Biol. Chem.* **241**, 2084.
Richardson, C. C., Inman, R. B. & Kornberg, A. (1964). *J. Mol. Biol.* **9**, 46.
Richardson, C. C. & Kornberg, A. (1964). *J. Biol. Chem.* **239**, 242.
Richardson, C. C., Lehman, I. R. & Kornberg, A. (1964). *J. Biol. Chem.* **239**, 251.
Ritchie, D. A., Thomas, C. A., Jr., MacHattie, L. A. & Wensink, P. C. (1967). *J. Mol. Biol.* **23**, 365.
Séchaud, J., Streisinger, G., Emrich, J., Newton, J., Lanford, H., Reinhold, H. & Stahl, M. M. (1965). *Proc. Nat. Acad. Sci. Wash.* **54**, 1333.
Streisinger, G., Edgar, R. S. & Denhardt, G. H. (1964). *Proc. Nat. Acad. Sci., Wash.* **51**, 775.
Thomas, C. A., Jr. (1966). *J. Gen. Physiol.* **49**, no. 6, pt. 2, 143.
Thomas, C. A., Jr. & MacHattie, L. A. (1964). *Proc. Nat. Acad. Sci., Wash.* **52**, 1297.
Thomas, C. A., Jr. & Rubenstein, I. (1964). *Biophys. J.* **4**, 93.

ARTICLE 34

From the *Proceedings of the National Academy of Sciences*, Vol. 59, No. 1, 131–138, 1968. Reproduced with permission of the author.

EVIDENCE FOR LONG DNA STRANDS IN THE REPLICATING POOL AFTER T4 INFECTION*

By Fred R. Frankel

DEPARTMENT OF MICROBIOLOGY, SCHOOL OF MEDICINE,
UNIVERSITY OF PENNSYLVANIA, PHILADELPHIA

Communicated by A. D. Hershey, October 30, 1967

T4 DNA extracted from phage-infected bacteria sediments faster and differs in other ways from the DNA found in phage particles.[1, 2] The rapid sedimentation could result from an increased molecular length, from association with other molecules, DNA or otherwise, from an altered conformation, or from any combination of these. Many properties of the DNA suggest increased length, but other possibilities have not been excluded.

If replicating DNA contains any single strands of exceptional length, the rapid sedimentation must signify, at least in part, native molecules of exceptional length. More important, the generation of such strands would furnish a clue to mechanisms of replication or recombination or both. Initial attempts to demonstrate long strands failed.[2, 3] We report here that part of the replicating DNA sediments in alkali at a rate appropriate to single strands of two or more times the normal length of T4 DNA.

Methods.—Lysates containing tritium-labeled DNA were prepared from cultures of *Escherichia coli* B3 (thymine-requiring) grown at 36°C to 2×10^8 cells/ml in a Tris-glucose-ammonium medium[4] containing 5 μg/ml of thymidine. The cells were transferred to 0.1 vol of an adsorption medium[5] containing 2 μg/ml of thymidine, and the suspension was aerated for 10–20 min at 36°C. Five particles per bacterium of phage T4 D (from R. S. Edgar) and 10 μg/ml of L-tryptophan were added. After 4 min, the infected cells were diluted to 2×10^8/ml in warm Tris-glucose-ammonium medium containing 2 μg/ml of thymidine. Four min later, 20–80 μc/ml of H³-thymidine was added. Samples of the culture were lysed at various times thereafter by first chilling in ice and then adding an equal volume of an ice-cold solution containing 2 mg/ml of lysozyme in 0.1 M KCN and 0.05 M ethylenediaminetetraacetate (EDTA), pH 8. After 20 min, sodium lauryl sarcosinate[6] was added to 3%. Five min later the tube contents were gently mixed by rotating the tube for 10 min at 5°C. This lysis procedure is a modification of that used by Bode and Kaiser.[7] The procedure that we used previously[2] seldom yields denatured DNA that sediments faster than single strands from phage particles, possibly because of nuclease action during a 65°C incubation step. Both procedures yield native DNA that sediments rapidly.

Phage particles labeled with radiophosphorus were prepared in the usual way by growth in synthetic medium containing P³²-phosphate. DNA used as a marker in sedimentation measurements contained 0.2 c/gm P. Phage particles used to infect cultures in which parental DNA was to be analyzed contained 5 c/gm P.

The DNA in lysates of infected cultures was denatured by adding NaOH to 0.3 N and mixing gently at 5°C for 10 min. The uncorrected pH of this solution was 12.75 (Leeds and Northrup pH meter; radiometer electrode-type GK 2021 C; manufacturer's correction for sodium ion concentration at this pH is + 0.36). P³²-labeled phage particles were then added and gently mixed with the alkaline lysate. This results in dissolution of the phage particles and the release of denatured DNA,[8] henceforth designated reference or marker DNA. Samples of this mixture were layered onto 4.5-ml or 25-ml linear gradients prepared from 5% and 20% sucrose solutions containing 0.9 M NaCl, 0.1 M NaOH, and 0.01 M EDTA, and the tubes were spun at 15×10^8 rpm²-hr (4.5 ml) or

22×10^8 rpm²-hr (25 ml) at 15°C. Increasing the NaOH concentration sixfold had no striking effect on the sedimentation patterns obtained. The tube contents were analyzed in the way already described.[2]

The hybridization technique used was that of Richards.[9]

Results.—Labeled parental phage: Shortly after T4 infection, the DNA originating from parental phage particles can be recovered as normal phage DNA molecules in high yield, but at later times the recovery is somewhat reduced.[10, 1] When the parental DNA was examined by sedimentation after alkaline denaturation (Fig. 1), the sample taken early after infection (4 min) resembled the marker DNA. At later times, however, the proportion of normal molecules decreased, and a faster-sedimenting component appeared.

A sample taken at 14 minutes after infection was denatured and sedimented through a 25-ml alkaline sucrose gradient (see inset, Fig. 2). Fractions of this DNA were then mixed with a peak fraction of marker DNA which had also been denatured and sedimented in an alkaline gradient. Figure 2 shows the results of recentrifugation of these mixtures. Fractions 5 and 6 showed higher sedimentation coefficients than the marker. The peak parental fraction was indistinguishable from the marker. There is clearly a marked heterogeneity of the strands.

The rapid sedimentation of the parental label probably does not signify cosedimentation with host DNA or cytoplasmic components. When a culture was infected with a P^{32}-labeled mutant of T4 (T4amA453, gene 32)[11] that is unable to synthesize DNA but carries out other early functions,[12] no fast com-

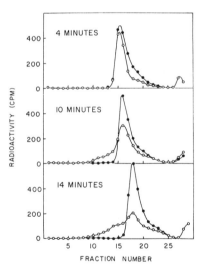

Fig. 1.—Sedimentation of precursor DNA containing P^{32} of parental origin (*open circles*) and admixed H^3-labeled marker DNA (*filled circles*) in alkaline sucrose. Lysates were prepared at times indicated. Sedimentation from right to left.

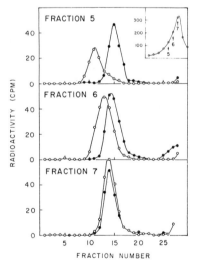

Fig. 2.—Recentrifugation of precursor DNA fractions in alkaline sucrose. *Inset:* Sedimentation in alkaline sucrose of a lysate prepared from a culture infected 14 min earlier with P^{32}-labeled phage. The fractions indicated by arrows were mixed with H^3-labeled reference DNA (also recovered from an alkaline sucrose gradient) and recentrifuged with the results shown. *Open circles*, P^{32}-labeled DNA; *filled circles*, H^3-labeled DNA.

ponent appeared at 15 minutes. However, if cells were mixedly infected with this P^{32}-labeled mutant and unlabeled T4 or T4amB22 (gene 43),[13] the fast component did appear.

Newly synthesized DNA: We previously reported that DNA continuously labeled after infection exhibits a gradually increasing sedimentation coefficient for about 15 minutes, when phage particles begin to appear.[2] Such DNA was examined after alkaline denaturation, with the results shown in Figure 3. At seven minutes after infection, the denatured strands sedimented slowly. With increasing time after infection the sedimentation coefficient of the DNA gradually increased, reaching the rate characteristic of the marker at about 9 or 10 minutes after infection, and then exceeding it at 12 minutes. By 17 minutes, DNA resembling the marker had begun to accumulate.

The DNA strands from a culture labeled after infection with H^3-thymidine and lysed at 14 minutes were isolated from a preparative alkaline sucrose gradient (Fig. 4A). Fractions of denatured marker DNA were also isolated (Fig. 4B). The results of centrifugation of several mixtures of these fractions are shown in Figure 5. A slowly sedimenting fraction (Fig. 5A) and a rapidly sedimenting fraction (Fig. 5B) of the newly synthesized DNA both retained their characteristic rate of sedimentation when compared with the peak fraction of the marker. However, a rapidly sedimenting fraction of the marker (containing less than 4% of the DNA) did not show rapid sedimentation after recentrifugation (Fig. 5C). It appears that the distribution of marker DNA in an alkaline gradient mainly reflects diffusion and mechanical mixing, whereas the distribution of the newly synthesized DNA reflects a heterogeneity of sedimentation coefficients.

When infected cells were labeled from the fourth to the seventh minute after infection, and then transferred to unlabeled medium until 15 minutes, the labeled DNA that moved slowly at 7 minutes was converted, by 15 minutes, into a mixture of normal and rapidly sedimenting strands. When infected cells were labeled at 15 minutes for 30 seconds or less, most of the labeled DNA sedimented heterogeneously and more slowly than the marker; however, a small fraction still sedimented faster. Thus, although pulse-labeled DNA adds directly to a rapidly sedimenting native structure in the replicating pool,[2] it does so mainly, but not entirely, in the form of short strands.

The effect of chloramphenicol and certain phage mutations: When chloramphenicol was added to a culture at 10 minutes after infection, and the culture was lysed at 30 minutes and analyzed after alkali treatment, the bulk of the newly synthesized DNA was found to sediment faster than the marker (Fig. 6, bottom). A similar result was obtained in the absence of chloramphenicol if the infecting phage was T4amB17 (gene 23) or T4tsL 147 (gene 22), mutants unable to synthesize head protein under restrictive conditions.[11, 14]

As the figure shows, different results were obtained if chloramphenicol was added at earlier times. Under these conditions the component that accumulated in greatest amount had a sedimentation coefficient somewhat lower than marker.

Test for cross-linking between DNA strands: If the polynucleotide chains of a DNA molecule are cross-linked, the molecular weight cannot halve on denaturation, and the structure will sediment faster than the free single strands. To

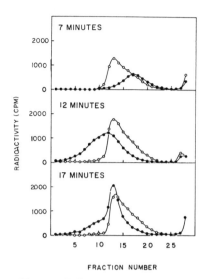

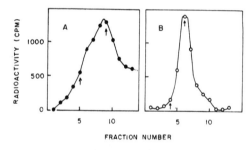

FIG. 3.—Sedimentation of H³-labeled newly synthesized DNA (*filled circles*) and admixed P³²-labeled reference DNA (*open circles*) in alkaline sucrose. The cultures were continuously labeled after infection and samples were lysed at the times indicated.

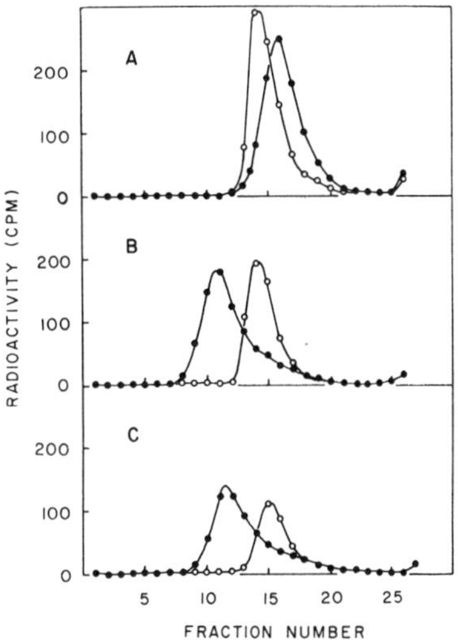

FIG. 5.—Recentrifugation of fractions of H³-labeled newly synthesized DNA (*filled circles*) and fractions of P³²-labeled marker (*open circles*) in alkaline sucrose. (*A*) Fraction 9 of Fig. 4*A* and fraction 6 of Fig. 4*B*; (*B*) fraction 5 of Fig. 4*A* and fraction 6 of Fig. 4*B*; (*C*) fraction 5 of Fig. 4*A* and fraction 4 of Fig. 4*B*.

FIG. 4.—Preparative alkaline sucrose gradients. (*A*) DNA from a culture labeled after infection with H³-thymidine and lysed at 14 min. (*B*) P³²-labeled phage particles treated with alkali.

determine whether interstrand linkages are present in the rapidly sedimenting T4 DNA, the possibility of reversible denaturation was examined.[15]

A lysate was prepared at 20 minutes after infection with T4*am*B17, denatured with 0.3 N NaOH, and marker added. One sample was analyzed by sedimentation and found to contain predominately fast-sedimenting DNA. A second sample was carefully adjusted to pH 7.5 with 0.5 M KH$_2$PO$_4$, layered onto a solution of CsCl, and sedimented to equilibrium. The rapidly sedimenting strands and marker both formed single bands indicating a buoyant density 0.015 gm/ml greater than that of native marker. Thus, the rapidly sedimenting DNA does not show reversible denaturation, and we conclude that its high rate of sedimentation does not result from interstrand linkages.

Association with protein also can result in rapid sedimentation of DNA. Therefore, rapidly sedimenting strands were isolated from a lysate prepared at 20 minutes after infection with T4*am*B17, neutralized, and incubated at 36°C for

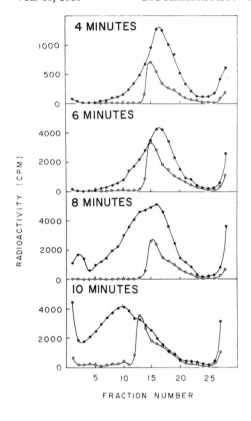

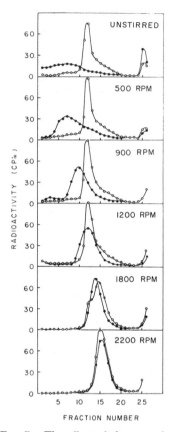

Fig. 6.—Sedimentation in alkaline sucrose of H³-labeled DNA synthesized in the presence of chloramphenicol (100 μg/ml) (*filled circles*), and admixed P³²-labeled marker DNA (*open circles*). Chloramphenicol was added at the times indicated. Lysates were prepared at 30 min after infection.

Fig. 7.—The effect of shear on single strands of precursor DNA. A lysate was prepared at 30 min from an infected culture to which chloramphenicol (100 μg/ml) had been added at 10 min and H³-thymidine at 11 min. After addition of alkali and P³²-labeled phage particles, the mixture was stirred at 25° for 5 min each at the indicated speeds.

30 minutes with 1 mg per milliliter of pronase. Sedimentation after treatment with detergent and alkali showed that the enzyme had little effect.

T4 specificity of the rapidly sedimenting DNA: We have found that denatured DNA obtained from uninfected cultures of *E. coli* according to our procedure sediments in alkali about 1.6 times faster than denatured marker DNA. At ten minutes after infection, little of the prelabeled bacterial DNA sediments faster than the marker. Although bacterial DNA synthesis ceases after phage infection,[16] it seemed possible that some incorporation of labeled precursors into fast-sedimenting host DNA strands might occur. Therefore, rapidly sedimenting strands were isolated from cells infected with either T4 or T4*am*B17 and the labeled DNA was tested for base-sequence homology with *E. coli* DNA and T4 DNA. It was found to resemble authentic T4 DNA (Table 1).

TABLE 1.   *Specific hybridization tests of mixtures containing precursor DNA and reference DNA.*

DNA test mixture	DNA on filter	Competing DNA	Per Cent Bound Precursor DNA	Reference DNA
	—	—	0.1	0.2
T4 precursor DNA	T4 DNA	—	51	58
(93S) plus reference DNA	T4 DNA	T4 DNA	21	23
	T4 DNA	E. coli DNA	58	60
	E. coli DNA	—	0.4	0.7
	—	—	0.2	0.0
T4amB17 precursor	T4 DNA	—	54	64
DNA (80S to 113S) plus	T4 DNA	T4 DNA	22	26
reference DNA	T4 DNA	E. coli DNA	59	62
	E. coli DNA	—	0.3	0.9

The *DNA test mixture* was prepared by mixing an alkaline sucrose fraction containing H^3-labeled precursor DNA isolated at 15 min after infection with P^{32}-labeled reference DNA. After neutralization, the mixture was sonicated for 4 × 10 sec in ice and heated at 100°C for 10 min; aliquots were added to the various membrane filters in vials and incubated for 20 hr at 63°C to 65°C. *DNA on filter:* The filters received 5 µg of heat-denatured T4 DNA or E. coli DNA. *Competing DNA* was added at 50 µg/ml to the test mixtures before neutralization.

Fragility under shear: Hershey, Burgi, and Ingraham[17] found a direct relationship between the length of DNA molecules and their fragility, and Davison and Freifelder[18] presented some evidence that the same may be true for single-stranded DNA.   We have examined the behavior under shear of an alkaline solution containing rapidly sedimenting DNA and marker.

As shown in Figure 7, the effect of shear on the denatured marker was similar to its effect on native DNA.   No change in the sedimentation pattern of the DNA was observed until a critical speed of stirring was exceeded (1800 rpm).   At this point, the marker gave rise to a slower-sedimenting species with concomitant loss of the original molecules.   The rapidly sedimenting DNA in the mixture showed very different behavior.   With each increment in stirring speed, its sedimentation rate decreased.   However, strands with the sedimentation rate of the marker were not produced until a stirring speed just below the critical speed for the marker was reached.   After being stirred at the critical speed, the DNA sedimented more slowly than the original marker strands but faster than their breakage product.   This is the behavior expected of molecules that are longer and more heterogeneous in length than marker DNA, if fragility depends on length.

Native, rapidly sedimenting DNA contains rapidly sedimenting strands: The experiments described up to this point have been concerned with the nature of alkali-denatured DNA in unfractionated lysates of infected cells.   The rapidly sedimenting strands described above may arise from a different fraction of the phage-specific DNA pool than the native, rapidly sedimenting DNA that we described previously.[2]   Therefore, native rapidly sedimenting DNA was isolated from cultures at 15 minutes after infection with T4 or T4amB17.   Re-analysis of selected fractions on neutral sucrose gradients showed that the isolated DNA's had high sedimentation rates (in one experiment, 6, 4, and 3.2 times faster than the marker) that depended on their original positions in the preparative gradients (fractions 6, 9, and 11, respectively).   Analysis on alkaline sucrose gradients showed the fractions to be similar to one another in that 40–60 per cent of the DNA sedimented faster than marker DNA.   We have been unable to increase the proportion of faster-sedimenting strands either by more rapid re-analysis or by gentler manipulations.   Evidently, the rapidly sedimenting native DNA contains both slowly and rapidly sedimenting strands.

Discussion.—DNA extracted from bacteria infected with phage T4 contains a considerable fraction of strands that sediment more rapidly in alkali than reference DNA strands.   These strands incorporate isotopic labels both from the parental phage particle and from the culture medium.

Phage particles treated with alkali released DNA whose major component sedimented homogeneously and reproducibly.   We took this material to be intact single strands of T4 DNA, and used it as our reference DNA.   Occasionally, a small amount of faster-sedimenting material was seen in these preparations[18] (see Fig. 7).   The latter may result from cross-links preventing separation of the DNA strands[19] or from incomplete dissociation of protein from the DNA.

We found no indication that the rapid sedimentation of alkali-treated DNA extracted from infected bacteria was due to cross-linking, bound protein, or cosedimentation with host materials.   In addition, we showed that the rapid sedimentation could not result from synthesis of long strands of host DNA.

This appears to leave two alternatives: that the rapidly sedimenting strands are longer than strands from phage DNA, or that they are altered conformations of phage DNA strands.   The only alternate stable conformation that a single polynucleotide chain might be expected to achieve is one in which the ends are joined to form a circle.   A unique circular molecule should exhibit a unique sedimentation coefficient.   However, heterogeneous sedimentation was observed in the studies reported here, even under conditions in which the alkali concentration of the sedimentation medium was varied from 0.1 N to 0.6 N.   Therefore there is no reason to suspect circular strands at present.

The effect of shear on the rapidly sedimenting strands was like that seen for long double-helical DNA molecules.   Very low rates of shear caused the sedimentation rate of the DNA to decrease without generating strands of any characteristic length; this occurred without breakage of marker strands in the mixture.   We therefore conclude that the rapidly sedimenting strands extracted from infected cells represent unusually long molecules of phage-specific DNA.   These strands sediment up to 1.5 times faster than reference strands and thus appear to be more than twice the length of the reference DNA.[20]   This is not to say that all molecules in the pool have this property, or that their rapid sedimentation in the native state may not be due in part to other structures.   Rapidly sedimenting strands have previously been observed in cells infected with phage lambda.[21]

Addition of chloramphenicol at a late time after infection, or infection by phage mutants unable to synthesize head protein, resulted in accumulation of rapidly sedimenting strands.   These strands showed sedimentation rates appropriate to polynucleotide chains about four times the length of the reference DNA. An analogous effect of chloramphenicol[2] and infection with the phage mutants[22] was observed on the native DNA in the replicating pool.   The production of normal phage DNA molecules appears to be a late step in maturation dependent on head formation, as already suggested by others.[23, 24]

Long strands of DNA could arise by recombination between molecules of ordinary size bearing terminal repetitions of nucleotide sequences. Formation of long strands in this way has been suggested by Streisinger and co-workers.[25] Alternatively, these molecules could be synthesized directly. A mechanism that could generate molecules longer than the template upon which they are replicated has been suggested for chromosome transfer during conjugation.[26] The short strands synthesized both early and late after infection may represent synthesis at new growing points since they are preferentially labeled by a very short pulse of H^3-thymidine.

Very early addition of chloramphenicol to an infected culture resulted in the accumulation of strands that sedimented somewhat more slowly than reference strands. This might result from failure of the appearance of enzymes concerned with the joining of DNA fragments[27] or from abortive DNA replication.

Summary.—Evidence is presented that some DNA in the replicating pool of T4-infected cells contains strands that are longer than those found in the mature phage particle.

The author thanks S. H. Mayne and A. Stamatelaky for excellent technical assistance, D. M. Frankel for performing several important experiments, and H. S. Smith and S. Goodgal for useful discussions.

* Aided by grant AI 05722 from the National Institutes of Health, U.S. Public Health Service.

[1] Frankel, F. R., *J. Mol. Biol.*, **18**, 109 (1966).
[2] Frankel, F. R., *J. Mol. Biol.*, **18**, 127 (1966).
[3] Korn, D., *J. Biol. Chem.*, **242**, 160 (1967).
[4] Burgi, E., these Proceedings, **49**, 151 (1963).
[5] Hershey, A. D., and M. Chase, *J. Gen. Physiol.*, **36**, 39 (1952).
[6] Davern, C. I., these Proceedings, **55**, 792 (1966).
[7] Bode, V. C., and A. D. Kaiser, *J. Mol. Biol.*, **14**, 399 (1966).
[8] Davison, P. F., D. Freifelder, and B. W. Holloway, *J. Mol. Biol.*, **8**, 1 (1964)
[9] Richards, O. C., these Proceedings, **57**, 156 (1967).
[10] Kozinski, A. W. and T. H. Lin, these Proceedings, **54** 273 (1965).
[11] Epstein, R. H., A. Bolle, C. M. Steinberg, E. Kellenberger, E. Boy de la Tour, R. Chevalley, R. S. Edgar, M. Susman, G. H. Denhardt, and A. Lielausis, in *Cold Spring Harbor Symposia on Quantitative Biology*, vol. 28 (1963), p. 375.
[12] Tomizawa, J., N. Anraku, and Y. Iwama, *J. Mol. Biol.*, **21**, 247 (1966).
[13] de Waard, A., A. V. Paul, and I. R. Lehman, these Proceedings, **54**, 1241 (1965).
[14] Sarabhai, A. S., A. O. W. Stretton, S. Brenner, and A. Bolle, *Nature*, **201**, 13 (1964).
[15] Geiduschek, E. P., these Proceedings, **47**, 950 (1961).
[16] Cohen, S. S., *J. Biol. Chem.*, **174**, 281 (1948).
[17] Hershey, A. D., E. Burgi, and L. Ingraham, *Biophys. J.*, **2**, 423 (1962).
[18] Davison, P. F., and D. Freifelder, *J. Mol. Biol.*, **16**, 490 (1966).
[19] Alberts, B., thesis, Harvard University (1965).
[20] Abelson, J., and C. A. Thomas, Jr., *J. Mol. Biol.*, **18**, 262 (1966).
[21] Smith, M. G., and A. Skalka, *J. Gen. Physiol.*, **49**, no. 6, part 2, 127 (1966).
[22] Frankel, F. R., and S. H. Mayne, *Abstracts*, 66th Meeting of the American Society for Microbiology (1966), p. 125.
[23] Streisinger, G., in *Phage and the Origins of Molecular Biology*, ed. J. Cairns, G. S. Stent, and J. G. Watson (Cold Spring Harbor Laboratory of Quantitative Biology, 1966), p. 335.
[24] Ikeda, H., and J. Tomizawa, *J. Mol. Biol.*, **14**, 120 (1965).
[25] Streisinger, G., J. Emrich, and M. M. Stahl, these Proceedings, **57**, 292 (1967).
[26] Fulton, C., *Genetics*, **52**, 55 (1965).
[27] Kozinski, A., P. Kozinski, and P. Shannon, these Proceedings, **50**, 746 (1963).

Questions

ARTICLE 24   Davison, Freifelder, and Holloway

1. Why was formaldehyde used?

2. What was the reason for performing the sucrose gradient shown in Fig. 5 of the article?

3. State the argument given that single-strand breaks are present in double-stranded DNA.

4. Draw the sedimentation diagram expected if 30% of the molecules had one single-strand break at 25% of the distance from one end.

•5. Suppose the sedimentation pattern indicated that the fragments were roughly half unit size. Design an experiment to distinguish these two possibilities: (1) all double-stranded molecules have one break in one strand; (2) half have two breaks, one in each strand, and half have no breaks.

ARTICLE 25   Burgi, Hershey, and Ingraham

•1. How would a single-strand break 10% from one end be detected by shearing experiments?

•2. Suppose each molecule in a population contained one single-strand break but at random positions. What effect would this have on a shearing experiment? Suppose only 10% had a randomly situated break and 90% had no break.

ARTICLE 26   Hershey, Burgi, and Ingraham

•1. The authors realized that the folded molecules were circular. Why do you think they did not state it explicitly?

2. One possible explanation for the cohesive ends is that they are complementary single strands. If so, is there anything in the paper that suggests they are short? What?

3. Why should a circular molecule have a higher s-value than a linear one would?

4. What effect does DNA concentration have on the formation of circles, dimers, and circular dimers?

ARTICLE 27   Hershey and Burgi

1. What would have been the result of Hershey and Burgi's annealing experiment if the left and right cohesive ends had been the same?

ARTICLE 28   Strack and Kaiser

1. Why was it important for Strack and Kaiser to show that exonuclease III restores the activity of DNA inactivated by polymerase?

2. In Fig. 2 of the article why does exonuclease III decrease the activity after an initial rise? What effect would exonuclease III have on DNA that had not been exposed to polymerase?

3. Recalling the information in article 26 by Hershey, Burgi and Ingraham, give the s of λ DNA treated with polymerase and stored in the cold for several days. Explain.

•4. Would the biological activity of λ DNA stored in the cold for several days be inactivated by polymerase? If the DNA were heated before polymerase treatment, would the polymerase reduce the activity? Would the activity of untreated DNA stored in the cold be affected by heating alone?

ARTICLE 29   Streisinger, Edgar, and Denhardt

•1. Explain how a circular map can be generated from a nonpermuted rod if genetic exchanges frequently occur in pairs.

•2. What is the minimum number of genetic markers needed to demonstrate circularity?

ARTICLE 30   Sechaud et al.

•1. Draw a DNA molecule containing a point mutant, terminal redundancy heterozygote and one containing a point mutant, internal heterozygote. Draw the structure of an internal deletion heterozygote and explain why the authors think its existence is unlikely. Why is a deletion heterozygote more understandable in the terminally redundant regions?

•2. Why were the authors concerned about clumps?

ARTICLE 31   Streisinger, Emrich, and Stahl

•1. Is the Watson-Crick replication scheme consistent with cyclic permutation *if a single phage particle can give rise to a permuted population*? Explain.

2. Mutations are known in the T4 head protein that result in the formation of both normal phage particles and those with smaller heads. Some of these smaller particles contain DNA, although the amount is less than in a normal phage. Under what conditions might the small phages be able to produce a plague? What would be the smallest DNA content that would allow a successful infection of a single bacterium by *two* small particles?

ARTICLE 32   Thomas and MacHattie
ARTICLE 33   MacHattie, Ritchie, Thomas, and Richardson

1. Draw the expected result of annealing denatured DNA if there is (1) cyclic permutation but not terminal redundancy and (2) terminal redundancy but not cyclic permutation.

2. Draw the result of the exonuclease experiment in (1) and (2) of the preceding question.

3. Why did these workers see "bushes" and not extended single strands?

4. What should be the relative lengths of the circles in the exonuclease experiment, the circles in the annealing experiment, and a normal, linear molecule? Assume that the redundant region is 1% of the length of the linear molecule.

5. In the exonuclease experiment would you ever expect to see linear dimers or circular dimers? If so, estimate the expected frequency of each.

1. Are all the long molecules in a cell infected with T4 the same size? Explain.

2. Since the long molecules may have contained single-strand breaks, would you say that Frankel's size estimate (2–3 times the normal length of the T4 DNA) is a maximum or minimum value?

3. What is the argument that the long molecules are not unit size or dimer circles?

4. Is there any evidence suggesting that the long molecules might be products of replication?

•5. Why should the molecules be longer if chloramphenicol is added 10 minutes after infection, or if a head mutant is used? (Chloramphenicol prevents protein synthesis.)

___

References

Sinsheimer, R. L. 1954. "Nucleotides from T2r⁺ Bacteriophage." *Science* **120**, 551–552. The discovery of glucosylated DNA.

Wyatt, G. R. and S. S. Cohen. 1952. "A New Pyrimidine Base from Bacteriophage DNA." *Nature* **170**, 1072–1073.

Lehman, I. R., and E. A. Pratt. 1960. "On the Structure of Glucosylated Hydroxymethylcytosine Nucleotides of Coliphage T2, T4, and T6." *J. Biol. Chem.* **235**, 3254–3259. The chemistry of glucosylation is described.

Stent, G., and R. Calendar. 1978. *Molecular Genetics*, 2nd ed. W. H. Freeman and Company.

Thomas, C. A. 1966. "The Arrangement of Information in DNA Molecules." *J. Gen. Physiol.* **49**, 143–169.

Watson, J. D. 1976. *Molecular Biology of the Gene.* W. A. Benjamin.

Wu, R., and E. Taylor. 1971. "Nucleotide Sequence Analysis of DNA. II. Complete Nucleotide Sequence of the Cohesive Ends of Bacteriophage λ DNA." *J. Mol. Biol.* **57**, 491–511.

Additional Readings

Foss, H. M., and F. W. Stahl. 1963. "Circularity of the Genetic Map of Bacteriophage T4." *Genetics* **48**, 1659–1672. Another demonstration of the circularity of the T4 genetic map.

NATURALLY OCCURRING SINGLE-STRANDED DNA

The Phage ϕX174 Experiments

That naturally occurring DNA is a double-stranded helix was clear from 1954 on. However, in a 1959 study of *E. coli* phage ϕX174, a phage that seemed of interest because of its extremely small size, Robert Sinsheimer made two striking observations about the phage DNA: the base composition did not reflect the base pairing rule, and the amino groups of the bases were available to react rapidly with formaldehyde; neither phenomenon applied to bacterial or mammalian DNA. These and other findings—reported by him in Article 35—led Sinsheimer to conclude that the DNA of ϕX174 is single-stranded, as was later shown to be true for several other phages.

Important findings in DNA biology in general resulted from various studies of ϕX174. For example, as was pointed out in Section VII, the discovery of the cohesive ends of λ DNA and the cyclic permutation and terminal redundancy of T4 DNA led to the idea that circular DNA molecules might be important intermediates in the life cycle of a phage. This idea was actually predated by studies of ϕX174 DNA, which was found to be not only single-stranded but also a circular DNA molecule. Furthermore the initially baffling problem of how a single-stranded DNA molecule could replicate (to produce a DNA molecule with the same—rather than the complementary—base sequence) was solved by the discovery (also in Sinsheimer's laboratory) of a double-stranded, circular, replicative intermediate. The studies of circularity are contained in Section IX; details of the biology of ϕX174 can be found in the additional readings listed for this section.

From *Journal of Molecular Biology*, Vol. 1, 43–53, 1959.
Reproduced with permission of the author and publisher.

A Single-Stranded Deoxyribonucleic Acid from Bacteriophage φX174†

ROBERT L. SINSHEIMER

Division of Biology, California Institute of Technology, Pasadena, California

(*Received 16 January 1959*)

The deoxyribonucleic acid (DNA) of bacteriophage φX174 can be extracted by phenolic denaturation of the virus protein. The DNA thus obtained has a molecular weight of $1 \cdot 7 \times 10^6$, indicating that there is one molecule per virus particle.

This φX DNA does not have a complementary nucleotide composition. The ultraviolet absorption of this DNA is strongly dependent upon temperature in the range 20° to 60°C, and upon NaCl concentration in the range 10^{-3} to 10^0 M. This DNA reacts with formaldehyde at 37°C and is precipitated by plumbous ions. This evidence is interpreted to mean that the purine and pyrimidine rings are not involved in a tightly hydrogen-bonded complementary structure.

Light scattering studies indicate that this DNA is highly flexible and that its configuration is strongly dependent upon the ionic strength of the solution. Upon treatment with pancreatic deoxyribonuclease, the weight-average molecular weight decreases in accordance with the function expected for a single-stranded molecule.

It is concluded that the DNA of bacteriophage φX174 is single-stranded.

In a previous paper (Sinsheimer, 1959) describing the properties of bacteriophage φX174, evidence was presented that the deoxyribonucleic acid (DNA) of this virus appeared to be unusual, in that the amino groups of the purine and pyrimidine rings were accessible to reaction with formaldehyde, and in that the atomic efficiency of inactivation of the virus by disintegration of ^{32}P atoms, incorporated into the viral DNA, was $1 \cdot 0$ (Tessman, 1959). In the present paper chemical and physical studies of the DNA extracted from this virus are presented. The evidence obtained indicates that each φX174 particle contains one molecule of a single-stranded DNA.

Materials and Methods

Bacteriophage φX174 was prepared and purified as previously described.

Preparation of φX DNA

To a cold suspension of φX174 containing 10 to 15 mg/ml. in borate buffer ‡ is added an equal volume of cold phenol § (Gierer & Schramm, 1956) previously equilibrated with borate buffer by shaking. The resultant emulsion is shaken for 5 min in the cold room and then spun at 1800 *g* for 5 min to separate the phases. The aqueous phase is removed, added to an equal volume of cold phenol, and the process repeated two more times. The final aqueous layer which contains the DNA is removed and allowed to stand.

---

† This research has been supported by grants from the U.S. Public Health Service, the American Cancer Society, and the California Division of the American Cancer Society.

‡ Borate buffer is a solution of sodium tetraborate, saturated at 4°C.

§ Mallinckrodt liquified phenol, AR " intended for chromatographic purposes," is used.

To improve the yield of DNA a volume of borate buffer, equal to one-half the initial volume of ϕX suspension, is added to the phenol phase from the first extraction, shaken and spun as above. The aqueous layer is removed and then used to extract in turn the DNA from the phenol layers of the second and third extractions. The final aqueous layer is pooled with the DNA solution obtained in the first series of extractions, and the combined solutions are shaken with an equal volume of cold ether. After shaking the emulsion is allowed to stand in a separatory funnel for 1 hr to allow the phases to separate. The clear aqueous layer is drawn off. The turbid ether layer is then extracted with a few ml. of borate, which, after separation of the phases, is added to the DNA solution. Two more ether extractions are employed to remove most of the residual phenol.

At this stage the DNA content of the extract accounts for 65 to 70 % of the DNA in the initial suspension of virus.

The DNA is now dialyzed against two changes of 1·3 M-potassium phosphate buffer, pH 7·5 (Kirby, 1957). There is no loss of ultraviolet absorbing material upon dialysis. To complete the removal of residual denatured protein, the DNA is extracted from the salt solution by shaking with one-half volume of either 2-methoxyethanol or N,N-dimethyl-formamide (Porter, 1955). After vigorous shaking the two phases are separated by centrifugation at 1800 g for 8 min. The upper organic layer contains all the DNA. A thin film of denatured material appears at the interface of the two phases.

The DNA solution is then dialyzed against three changes of 0·2 M-sodium chloride plus 0·001 M-phosphate buffer, pH 7·5. 20 to 30 % of the ϕX DNA dialyses from the organic phase through the membrane into the first two changes.†

Methods of purine and pyrimidine analysis

Degradation to purines and pyrimidines was carried out with either 12 N-HClO$_4$ at 100°C (Marshak & Vogel, 1951) or with glass distilled 6 N-HCl at 100°C (Hershey, Dixon & Chase, 1953). With either method, the purines and pyrimidines recovered accounted for over 90 % of the phosphorus of the DNA preparation. Phosphorus analyses were made by a modification of the method of Allen (1940). Purines and pyrimidines were separated by descending chromatography in the isopropanol-water-HCl solvent of Wyatt (1951) using S. and S. 597 paper. In two instances this separation was checked, with good agreement, by a paper electrophoretic separation of the same digest, using 20 volts/cm at pH 3·2, 0·1 M-tris formate buffer, and Whatman no. 3 paper (Nutter & Sinsheimer, 1959). In all cases the purines and pyrimidines were eluted from the ultraviolet absorbing spots in 0·1 N-HCl and the quantities determined by measurement of the ultraviolet absorption, after subtraction of the absorption of appropriate blanks. The extinction coefficients employed were: adenine, $A_M = 13,100$ at 260 mμ (Beaven, Holiday & Johnson, 1955); guanine, $A_M = 11,400$ at 248 mμ (ibid.); thymine, $A_M = 7,890$ at 264·5 mμ (Shugar & Fox, 1952); and cytosine, $A_M = 10,400$ at 275 mμ (ibid.).

Degradation to nucleotides was accomplished by successive use of pancreatic deoxyribonuclease (DNase) (Worthington Biochemical Company) and venom phosphodiesterase (Koerner & Sinsheimer, 1957). Ion exchange fractionation of the nucleotides with 98 % recovery of the ultraviolet absorption was carried out as previously described (Sinsheimer & Koerner, 1951). Extinction coefficients for the nucleotides have been previously presented (Sinsheimer, 1954).

All spectra were obtained on a Beckman DK2 spectrophotometer.

Light scattering

All light scattering measurements were made in a Brice-Phoenix light scattering photometer, model 1000 D, made by Phoenix Precision Instruments, Inc. ϕX DNA solutions were cleaned of dust by filtration through type AA Millipore filters.

† This DNA can be recovered by evaporation of the pooled dialysate and solution in 0·2 M-NaCl plus 6 M-urea plus 0·001 M-versene, pH 7·0, followed by dialysis versus 0·2 M-NaCl. This DNA which has passed through the membrane has the same physical properties and composition as the DNA which remains within. It was at first thought that the escape of the ϕX DNA was a consequence of attack upon the membrane by methoxyethanol but the same result was observed with N,N-dimethyformamide and subsequent studies have indicated that the membranes are not permanently changed. Apparently ϕX DNA, in a largely organic medium, assumes a configuration in which it can slowly pass through the cellophane membrane.

Results

Composition of ϕX DNA

The results of several analyses of the purine and pyrimidine composition of two preparations of ϕX DNA are presented in Table 1. The nucleotide analysis achieved by enzymatic degradation is considered to be the most reliable. It is clear that the molar equalities of adenine and thymine and of guanine and cytosine observed in nearly all other DNA (Chargaff, 1955) are not observed. This DNA thus cannot have the complementary structure formulated by Watson & Crick (1953a). The only regularity that can be observed is that thymine/adenine = guanine/cytosine, but the significance of this is not immediately evident.

It appears that the methods of acid hydrolysis which have proven reasonably satisfactory for conventional forms of DNA are not entirely satisfactory for this DNA. A tendency to loss of thymine in $HClO_4$ hydrolysis (Wyatt, 1952) is exaggerated, as in the loss of adenine (Hershey, Dixon & Chase, 1953) in HCl hydrolysis.

Hydrogen bonding in ϕX DNA

The extracted DNA reacts with formaldehyde (Fig. 1) as does the native virus. This reaction is given by RNA (Fraenkel-Conrat, 1954; Staehelin, 1958) and by heat denatured DNA, but not by native DNA (Hoard, 1957). It is an indication that the amino groups of the purine and pyrimidine rings are accessible and are not involved in a tightly hydrogen bonded structure.

TABLE 1

Purine and pyrimidine composition of ϕX DNA

Preparation	Mode of degradation	Molar ratios			
		Adenine	Thymine	Guanine	Cytosine
A	6 N-HCl Hydrolysis	1·00	1·31	1·06	0·82
B	6 N-HCl Hydrolysis	1·00	1·31	1·06	0·82
B	12 N-$HClO_4$ Hydrolysis	1·00	1·06	0·99	0·76
B	Enzymatic digestion to nucleotides	1·00	1·33	0·98	0·75

Upon thermal denaturation of native DNA there is a disruption of the hydrogen bonded structure with a concomitant increase of ultraviolet absorption. With native DNA this transition takes place within a narrow temperature zone in the region of 80° to 90°C (Thomas, 1954; Lawley, 1956; Shack, 1958). Over the temperature range of 20° to 70°C there is very little influence of temperature upon the absorption of native DNA.

In contrast, the ultraviolet absorption of ϕX DNA is a marked function of temperature over a very wide range (Fig. 2). A temperature dependence of ultraviolet absorption of this nature is also observed with thermally denatured DNA (Shack, 1958), although the variation observed does not appear to be as great over the temperature region 20° to 50°C (Fig. 2). This variation of absorption with temperature is presumably the result, in part, of rupture of random intramolecular hydrogen bonds and, in part, a consequence of thermal expansion of the molecule.

Similarly, the ultraviolet absorption of ϕX DNA is strongly dependent upon ionic strength at values of ionic strength in which the absorption of native DNA is unaffected (Thomas, 1954; Lawley, 1956; Shack, 1958) (Fig. 3). Again the increase of absorption at lower ionic strength is presumably the result of decreased intramolecular hydrogen bonding and molecular expansion (*vide infra*) brought about by an increase of net charge upon the DNA.

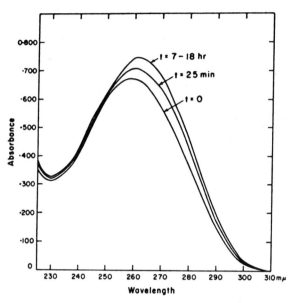

Fig. 1. Effect of formaldehyde (1·8 %) upon the ultraviolet absorption of φX DNA in 0·2 M·NaCl at 37°C. Concentration of DNA = 25·5 μg/ml.

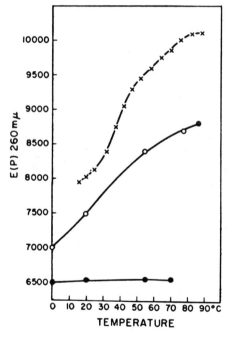

Fig. 2. Variation of the ultraviolet absorption of φX DNA at 260 mμ with temperature. *E(P)* = absorbance of a solution containing one mole of phosphorus per liter.

    x — x   DNA from φX in 0·2 M·NaCl + 10⁻³ M·phosphate buffer, pH 7·5.

    . — .   Native thymus DNA in 0·1 M·NaCl (from Shack, 1958).

    0 — 0   Heat denatured thymus DNA in 0·1 M·NaCl (from Shack, 1958).

It may be noted that $E(P)$, the absorbance per mole of phosphorus (Chargaff & Zamenhof, 1948), is 8700 for ϕX DNA in 0·2 M-NaCl at 37°C. This is considerably higher than has been observed with native or denatured DNA under comparable conditions. In accord with this observation, there is only an 11 % increase in the ultraviolet absorption of ϕX DNA upon degradation with pancreatic DNase in 0·2 M-sodium chloride at 37°C, as compared to the 30 to 35 % increase observed with native thymus DNA (Kunitz, 1950).

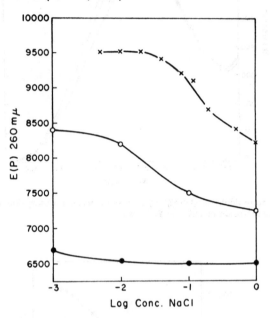

FIG. 3. Variation of the ultraviolet absorption at 260 mμ of ϕX DNA with salt concentration.

     x — x   DNA from ϕX, at 37°C.

     . — .   Native thymus DNA, at 22°C (from Shack, 1958).

     0 — 0   Heat denatured thymus DNA, at 22°C (from Shack, 1958).

Stevens & Duggan (1957) showed that thermally denatured thymus DNA is precipitated by Pb^{++} ion under conditions such that native DNA remained soluble. Similarly, ϕX DNA is completely precipitated from a solution in 0·1 M sodium chloride containing 165 μg per ml. by addition of Pb^{++} ions to 0·017 M.

Light scattering observations

The particle weight of ϕX DNA has been measured by light scattering † (Figs. 4 and 5). Several preparations have been used under various ionic conditions and the weights obtained have always been within the range of 1·6 to 1·8 × 10^6. This is in good agreement with the weight of DNA per ϕX particle calculated from the particle weight, 6·2 × 10^6, and the DNA content, 25·5 %.‡ Hence there is one molecule of DNA per ϕX particle.

   † In this calculation, the refractive increment of the DNA has been assumed to be 0·201 cm^3/g at 4358 Å (Northrop, Nutter & Sinsheimer, 1953), although this value may not be quite correct for a DNA of this type.

   ‡ The DNA content of ϕX was calculated from colorimetric analysis by the diphenylamine and p-nitrophenylhydrazine methods. In ordinary DNA these methods are believed to measure purine deoxyribosides (Dische, 1955). If this is still true with ϕX DNA, the total DNA content of ϕX will be slightly greater than 25·5 %, because the purine deoxyribosides comprise less than 50 % of the total DNA.

In 0·02 M-NaCl at 37°C the ϕX DNA has a radius of gyration of 1140 Å. Upon addition of salt to 0·2 M, this decreases to 440 Å (Fig. 6), while in 0·02 M-NaCl plus 4×10^{-3} M-Mg^{++} the radius of gyration is 325 Å.† ϕX DNA is thus a highly flexible

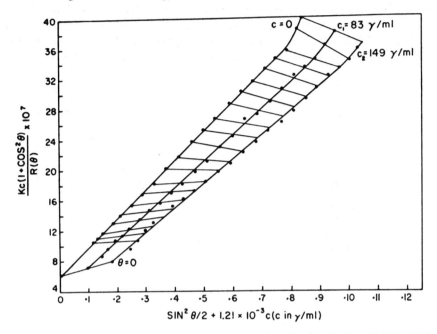

FIG. 4. Light scattering of ϕX DNA in 0·02 M-NaCl + 10^{-3} M-phosphate buffer, pH 7·5 at 37°C.

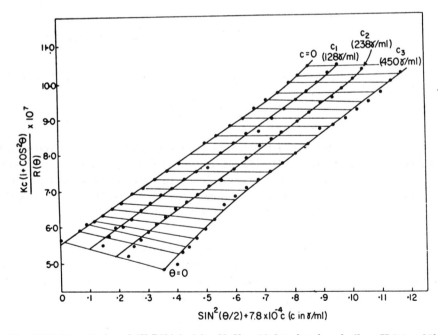

FIG. 5. Light scattering of ϕX DNA in 0·2 M-NaCl + 10^{-3} M-phosphate buffer, pH 7·5 at 37°C.

† At the same time, the ultraviolet absorption decreases by 14 % upon addition of Mg^{++}, and the sedimentation velocity increases (*vide infra*).

molecule. With native DNA, almost no change of radius of gyration is observed with change of ionic strength (Ehrlich & Doty, 1958). With alkali denatured DNA, a considerable variation is observed, nearly equal to that of ϕX DNA.

In 0·2 M-NaCl, the scattering envelope agrees well with that of a random coil (Doty & Steiner, 1950). For a random coil the radius of gyration should vary as $M^{0.5}$ (Tompa, 1956). For ϕX DNA in 0·2 M-sodium chloride, the ratio of $\sqrt{M}/\rho_g$ (M is the molecular weight in molecular weight units, ρ_g is the radius of gyration in Å) is 3·0. This may be

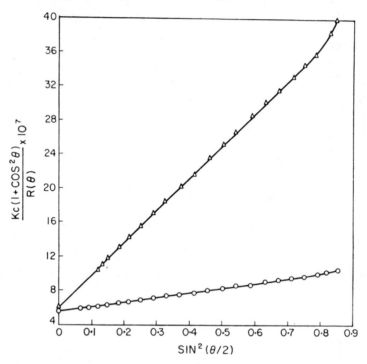

Fig. 6. Comparison of the scattering envelopes (extrapolated to zero concentration) of ϕX DNA in 0·02 M and 0·2 M-NaCl.

$$\triangle \text{——} \triangle \quad \text{in } 0\cdot02 \text{ M-NaCl.}$$
$$0 \text{——} 0 \quad \text{in } 0\cdot2 \text{ M-NaCl.}$$

compared with the ratio of 1·1 observed for native and sonicated thymus DNA (Reichmann, Rice, Thomas & Doty, 1954; Doty, McGill & Rice, 1958) and with a ratio of 2·5 to 2·6 observed with heat, acid, or alkali denatured thymus DNA in 0·2 M-sodium chloride (Rice & Doty, 1957; Thomas & Doty, 1956; Ehrlich & Doty, 1958). ϕX DNA is evidently more flexible than either native DNA or denatured DNA of thymus.

Kinetics of degradation by pancreatic deoxyribonuclease

As shown by Charlesby (1954) (also Tobolsky, 1957) the decline of the weight average molecular weight of a monodisperse single chain polymer upon random degration of the inter-monomeric links will be described by

$$\frac{M(p)}{M(O)} = \frac{2}{\gamma^2}[e^{-\gamma} + \gamma - 1]$$

where $M(p)$ is the weight average molecular weight when the probability of rupture of any link is p, $M(O)$ is the initial molecular weight and γ is proportional to p.

D

For a double chain structure, with cross linkage between the chains, the same equation applies except that γ is proportional to p^2, as shown by Thomas (1956).

Thomas and also Schumaker, Richards & Schachman (1956) have shown that the rate of splitting of bonds by pancreatic DNase is initially uniform with time ($p = kt$). Therefore, measurements were made of the decline with time (and thus with p) of the

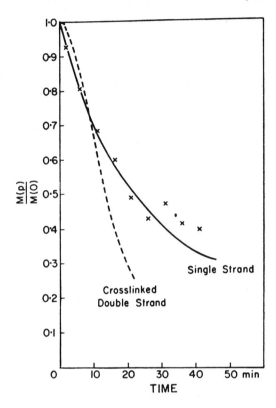

FIG. 7. Decline of weight-average molecular weight, as measured by light scattering, of ϕX DNA acted upon by pancreatic deoxyribonuclease. At $t = 0$, DNase in a final concentration of 3.7×10^{-7} μg/ml. was added to a solution containing 60 μg/ml. of ϕX DNA in 0.2 M-NaCl + 0.001 M-phosphate buffer, pH 7.5 + 0.02 M-magnesium acetate. The temperature was maintained at 34°C.

—— Function expected for a single-stranded molecule.

· · · · Function expected for a cross-linked double-stranded molecule.

xxxx Experimental observations.

The theoretical functions were fitted to intersect a smooth curve representing the experimental data at one point, at $\dfrac{M(p)}{M(O)} = 0.74$.

weight average molecular weight of ϕX DNA under attack by DNase. The form of the curve thus obtained (Fig. 7) agrees well with that expected for a single-stranded molecule and is clearly distinct from that expected for a two-stranded molecule with linked strands, such as native DNA.†

† On some occasions, with considerably lower concentrations of DNase, a slight curvature (concave downwards) has been observed in the initial stages of the digestion. This curvature, which was evident as an increase in slope to a value greater than the always finite initial slope, was not reproducible and has never been observed at the enzyme concentration employed in Fig. 7.

Ultracentrifuge studies

When banded by the cesium chloride density gradient equilibrium method of Meselson, Stahl & Vinograd (1957), φX DNA appears at a density of 1·72. This is distinctly higher than the density of native DNA (1·70) (Plate I (a)) and is very closely the density at which heat denatured salmon sperm DNA is observed to band (Meselson & Stahl, 1958).

In velocity centrifugation in 0·2 M-NaCl, φX DNA sediments as a single, rather sharp boundary, of $S = 23·8$ (Plate I (b)). At lower ionic strength, the sedimentation rate decreases, and below 0·08 M-NaCl the boundary splits in two indicating the presence of two centrifugal components (Plate I (c)).

A similar splitting of the boundary can be observed in 0·2 M-sodium chloride after treatment with formaldehyde (Plate I (d)). The sedimentation constants of the two components are 15·8 and 13·9 (measured in the presence of formaldehyde). The

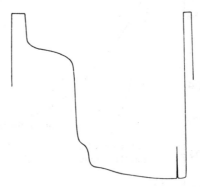

Fig. 8. Densitometer tracing of frame 11 of Plate I (d) indicating the relative proportions of the two centrifugal components.

relative proportions of the two components observed either in 0·04 M-sodium chloride or after formaldehyde treatment are the same for a given φX DNA preparation, although variations are observed in different preparations. The boundaries obtained after formaldehyde treatment are very sharp. A densitometer tracing is shown in Fig. 8.

The splitting of the boundary observed in dilute salt is completely reversible; upon addition of salt or Mg^{++} a single boundary is again observed.

Discussion

The chemical reactivity of φX DNA, its ultraviolet absorption properties, its density, and its flexibility as demonstrated by light scattering, all serve to distinguish the DNA of φX174 from conventional DNA. In several ways, such as the dependence of ultraviolet absorption upon temperature and ionic strength, and the dependence of the radius of gyration upon ionic strength, the properties of φX DNA appear to be an exaggeration of the properties of a denatured DNA, as contrasted to those of native DNA. The nucleotide composition rules out the possibility of a complementary structure, native or denatured. The simplest explanation of the properties observed is that the DNA of φX174 is a single-stranded structure. Such an explanation is also at least consistent with the high efficiency of inactivation of the φX virus by the decay of incorporated [32]P.

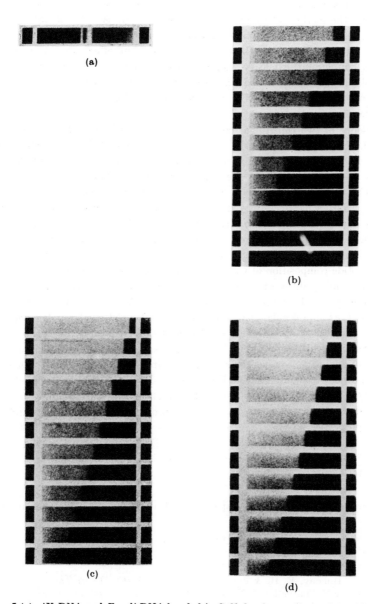

(a)

(b)

(c)

(d)

PLATE I (a). φX DNA and *E. coli* DNA banded in CsCl density gradient, after 20 hr at **44,770** rev/min. Density increasing to the right.

PLATE I (b). Ultracentrifuge sedimentation pattern of φX DNA (17 μg/ml.) in 0·2 M-NaCl at 20° C. Pictures taken every **4** min at 56,100 rev/min.

PLATE I (c). Ultracentrifuge sedimentation pattern of φX DNA (23 μg/ml.) in 0·04 M-NaCl at 20° C. Pictures taken every **4** min at 56,100 rev/min.

PLATE I (d). Ultracentrifuge sedimentation pattern at 20° C of φX DNA (32 μg/ml.) after **18 hr at 37°** C in 1·8 % formaldehyde in 0·2 M-NaCl. Pictures taken every 4 min at 56,100 rev/min.

In order to test the possibility that some of the properties of φX DNA were artifacts produced by the method of extraction and purification, samples of native thymus and of native *E. coli* DNA were taken through the entire extraction and purification procedures. It was shown that these procedures have no effect upon the density of these DNA samples in the density gradient equilibrium method, or upon the increase in ultraviolet absorption observed when they were degraded with deoxyribonuclease. In addition the reactivity of φX DNA with formaldehyde before extraction from the virus may be cited.

The presence of two centrifugal components in the φX DNA preparation is not understood. They appear when the molecule is caused to take an extended form, either by destruction of intramolecular hydrogen bonds by formaldehyde or by lowering of the ionic strength. At one time it was thought that these might represent complementary strands of a DNA which had been separated biologically before incorporation into separate virus particles. However, it has been observed that the proportions of the two components varied from 62/38 to 79/21 in different preparations. Despite these variations, the preparations had identical nucleotide composition. This observation would appear to rule out the hypothesis of complementary chains.

The nucleotide composition of this DNA fulfils neither the adenine, thymine and cytosine, guanine equalities of ordinary DNA, nor the guanine plus uracil (thymine) equals adenine plus cytosine relation proposed for RNA by Elson & Chargaff (1954). In this respect, as in molecular size, this DNA is similar to the ribonucleic acid of the small plant viruses. It is, however, a DNA and differs notably from the RNA of TMV, for instance, with respect to stability and to several physical properties.

DNA fractions with compositions deviating from complementarity have been reported on other occasions (Lucy & Butler, 1954; Bendich, Pahl & Beiser, 1956). It might be very worth while to examine such fractions for the possible presence of single-stranded DNA components, by comparison of their properties with any of the distinctive properties of φX DNA. The existence of a single-stranded DNA would seem to imply the existence of a distinct mode of DNA replication (Watson & Crick, 1953b; Meselson & Stahl, 1958). It would be surprising if this were limited to a special class of bacterial viruses.

It is a pleasure to acknowledge the capable technical assistance of Miss Sharon Palmer.

REFERENCES

Allen, R. J. L. (1940). *Biochem. J.* **34**, 858.

Beaven, G. H., Holiday, E. R. & Johnson, E. A. (1955). Ch. 14 in *The Nucleic Acids*, Vol. I, ed. by E. Chargaff & J. N. Davidson. New York: Academic Press.

Bendich, A., Pahl, H. B. & Beiser, S. M. (1956). *Cold Spr. Harb. Symp. Quant. Biol.* **21**, 31.

Chargaff, E. (1955). Ch. 10 in *The Nucleic Acids*, Vol. I, ed. by E. Chargaff & J. N. Davidson. New York: Academic Press.

Chargaff, E. & Zamenhof, S. (1948). *J. Biol. Chem.* **173**, 327.

Charlesby, A. (1954). *Proc. Roy. Soc.* A **224**, 120.

Dische, Z. (1955). Ch. 9 in *The Nucleic Acids*, Vol. I, ed. by E. Chargaff & J. N. Davidson. New York: Academic Press.

Doty, P., McGill, B. B. & Rice, S. A. (1958). *Proc. Nat. Acad. Sci., Wash.* **44**, 432.

Doty, P. & Steiner, R. F. (1950). *J. Chem. Phys.* **18**, 1211.

Ehrlich, P. & Doty, P. (1958). *J. Amer. Chem. Soc.* **80**, 4251.

Elson, D. & Chargaff, E. (1954). *Nature*, **173**, 1037.

Fraenkel-Conrat, H. (1954). *Biochim. biophys. Acta*, **15**, 307.

Gierer, A. & Schramm, G. (1956). *Z. Naturf.* **11b**, 138.

Hershey, A. D., Dixon, J. & Chase, M. (1953). *J. Gen. Physiol.* **36**, 777.

Hoard, D. E. (1957). Ph.D. Thesis, University of California, Berkeley.

Kirby, K. S. (1957). *Biochem. J.* 495.

Koerner, J. F. & Sinsheimer, R. L. (1957). *J. Biol. Chem.* 1049.

Kunitz, M. (1950). *J. Gen. Physiol.* **33**, 349.

Lawley, P. D. (1956). *Biochim. biophys. Acta,* **21**, 481.

Lucy, J. A. & Butler, J. V. (1954). *Nature,* **174**, 32.

Marshak, A. & Vogel, H. J. (1951). *J. Biol. Chem.* **189**, 597.

Meselson, M. & Stahl, F. W. (1958). *Proc. Nat. Acad. Sci. Wash.* **44**, 671.

Meselson, M., Stahl, F. W. & Vinograd, J. (1957). *Proc. Nat. Acad. Sci., Wash.* **43**, 581.

Northrop, T. G., Nutter, R. L. & Sinsheimer, R. L. (1953). *J. Amer. Chem. Soc.* **75**, 5134.

Nutter, R. L. & Sinsheimer, R. L. (1959). *Virology,* in press.

Porter, R. R. (1955). pp. 98-112 in *Methods in Enzymology,* Vol. I, ed. by S. P. Colowick & N. O. Kaplan. New York: Academic Press.

Reichmann, M. E., Rice, S. A., Thomas, C. A. & Doty, P. (1954). *J. Amer. Chem. Soc.* **76**, 3047.

Rice, S. A. & Doty, P. (1957). *J. Amer. Chem. Soc.* **79**, 3937.

Schumaker, V. N., Richards, E. G. & Schachman, H. K. (1956). *J. Amer. Chem. Soc.* **78**, 4230.

Shack, J. (1958). *J. Biol. Chem.* **233**, 677.

Shugar, D. & Fox, J. J. (1952). *Biochim. biophys. Acta,* **9**, 199.

Sinsheimer, R. L. (1954). *J. Biol. Chem.* **208**, 445.

Sinsheimer, R. L. (1959). *J. Mol. Biol.* **1**, 37.

Sinsheimer, R. L. & Koerner, J. F. (1951). *Science,* **114**, 42.

Staehelin, M. (1958). *Biochim. biophys. Acta,* **29**, 410.

Stevens, V. L. & Duggan, E. L. (1957). *J. Amer. Chem. Soc.* **79**, 5703.

Tessman, I. (1958). Personal communication.

Thomas, Jr., C. A. (1956). *J. Amer. Chem. Soc.* **78**, 1861.

Thomas, Jr., C. A. & Doty, P. (1956). *J. Amer. Chem. Soc.* **78**, 1854.

Thomas, R. (1954). *Biochim. biophys. Acta,* **14**, 231.

Tobolosky, A. (1957). *J. Polym. Sci.* **26**, 247.

Tompa, H. (1956). Polymer Solutions. London: Butterworths Scientific Publications.

Watson, J. D. & Crick, F. H. C. (1953*a*). *Nature,* **171**, 737.

Watson, J. D. & Crick, F. H. C. (1953*b*). *Nature,* **171**, 964.

Wyatt, G. R. (1951). *Biochem. J.* **48**, 584.

Wyatt, G. R. (1952). *J. Gen Physiol.* **36**, 201.

Questions

ARTICLE 35 Sinsheimer

1. Explain how the following findings support the view that the DNA is single-stranded.
    (a) increasing ultraviolet absorption with decreasing ionic strength;
    (b) high value of $E(P)$;
    (c) small increase in absorbance when the DNA is heated;
    •(d) change in radius of gyration with ionic strength;
    •(e) kinetics of degradation by DNase.

2. At low ionic strength the sedimentation boundary splits. We now know that ϕX174 DNA is a single-stranded circle. Explain what the two boundaries are. Why is there a single boundary at high ionic strength? Why does the proportion of the two components vary in different DNA preparations?

Additional Readings

Kornberg, A. 1974. *DNA Synthesis.* W. H. Freeman and Company. This is the most complete description of the replication of ϕX174 DNA.

Sinsheimer, R. L. 1962. "Bacteriophage ϕX174 and Related Viruses." *Prog. Nucleic Acid Res.* 8, 115–170. An early review.

Watson, J. D. 1976. *Molecular Biology of the Gene.* W. A. Benjamin. This contains an excellent description of the biology of ϕX174.

CIRCULAR AND SUPERCOILED DNA MOLECULES

The Discovery of Circular DNA

In Section VII we discussed the finding that the genetic map of *E. coli* phage T4 is circular. This was actually not the first time that a circular map had been observed. Several years before, various groups (principally François Jacob and Elie Wollman at the Institut Pasteur in Paris, and Ed Adelberg, then at Berkeley) had proposed that the genetic map of *E. coli* was circular. The evidence in support of this suggestion is the following.

Genetic Evidence for a Circular *E. coli* Chromosome

Some strains of *E. coli* are capable of transferring genes to other strains: the donors are called male and the recipient strains female. Some male strains transfer many genes with high efficiency and are called "*high frequency of recombination*," or Hfr, strains. When an Hfr male is mixed with a female, DNA is transferred from male to female (Fig. IX-1). By interrupting this process at various times and determining by genetic tests which genes have been transferred as a function of time, a gene order (actually a temporal map) can be obtained, e.g., ABC ··· XYZ (Fig. IX-2). Each Hfr strain starts transferring from a single point in the DNA, but this point varies from one strain to another. When maps from many different Hfr strains are compared, it is found that all are cyclic permutations of one another—i.e., one map may be ABC ··· XYZ, another WXYZABC ··· TUV, and so forth. Hence the map generated from combining those of many Hfr strains is circular (Fig. IX-2).

If a genetic map is obtained for various female strains by linkage analysis (i.e., by determining which genes are adjacent), it is also found that A is near B, B is near C, and so forth, and that Z is near A. Furthermore, when linkage was determined in an Hfr strain by P1 transduction, a system by which a small con-

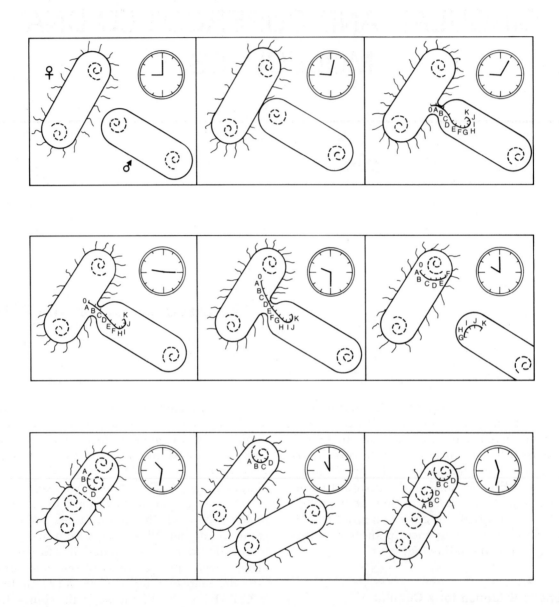

Figure IX-1
Diagrammatic representation of conjugation between Hfr (♂) and F-(♀) bacteria. The female recipient in this diagram is represented by the hairlike extensions. Conjugation begins shortly (at upper right) after the bacteria are mixed, and is soon followed by transfer of DNA. The letters represent the location of various genes along the donor chromosome. The clock shows the amount of time it takes for this DNA transfer to occur. [From F. Jacob and E. Wollman, *Sexuality and Genetics of Bacteria*, Academic Press, 1961.]

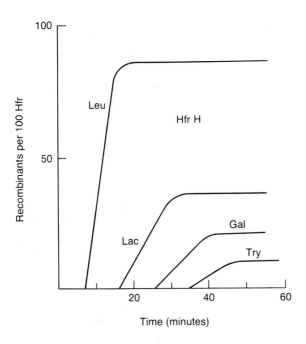

a. Time of entry experiment

Types of Hfr									
H	O	Leu	Lac	Try	His	Gly	Mal	Iso	Thi
2	O	Leu	Thi	Iso	Mal	Gly	His	Try	Lac
5	O	Thi	Leu	Lac	Try	His	Gly	Mal	Iso
AB311	O	His	Try	Lac	Leu	Thi	Iso	Mal	Gly
AB312	O	Mal	Iso	Thi	Leu	Lac	Try	His	Gly

b. Order of transfer of genes

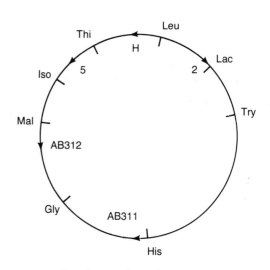

c. Genetic map of *E. coli* derived from **b**

Figure IX-2

The ordering of genes in *E. coli*. (a) A time-of-entry experiment. A donor cell (Hfr) whose genotype is Leu⁺ Lac⁺ Gal⁺ Try⁺ is mated with a recipient that is Leu⁻ Lac⁻ Gal⁻ Try⁻. At various times after mating, the male cells are destroyed and the presence of a transferred + gene is determined by genetic tests. The frequency of gene transfer is plotted as recombinants (i.e., females containing the indicated + gene) per 100 males. The intersection with the abscissa marks the time of first entry of the particular gene into the female. The genes can then be ordered according to their times of entry. (b) The order of transfer of genes for various Hfr strains. The symbol O refers to the origin of transfer. (c) The orders of (b) plotted on a single line. Examination of part (b) shows that the line must be a circle. This is called a genetic map.

tinuous fragment of bacterial DNA is packaged in a P1 phage coat (Fig. IX-3), it was found by Sanne DeWitt and Adelberg (1962) that the F factor (which in an Hfr mating is always transferred to the female last) is linked both to genes transferring soon after mating begins and to those transferring very late. As we saw earlier for T4 DNA, a circular map is not necessarily indicative of a circular DNA molecule. Indeed, on many occasions geneticists pointed out that there were several ways to generate a circular map from a linear chromosome. However, in *E. coli*, the DNA is in fact circular. A demonstration of this was long in coming because of the extraordinary difficulty in isolating such a large DNA molecule (the molecular weight was about 2.6×10^9 in comparison with 1.1×10^8 for T4 DNA). It was John Cairns who conceived of a workable scheme to avoid breaking this huge molecule and obtained the beautiful autoradiograms seen in Article 6 (Section III).

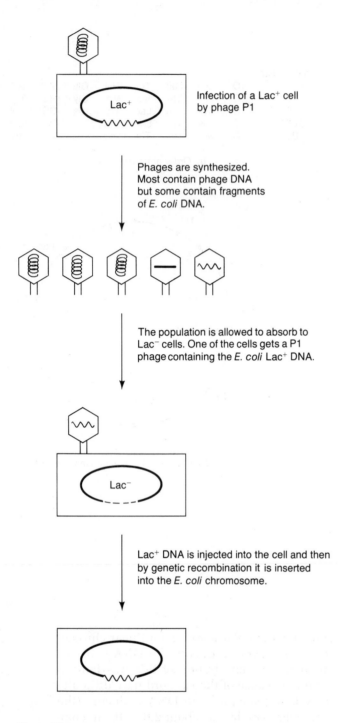

Infection of a Lac⁺ cell by phage P1

Phages are synthesized. Most contain phage DNA but some contain fragments of *E. coli* DNA.

The population is allowed to absorb to Lac⁻ cells. One of the cells gets a P1 phage containing the *E. coli* Lac⁺ DNA.

Lac⁺ DNA is injected into the cell and then by genetic recombination it is inserted into the *E. coli* chromosome.

Figure IX-3
A diagram of transduction of *E. coli* by phage Pl. A piece of Lac⁺ DNA is packaged into a Pl phage coat and transferred to a Lac⁻ strain of *E. coli*. It is incorporated into the recipient chromosome, thus converting the Lac⁻ cell to Lac⁺.

φX174 DNA is Circular

Cairns was actually not the first person to demonstrate circularity. Rather circularity was another of the surprises revealed by the tiny phage φX174 whose single-stranded DNA is also circular. In Articles 36 and 37, on the structure of the DNA of Bacteriophage φX174, Walter Fiers and Robert Sinsheimer describe the various tools they used to show that φX174 DNA has no free ends. First, their enzymatic studies with exonucleases and endonucleases* showed that no ends are available for reaction. This finding itself did not prove circularity since each end could be linked to some element that could protect it from enzymatic attack. However, their sedimentation studies provided the remaining necessary information by showing that endonucleolytic cleavage does not yield two units, but only one; hence if there were a terminal blocking element, both ends would have to be linked to one such element, and the same one. It is clear that a circular DNA is a more economical hypothesis.

Circular DNA Seen by Electron Microscopy

After the work described in Articles 36 and 37 by Fiers and Sinsheimer, an attempt at an electron microscopic demonstration of circularity was inevitable. Albrecht Kleinschmidt, who had developed the technique for visualizing DNA that was described earlier (page 82), had not yet succeeded in visualizing single-stranded DNA. However, in Sinsheimer's laboratory it had been shown also that the single-stranded φX174 DNA was converted to a double-stranded replicative form shortly after infection, and there was reason to believe that it too was circular. Therefore, as a start, Kleinschmidt looked at this replicative form and in fact observed that it was circular. This result was reported by him, Alice Burton, and Robert Sinsheimer in Article 38.

Kleinschmidt, continuing his efforts to see single-stranded DNA, tried various shadowing methods using boiled DNA but succeeded only in seeing tiny collapsed "puddles" of

---

*An exonuclease can remove nucleotides one at a time only from a terminus of a polynucleotide. An endonuclease can cut internal phosphodiester bonds also.

DNA. I hypothesized that the puddles were the result of random reformation of hydrogen bonds, as discussed in Section V. Having studied formaldehyde stabilization of single-stranded DNA, I suggested that the DNA be heated in the presence of formaldehyde. This idea was frankly a naive one, since I had not considered that formaldehyde might react with the cytochrome C used in the sample preparation for microscopy. Luckily it did not, and the single-stranded circles of φX174 DNA were seen. These rings are described in more detail in Article 39 by Kleinschmidt, Sinsheimer, and me.

Circles Are Twisted

For several years after these micrographs of circular φX174 DNA were obtained, it was assumed that except for φX174 DNA all DNA contained in a phage particle was linear. In fact more than five years passed before the next circular phage DNA was found (in the phage PM-2 whose host is a marine Pseudomonad). It was also assumed that the DNA of plant and animal viruses would contain linear DNA. However, as various laboratories turned to detailed study of the DNA molecules of avian and mammalian viruses, it became apparent that these molecules behaved in surprising ways. In an unnoticed report Waclaw Szybalski and his students (1963) observed that fowlpox viral DNA could not be irreversibly denatured. He suggested that the DNA might contain cross-links between the strands, thus preventing complete strand separation and providing a point of register for renaturation as we saw in Peter Geiduschek's study of reversible DNA (Article 21).

In another unnoticed paper James Watson and John Littlefield observed that the DNA isolated from rabbit papilloma virus sedimented as two components. In the next year three papers appeared that surprised almost everyone. Renato Dulbecco and Marguerite Vogt, then at the California Institute of Technology, observed that when the DNA isolated from polyoma virus was sedimented, it consisted of two components. They eliminated the possibility that the faster (F) component is an end-to-end dimer of the slower (S) component by showing that when chromatographed on MAK, F eluted at a lower salt concentration than S, contrary to what would be expected if F were a dimer. They proposed that F was a circle, but they felt sure that the distinction between F and S could not be simply that the one was a circular form and the other a linear form for three reasons: (1) the ratio of the sedimentation coefficient of F to that of S was greater than for the circular and linear forms of λ as reported by Hershey, Burgi, and Ingraham in Article 26; (2) F sedimented 2.5 times as rapidly at alkaline pH than at neutral pH, whereas 1.3 would be the expected ratio if F were a circle; (3) F was converted to S by one single-strand break produced by DNase treatment.

At the same time Roger Weil (1963) in Jerome Vinograd's laboratory at Cal Tech was studying the polyoma DNA isolated from infected mouse cells. Polyoma viral DNA can productively infect mouse cells, thus producing virus particles.* He found that when this intracellular DNA was heated to 95°C and rapidly cooled, the infectivity was not lost. This differed from the work of Marmur and Lane (Article 17), in which pneumococcal transforming DNA was inactivated about 100-fold by a similar treatment. Similarly the infectivity was not lost when the DNA was made alkaline and then neutralized. Furthermore, when the heated and rapidly cooled DNA was banded in CsCl, two bands were seen at the densities of native and denatured DNA. Weil drew an analogy between this DNA and the reversible DNA studied by Geiduschek (Article 21) since it was clear that a major fraction of the intracellular polyoma DNA was not irreversibly denaturable. (Note that Weil in fact rediscovered the phenomenon observed by Szybalski with fowlpox viral DNA).

Although Weil did not at first correctly understand the structure of the intracellular viral DNA, his work supported the idea that the F form of the viral DNA observed by Dulbecco and Vogt must have a special structure. The

---

*This process is similar to bacterial transformation in that the cells take up free DNA. Transformation however refers to the production of a new genotype, usually by genetic recombination. But in this process of infection with viral DNA the cell is killed and viruses are produced. There is no general term for this process; when phage DNA is used to infect a bacterium, the word *transfection* is used to describe it.

nature of this structure was elucidated in the paper by Weil and Vinograd demonstrating that the fast form is a *covalently closed* circle whose strands cannot be separated because both are closed circles intertwined with one another.* This explained the reversible denaturation but not the high sedimentation coefficient observed by Dulbecco and Vogt. Later in the year Weil and Vinograd demonstrated by both sedimentation data and electron microscopy that some of the DNA is circular, as suggested by Dulbecco and Vogt. However, several questions still remained, mainly how to explain the particular values of the sedimentation coefficient and the conditions that convert one form to another. The answer was actually present in their electron micrographs, one of which showed a new structure, the *twisted circle*. Walter Stoeckenius noticed this structure, but because electron microscopists at the time suspected that many of the odd structures that had been observed for years were the results of an uncontrolled and unknown type of protein-DNA interaction, he failed to conclude that this was a new structure and suggested instead that circular molecules were more prone to this uncontrolled protein-mediated twisting. The final solution came two years later from Vinograd's laboratory, where it was realized that the double-stranded, uninterrupted DNA is twisted or supercoiled and that an untwisted circle can be formed from a twisted one by the introduction of only one single-strand break. It is important at this point to understand the topology of supercoiling. If the two ends of a linear DNA molecule are attached, a covalent circle results. However, if one end is rotated 360° in either direction from the other before joining, the resulting circle will twist in the opposite direction to generate a figure-8. For *each* 360° rotation the circle will receive one twist in the opposite direction to relieve strain in the DNA. Hence supercoiling can be thought of as a response of the circle to either underwinding or overwinding of the double helix.*

Vinograd's laboratory then became the center for the study of supercoiled DNA molecules and, as part of this effort, researchers there developed an elegant method for detecting and identifying supercoiled DNA, utilizing the fact that, in the presence of the dye ethidium bromide, the density of supercoiled DNA differs from that of linear and untwisted DNA; this density could be detected by centrifuging the DNA to equilibrium in solutions of buoyant CsCl containing ethidium bromide. Their work is described in Article 40 by R. Dulbecco and M. Vogt, Article 41 by Weil and Vinograd, Article 42 by Vinograd and others, and Article 43 by Roger Radloff, William Bauer, and Vinograd.

The ethidium bromide-CsCl method made it possible to determine whether supercoiled DNA exists in other systems (Fig. IX-4). Vinograd had also provided earlier another method for detecting circular DNA that had no interruptions (i.e., covalent circles): it was sedimentation in alkali, a method based on the fact that the covalent circle has a much higher sedimentation coefficient in alkali than does an open circle, i.e., one with an interruption (Fig. IX-5). A third method, sedimentation at neutral pH in the presence of ethidium bromide, was developed several years later by Lionel Crawford and is based on three facts: supercoiled DNA sediments more rapidly than an untwisted covalent circle, naturally occurring supercoils contain negative twists (i.e., those which would be produced by underwinding of the helix before end joining), and ethidium bromide introduces positive twists. Therefore, as increasing amounts of ethidium bromide are added to supercoiled DNA, the negative twists are gradually eliminated until

---

*The terminology used to describe circles has been widely confused. The word *circle* is used to indicate a molecule lacking free ends and one for which no other properties are known. If the circle has one or more single-strand breaks, it is an *open circle*. An open circle can never be a *twisted circle*. If the circle contains no single-strand interruptions the terms *closed circle*, *covalently closed circle*, or *covalent circle* are used interchangeably. Such a circle need not be twisted but, because naturally occurring closed circles are always twisted, these terms are frequently used to mean a twisted circle. If the circle is twisted, the terms *supercoiled circle*, *supercoil*, *twisted circle*, *superhelix*, and *superhelical circle* are used interchangeably. One also sees such phrases as, "The circle has superhelical twists."

---

*During the final stages of this book's publication, a novel *E. coli* enzyme, DNA gyrase, was discovered by Martin Gellert and Howard Nash. This enzyme can convert an untwisted, covalently closed circle to a twisted circle. Thus, whereas it is possible in the laboratory to produce a twisted circle by joining the ends of a linear molecule, it is probable that intracellular twisting occurs by the action of enzymes like DNA gyrase.

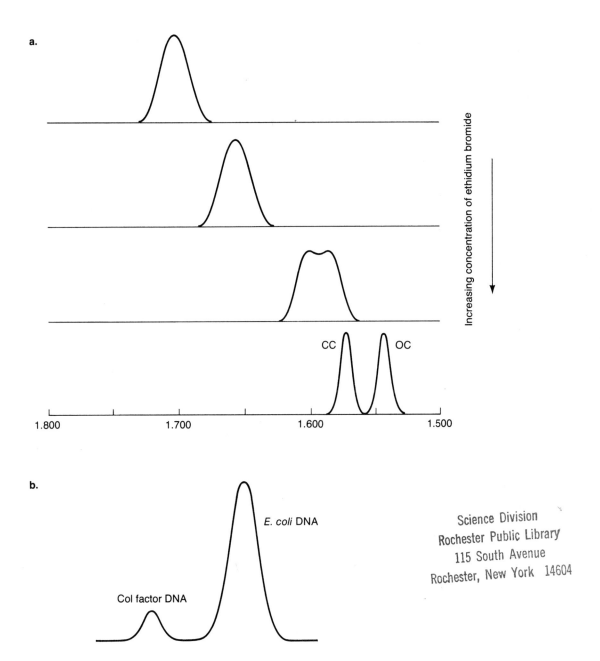

Figure IX-4
Effect of ethidium bromide on the density of DNA in CsCl. In part (a), a mixture of equal amounts of open circles (OC) and covalent circles (CC) is centrifuged in CsCl containing various amounts of ethidium bromide. The density decreases until, at saturation, the two components separate. The covalent circles bind less ethidium bromide and are therefore at a higher density. Part (b) shows a photometric tracing of the DNA extracted from an *E. coli* strain containing a colicinogenic factor, centrifuged in CsCl containing ethidium bromide. [From *Physical Biochemistry* by D. Freifelder. W. H. Freeman and Company. Copyright © 1976.]

Figure IX-5

Detection of covalent circles of *E. coli* F factor DNA. Bacteria were grown in ³H-thymidine, lysed, and sedimented through an alkaline sucrose gradient. The F supercoils comprise approximately 1% of the total DNA but are easily resolved because of their high *s* in alkali. [From *Physical Biochemistry* by D. Freifelder. W. H. Freeman and Company. Copyright © 1976.]

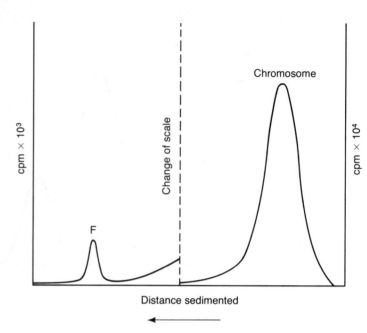

Figure IX-6

(a) Effect of ethidium bromide on the sedimentation coefficient of polyoma virus DNA. Samples of polyoma virus DNA, containing both fast (supercoiled) and slow (open-circle) forms of the DNA, were exposed to a range of concentrations of ethidium bromide. Each sample contained 9 μg of DNA, plus ethidium bromide to give the binding ratio shown. ○—Supercoiled molecules; △—open-circle (unsupercoiled) molecules; ●—molecules that formed only a single boundary.

(b) Diagrammatic representation (not to scale) of the removal and reversal of supercoiling turns. The Watson-Crick helix is represented as a single continuous line. The number of supercoil turns in the original molecule (A) decreases as ethidium bromide molecules (represented by bars perpendicular to the helix axis) bind to the DNA (B). The ethidium bromide molecules are inserted at random. At equivalence, the accumulated untwisting due to the number of ethidium bromide molecules bound exactly balances the initial number of supercoiling turns (C). The untwisting caused by the binding of more ethidium bromide molecules produces tertiary twists, this time in the opposite direction (D) and (E). [Redrawn from L. V. Crawford and M. J. Waring, *J. Mol. Biol.* **25** (1967), 23–30.]

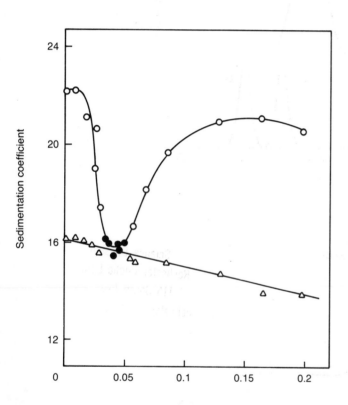

a.

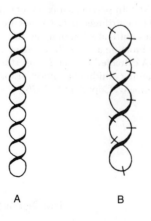

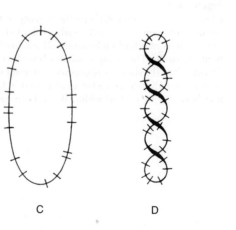

b.

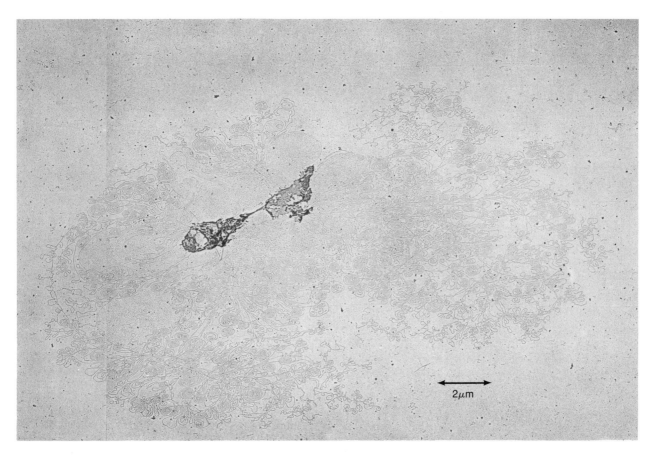

Figure IX-7
An electron micrograph of the chromosome of *E. coli* showing supercoiling. The dark structure is a portion of the cell membrane. [From H. Delius and A. Worcel, *J. Mol. Biol.* **82** (1974), 108–109.]

the circle is untwisted. However, since more dye can still be bound at this point, the circle becomes twisted again but in the opposite direction of the original DNA. Therefore, a test of supercoiling is a decrease in sedimentation coefficient followed by an increase, as ethidium bromide is added (Fig. IX-6).

These three methods were widely used and supercoiled DNA was found in other viruses, in some phages, in phage-infected bacteria and viral-infected eukaryotic cells, in bacteria containing plasmids, and in cell organelles such as mitochondria and chloroplasts.

Some years later Gordon Stonington and David Pettijohn developed a method for isolating the intact *E. coli* chromosome—in much larger amounts than yielded by the procedures of Cairns (Article 6) and of Davern (Article 7). Abraham Worcel and Elizabeth Burgi used this procedure to demonstrate that even *E. coli* DNA is supercoiled. These results are shown in the electron micrograph of Fig. IX-7, and Worcel and Burgi describe their findings in more detail in Article 44.

From *Journal of Molecular Biology*, Vol. 5, 408–419, 1962. Reproduced with permission of the authors and publisher.

The Structure of the DNA of Bacteriophage φX174

I. The Action of Exopolynucleotidases

Walter Fiers and Robert L. Sinsheimer

California Institute of Technology, Division of Biology, Pasadena, California, U.S.A.

(*Received 20 June 1962*)

A φX-DNA preparation cannot be inactivated, and can be hydrolysed only to a small extent, by *E. coli* phosphodiesterase. This result is a consequence of the unavailability of a free 3′-OH terminus. Studies of the action of spleen phosphodiesterase upon φX DNA similarly indicate the absence of a free 5′-OH terminus. Inactivation of φX DNA by spleen or venom phosphodiesterase is exclusively a consequence of endonuclease activities present in these enzyme preparations.

Pre-treatment of φX DNA with bacterial phosphomonoesterase does not increase its susceptibility to the action of *E. coli* or spleen phosphodiesterase other than by the production of fragments as a consequence of contaminating endonuclease action.

1. Introduction

The DNA isolated from the minute *Escherichia coli* bacteriophage φX174 has been characterized both by chemical and by physico-chemical methods (Sinsheimer, 1959). It consists essentially of a single-stranded chain with a mass corresponding approximately to 5500 nucleotides. The φX DNA carries all the information needed for the synthesis of the complete virus as shown by infection of *E. coli* protoplasts (Guthrie & Sinsheimer, 1960). It would be of interest to correlate precise regions of the molecule, identified in terms of chemical structure, with the biological information carried.

In principle, a definite and controllable modification of the DNA chain and, hopefully, of the viral genome, could be obtained by means of limited digestion with exopolynucleotidases, which attack nucleic acid chains exclusively or preferentially at one of the ends, liberating mononucleotides. *E. coli* phosphodiesterase (Lehman, 1960) and venom phosphodiesterase (Razzell & Khorana, 1959) were therefore used to attempt to modify the hypothetical 3′-OH end of φX DNA, and spleen phosphodiesterase was used for the corresponding 5′-OH end (Hilmoe, 1960; Razzell & Khorana, 1961). It has been found however that φX DNA lacks end groups susceptible to any of these enzymes.

2. Materials and Methods

(a) *Materials*

The DNA was isolated from a φX174 bacteriophage preparation in borate buffer by four successive treatments with freshly re-distilled phenol (Sinsheimer, 1959). Phenol dissolved in the aqueous phase was extracted with ether and traces of the latter removed by a N_2-stream. The DNA solution was dialysed against 3 changes of 0·1 M-NaCl plus 0·001 M-tris, pH 7·5 or 8·1, in the cold. The spectrum was determined at 37°C, and the concentration was calculated from the absorbancy taking $E(P)$ as 8700 at 260 mμ.

Calf thymus DNA was purchased from Worthington Biochemical Corp.

408

E. coli phosphodiesterase was a generous gift of Dr. I. R. Lehman. It contained 16,600 units/ml. and approximately 3 mg protein/ml. A 1:50 dilution in 0·1% gelatin, adjusted to pH 8·0, was used as a stock solution for all experiments. It was divided into a set of equal samples and stored at −10°C; each tube was frozen only once.

Venom phosphodiesterase was kindly provided by Dr. J. B. Hall. The preparation had been purified by acetone fractionation (Koerner & Sinsheimer, 1957) and by chromatography on a DEAE-cellulose column (Felix, Potter & Laskowski, 1960). A solution in 0·001 M-MgCl₂ plus 0·01 M-tris, pH 8·8, was kept frozen until use.

Spleen phosphodiesterase prepared according to Hilmoe (1960) was bought from Worthington Biochemical Corp. For some experiments the enzyme was further purified by adsorption to and elution from calcium phosphate gel (Razzell & Khorana, 1961), followed by dialysis for 8 hr against 2 changes of a 0·01 M-sodium succinate buffer, pH 6·5 plus 0·001 M-EDTA plus 0·001% Tween-80 solution.

Crystalline pancreatic DNase was purchased from Worthington Biochemical Corp. It was dissolved in 0·1% BSA† and further diluted in the presence of 0·02% BSA.

Bacterial alkaline phosphatase (Torriani, 1960; Garen & Levinthal, 1960) was purchased either as a partially purified preparation, or, later, as a chromatographically purified preparation. It was diluted 1 : 25 and heated for 10 min at 90°C in the presence of 0·1 M-NaCl containing 0·01 M-MgCl₂ and 0·005 M-tris, pH 8·1. Under these conditions 92% of the phosphatase activity was recovered. For some experiments the heated enzyme solution was dialysed against 0·02 M-glycine, pH 9·5 plus 0·002 M-MgCl₂. One phosphatase unit is defined as the amount of enzyme which liberates 1 mμmole inorganic phosphate/hr at 37°C (measured in the presence of 0·001 M-*p*-nitrophenylphosphate plus 0·01 M-MgCl₂ plus 0·01 M-tris, pH 8·0).

One Hilmoe unit of spleen phosphodiesterase activity (Hilmoe, 1960), defined by activity on an "RNA core", was found to be approximately equivalent to one Lehman unit (Lehman, 1960) of *E. coli* phosphodiesterase, when the activities of both enzymes were measured with denatured calf thymus DNA as a substrate. This useful result is a consequence of the definitions of the units and the much greater number of chain ends in the "RNA core".

All enzyme dilutions were made in the presence of a protective protein (usually 0·1% gelatin or 0·1% BSA); unless otherwise indicated, incubations with *E. coli* phosphodiesterase and with venom phosphodiesterase were carried out at 30°C, and with all other enzymes at 37°C.

(b) *Methods*

Determination of percentage hydrolysis to mononucleotides

Exopolynucleotidase action was measured by precipitating the polynucleotides with perchloric acid and measuring the acid soluble nucleotides, either by their absorbancy at 260 mμ, or, less often, by ³²P-count (Lehman, 1960). In a typical experiment, to 100 μl. enzymatic hydrolysate, originally containing 15 μg of φX DNA, were added 50 μl. "carrier" DNA (2·5 mg/ml. calf thymus DNA in 0·02 M-NaCl plus 0·002 M-tris, pH 7·5) and, with vigorous stirring, 100 μl. ice-cold 0·625 N-HClO₄. After standing for 5 min at 0°C, followed by centrifugation, 200 μl. of the supernatant was withdrawn and the absorbancy measured at 260 mμ (immediate reading was necessary as the absorbancy would decrease by 10% due to depurination, at such a rate that the reaction was half completed in approximately 15 min at room temperature). (For the experiments with spleen phosphodiesterase the carrier DNA was previously heated for 3 min at 100°C to provide a more compact precipitate.)

One unit of phosphodiesterase activity corresponds to the release of 10 mμmoles DNA-P in 30 min (Lehman, 1960). However, this unit depends not only on the enzyme concentration, but also on the amount and nature of the substrate as well as other experimental conditions. The $E(P)$ of the acid-soluble nucleotides was assumed to be 10,000 at 260 mμ.

† The following abbreviation is used: BSA is bovine serum albumin.

Sedimentation analysis

200 µl. of the enzymatic reaction mixture containing approximately 30 µg φX DNA was diluted to 1 ml. with a solvent which prevented further enzymatic reaction. The samples were centrifuged at 20°C in a Spinco model E analytical centrifuge in a rotor provided with a reference wedge (Robkin, Meselson & Vinograd, 1959). (We thank Dr. J. Vinograd for lending this rotor.) Photographs were taken with the ultraviolet optical system and densitometer tracings were made with a Joyce-Loebl double beam recording photo-densitometer. The reported S values are uncorrected.

Assay of biological activity

Protoplasts of *E. coli* K12, strain W6, were infected with φX DNA and the mature phages produced were titrated on the regular host, *E. coli*, strain C, according to Guthrie & Sinsheimer (1960). Usually, from each DNA sample, 5 fivefold dilutions were assayed, ranging from approximately $2 \cdot 5 \times 10^{10}$ to 4×10^{7} φX-DNA particles/ml. of infection mixture. All assays were compared to a standard φX-DNA preparation, and as such were independent of the non-linearity of the dilution curve and of the variability of the proto-plast preparations. One absorbancy unit of φX DNA, in 0·2 M-NaCl at 25°C, corresponds to $1 \cdot 3 \times 10^{13}$ molecules/ml.

3. Experiments and Results

(a) *The action of* E. coli *phosphodiesterase on* φX DNA

In preliminary experiments it was found that when as much as 20 to 30% of a φX-DNA preparation (preparation 1) was hydrolysed (henceforth meant as hydrolysis to mononucleotides), no detectable decrease of the sedimentation coefficient of the main component was observed and there was little, if any, decrease in biological activity. These results indicated that the mononucleotides produced could not have been removed at random from all the DNA chains present. Some chains must have been hydrolysed extensively while the majority remained intact. Two explanations appeared possible. Either the phosphodiesterase acted by an "all or none" mechanism, forming an undissociable enzyme–substrate complex and releasing mononucleotides as it progressed along the DNA chain, or alternatively, the φX DNA was hetero-geneous and only a fraction of the DNA could be degraded, while most—and presumably all the biologically active—DNA chains had no 3′-OH end group susceptible to the enzyme.

As the hydrolysis showed no initial lag period it followed from the undissociable enzyme substrate hypothesis that a titration (by rate of hydrolysis) of φX DNA with phosphodiesterase, or *vice versa*, would show a sharp end point. Instead, a function suggesting a normal equilibrium between substrate, enzyme, and enzyme–substrate complex was observed (Fig. 1 (a),(b)). Furthermore, if this hypothesis were correct, an increase in the enzyme-to-substrate ratio would involve more φX-DNA chains in complexes with the phosphodiesterase. This fraction could be estimated by the ratio of V, the hydrolysis rate, to V_m, the maximum hydrolysis rate found by plotting the reciprocal enzymatic rate *versus* the reciprocal enzyme concentration (similar to a Lineweaver-Burk plot). The results of such a calculation, shown in Table 1, indicated that even when a calculated 64% of the φX-DNA chains would be bound to the enzyme, no significant decrease in the relative amount of the homogeneous component or its S value could be demonstrated. In each case, the biological activity remained unchanged.

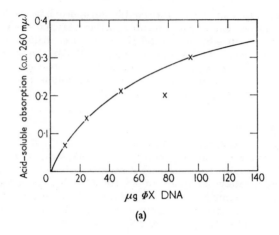

(a)

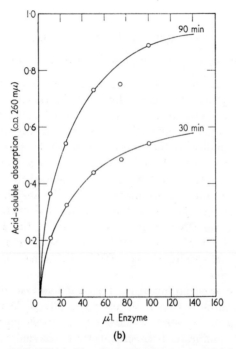

(b)

Fig. 1 (a). *Hydrolysis of ϕX DNA by* E. coli *phosphodiesterase at various concentrations of DNA.* The 200 μl. reaction mixture contained ϕX DNA (preparation 1) as indicated, 0·066 M-glycine buffer, pH 9·5, 0·001 M-MgCl$_2$, 0·1 M-NaCl, and 12·5 μl. phosphodiesterase. Incubation was for 90 min at 30°C.

Fig. 1 (b). *Hydrolysis of ϕX DNA by* E. coli *phosphodiesterase at various enzyme concentrations.* The 200 μl. reaction mixture contained 28 μg ϕX DNA (preparation 1), 0·066 M-glycine buffer, pH 9·5, 0·001 M-MgCl$_2$, 0·05 M-NaCl, and phosphodiesterase as indicated.

TABLE 1

Effect of digestion with E. coli *phosphodiesterase upon* φX DNA

Enzyme added (μl.)	% hydrolysis after 90 min	% φX DNA bound to enzyme†	Principal boundary‡ S_{20}	% of total	Infectivity
0	0	0	20·4	74	+ + +
3·3	5·5	16·1	20·3	77	+ + +
10	8·7	25·4	21·1	77	+ + +
40	16·1	46·9	20·6	72	+ + +
100	22·0	64·0	20·3	72	+ + +

† Based upon the assumption that all DNA molecules behave identically with respect to formation of an active complex with the enzyme.

‡ Centrifugation performed in 0·1 M-phosphate buffer, pH 7·2 plus 0·002 M-EDTA. All densitometer tracings show a major, sharp boundary (principal component) and trailing material of varying sedimentation coefficient.

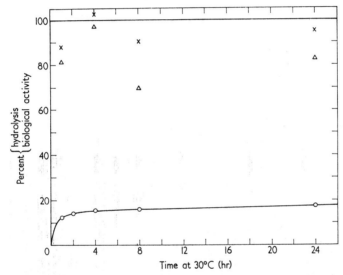

Fig. 2. *Hydrolysis of φX DNA by* E. coli *phosphodiesterase; effect upon biological activity.* The reaction mixture contained 150 μg φX DNA (preparation 3)/ml., 0·066 M-glycine buffer, pH 9·5, 0·001 M-MgCl₂, 0·014 M-NaCl, and 70 μl. phosphodiesterase/ml. Samples of 100 μl. and 10 μl. were taken at various times for determination of, respectively, percent hydrolysis and infectivity (expressed as percent of an unincubated standard). Enzyme was omitted in the control. ○——○, acid-soluble nucleotides. ×——×, biological activity. △——△, biological activity of control.

These data also indicate that the *E. coli* phosphodiesterase preparation is remarkably free of endonuclease activity: 22% hydrolysis corresponds to an average removal by exonuclease action of 1200 nucleotides per chain length of 5500 without any evidence of a single phosphodiester scission by endonuclease action (it is estimated that 0·1 hits per molecule would have been detected by the change in sedimentation pattern).

The alternative hypothesis, that of a heterogeneous substrate containing a majority of DNA chains resistant to the enzyme (including those biologically active), and a small fraction of susceptible molecules, could be confirmed with a better φX DNA preparation. With preparation 3 the enzymatic activity became negligibly small after approximately 16% hydrolysis (Fig. 2). It could be demonstrated that under these conditions the enzyme was neither inactivated nor inhibited. The biological activities of the treated samples were even slightly above the control values indicating no loss due to the addition of the enzyme. Carrying out the reaction at 37°C instead of 30°C had little effect (after 8 hours, respectively 18·4% and 16·1% hydrolysis); neither did addition of an equal amount of fresh enzyme after 8 hours (an additional 5% hydrolysis in 3 hours).

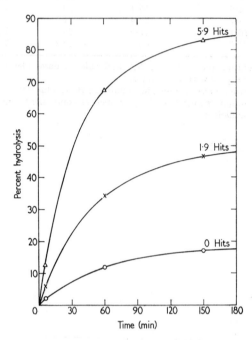

Fig. 3. *Effect of prior treatment with pancreatic DNase on the susceptibility of φX DNA to hydrolysis by E. coli phosphodiesterase.* The 200 μl. reaction mixture contained 95 μg φX DNA (preparation 1), 0·0015 M-tris, pH 7·5, 0·005 M-MgCl₂, 0·1 M-NaCl, 0·005% BSA, and 8×10^{-3} μg pancreatic DNase. After incubation for 10 min (1·9 hits) or 100 min (5·9 hits), reaction was stopped by addition of 150 μl. 0·2 M-glycine buffer, pH 9·8. The control lacked enzyme. After removal of 10 μl. samples for biological assay, 50 μl. of phosphodiesterase was added and equal portions removed for determination of percent hydrolysis at the indicated times.

Brief pre-treatment with pancreatic deoxyribonuclease, which is known to release 3′-OH ends, results in the susceptibility of at least a large fraction of the φX-DNA chain (Fig. 3). Therefore the lack of susceptibility of the φX DNA seems only to be due to the lack of a suitable end group (presumably a free 3′-OH terminus). [Evidence has been presented that in TMV RNA the corresponding end is free (Sugiyama & Fraenkel-Conrat, 1961).]

The 40% hydrolysis of φX DNA found by Lehman (1960) was probably an observation on a more degraded preparation.

(b) *The action of spleen phosphodiesterase on φX DNA*

As the hypothetical 3′-OH end of φX DNA appeared to be resistant to attack, an attempt was made to degrade selectively the potential 5′-OH end by means of spleen phosphodiesterase.

Incubation of ϕX DNA with spleen phosphodiesterase resulted in a gradual, approximately first-order loss of biological activity (Fig. 4) accompanied, after an initial lag, by the release of acid-soluble fragments. However, as can be seen in Fig. 5, at 50% inactivation, an average of 347 mononucleotides (6%) had been released per chain. If the observed inactivation were a consequence of terminal hydrolysis until a vital part of the ϕX-DNA molecule were reached, some 350 nucleotides from the original end, the biological activity would then be expected to

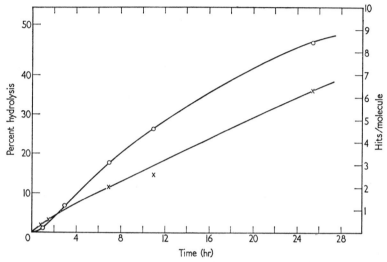

FIG. 4. *Action of spleen phosphodiesterase on ϕX DNA.* Biological inactivation (×——×) (expressed as hits/molecule) and percent hydrolysis (○——○) *versus* time. One ml. reaction mixture contained 150 μg ϕX DNA (preparation 6), 0·033 M-succinate buffer, pH 6·5, 0·005 M-EDTA, 0·035 M-NaCl, and 1 unit spleen phosphodiesterase.

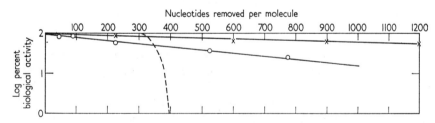

FIG. 5. *Action of spleen phosphodiesterase on ϕX DNA.* Residual biological activity plotted against average number of nucleotides released per molecule.

Experiment 1 (○——○): 500 μl. reaction mixture contained 69 μg ϕX DNA (preparation 5), 0·033 M-succinate buffer, pH 6·5, $3\cdot3\times10^{-4}$ M-EDTA, 0·02 M-NaCl, 0·03% gelatin, and 0·6 units spleen phosphodiesterase.

Experiment 2 (×——×): effect of pre-treatment with phosphomonoesterase. 100 μg ϕX DNA (preparation 5) was incubated for 24 hr at 37°C with $4\cdot5\times10^{4}$ units of chromatographed and heated bacterial alkaline phosphatase in 300 μl. 5×10^{-3} M-glycine buffer, pH 9·5 plus 5×10^{-4} M-MgCl$_2$ plus 0·05 M-NaCl. 46% of the biological activity survived. The solution was diluted to 600 μl. with succinate buffer, pH 6·5 (final concentration 0·033 M), EDTA (final concentration 5×10^{-3} M), and 0·4 units spleen phosphodiesterase. Samples were taken at various times for measurement of extent of hydrolysis and assay of infectivity (relative to a control containing phosphatase but no spleen phosphodiesterase).

(- - -): theoretical inactivation function if inactivation were due to exonuclease action and there were 347 superfluous nucleotides at the 5'-OH end of ϕX DNA (intersects inactivation curve of experiment 1 at 50% survival). Removal of the 348th nucleotide is assumed to inactivate.

drop drastically upon further action of the exonuclease (Fig. 5, dashed line). This is not observed. Even a sample hydrolysed until 48% was released as mononucleotides still possessed 1·4% of the original biological activity. Hence it appears more likely that all of the inactivation is produced by an independent endonuclease activity in the enzyme preparation.

The presence of endonuclease activity is also evident from ultracentrifuge studies. The number of chain breaks that can be estimated from the decrease in the amount of the principal, homogeneous component or from the average sedimentation coefficient agrees well with the observed loss in biological activity (Table 2).

TABLE 2

The effect of spleen phosphodiesterase on the biological activity and the sedimentation pattern of ϕX DNA

Experiment	% hydrolysis	Hits/molecule (bio-assay)	Ultracentrifugal analysis S	Hits/molecule[†]
1	Control	—	21·5[‡]	—
	0·11	0·1	22·5[‡]	0·1[§]
2	Control	—	21·9[‡]	—
	48·1	4·3	10·7[‡]	4·2–7·7
3	Control	—	12·1[‖¶]	—
	13·8	1·4	9·5[‖]	1·6–2·0
4	Control	—	12·1[‖¶]	—
	25·9	2·6	8·9[‖]	1·8–2·4

Reactions were carried out as in experiment 1 of Fig. 5.

[†] The number of chain breaks is estimated from the average (50%) sedimentation coefficient S, assuming that $M = k S^\alpha$ and $0·35 < \alpha < 0·50$. It is also assumed that the first hit does not decrease M (molecular weight) (Fiers & Sinsheimer, 1962).

[‡] Sedimentation in 0·08 M-phosphate, pH 9·0.

[§] From quantitative recovery of the principal centrifugal component.

[‖] Sedimentation after treatment with formaldehyde (Fiers & Sinsheimer, 1962).

[¶] Sedimentation coefficient of S_2 component (Fiers & Sinsheimer, 1962).

The lag in release of acid-soluble material supports the hypothesis that the exonuclease action needs the chain ends produced by the endonuclease activity. Under identical conditions the hydrolysis of calf thymus DNA proceeds without lag.

Taken together with the previously described results of the action of the *E. coli* phosphodiesterase, these findings suggest that the smaller fragments present in ϕX-DNA preparations have a free 3'-OH end, but not a free 5'-OH end. The endonuclease activity present in the spleen phosphodiesterase preparation must lead to free 5'-OH ends.

(c) *Influence of pre-treatment with phosphomonoesterase*

A possible explanation for the resistance of both hypothetical chain ends to specific exonucleases would be that they were blocked by phosphate groups. The resistance of chains terminated by 5'-phosphate to spleen phosphodiesterase has been demonstrated by Razzell & Khorana (1961). Although the specificity of the *E. coli* phosphodiesterase has not been investigated in detail, its similarity to venom phosphodiesterase in mode of action suggests that it may likewise be inactive on a 3'-phosphate terminated end (Lehman, 1960; Razell & Khorana, 1959).

These possibilities were tested by pre-treatment of the ϕX DNA with bacterial alkaline phosphomonoesterase. The parameters, V_M and K_M, of this enzyme, are remarkably similar for most phosphate monoesters studied (Torriani, 1960; Garen & Levinthal, 1960; Heppel, Harkness & Hilmoe, 1962) and oligonucleotides are readily dephosphorylated, so it seemed likely, though not certain, that this enzyme could remove a postulated terminal phosphate from a chain as long as that of ϕX DNA.

An increased susceptibility to the action of *E. coli* phosphodiesterase was observed if the ϕX DNA was pre-incubated with the heated phosphomonoesterase (Fig. 6, curve A). However, this effect was markedly reduced if a comparable activity of a chromatographically purified monoesterase preparation was used (Fig. 6, curve B).

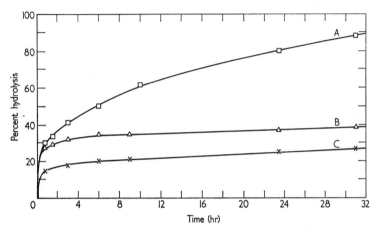

FIG. 6. *Effect of pre-treatment with bacterial alkaline phosphomonoesterase upon the susceptibility of ϕX DNA to* E. coli *phosphodiesterase.*

Curve A ($\square$——$\square$): 600 μl. reaction mixture containing 120 μg ϕX DNA (preparation 4), 0·066 M-glycine buffer, pH 9·5, 0·0035 M-MgCl$_2$, 0·05 M-NaCl, and 3·3 × 10^4 units phosphomonoesterase (partially purified) were incubated for 24 hr at 37°C. Survival of biological activity was 12%. 100 μl. 0·08 M-K$_2$HPO$_4$ and 100 μl. *E. coli* phosphodiesterase were added and samples taken for determination of the extent of hydrolysis at the indicated times.

Curve B ($\triangle$——$\triangle$): same reaction as in (A), but with 4·5 × 10^4 units chromatographed phosphomonoesterase. The survival of biological activity after 24 hr incubation was 60%.

Curve C ($\times$——$\times$): control for (A) and (B), without addition of phosphomonoesterase.

The biological inactivation accompanying the action of the monoesterase was also greatly reduced in the latter experiment. These results suggested that the observed increase in susceptibility to *E. coli* phosphodiesterase action was a consequence of a contaminating nuclease activity (which liberated 3'-OH end groups) present in the monoesterase preparations and not a consequence of removal of a 3'-phosphate end group. The presence of residual nuclease activity in the chromatographed and heated phosphomonoesterase preparation has also been demonstrated by ultracentrifugal experiments (Fiers & Sinsheimer, 1962, Table 2). (It should be emphasized that in these experiments a relatively great amount of phosphomonoesterase was employed in order to hydrolyse completely even a very unreactive phosphomonoester).

Pre-treatment of ϕX DNA with phosphomonoesterase does not produce any increase in susceptibility to inactivation by the spleen phosphodiesterase (Fig. 5, experiment 2); suggesting that the active molecules remain resistant to exonuclease action.

(d) *Action of venom phosphodiesterase on ϕX DNA*

A possible explanation for the resistance of the ends of ϕX DNA to enzymatic attack could be that these ends are involved in short double-stranded structures, e.g. by being folded back over a limited length. Lehman (1960) has clearly shown that the *E. coli* phosphodiesterase can hydrolyse only single-stranded DNA.

The action of venom phosphodiesterase, however, does not seem to be prevented by the secondary structure of its substrate (Boman & Kaletta, 1957; Williams, Sung & Laskowski, 1961). Unfortunately endonuclease activity can be detected even in the best venom phosphodiesterase preparations (Razzell & Khorana, 1959).

ϕX DNA is readily hydrolysed by the venom phosphodiesterase preparation with loss of biological activity and production of acid soluble fragments. When 2·0% has been hydrolysed to acid solubility, there are 3·7 lethal hits per molecule, by biological assay. By the same argument as given above for the spleen phosphodiesterase, one might then expect that if inactivation is a consequence of hydrolysis from the end into a critical region (about 100 nucleotides from the 3'-OH terminus) further hydrolysis would drastically reduce the biological activity. This is not observed, as 3·9% hydrolysis corresponds to 6·4 hits per molecule.

It thus appears more plausible to attribute the inactivation to the endonuclease activity, with a ratio of approximately 30 terminal nucleotides released per endonuclease action. Apparently ϕX DNA cannot be inactivated by removal of a few nucleotides from the end by venom phosphodiesterase. It therefore seems likely that all mononucleotides are released from broken chains, and that the hypothetical 3'-OH end of the biologically active ϕX DNA is as resistant to the venom phosphodiesterase as to the *E. coli* phosphodiesterase.

The endonuclease nature of the lethal hits is confirmed by ultracentrifugal data. After 3·9% hydrolysis the mean sedimentation coefficient had dropped from 20·0 s to 8·3 s. Dependent upon the assumed exponent of the relationship between S and the molecular weight, this decrease would correspond to between 5·8 and 12·2 chain breaks per molecule, to be compared with the 6·4 hits per molecule found by biological assay.

(e) *Effect of exposure to denaturing conditions upon enzymatic susceptibility of ϕX DNA*

The possibility of a limited hydrogen bonded structure was further explored by studying the effect of a denaturing treatment. ϕX DNA was treated with formamide under conditions which according to Marmur & Ts'o (1961) would insure complete melting out of any hydrogen-bonded region in DNA. One ml. of a reaction mixture containing 105 μg ϕX DNA (preparation 3), 95% (v/v) redistilled formamide, 50 mM-NaCl and 0·05 mM-tris buffer, pH 7·5, was incubated for 20 minutes at 37°C. The solution was cooled to 0°C and 650 μl. 0·2 M-NaCl plus 0·001 M-tris buffer was added. The formamide was removed by six extractions with isobutanol previously equilibrated with the NaCl–tris solution, and each isobutanol extract was washed once with 200 μl. of 0·1 M-NaCl (the procedure is similar to the phenol extraction described by Sinsheimer, 1959). 80·2% of the DNA was recovered in a total volume of 600 μl. The residual formamide was approximately 0·3% (estimated from the absorbancy at 260 mμ and at 235 mμ). In the control experiment no formamide was added, but the solution was similarly incubated at 37°C and extracted with isobutanol. 94% of the DNA was recovered.

The recovery of the biological activity in the formamide-treated sample was 89% of the control, indicating either that a specific hydrogen bonding is not necessary for infectivity, or that it is readily reformed upon removal of the denaturing agent. This formamide treatment was unable to remove the enzymatic block from the hypothetical 3′-OH end (Table 3). The absence or reversibility of any essential secondary structure is also evident from the stability of ϕX DNA in alkaline solution (Fiers & Sinsheimer, 1962).

TABLE 3

Hydrolysis of formamide-treated ϕX DNA by E. coli phosphodiesterase

Time of incubation (hr)	% hydrolysis Formamide-treated	Control
0·75	19·4	19·3
1·5	21·8	22·4
3	26·8	28·0
7	33·1	31·4

Reaction mixture contained, in 600 μl., 62 μg formamide-treated ϕX DNA, 0·066 M-glycine buffer, pH 9·5, 0·001 M-MgCl$_2$, 0·08 M-NaCl, and 37 μl. phosphodiesterase. Control contained untreated ϕX DNA.

It should be noted, however, that the recovery of biological activity in the control experiment was only 23·5%. This inactivation is a result of the isobutanol extraction (Fiers & Sinsheimer, 1962).

4. Discussion

The biological activity of a ϕX DNA preparation cannot be destroyed by treatment with *E. coli* phosphodiesterase. The action of this enzyme is limited to degradation of a small proportion, presumably ϕX DNA fragments, of the preparation. However, after a few hits by pancreatic deoxyribonuclease, the bulk of such preparations is readily susceptible. This evidence indicates that the block to the action of this exonuclease is at the 3′-OH end of the ϕX DNA chain.

Similarly, the inactivation of ϕX DNA by spleen or venom phosphodiesterase cannot be ascribed to exonuclease actions of these enzymes from the 5′-OH or the 3′-OH ends respectively, but rather appears to be a consequence of contaminating endonuclease activities in both enzyme preparations.

It is possible that the actions of these enzymes are blocked by the presence of terminal phosphate groups at both ends of the ϕX DNA. Attempts to eliminate this possibility by prior use of phosphomonoesterase were complicated by residual endonuclease activity in the phosphomonoesterase preparation. It appears that any enhancement of the susceptibility of ϕX DNA to exonuclease action by prior use of phosphomonoesterase is a consequence of this endonuclease action. However, this result does not eliminate the possibility of terminal phosphate groups, because of uncertainty as to whether they would be susceptible to the monoesterase.

The possibility that exonuclease action at the 3′-OH end could be blocked by hydrogen bonding of the terminal nucleotides would appear to be unlikely because of the failure of the venom phosphodiesterase to attack this end. A similar argument cannot be made for the other end. The formamide experiments indicate that any such hydrogen-bonded structure, if essential to biological activity, must be re-formed after denaturation.

Other possible explanations of these results include the presence of non-nucleotide substituents upon both ends of the φX DNA, or a closed ring structure (Fiers & Sinsheimer, 1962).

This research was supported by U.S. Public Health Service grant RG6965. One of the authors (W. F.) is a Rockefeller Foundation Fellow; "Chargé de Recherches du Fonds National Belge de la Recherche Scientifique".

REFERENCES

Boman, H. G. & Kaletta, U. (1957). *Biochim. biophys. Acta*, **24**, 619.
Felix, F., Potter, J. L. & Laskowski, M. (1960). *J. Biol. Chem.* **235**, 1150.
Fiers, W. & Sinsheimer, R. L. (1962). *J. Mol. Biol.* **5**, 424.
Garen, A. & Levinthal, C. (1960). *Biochim. biophys. Acta*, **38**, 470.
Guthrie, G. D. & Sinsheimer, R. L. (1960). *J. Mol. Biol.* **2**, 297.
Heppel, L. A., Harkness, D. R. & Hilmoe, R. J. (1962). *J. Biol. Chem.* **237**, 841.
Hilmoe, R. J. (1960). *J. Biol. Chem.* **235**, 2117.
Koerner, J. F. & Sinsheimer, R. L. (1957). *J. Biol. Chem.* **228**, 1049.
Lehman, I. R. (1960). *J. Biol. Chem.* **235**, 1479.
Marmur, J. & T'so, P. O. P. (1961). *Biochim. biophys. Acta*, **51**, 32.
Razzell, W. E. & Khorana, H. G. (1959). *J. Biol. Chem.* **234**, 2114.
Razzell, W. E. & Khorana, H. G. (1961). *J. Biol. Chem.* **236**, 1144.
Robkin, E. M., Meselson, M. & Vinograd, J. (1959). *J. Amer. Chem. Soc.* **81**, 1305.
Sinsheimer, R. L. (1959). *J. Mol. Biol.* **1**, 43.
Sugiyama, T. & Fraenkel-Conrat, H. (1961). *Proc. Nat. Acad. Sci., Wash.* **47**, 1393.
Torriani, A. (1960). *Biochim. biophys. Acta*, **38**, 460.
Williams, E. J., Sung, S. & Laskowski, M. (1961). *J. Biol. Chem.* **236**, 1130.

From *Journal of Molecular Biology*, Vol. 5, 424–434, 1962. Reproduced with permission of the authors and publisher.

The Structure of the DNA of Bacteriophage φX174

III. Ultracentrifugal Evidence for a Ring Structure

WALTER FIERS AND ROBERT L. SINSHEIMER

California Institute of Technology, Division of Biology, Pasadena, California, U.S.A.

(*Received 20 June 1962*)

By velocity sedimentation, in appropriate solvents, the presence of two discrete components can be demonstrated in preparations of the DNA of bacteriophage φX174. The major, faster-moving S_1 and the slower S_2 are present under conditions which exclude the possibility of hydrogen-bond formation. It can be shown, either by treatment with pancreatic deoxyribonuclease or by thermal inactivation, that S_2 is the first degradation product of S_1, formed by scission of S_1, without significant decrease in molecular weight. A subsequent chain scission in S_2, which occurs with equal likelihood, results in random fragmentation. These results are interpreted to mean that the S_1 component is a covalently linked ring structure and the S_2 component is the corresponding open-chain degradation product.

Under certain conditions the S_2 component can be selectively degraded by *E. coli* phosphodiesterase. The digestion is not complete and there appears to be a single discontinuity, resistant to phosphodiesterase, present in the φX-DNA ring.

1. Introduction

One of us (Sinsheimer, 1959) has reported that when φX DNA was centrifuged under conditions which produced an extension of the molecules, two components could be demonstrated. The minor component (henceforth called S_2) comprised about 20 to 40 % of the total, and under the conditions employed, moved at about a 10% slower rate than the major component (S_1), which usually comprised 50 to 70%. This heterogeneity was rather surprising in a preparation derived from a supposedly homogeneous virus preparation. The difference was suspected to be configurational, as the separation was not observed during centrifugation in a 0·2 M-salt solution, and because the nucleotide composition seemed to be independent of the relative amounts of S_1 and S_2.

In a previous paper (Fiers & Sinsheimer, 1962a) evidence has been presented which indicates that infective φX-DNA molecules lack both a free 3′-OH and a free 5′-OH terminus. The following relation explains both this chemical peculiarity and the physical duality: the S_1 seems to be identical with the infective form and has a ring structure, while S_2 is the first breakdown product of S_1 and corresponds (after treatment with formaldehyde, or in alkali) to an open chain form.

2. Materials and Methods

The sources of the materials have been described in a previous paper (Fiers & Sinsheimer, 1962a).

The "standard treatment" with formaldehyde consisted of adding 1/19 vol. of 36% formaldehyde (Baker and Adamson, reagent grade) followed by 20 hr incubation at 37°C.

Centrifugation was performed in the Spinco model E ultracentrifuge. Alkaline solutions were centrifuged in Kel-F centerpieces.

Quantitation of ultracentrifuge patterns was routinely performed by means of the reference-wedge method of Robkin, Meselson & Vinograd (1959). No discrepancy was found between these results and results obtained by direct evaluation from the densitometer tracings. This indicates that the film response was linear with DNA concentration, and that the photographic procedure was reproducible. Values reported are the averages from two frames, usually the no. 5 frame (40 min) and the no. 7 frame (56 min).

In the digestions with pancreatic DNase (Fig. 2) each enzymic reaction mixture (200 μl.) contained 30 μg ϕX DNA (preparation 6), $1 \cdot 25 \times 10^{-3}$ M-tris buffer, pH 7·5, 5×10^{-3} M-MgCl$_2$, 5×10^{-2} M-NaCl, $5 \times 10^{-3}\%$ BSA,† and 7×10^{-5} μg pancreatic DNase. To terminate the reaction the digest was diluted fivefold with potassium phosphate buffer and EDTA to yield final concentrations respectively of 0·08 M and 0·002 M at pH 9·0. A sample was taken for assay of the biological activity and the remainder was analysed in the ultracentrifuge after the standard treatment with formaldehyde.

For the thermal inactivation study (Fig. 3), each sample (200 μl.) containing 34 μg ϕX DNA (preparation 5), 0·025 M-glycine buffer pH 9·0, 5×10^{-5} M-EDTA, and 0·025 M-NaCl, was heated at 98·2°C for a specified time and fast-cooled. A sample was taken for assay of biological activity and the remainder was analysed in the ultracentrifuge after the standard treatment with formaldehyde.

The combined digestions of ϕX DNA with pancreatic DNase and *E. coli* phosphodiesterase (Fig. 5) were performed as follows: a 50 μl. reaction mixture containing 52 μg ϕX DNA (preparation 3), 0·1 M-NaCl, 0·0021 M-tris buffer pH 7·5, 0·008 M-MgCl$_2$, 0·01% BSA, and 62 $\mu\mu$g pancreatic DNase was incubated at 37°C for an appropriate time. 100 μl. 0·05 M-sodium carbonate buffer, pH 11·0, was added and the sample incubated 2 hr at 30°C. The sample was then diluted to 300 μl. by addition of 4·5 μmoles of HCl–glycine buffer, pH 9·5, to a final concentration of 0·066 M and 50 μl. of phosphodiesterase. After 24 hr incubation at 30°C a 100 μl. sample was taken for measurement of per cent hydrolysis (to acid solubility) while the remainder was used for assay of biological activity and for ultracentrifugal analysis (Table 6).

3. Results

(a) *Observation of sedimentation heterogeneity in ϕX DNA*

In the earlier study, heterogeneity of ϕX DNA was observed during sedimentation in formaldehyde, or in a medium of low ionic strength. In the former medium, ϕX DNA is irreversibly inactivated; in the latter its biological activity is unstable and reproducible observation of heterogeneity is difficult, probably because of convection during sedimentation.

A quantitatively similar pattern can be obtained during sedimentation in alkaline medium at pH 11·0 (Table 1, experiment 1). ϕX DNA can be recovered from this medium with full biological activity (after 2 hours at 30°C in 0·033 M-sodium carbonate, pH 11·0, plus 0·01 M-NaCl, plus $1 \cdot 3 \times 10^{-4}$ M-EDTA).

The S$_1$ and S$_2$ components are also present during sedimentation at pH 12·0 (Table 1, experiment 2) and at pH 12·5, in the usual proportions. As pH 12·5 is well above the denaturation point of any DNA studied (Vinograd, Morris, Dove & Davidson at the sixth annual meeting of the American Biophysical Society, 1962; Dove 1962), this observation would seem to exclude the possibility that the distinction between S$_1$ and S$_2$ is produced by a hydrogen-bonded configuration. This conclusion is supported by the persistence of the S$_1$, S$_2$ pattern after heating of the ϕX DNA to 80°C in formaldehyde (Table 1, experiment 3), a condition known to denature calf thymus DNA irreversibly.

† Bovine serum albumin.

W. FIERS AND R. L. SINSHEIMER

TABLE 1

Sedimentation heterogeneity of ϕX DNA

Experiment	Conditions of centrifugation	Sedimentation coefficient		% Composition		
		S_1	S_2	S_1	S_2	Slow†
1‡	CH₂O (control)	13·5 s	12·1 s	72·2	18·9	8·9
	OH⁻ (pH 11)	12·8	12·1	67·7	16·8	15·5
	OH⁻ (pH 11)	13·1	12·0	74·4	20·4	9·2
2§	OH⁻ (pH 12)	11·5	10·7	62·4	25·7	11·9
3‖	CH₂O (control)	13·1	12·1	60·4	17·0	22·6
	CH₂O (heated)	13·8	12·3	57·0	25·3	17·7

† All centrifuge patterns contain a certain amount of slower moving degradation products with a distribution of sedimentation coefficients.

‡ The control contained 32 μg ϕX DNA (preparation 6) in 1 ml. 0·1 M-sodium phosphate buffer, pH 7·0 plus 0·01 M-NaCl. It was treated with formaldehyde as described. The pH 11 experiments contained 30 μg ϕX DNA (preparation 6) in 1 ml. 0·04 M-sodium carbonate buffer, pH 11·0 plus 6×10^{-4} M-EDTA plus 0·01 M-NaCl.

§ The pH 12 experiment contained 30 μg ϕX DNA (preparation 6) in 0·032 M-sodium arsenate buffer, pH 12·0.

‖ Control, as in experiment 1, but with ϕX DNA preparation 3. The heated sample was held at 80°C for 10 min in the presence of the formaldehyde, fast-cooled, and centrifuged.

(b) *Correlation of biological activity and S_1 content*

In the course of various treatments of ϕX DNA (Fiers & Sinsheimer, 1962*a,b*) it was observed that after exposure to enzyme preparations which presumably produced chain scission, no quantitative correlation could be made between the residual biological activity and the proportion of the total DNA in the leading boundary when ultracentrifugal analysis was performed in 0·1 to 0·2 M-salt solution (Table 2).

TABLE 2

Ultracentrifuge patterns of ϕX DNA partially inactivated by various endonucleases

Enzyme	Residual biological activity (%)	Solvent for ultracentrifugation	Proportion in principal boundary (% of control)	Calculated† proportion of S_1 plus S_2 (%)
E. coli phosphatase	42	A	62·8	78·4
Pancreatic DNase	37	B	80·1	74·0
Venom diesterase	17	A	49·8	43·5
Spleen diesterase	1·4	C	0	7·5

Solvents: A is 0·1 M-sodium phosphate, pH 7·0 plus 0·002 M-EDTA.
B is 0·22 M-NaCl plus 0·01 M-tris, pH 8·1.
C is 0·08 M-potassium phosphate, pH 9·0.

† Calculated from residual biological activity assuming a Poisson distribution of "hits".

However, sedimentation of several variously treated ϕX DNA preparations under conditions which permitted observation of sedimentation heterogeneity demonstrated a direct correlation between infectivity and the content of the S_1 component (Fig. 1).

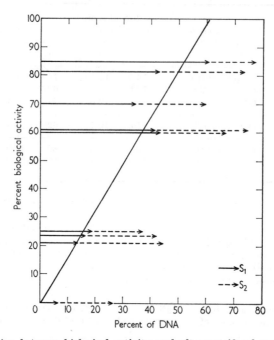

FIG. 1. Correlation between biological activity and ultracentrifugal components. Results of samples with different histories are collected. Treatments include enzymic, thermal and spontaneous inactivation. The ordinate is the percentage residual biological activity, compared to a standard ϕX DNA (preparation 3) of equal ultraviolet absorption.

(c) *Effect of action of pancreatic DNase upon infectivity and sedimentation pattern of ϕX DNA*

If ϕX DNA is treated with pancreatic DNase for various periods of time, it is found (Fig. 2) that the proportion of the S_1 component decreases exponentially, as would be expected for the infective component (Guthrie & Sinsheimer, 1960). Using the residual biological activity to calculate the number of DNase "hits" upon S_1, it is found that the proportion of S_2 follows closely the expected function for the proportion of the component with one hit. First the proportion of S_2 rises to nearly the theoretical value (37% at one hit), indicating that it is formed from S_1 without appreciable loss of substance or decrease in molecular weight (excluding a specific, median scission, *vide infra*). After an exposure to the enzyme sufficient to produce an average of one hit per molecule, the proportion of S_2 decreases, always remaining larger, however, than the proportion of S_1. The good agreement between the experimental points and the calculated function also indicates that the probability of destruction of S_2 (by a second hit) is similar to the probability of formation of S_2 from S_1 by the first hit (thus excluding the possibility of a special site of attack for conversion of S_1 to S_2). No discrete boundary other than S_1 and S_2 was observed in any of these experiments. A few samples, centrifuged in alkali rather than in formaldehyde, gave quantitatively similar results.

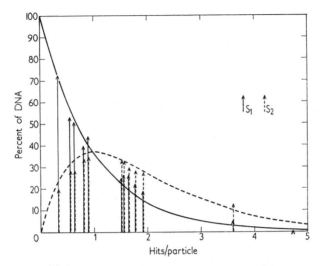

FIG. 2. Ultracentrifugal composition of ϕX DNA after varying extent of hydrolysis by pancreatic DNase. The ordinate is the proportion of ultraviolet absorption contributed by S_1 and S_2 to the ultracentrifuge pattern. The abscissa is the number of "hits" calculated from the residual biological activity. The control (no enzyme added) was considered to contain 72% (the same as its S_1 content) of the biological activity of a theoretical fully-active sample (i.e. we assume 28% of the DNA has already received one or more "hits" during its preparation).

The smooth curve is the theoretical function for the no-hit (e^{-m}, where m is the mean number of hits per molecule) fraction. The dashed curve corresponds to the one-hit function ($m.\ e^{-m}$).

The sedimentation coefficients of the S_1 component were between 13·5 and 14·1 s: of S_2 component between 11·5 and 12·5 s.

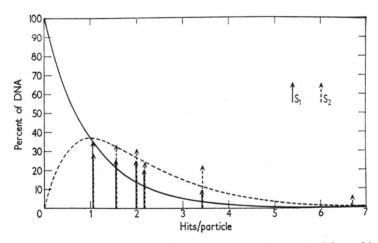

FIG. 3. Ultracentrifugal composition of ϕX DNA after varying extent of thermal inactivation. Ordinate, abscissa and curves as in Fig. 2. The S_1 content of the unheated control was 35% for this preparation.

The sedimentation coefficients of the S_1 component were between 12·8 and 13·2 s; of the S_2 component between 11·2 and 11·5 s.

(d) *Effect of heat upon infectivity and sedimentation pattern of ϕX DNA*

A similar pattern of centrifugal observations is obtained (Fig. 3) if inactivation of ϕX DNA is brought about by heat rather than by enzymic action. The results in Fig. 3 are somewhat more erratic and in poorer agreement with the theoretical functions. The proportions of the S_1 and S_2 components exceed those expected from the degree of inactivation. However, it is evident that the application of heat, as that of DNase, results in the conversion of S_1 to S_2 and in the degradation of S_2. The chance of the second event is comparable with the chance of the first, thus indicating that there is no especially heat-sensitive site for chain scission.

(e) *Masked or potential chain scission in ϕX-DNA structure*

The results just described indicate that thermal inactivation (at pH 9·0) can result in a conversion of S_1 to S_2 (and further) and thus by analogy with the results obtained with DNase, can be accompanied by chain scission. However, in a previous paper (Fiers & Sinsheimer, 1962b) we have demonstrated that thermal inactivation of ϕX DNA is not immediately accompanied by a susceptibility to the action of *E. coli* phosphodiesterase. Further centrifugal analyses, under conditions which do not rupture hydrogen bonds, of ϕX-DNA preparations inactivated by heat or acid (Table 3) do not reveal any simple correlation between extent of inactivation and that of molecular degradation. When the same preparations are centrifuged in the presence of formaldehyde, a distinctly lesser, but still unexpectedly large amount of macromolecular components is observed.

TABLE 3

Ultracentrifuge patterns of variously inactivated ϕX DNA

Treatment	Residual biological activity (%)	Proportion in leading boundary† (% of control)	Calculated¶ proportion of S_1 plus S_2 (%)	Observed proportion of S_1 plus S_2 in formaldehyde (%)
Heat‡	<0·1	56·7	<0·8	22·6
pH 2·9§	0·03	104·5	0·3	28·3
Isobutanol‖	23·5	100·6	57·6	47·5
Isobutanol	20·9	101·2	53·7	51·6

† Centrifuged in 0·1 M-sodium phosphate plus 0·002 M-EDTA.
‡ 10 min at 100°C, at pH 7·5.
§ 15 hr in 0·1 N-acetic acid at 30°C.
‖ Repeated extraction of solution with isobutanol as described in Fiers & Sinsheimer (1962a).
¶ Calculated from residual biological activity, assuming a Poisson distribution of hits.

Evidently the inactivating event upon exposure of ϕX DNA to either heat or acid can leave the macromolecular structure intact. Some degree of chain scission can, however, be demonstrated after exposure of the inactivated DNA to more disruptive conditions (higher temperature, formaldehyde, alkali). Either some chain scission has occurred during inactivation and its effect is masked by hydrogen-bonding of adjacent nucleotides† or the chain scission subsequently observed is not an immediate

† A masking of scissions in this way does not seem to occur if the scissions are produced by endonuclease action (Table 2). Following endonuclease action the proportion of DNA in fragments of S less than that of the principal component is very close to that expected from the number of hits.

consequence of the inactivating event (presumably depurination) but occurs at the weakened site upon molecular extension. The persistence of macromolecular components (Fig. 3, Table 3) in amounts considerably in excess of those to be expected from the number of hits, suggests that scission does not occur at all such sites under conditions employed here.

The nature of inactivation by extraction with isobutanol is unknown, but evidently the process results in direct or potential chain scissions which are quantitatively revealed in the presence of formaldehyde.

(f) The endogenous S_2 component

All ϕX-DNA preparations thus far examined contained proportions of S_1, S_2 and slower moving fragments which approximately correspond to the proportions to be expected of zero, one, and more-than-one hit components in a population which has been subjected to random degradation (Table 4). The nature of these hits may be heterogeneous (nucleolytic, depurination) and may vary from one preparation to another.

TABLE 4

Sedimentation heterogeneity of ϕX-DNA preparations

Preparation	%S_1	%S_2		% Slow component	
		Found	Expected	Found	Expected
3	64·5	19·7	28·3	15·8	7·2
5	36·3	34·2	36·8	29·5	26·8
6	72·2	18·9	23·5	8·8	4·3

Each preparation was analysed after formaldehyde treatment. Expected amounts of S_2 (one-hit) and slow (more-than-one-hit) components were calculated from the proportion of the S_1 (zero-hit) component.

In one preparation (preparation 6) the S_2 component could be completely degraded by the action of *E. coli* phosphodiesterase (Fig. 4). However, the S_2 component of other preparations has been largely resistant to *E. coli* phosphodiesterase attack (Table 5). Prior treatment with *E. coli* phosphomonoesterase, while resulting in some degradation because of endonuclease action, did not significantly increase the susceptibility of S_2 to *E. coli* phosphodiesterase action.

The possibility that the termini of the S_2 component molecules could be protected from phosphodiesterase attack by coiling and hydrogen bond formation would seem to be excluded by the evident susceptibility of S_2 component formed by pancreatic DNase action (cf. Fig. 5 of this paper and Fig. 3 of Fiers & Sinsheimer, 1962a). A more plausible explanation of the resistance of endogenous S_2 component is that—by analogy to the depurinated molecules described above—it remains in a ring form until exposed to conditions producing molecular extension. If this is so, a prior exposure to such conditions (e.g. alkali) should render at least a portion of the potential S_2 component susceptible to phosphodiesterase action. Preliminary experiments indicate that this hypothesis is correct.

TABLE 5

Effect of E. coli *phosphodiesterase upon sedimentation pattern of* ϕX *DNA*

% Hydrolysis† (acid soluble)	% Composition			
	S_1	S_2	Slow	Total
0 (control)	48·2	30·7	20·1	100
12·5	45·4	25·6	10·6	81·6
0 (control)	81·8 ($S_1 + S_2$)		18·1	100
16·1	72·4 ($S_1 + S_2$)		8·0	80·4

† Experimental procedure identical to that described in Fig. 2, Fiers & Sinsheimer (1962a).

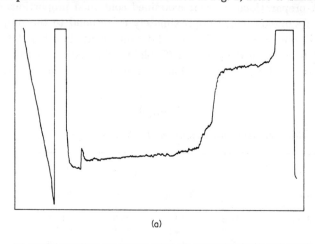

(a)

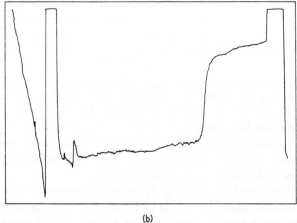

(b)

Fig. 4. Selective degradation of the S_2 component by *E. coli* phosphodiesterase. Each sample contained 30 μg/ml. ϕX DNA (preparation 6). Sedimentation from left to right at 56,100 rev./min. Pictures taken after 56 min of centrifugation. Densitometer traces from Joyce-Loebl densitometer.

(a) Control, centrifuged in 0·04 M-sodium carbonate buffer, pH 11·0 plus 0·01 M-NaCl plus 1·5 × 10⁻⁴ M-EDTA.

(b) The sample was pre-incubated for 2 hr at 30°C in the solvent described in (a). Glycine buffer, pH 9·5, MgCl₂, and *E. coli* phosphodiesterase (final concentration, 166 μl./ml.) were added and digestion carried out for 5 hr at 37°C. After addition of EDTA to halt reaction, the sample was dialysed against the carbonate buffer solvent described in (a) and then analysed in the ultracentrifuge.

(g) A discontinuity in the φX-DNA ring

The extent of hydrolysis possible by *E. coli* phosphodiesterase after varying degrees of pancreatic DNase action can be measured both by determination of the amount of acid-soluble nucleotides produced and by measurement of the decline in the ultra-violet absorbing sedimentable material in the ultracentrifuge. The results of both measurements (Table 6) are in good agreement. When the data are plotted (Fig. 5)

TABLE 6

Biological inactivation (hits/molecule)	% hydrolysis (acid soluble)	Sedimentation analysis	
		Hits/molecule†	% hydrolysis‡
0 (control)§	0	0	0
0 (control)§	0	0	0
0·20	20·0	0·43	14·3
1·2	38·0	1·2	33·9
1·4	48·4	1·5	38·9
2·2	55·0	1·8	46·5
3·8	75·1	>2	63·0
4·9	79·0	>2	69·4

Biological activity, percent hydrolysis, and sedimentation data after successive action of pancreatic DNase and *E. coli* phosphodiesterase, as in Fig. 5. For sedimentation, the 200 μl. reaction mixture was diluted fivefold with sodium carbonate and EDTA (final concentrations 0·04 M and 6·7 × 10⁻⁴ M respectively) and 5 μmole NaOH added to adjust the pH to 11·0.

  † Calculated from decrease in S_1.

  ‡ Decrease in sedimentable, ultraviolet absorbing material.

  § Recovery of biological activity in the controls was 109 and 95%, respectively.

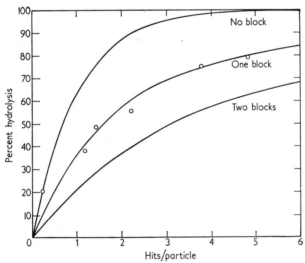

FIG. 5. Extent of hydrolysis by *E. coli* phosphodiesterase following limited digestion by pancreatic DNase. The ordinate is the percentage of φX DNA hydrolysed, as measured by the amount of acid-soluble ultraviolet absorption, expressed relative to the ultraviolet absorption of the non-hydrolysable material of the control (φX DNA treated with *E. coli* diesterase, but not pre-treated with pancreatic DNase). The abscissa is the number of "hits" due to pancreatic DNase, as measured by the residual biological activity.

The theoretical functions are computed on the assumption of random DNase "hits" in a ring structure containing no, one, or two maximally separated blocks to *E. coli* phosphodiesterase action. The functions are, m hits: for no block $[1-e^{-m}]$; for one block, $[(m+e^{-m}-1)/m]$; for two blocks, $[(m+2\ e^{-m/2}-2)/m]$.

against the number of hits produced by the nuclease (calculated, with good agreement, from either the decline in biological activity or in the amount of the S_1 component) they are found to fit the function expected if, after opening of the ring, there is a single discontinuity or block to *E. coli* phosphodiesterase action, randomly located with respect to the first nuclease hit. (The hypothesis of random location is supported by the failure to observe a discrete boundary, other than residual S_1, after completion of the *E. coli* phosphodiesterase action.)

That the action of the *E. coli* phosphodiesterase in these experiments was maximal is supported by the observation that the ratio of per cent hydrolysed to the theoretical amount of hydrolysable material (100% minus the proportion of residual S_1) increased with increasing hits per molecule, and by direct demonstration, in the case of the 2·2 hit sample, that dialysis, lyophilization, and re-incubation with *E. coli* phosphodiesterase resulted in no further degradation.

These observations account for the consistent failure in earlier experiments to remove completely the slow-moving components by enzymic action (Table 5).

4. Discussion

The enzymic and centrifugal experiments indicate that the infective ϕX-DNA chains are chemically and physically indistinguishable from the S_1 component, which has a ring structure. The principal evidence for a ring structure is that the first degradation product has an unchanged, or nearly unchanged, molecular weight, and constitutes physically an homogeneous component, while further chain breaks, which occur at a comparable rate to the first, result in random fragmentation.

Other models might be invented to explain the homogeneity of the first degradation product, which involved an exclusive site for the first either enzymic or thermal hit, resulting in a major component, the S_2, and a small fragment which has been overlooked. Such an explanation seems unlikely considering the known broad specificity of pancreatic DNase and the lack of secondary structure in DNA at 100°C. Furthermore, in both types of degradation, the rate of occurrence of the second hit is as large as that of the first, suggesting the similarity of both events. It is also pertinent that the results in the last section demonstrate that S_2 is physically, but not chemically, homogeneous. After phosphodiesterase action a distribution of sizes is obtained, indicating that the distance between the chain scission and the enzymic block is random. This result again argues against the possibility of an exclusive first-hit site.

The ring structure explains the earlier failure to find end-groups by enzymic methods. The covalent nature of the ring follows from the persistence of the S_1 and S_2 boundaries under conditions of complete exclusion of hydrogen bond formation. The ring structure of this viral DNA offers interesting possibilities for understanding the molecular basis of cyclic genomes.

It seems significant that the infective ring molecule contains one discontinuity (resistant to *E. coli* phosphodiesterase) along the primary DNA chain. The nature of this discontinuity is at present unknown. It may be an unusual nucleotide, an unusual linkage, a small hairpin-like tightly hydrogen-bonded region, or the ring may be closed by non-nucleotide components. It should be noted, however, that the two bonds which link this discontinuity to the 3′-OH and 5′-OH ends of the DNA chain, are at least as stable as ordinary phosphodiester bonds under a variety of conditions which include heating in neutral solution and at pH 9, treatment with 0·1 N-sodium hydroxide at 37°C, and with 0·1 N-acetic acid at 30°C. It is tempting to speculate that this special region may play an important part in the initiation of infection and in the replication of the viral DNA (Sinsheimer, Starman, Nagler & Guthrie, 1962).

We thank Dr. J. Vinograd for showing us prior to publication his data on the behavior of ϕX DNA in alkaline solution. Dr. E. Carusi assisted on several occasions with the ultracentrifuge studies. This research was supported by United States Public Health Service Grant RG6965. One of us (W. F.) is a Rockefeller Foundation Fellow; "Chargé de Recherches du Fonds National Belge de la Recherche Scientifique".

REFERENCES

Dove, W. F. (1962). Ph.D. Thesis, California Institute of Technology, Pasadena.
Fiers, W. & Sinsheimer, R. L. (1962a). J. Mol. Biol. **5**, 408.
Fiers, W. & Sinsheimer, R. L. (1962b). J. Mol. Biol. **5**, 420.
Guthrie, G. D. & Sinsheimer, R. L. (1960). J. Mol. Biol. **2**, 297.
Lehman, I. R. (1960). J. Biol. Chem. **235**, 1479.
Robkin, E. M., Meselson, M. & Vinograd, J. (1959). J. Amer. Chem. Soc. **83**, 2926.
Sinsheimer, R. L. (1959). J. Mol. Biol. **1**, 43.
Sinsheimer, R. L., Starman, B., Nagler, C. & Guthrie, S. (1962). J. Mol. Biol. **4**, 142.

Electron Microscopy of the Replicative Form of the DNA of the Bacteriophage φX174

Abstract. *Electron micrographs of surface films containing the replicative form of the DNA of bacteriophage φX174 show ring structures, whose average contour length is 1.64 μ, which have the characteristic appearance of double-stranded DNA throughout most of their length.*

The DNA of bacteriophage φX174 has been shown to be single-stranded (*1*) and to have a ring structure (*2*). Evidence has been presented (*3*) that during the intracellular reproduction of this virus, the viral DNA is converted to a double-stranded form, referred to as a "replicative" form, or RF, which is then multiplied. A purification of the replicative form has recently been described (*4*).

Our purification of the RF has included the use of fractional precipitation by cetyltrimethylammonium bromide (*5*) and of column chromatography, as described by Mandell and Hershey (*6*). The preparation obtained in this way is infective to protoplasts (*7*), and this infectivity has the resistance to inactivation by ultraviolet light and the buoyant density characteristic of the replicative form (*3*).

This material has been examined in the electron microscope by the monolayer technique of Kleinschmidt *et al.* (*8*). A solution of 2×10^{-6} g DNA per ml is mixed with 10^{-4} g/ml cytochrome *c* in $1M$ ammonium acetate, and 0.2 ml of the mixture is spread upon a clean surface of $0.1M$ ammonium acetate as hypophase in a Langmuir trough.

The DNA threads diffuse and part of them become adsorbed to the protein film. After full expansion to

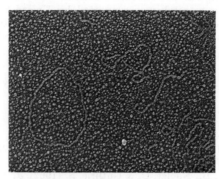

Fig. 1. Opened and twisted rings of φX-RF DNA with filamentous DNA of *E. coli*. Uranium contrast, negative. × 75,000.

about 0.85 m² per mg of protein, the film is transferred to carbonized support films (*9*) and rotary shadowed with uranium (*10*). An appreciable fraction of the DNA threads are seen in the form of rings, either open or twisted to varying extent (Fig. 1). The appearance of these circular structures throughout most of their length is that characteristic of double-stranded DNA. Measurement of the length of the DNA in over 200 of these rings, both open and twisted, has given an average length of 1.64 ± 0.11 μ. Many of the DNA threads in the preparation appear to have two ends, but only a small fraction of these are comparable in length with the rings.

If a Watson-Crick structure (*11*) is assumed for a double-stranded φX174 DNA, it should have a weight of 1.96×10^6 avograms (1 avogram = 1g/avogadro number) per micron (Na⁺ salt). Thus the observed mean length corresponds to a molecular weight of 3.2×10^6, in remarkably good agreement with the calculated value of twice the weight of the viral DNA (1.7×10^6 avograms) (*1*).

It thus appears plausible to associate these structures with the replicative

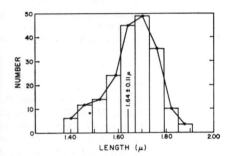

Fig. 2. Length distribution of φX-RF DNA of over 200 rings.

form of φX174 DNA and to conclude that the RF, like the viral form, occurs in a ring structure during the vegetative stage of the virus (*12*).

. ALBRECHT K. KLEINSCHMIDT
Virus Laboratory, University of California, Berkeley

ALICE BURTON
ROBERT L. SINSHEIMER
Division of Biology, California Institute of Technology, Pasadena

References and Notes

1. R. L. Sinsheimer, *J. Mol. Biol.* **1**, 43 (1959).
2. W. Fiers and R. L. Sinsheimer, *ibid.* **5**, 424 (1962).
3. R. L. Sinsheimer, B. Starman, C. Nagler, S. Guthrie, *ibid.* **4**, 142 (1962).
4. M. Hayashi, M. N. Hayashi, S. Spiegelman, *Science* **140**, 1313 (1963).
5. S. K. Dutta, A. S. Jones, M. Stacey, *Biochim. Biophys. Acta* **10**, 613 (1953).
6. J. D. Mandell and A. D. Hershey, *Anal. Biochem.* **1**, 66 (1960).
7. G. D. Guthrie and R. L. Sinsheimer, *J. Mol. Biol.* **2**, 297 (1960).
8. A. K. Kleinschmidt, D. Lang, D. Jacherts, R. K. Zahn, *Biochim. Biophys. Acta* **61**, 857 (1962).
9. A. Kleinschmidt, D. Lang, R. K. Zahn, *Z. Naturforsch.* **16b**, 730 (1961).
10. R. C. Williams and R. W. G. Wyckoff, *J. Appl. Phys.* **15**, 44 (1944).
11. R. Langridge, H. R. Wilson, C. W. Hooper, M. H. F. Wilkins, L. D. Hamilton, *J. Mol. Biol.* **2**, 19 (1960).
12. Supported in part by U.S. Public Health Service grants RG-6965 (California Institute of Technology) and by grant CA-02245 (University of California).

26 August 1963

From *Science*, Vol. 146, No. 3641, 254–255. Reproduced with permission of the authors and publisher. Copyright 1964 by the American Association for the Advancement of Science.

Electron Microscopy of Single-Stranded DNA: Circularity of DNA of Bacteriophage φX174

Abstract. *The single-stranded DNA of coliphage φX174 has been examined with the electron microscope by a modification of the protein-monolayer-adsorption technique. The molecules were found to be circular with a total length of 1.77 ± 0.13 microns.*

The DNA of coliphage φX174 is single-stranded (*1*) and also circular (*2*). The latter conclusion, which has been based upon the resistance of the DNA to digestion by exonucleases and upon an analysis of the change in the ultracentrifugal pattern after the introduction of chain scissions by deoxyribonuclease, is confirmed by electron microscopy.

Direct observation of double-stranded DNA molecules by electron microscopy is easily performed if the DNA is adsorbed onto protein monolayers (*3*). However, making this observation has been complicated by the entanglement of a variable fraction of the molecules, although it is usually possible to find a sufficient number of extended, untangled filaments to obtain a length distribution.

When φX174 viral DNA was examined, exceedingly severe tangling eliminated any possible conclusion concerning length or configuration, although the success with the double-stranded replicative form (*4*), that is, the clear demonstration of ring molecules (*5, 6*), made it seem likely that the single-stranded form is also circular.

We have assumed that intramolecular hydrogen bonding is principally responsible for entangling and have therefore modified the preparative procedure to minimize such bonding. This modification consists of denaturing the DNA by one of the methods of Freifelder and Davison (*7, 8*) in order to break all hydrogen bonds and of subsequently handling the DNA in the presence of formaldehyde at concentrations sufficient to prevent re-formation of hydrogen bonds.

The DNA of φX174 was prepared according to the method of Sinsheimer (*1*) and adjusted to a concentration of 70 μg/ml in a solution of 0.1*M* NaCl and 0.05*M* tris (*p*H 7.7); this served as a stock. The solution for spreading monofilms was made by mixing the following components in sequence for the indicated times: 0.04 ml DNA + 0.04 ml 1*M* NaOH, 20 seconds; 0.4 ml 37 percent formaldehyde (adjusted to *p*H 11 with 1*N* NaOH), 30 seconds; 0.1 ml 1*M* KH_2PO_4; 0.6 ml 0.01*M* PO_4–0.005*M* EDTA (*p*H 7.8); and 0.4 ml 4*M* NaCl.

One-tenth milliliter of the denatured DNA solution was diluted to contain 0.5 μg of DNA per milliliter by adding 1*M* ammonium acetate containing 0.5 percent formaldehyde (neutralized and boiled for 10 minutes). Finally, 0.1 ml of 0.01 percent cytochrome *c* in 1*M* ammonium acetate was added. This solution was spread, 4 minutes later, by the standard method (*3*) except that the spreading was done on 0.5 percent formaldehyde (pretreated as described) instead of water. Lower concentrations of formaldehyde were not satisfactory. The viscoelastic properties and the spreading speed of the film on this solvent are comparable with the properties and speed obtained with spreading the film on water. The film was transferred to carbonized support films on platinum grids by touching the surface and the adhering drople was removed by touching an ethanol surface (*6*). After drying with filter paper the preparation was rotary shadowed with uranium at an angle of 6 to 10 degrees. Undenatured φX174 DNA was also examined by the standard technique (without treatment with alkali and formaldehyde).

Figure 1 shows representative single-stranded rings of φX174 DNA. In a

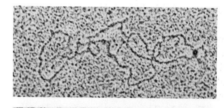

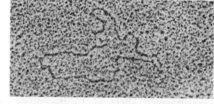

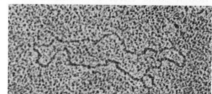

Fig. 1. Electron micrographs of φX174 DNA rings; 7.8 cm is equivalent to 1 μ. Total magnification, 78000.

typical field about one-half of the DNA is easily seen as circular molecules; filaments account for fewer than 5 percent. The remainder of the DNA consists of molecules overlapping, aggregated, or tangled, the number of which can be substantially reduced by spreading the DNA at lower concentrations. When they are mixed with other material which is double-stranded (9) the single strands appear thinner, are more difficult to see because of lower contrast with the background and, in general, show sharp kinks, an observation consistent with the expected greater flexibility of the single strands. Undenatured preparations (untreated with alkali and formaldehyde) show badly tangled masses, without any visible free ends.

The lengths of most of the rings were measured by tracing a projected enlarged image on paper and placing 0.08-cm diameter polyethylene tubing along the path (6). A histogram of the measured lengths of 186 rings is shown in Fig. 2. The mean contour length is $1.77 \pm 0.13 \mu$ (7.3 percent), and it is greater than that found for the replicative form ($1.7 \mu \pm 7$ percent) (5). The greater mean length probably reflects the greater flexibility and less degree of coiling of single-stranded DNA, although some effect of formaldehyde is certainly possible. If the molecular weight is 1.7×10^6 (1) the mass per unit length of molecules prepared in this way is 0.95×10^6 dalton/micron.

The small number of filaments is an indication of the safety of the alkali treatment for denaturing DNA (7) and furthermore reflects the stability

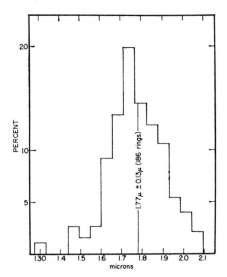

Fig. 2. Histogram of length distribution of 186 ϕX174 DNA rings. The error shown for the mean is the standard deviation.

of the linkage that causes ring closure. We assume that this linkage is probably a covalent bond. This is consistent with the resistance of infectious ϕX174 DNA to alkali (10).

DAVID FREIFELDER
Donner Laboratory, University of California, Berkeley
ALBRECHT K. KLEINSCHMIDT
Virus Laboratory, University of California, Berkeley
ROBERT L. SINSHEIMER
Division of Biology, California Institute of Technology, Pasadena

References and Notes

1. R. L. Sinsheimer, *J. Mol. Biol.* **1**, 43 (1959).
2. W. Fiers and R. L. Sinsheimer *ibid.* **5**, 408 (1962).
3. A. K. Kleinschmidt and R. K. Zahn, *Z. Naturforsch.* **14b**, 770 (1959).
4. R. L. Sinsheimer, B. Starman, C. Nagler, S. Guthrie, *J. Mol. Biol.* **4**, 142 (1962).
5. A. K. Kleinschmidt, A. Burton, R. L. Sinsheimer, *Science* **142**, 961 (1963).
6. B. Chandler, M. Hayashi, M. N. Hayashi, S. Spiegelman, *ibid.* **143**, 47 (1964).
7. D. Freifelder and P. F. Davison, *Biophys. J.* **3**, 49 (1963).
8. P. F. Davison, D. Freifelder, B. W. Holloway, *J. Mol. Biol.* **8**, 1 (1964).
9. D. Lang, R. K. Zahn, A. K. Kleinschmidt, *Biophysik*, in press.
10. R. L. Sinsheimer, unpublished.
11. Supported in part by grants from the Atomic Energy Commission and the National Aeronautics and Space Administration, by grant AI 01267 from the National Institute of Allergy and Infectious Diseases, by grant CA 02245 from the National Cancer Institute, and by grant RG 6965 from the National Institutes of Health.
12. We thank Susan G. Johnson who took the electron micrographs, and Katherine Le Blanc, Kristin Chung, and John Polacheck, who traced and measured lengths.

20 July 1964

From the *Proceedings of the National Academy of Sciences*, Vol. 50, No. 2, 236–243, 1963. Reproduced with permission of the authors.

EVIDENCE FOR A RING STRUCTURE OF POLYOMA VIRUS DNA*

BY R. DULBECCO† AND M. VOGT†

CALIFORNIA INSTITUTE OF TECHNOLOGY

Communicated June 11, 1963

The DNA of polyoma (PY) virus has two interesting features: it is unusually resistant to heat or formamide denaturation,[1] and it has two distinct components, of sedimentation coefficient 14 and 21, respectively.[2] A two-component sedimentation was formerly shown to be a characteristic of the DNA of papilloma virus,[3] which is considered a member of the same group of viruses.

Evidence presented in this article shows that both these features of polyoma virus DNA are the consequence of special configurational properties.

Material and Methods.—Purified polyoma virus was prepared according to Winocour;[4] DNA was extracted from the purified virus with phenol[5] and stored at −70°C. Virus labeled with either P^{32} or H^3 was obtained by adding label (P^{32} orthophosphate or H^3 thymidine) 24 hr after infecting the cultures. Sedimentation studies were carried out by the band sedimentation method:[6] 3 ml of CsCl solution of density 1.50 were placed in a centrifuge tube and covered with paraffin oil; the sample, of a volume 0.05–0.25 ml. was added dropwise on top of the oil layer. The tube was spun in an SW-39 rotor in a model L Spinco ultracentrifuge at 35 K rpm; in most experiments, spinning time was 4 hr. For shorter runs, the tube containing the CsCl solution and oil was spun for 4 hr before adding the sample, in order to preform the gradient. The CsCl solution was buffered either at pH 7.5 with 0.01 M Tris, or at alkaline pH (11.0, 11.8, and 12.5) with 0.01 M phosphate buffer.[7] For density gradient equilibrium sedimentation, 3 ml of a CsCl solution of density 1.71 or 1.72 at pH 7.5, and 1.73 at pHs 11–12.5 were used. The solution was covered with oil, spun in the SW-39 head of the model L ultracentrifuge at 29 or 30 K for 70–80 hr. In all cases, fractions were collected from the bottom of the tube. Columns of methylated albumin were prepared according to Mandell and Hershey.[8] Infectivity of polyoma DNA was measured as described by Weil.[5] Pancreatic DNAase Worthington 1x crystallized was used. *E. coli* phosphodiesterase (Lehman enzyme)[9] was a gift of Dr. R. L. Sinsheimer. DNAase and phosphodiesterase digestion were carried out according to Fiers and Sinsheimer.[16]

Results.—In brief, the following results were obtained. (1) PY DNA always sediments in two bands, a fast one (*F*) and a slow one (*S*). The *F* and *S* components are stable under a variety of conditions; both are infectious. (2) In the column of methylated albumin the *S* component is eluted at *higher* salt concentration than the *F* component. (3) The following observations were made by density gradient equilibrium centrifugation. At pH 7.5 the bands formed by the *F* and by the *S* component have the same density and do not differ greatly in width; at pH 11.8 most of the *F* component has the density of native DNA, whereas the *S* component has the density of denatured DNA. At pH 12.5 both components are denatured; the *S* component forms a wider band. (4) In band sedimentation at pH 12.5 the *F* component sediments about 2.5 times faster than at pH 7.5; after neutralization and annealing it reacquires the original sedimentation characteristics. The *S* component, on the contrary, migrates at pH 12.5, about the same speed as at pH 7.5. (5) The infectivity of the *F* component is increased several times after treatment at pH 12.5 followed by neutralization; the infectivity of the *S* component is lost after this treatment. (6) Pancreatic DNAase converts the *F* component into the *S* component, with loss of infectivity. The kinetics of conversion is unusual, since it is of first order. These results show that molecules of polyoma virus DNA in different states exist in the F and in the S component. The data are consistent with the hypothesis that the F component is made up of molecules in ring form, and the S component of molecules in linear, or open, form. The results will now be given in greater detail, in terms of this hypothesis.

Properties of the two components of different sedimentation velocity: Many preparations of PY DNA, of both small plaque and large plaque type, were examined by band sedimentation in CsCl at pH 7.5. They all contained the F and the S band. The band pattern observed with this method was found to be similar to that observed with the sucrose gradient method. The major component was the F component; it represented 60–80 per cent of the total in DNA preparations which had been extracted only once with phenol, and 90–95 per cent in preparations which had been extracted two or three times.

The ratio of the uncorrected distances traveled by the two components was 1.31 in sucrose gradient with $5 \times 10^{-3} M$ Mg^{++}, $2.5 \times 10^{-2} M$ K^+, and $5 \times 10^{-2} M$ Tris buffer, pH 7.5; the ratio was 1.22 in CsCl of density 1.50 at pH 7.5.

Both components were infectious and produced plaques of equal type. The average specific infectivity (ratio: infectivity to OD_{260}) was two to three times larger in the F band (Fig. 1). In the F band the infectivity was distributed almost uniformly, whereas in the S band infectivity was present only in the frontal part. The S band is thus heterogeneous; this band is assumed to consist in its frontal part of infectious linear molecules, obtained by opening the ring, and in its trailing part, of smaller noninfectious fragments.

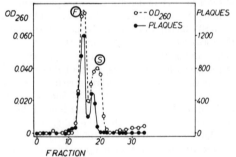

FIG. 1.—Band sedimentation of polyoma DNA at pH 7.5. Polyoma DNA extracted once with phenol and sedimented at 35 K rpm in CsCl of density 1.50 for 4 hr.

Stability: Isolated F and S components were prepared, either unlabeled or labeled with different labels, by collecting fractions corresponding to the peak part of each band. The F and S components prepared in this way maintained their characteristic sedimentation velocities after prolonged storage at $-70°C$. Neither conversion from one to the other nor alterations of their sedimentation properties were obtained by the following treatments: exposure to pHs from 4.5 to 11, or to $6 M$ CsCl; treatment with 2 per cent Na-dodecylsulfate for 30 min at room temperature; annealing separately or together at 70°C for up to 1 hr; chromatography in a column of methylated albumin; treatment with the Lehman enzyme.

Chromatographic properties: P^{32}-labeled isolated F and S components mixed with unfractionated H^3-labeled PY DNA were chromatographed in the column of methylated albumin (Fig. 2). The F component was distributed almost uniformly throughout the whole elution band; the S component eluted at higher salt concentration. In turn, the frontal part of the elution band was rich in the fast component, whereas the trailing part was richer in the slow component. The infectivity was distributed almost uniformly throughout the chromatographic band.

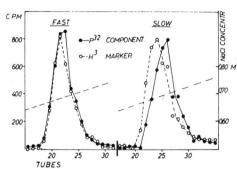

Fig. 2.—Chromatography of *F* and *S* components on a column of methylated albumin. 0.5 μg of DNA was used in each run. A column of 1 cm diameter, 11 cm long was used. The DNA was applied in 0.55 *M* NaCl, 0.05 *M* phosphate buffer pH 6.9. The elution gradient was linear. Fractions of 2 ml each were collected. To each fraction 0.03 ml of a 4% serum albumin solution and trichloroacetic acid to a final concentration of 5% were added in the cold. After centrifugation the pellets were dissolved in NH₄OH and counted.

These results contradict the hypothesis that the *S* band is made up of monomer DNA molecules and the *F* band of end-to-end dimers, as suggested by Watson and Littlefield[3] for the two components of papilloma virus DNA of different sedimentation velocity: the dimers should in fact elute at *higher*, rather than lower, salt concentration. The results are compatible with the ring hypothesis, since the linear molecules may well appear longer to the column than the ring molecules.

Density gradient equilibrium centrifugation: Mixtures of *F* and *S* components, labeled with different isotopes, were banded at pHs 7.5, 11, 11.8, and 12.5. At pH 7.5 the two components gave rise to essentially coincident bands at equilibrium (Fig. 3); the band of the *S* component was wider by a factor of 1.2, perhaps owing to the presence, in this component, of fragments of DNA molecules. The *S* component reached equilibrium more slowly than the *F* component. At pH 11 both components had the buoyant density of native PY DNA, although

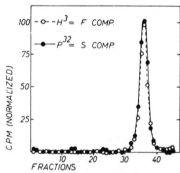

Fig. 3.—Density gradient equilibrium centrifugation of *F* and *S* components at pH 7.5. H³-labeled DNA containing *ca.* 90% *F* component, and P³²-labeled DNA containing *ca.* 80% *S* component, in amounts of 0.2 μg each, were centrifuged in CsCl of density 1.72 in 0.01 *M* Tris buffer of pH 7.5 for 80 hr at 29 K rpm.

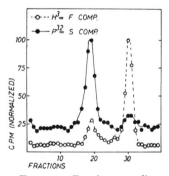

Fig. 4.—Density gradient equilibrium centrifugation of *F* and *S* components at pH 11.8. H³-labeled *F* and P³²-labeled *S* component in 0.1 *M* phosphate buffer at pH 11.8 were centrifuged in CsCl solution of density 1.73 in 0.01 *M* phosphate buffer, pH 11.8, for 80 hr at 30 K rpm.

the band made up of the *S* component tended to spread toward higher densities, owing to incipient denaturation. At pH 12.5 both components were completely denatured and had a correspondingly higher density;[7] the band of the *S* component was 1.5 times wider than that of the *F* component. At pH 11.8 the *S* component had a denatured density, whereas the *F* component separated into two bands, with most of the DNA in a band of native density, and the remainder in a band of denatured density (Fig. 4). The proportion of F DNA in these bands differed in various preparations.

Band sedimentation at pH 12.5: The S component sedimented at a speed not very different from that which it had at pH 7.5, whereas the F component gave rise to two bands, one (*SF band*) sedimenting like the S component at the same pH, the other (*FF band*) sedimenting about 2.5–2.8 times faster (Fig. 5). The proportion of F DNA which appeared in the FF band was similar to that resisting denaturation in the equilibrium centrifugation. The DNA of the FF band reacquired the sedimentation velocity of F DNA when annealed at pH 11.8 in 6 M CsCl: when rapidly neutralized, most of the FF DNA retained the high sedimentation velocity.

The results obtained at alkaline pH are interpreted as follows: the ring molecules (F component) are resistant to denaturation, presumably because they do not have free ends; when denatured, they remain topologically unchanged, i.e., still coiled together, and sediment very rapidly, like incompletely denatured DNA.[11] Upon annealing, they reconstitute the ring molecules characteristic of the F component. The linear molecules (S component) denature more readily, presumably because they have free ends; upon denaturation they give rise to separate single strands, which sediment at a rate similar to that of the native S component. The molecules of the F component, which sediment slowly at pH 12.5 (SF band) and denature readily, are thought to originate from F molecules either with single-chain cuts, or with alkali-labile points, such as pre-existing depurinations. Upon denaturation, such molecules would give rise both to single-stranded rings and single-stranded linear chains, which would move close to each other, like the analogous ϕX forms.[12]

Behavior of the infectivity at alkali pH: Treatment at pH 11.8 or 12.5 abolished the infectivity of the S component. The infectivity of the F component was not affected by treatment at pH 11.8; after treatment at pH 12.5 and rapid neutralization it was increased about four times. This increase can be probably attributed to increased uptake of partially denatured DNA by the cells.[1]

The two bands produced by the F component when sedimented at pH 12.5 were both infectious; in the SF band the specific infectivity (ratio: infectivity to radioactivity) was maximal in the frontal and absent in the trailing part of the band.

These results show that the SF band is heterogeneous. The heterogeneity is also shown by the different sensitivity to the Lehman enzyme of the various fractions composing the band. The DNA present in the frontal part of the band was resistant to the enzyme, whereas that of the trailing part was partly hydrolyzed. The infectivity was not affected by the enzyme. A possible interpretation of these results is that the infectious DNA present in the frontal part of the SF band is made up of single-stranded rings, and the DNA of the trailing part, mostly noninfectious, of single-stranded linear chains. Other interpretations are, however, possible.

DNAase digestion: Samples of the labeled isolated components were treated with the enzyme at constant concentration for various lengths of time. Then the sedimentation pattern of each sample was determined, after mixing it with differentially labeled unfractionated PY DNA as marker. *F component* (Fig. 6): As a function of the time of digestion the height of the F band decreased, and a band coincident with the native S band was formed. The change affected the height of the F band but not its position. No component migrating at a velocity intermediate between

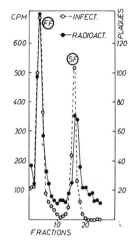

FIG. 5.—Band sedimentation of the F component at pH 12.5. 0.2 μg of H³-labeled PY DNA, made up almost exclusively of F component, was mixed with 1.5 μg of unlabeled isolated F component, and brought to pH 12.5 with 0.1 M phosphate buffer. The mixture was sedimented at 35 K rpm for two hr in a tube containing CsCl of density 1.50, in 0.01 M phosphate buffer, pH 12.5. The tube had been prespun for 4 hr at 35 K rpm in order to preform the gradient. Both radioactivity and infectivity were determined in all fractions.

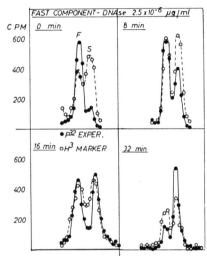

FIG. 6.—Action of DNAase on the F component. Isolated P³²-labeled F component DNA (0.5 μg/ml), unfractionated unlabeled PY DNA (4 μg/ml), and DNAase, pre-equilibrated at 37°C, were mixed at time zero and incubated at 37°C for various lengths of time. The incubation medium was: 0.04 M NaCl, 0.01 M Tris buffer, pH 7.5, 0.004 M MgCl₂, 5 × 10⁻⁵ gr/ml of bovine serum albumin. DNAase action was interrupted by 0.008 M Versene and 0.2 M phosphate buffer of pH 9.0. Each sample was then mixed with unfractionated H³-labeled PY DNA, and sedimented at 35 K rpm for 4 hr in CsCl of density 1.50.

those of the F and of the S band was formed. In the further study of this phenomenon, a DNAase treatment which converts 37 per cent of the F component into S will be designated as a "converting hit." *S component* (Fig. 7): As a function of the time of digestion the S band at first increased in width, and showed subsequently a progressively lower velocity of sedimentation without generating a new separate band.

The results obtained with the F component can be interpreted as those obtained by Fiers and Sinsheimer[13] with the DNA of phage φX 174: that a single scission in the ring molecule converts it to the linear form, thus causing a discrete difference in sedimentation velocity.

Kinetics of DNAase digestion: In repeated experiments the kinetics of disappearance of the F band as a function of time, at constant DNAase concentration, was found of first order. This result suggested at first that the ring contains an inserted single-stranded segment. To search for such a segment, labeled F DNA was exposed to several converting hits with DNAase and then to the Lehman enzyme. No measurable hydrolysis was observed under conditions where 1 per cent hydrolysis could have been detected. We conclude that the DNA of the *F* component does not have a single-stranded segment of sufficient size to explain the kinetics of DNAase action.

In order to interpret the kinetics of *F* → *S* conversion, the kinetics of DNAase action on either component was studied by sedimenting the enzyme-treated DNA both at pH 7.5 and at pH 12.5, together with untreated DNA, differentially labeled. The changes in sedimentation velocity at pH 7.5 reflect the consequences of *complete scissions* of the DNA molecules; the changes observed at pH 12.5 reflect the consequences of *single-chain scissions*. The following observations were made. *F component:* The effect of DNAase, whether studied at pH 7.5 or 12.5, was similar: at pH 12.5 the *FF* component was converted into the *SF* component at

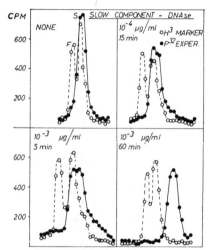

Fig. 7.—DNAase action on the *S* component. Same conditions as in Fig. 6, except that isolated *S* component was used.

exactly the same rate at which the *F* component was converted into the *S* component at pH 7.5. This result shows that DNAase did not produce single-chain cuts in the ring molecules, which would have resulted in faster disappearance of the *FF* component at pH 12.5, compared to the *F* component at pH 7.5. Thus, the enzyme broke both chains at once. *S component* (either native or produced from the action of DNAase on the *F* component): At either pH the band formed was modified in the same qualitative way by DNAase; the band became wider, and then its average velocity of sedimentation progressively decreased, as digestion proceeded. The effect was, however, much more pronounced at pH 12.5.

To compare the effects observed at the two pHs, the number of scissions was evaluated from the sedimentation behavior of the DNA molecules. When the number of scissions was low, it was evaluated by determining the proportion of material remaining in a position coincident with the *S* band (pH 7.5) or with the *SF* band (pH 12.5). When the number of scissions was higher, it was evaluated from the change in average velocity of sedimentation,[14] with the assumption that the sedimentation velocity is proportional to $MW^{0.37}$ at pH 7.5, and to $MW^{0.5}$ at pH 12.5.[15] In repeated experiments it was found that single-chain cuts, demonstrable at pH 12.5, were produced at a much faster rate than complete scissions, demonstrable at pH 7.5; and furthermore, that complete scissions occurred according to a kinetics which appeared to be of higher order (Fig. 8).

These results show that there is a profound difference in the effect of DNAase on the *F* and on the *S* component.

The interpretation of these results is that the ring molecules are labile for reasons related to their shape. Perhaps the ring form imposes a statistically higher stress on the phosphodiester bonds; hydrolysis of one bond may increase considerably the stress on the complementary bond, making it highly susceptible to hydrolysis by the same enzyme molecule, or to breakage. As soon as the ring opens, the stress disappears.

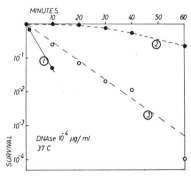

Fig. 8.—Kinetics of DNAase digestion on H³-labeled PY DNA containing almost exclusively *F* component. All rates were determined within a single experiment, by using a single sample which contained DNA, 15 μg/ml and DNAase, 10⁻⁴ μg/ml. Survival = proportion of the respective form that remained unchanged. *Curve 1:* kinetics of disappearance of *F* (pH of sedimentation 7.5) or *FF* (pH of sedimentation 12.5). *Curve 2:* kinetics of degradation of *S* (derived from *F*); pH of sedimentation 7.5. *Curve 3:* kinetics of degradation of *SF*; pH of sedimentation 12.5. In Curves 2 and 3 low survivals were calculated from the average number of scissions per molecule, according to the equation: survival = exp (− average number of scissions).

Effect of DNAase on infectivity: These experiments yielded two types of results. The first result is that the infectivity of the DNA of the *F* band decreased as a function of the time of exposure to DNAase at exactly the same rate as the radioactivity. This shows that the infectious molecules have the same properties as the bulk of the DNA molecules. Experiments carried out with pure *F* component showed that not more than 10 per cent of the infectivity lost by the *F* component appeared in the *S* band. Thus, *F* → *S* conversion by DNAase entails loss of infectivity in most, if not all, molecules.

The other result is that the infectivity of the *S* component was more resistant to the action of DNAase than that of the *F* component. The number of hits in the *S* component determined from infectivity loss corresponded to the average number of complete scissions per molecule, as determined from the sedimentation behavior. With increasing exposure to DNAase, as the radioactivity band increased in width, the residual infectivity remained associated with molecules which sedimented like those of the frontal part of the original *S* band.

These results show that in both the *F* and the *S* components loss of infectivity was caused by complete scissions, whereas single-chain breaks were inconsequential. Single-chain cuts had, however, an effect on infectivity, if DNA which had experienced a few DNAase hits was heated at 100°C for 5 min in 0.15 *M* NaCl. This DNA lost all its residual infectivity, in contrast to DNA not exposed to DNAase which increased its infective titer two or three times after heating.[1]

Conclusions.—The results here reported are in agreement with the hypothesis that the DNA of polyoma virus is a ring molecule. Many other possible structures which may account for the existence of two components of different sedimentation velocity are contradicted by the results.

According to the results, the ring should be made up of two complete complementary strands, each one separately closed on itself, plectonemically coiled, and not cross-linked. It is likely that the ends of each polynucleotide chain are connected by a special "linker," as in φX DNA.[13] This is suggested by the presence of infectivity in the native linear molecules, in contrast to its absence in at least most of the linear molecules obtained by DNAase treatment. These results also show that infectivity requires the information of the whole DNA molecule and the integrity of its functional units.

Single-chain breaks, as produced by DNAase in infectious *S* molecules, do not abolish infectivity. Thus, either these breaks are repaired after the DNA enters the cell, or the reading of the DNA molecules for RNA synthesis or replication is undisturbed by the breaks.

The behavior of infectivity at alkali pH suggests that single-stranded rings are infectious; only further experiments may, however, decide whether this is true.

Two results remain to be explained: (1) the difference in sedimentation velocity of the ring and of the linear molecules is larger than for other ring-shaped molecules, such as those of ϕX DNA[12] or λ DNA;[16] (2) the ring is unstable, as suggested by the kinetics of DNAase digestion. It can be tentatively suggested that they are the consequence of the combination of double strandedness and of the size of the ring.

We are grateful to Prof. R. L. Sinsheimer for useful discussions. The competent and dedicated assistance of Miss Maureen Muir is gratefully acknowledged.

This work will be supplemented by studies with the analytical ultracentrifuge.[17]

* Aided by grants from the National Institutes of Health, U.S. Public Health Service, the American Cancer Society, and The National Foundation.

† Present address: The Salk Institute for Biological Studies, La Jolla, California.

[1] Weil, R., these Proceedings, **49**, 480 (1963).

[2] Crawford, L. V., *Virology*, **19**, 279 (1963).

[3] Watson, J. D., and J. W. Littlefield, *J. Mol. Biol.*, **2**, 161 (1960).

[4] Winocour, E., *Virology*, **19**, 158 (1963).

[5] Weil, R., *Virology*, **14**, 46 (1961).

[6] Vinograd, J., R. Bruner, R. Kent, and J. Weigle, these Proceedings, **49**, 902 (1963).

[7] Vinograd, J., J. Morris, N. Davidson, and W. F. Dove, Jr., these Proceedings, **49**, 12 (1963).

[8] Mandell, J. D., and A. D. Hershey, *Anal. Biochem.*, **1**, 66 (1960).

[9] Lehman, I. R., *J. Biol. Chem.* **235**, 1479 (1960).

[10] Fiers, W., and R. L. Sinsheimer, *J. Mol. Biol.*, **5**, 408 (1962).

[11] Freifelder, D., and P. F. Davidson, *Biophys. J.*, **3**, 49 (1963).

[12] Sinsheimer, R. L., *J. Mol. Biol.*, **1**, 43 (1959).

[13] Fiers, W., and R. L. Sinsheimer, *J. Mol. Biol.*, **5**, 424 (1962).

[14] Thomas, C. A., *J. Am. Chem. Soc.*, **78**, 1861 (1956).

[15] Doty, P., J. Marmur, J. Eigner, and C. Schildkraut, these Proceedings, **46**, 461 (1960).

[16] Hershey, A. D., E. Burgi, and L. Ingraham, these Proceedings, **49**, 748 (1963).

[17] Weil, R., and J. Vinograd, in preparation.

From the *Proceedings of the National Academy of Sciences*, Vol. 50, No. 4, 730–738, 1963. Reproduced with permission of the authors.

THE CYCLIC HELIX AND CYCLIC COIL FORMS OF POLYOMA VIRAL DNA

By Roger Weil and Jerome Vinograd

GATES AND CRELLIN LABORATORIES OF CHEMISTRY* AND THE NORMAN W. CHURCH LABORATORY OF CHEMICAL BIOLOGY, CALIFORNIA INSTITUTE OF TECHNOLOGY

Communicated by R. Dulbecco, August 19, 1963

The DNA extracted from polyoma virus exhibits certain properties which have not been reported for other viral base-paired DNA's.[1] The DNA renatures monomolecularly. The loss of helical configuration does not impair biological activity. Heating at 100° for 10–20 min followed by rapid cooling does not reduce the infective titer. Dulbecco has given evidence suggesting that a fraction of the polyoma DNA molecules are cyclic.[2] This form would in part account for the above properties.

The results reported in this communication give strong evidence for cyclic polyoma DNA molecules; they also define certain properties of the cyclic and open forms. Some of the evidence reported here is similar to that obtained in an independent investigation by Dulbecco and Vogt.[3]

Materials and Methods.—Isolation and purification of the virus and extraction of the DNA: The virus was isolated and purified by the procedure described by Winocour.[4] The main visible band from the CsCl density gradient was collected, and the DNA extracted by a modified phenol procedure.[1] The homogenization step was omitted. The CsCl, Harshaw Chemical Company, Optical Grade, formed solutions which remained clear upon addition of alkaline buffers. Reagent grade inorganic chemicals and Merck's reagent grade formaldehyde were used.

Analytical ultracentrifugation: Sedimentation velocity analyses were performed in the Spinco analytical ultracentrifuge by band-centrifugation.[5] Boundary-velocity experiments were precluded by the limited supply of the virus. Buoyant density experiments were performed at 44,770 rpm at 25°C and analyzed by methods previously described.[6]

Preparation of the alkaline solutions: The alkaline CsCl solutions were prepared with the buffers described by Vinograd et al.[7] The alkaline NaCl solutions were prepared by mixing 1 volume of 4 M NaCl with 3 volumes of 0.11 M KOH.[7a] The pH measured with a small general-purpose Beckman glass electrode was 12.2 ± 0.1. DNA samples were made alkaline in the sample hole by addition of 1 volume of 1.1 M KOH[7a] to 9 volumes of DNA solution, and mixed by drawing the solutions back into the Kel-F tubing several times. The final pH was 12.5 ± 0.2. Preincubation for longer periods was performed in 0.032-in. I.D. Kel-F tubing sealed at both ends with 4-mm plugs of silicone grease.

Denaturation and renaturation: These experiments were performed as previously described.[1] Formaldehyde denaturations[8] were performed with 25 μl samples in 11% CH_2O, pH 8.3, ionic strength 0.01, in sealed glass melting-point tubes at 70–75°C 2–3 min followed by chilling at 0°C.

Sonication: The DNA solutions in thin 1-ml nitrocellulose test tubes were sonicated in a 9,000 cps Raytheon Sonicator Model no. S102A at the indicated power level.

Calculations: The molecular weight of anhydrous NaDNA was calculated from the standard deviation of the band.[6] The additional cesium ions and preferential hydration[7] introduced at high pH were included in the calculations. Buoyant densities were measured with *M. lysodeikticus* DNA, $\rho_0^0 = 1.732$ gm cm^{-3}, as a reference, a value based on $\rho_0^0 = 1.710$ gm cm^{-3} for DNA from *E. coli.*[9] This value disagrees with a recent redetermination by +0.006 gm cm^{-3}.[7]

Results.—Sedimentation analysis of DNA extracted from the virus: Band-centrifugation analyses of the DNA from six different virus batches all showed three components which we refer to as I, II, and III, in order of decreasing sedimentation coefficients. The values, $s_{20,w}$, in 1 M NaCl 0.01 M Na$_2$HPO$_4$, pH 8.0 were 20.3 S ± 0.4 S.D., 15.8 S ± 0.4 S.D., and 14.4 S[10] (Fig. 1A). The component composition was determined in 3.5 M CsCl (Fig. 1B). In this solvent the band profiles are more symmetrical than in 1 M NaCl or 1 M CsCl. In all preparations

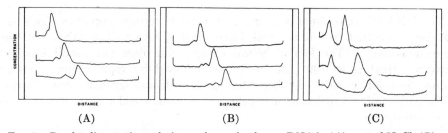

Fig. 1.—Band sedimentation velocity analyses of polyoma DNA in (*A*) neutral NaCl, (*B*) neutral CsCl, (*C*) alkaline NaCl. (*A*) Lamella: 15 μl, 24 μg/ml DNA; bulk solution: 1 M NaCl, 0.01 M Na$_2$HPO$_4$, pH 8.0. Densitometer records are from photographs taken at 16-min intervals. 35,600 rpm, 20°. (*B*) Lamella: 20 μl, 60 μg/ml DNA; bulk solution: CsCl, $\rho = 1.36$, 0.01 M Tris, pH 8.1. 16-min intervals, 35,600 rpm, 20°. (*C*) Lamella: 25 μl, 24 μg/ml DNA, pH = 12.75; bulk solution: 1.0 M NaCl, 0.10 M KOH, pH 12.2, 8-min intervals. 35,600 rpm.

I comprised 80–90 per cent of the DNA. Component II varied between 1 and 20 per cent and III between 1 and 10 per cent. DNA released from the virus in the sample hole, pH 11.2 or pH 12.3, gave patterns indistinguishable from those obtained with the phenol extract. The sedimentation pattern at neutral pH did not change upon storage at 4° for periods up to 20 weeks, or upon freezing and thawing. The DNA formed a single band at equilibrium in CsCl with a buoyant density of 1.709 gm cm^{-3}. All preparations formed bands which were skewed on the light side. It is shown later that this skewness is due to III.

The renaturation behavior of polyoma DNA suggested that the strands did not physically separate when heated under conditions in which strand separation normally occurs. To examine this problem we employed the procedure described by Freifelder and Davison.[8] Solutions of polyoma DNA were denatured in CH_2O and examined in sedimentation velocity experiments in 11 per cent CH_2O, 1 M NaCl. The DNA now sedimented in two components (80 per cent fast) with a ratio of sedimentation coefficients of 2.2. This result was not compatible with the hypothesis that the fast material represented singly cross-linked double strands, and the slow material single strands[8] because a ratio of $\sqrt{2}$ is to be expected. Strand separation was also examined at high pH, 12.2 ± 0.1, in 1 M NaCl. In this solvent, complete strand separation is to be expected.[7] All fresh polyoma DNA preparations, including those stored at −70°C, sedimented in two well-resolved components[11] (about three-fourths fast) with sedimentation coefficients, $s^\circ_{20,w}$ of 53.1 S ± 0.3 S.D. and 14.6 S ± 0.5 S.D. (Fig. 1C). We will refer to the alkali components as the 53 S and 15 S DNA's and will show below that these are double-stranded cyclic coils and single-stranded random coils, respectively. That both of these components can derive from I was shown in experiments in which the DNA was so dilute that II and III could not be detected at pH 8; in alkali, however, both the 53 S and 15 S components were observed.

The sedimentation velocity of the 15 S DNA was compared with the sedimentation velocity of ϕX-174 DNA (16.0 S) in the same solvent, and its molecular weight calculated with the relation for random coils, $S_1/S_2 = (M_1/M_2)^{0.5}$. With $M = 1.7 \times 10^6$ for ϕX-174 DNA[12] we obtain $M = 1.4 \times 10^6$ for the 15 S component. The same value is obtained if we use single-stranded T-7 DNA as a reference in this calculation, $s^\circ_{20,w} = 39$ S, $M = 9.5 \times 10^6$.[13]

Fresh polyoma DNA forms two well-separated bands in buoyant density experiments in alkaline CsCl pH 12.3. In *band-buoyancy* experiments (band-centrifugation performed with buoyant bulk solutions[5]) (Fig. 2), the 53 S DNA forms a band at a higher density than the 15 S DNA. The buoyant densities are 1.784 and 1.766 gm cm^{-3}, referred to 1.710 for *E. coli* DNA. The molecular weight calculated from several band profiles for the 53 S DNA was 2.4 ± 0.3 $\times 10^6$. The band formed by the 15 S DNA was skewed on the light side because of slowly sedimenting DNA and a small amount of III.

With a ratio of molecular weights $M_{53\ S}/M_{15\ S} \approx 2$, as indicated in the arguments in the above two paragraphs, we expect for random coils a ratio of sedimentation coefficients of 1.4. The high velocity, 53 S, in alkali of the double-stranded DNA has thus to be attributed to a drop in the frictional coefficient by a factor of 2.5 relative to a random coil containing two strands (Fig. 3). We therefore propose that the double-stranded cyclic helix denatures to form a double-stranded cyclic coil in which the turns originally present in the helix are conserved. The topological constraint imposes a compact structure on this molecule (Fig. 3) and gives rise to an unusually high sedimentation coefficient and buoyant density.

Several physical and chemical agents can convert the cyclic helix, the linear helix and the double-stranded cyclic coil to single strands, as well as the cyclic helix to the double-stranded cyclic coil and to the linear helix.

(1) *Exposure to high pH:* After the initial rapid conversion of about one fourth of component I into single strands, a further conversion into single-stranded 15 S DNA occurs at a slow rate. For example, after 72 hr incubation at pH 12.1 in 0.01 M NaCl, the amount of 53 S DNA decreased to about one third with a concomitant increase in 15 S DNA. Unlike the usual 15 S

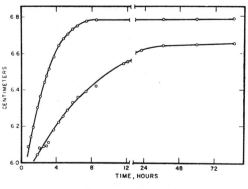

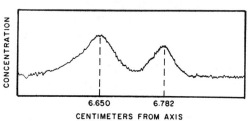

FIG. 2.—Band-buoyancy analysis of polyoma DNA in alkaline CsCl. Lamella: 20 μl, 24 μg/ml DNA, pH 12.75. Bulk solution: CsCl, ρ = 1.76, 0.04 M K₃PO₄, pH = 12.35. 44,770 rpm. *Upper:* position of 15 S and 53 S components at various times. *Lower:* densitometer records of bands at 72 hr, 25°.

DNA, this material contained two velocity components, a small fraction, presumably cyclic single strands, sedimenting about 10 per cent faster than the main component.[11] The patterns were similar to those observed in experiments with φX-174 DNA, which contained both linear and circular molecules.[14]

(2) *Exposure to high CsCl concentrations:* The sedimentation velocity patterns of polyoma DNA in dilute CsCl solutions are skewed because of the effects of concentration-dependent sedimentation. At a density of 1.35 gm cm⁻³ the bands become symmetrical, presumably because of a reduction in the equivalent hydrodynamic volume.

At still higher concentrations of CsCl, ρ = 1.5 and ρ = 1.6, the relative amounts of the components changed in two of the three preparations studied. Component I became smaller and II larger. That this conversion is not rapidly reversible was shown in experiments in which polyoma DNA was incubated at 23°C in CsCl, ρ = 1.72, 0.006 M Na citrate pH 7.0 for 1, 24, and 44 hr, and subsequently analyzed, after dilution, in CsCl, ρ = 1.35 gm cm⁻³. Analyses in neutral CsCl showed that irreversible conversion of I in all three DNA preparations had occurred to the extent of 50, 75, and 90 per cent, respectively. In one of the preparations of DNA it was confirmed that the conversion noted after 44 hr by analysis in neutral CsCl represented a ring cleavage. This was shown by a change in the relative amounts of fast and slow forms in alkaline CsCl, ρ = 1.36, 0.04 M K₃PO₄, pH = 12.1. The ratio of amounts changed from 0.6 to 0.1. The corrected sedimentation coefficients, $s_{20,w}^{0} \eta_{rel}$, for the double-stranded cyclic coils and the single-stranded coils were 37 S and 11.6 S. These results indicate that preparative equilibrium sedimentation in CsCl at room temperature can lead to a substantial loss of the cyclic helical form.

(3) *Exposure to high CsCl concentration at high pH:* A series of buoyant density experiments were performed to examine the effects of pH. At pH 11,

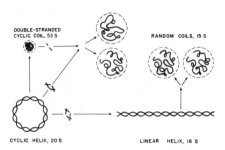

Fig. 3.—Diagrammatic representation of the two forms of polyoma DNA. The double-stranded cyclic helix denatures to form a double-stranded cyclic coil. The linear helix denatures to form two single-stranded random coils. The dashed circles indicate the relative equivalent hydrodynamic diameters. Possible links between chains or special bonds connecting chain ends are not considered. The sedimentation coefficients of the helices and coils were measured in neutral and alkaline NaCl solutions, respectively.

as at pH 8.1, the buoyant density was 1.709 gm cm^{-3}, and the band was skewed at the light side. At pH 11.8 a substantial fraction of the DNA titrated to form a band at 1.766 gm cm^{-3} that was skewed at the light side while the remainder, the intact cyclic helical DNA, formed a symmetrical band at 1.714 gm cm^{-3}.[11] At pH 12.2 the latter material denatured and titrated to form a band of double-stranded cyclic coils at 1.784 gm cm^{-3}. The results show that the cyclic helix is more stable than the linear helix to alkali denaturation. Both at pH 11.8 and 12.2 the relative amounts of the double-stranded cyclic DNA were smaller than the relative amount of 53 S DNA initially present in the preparations. A portion of the double-stranded cyclic DNA was therefore converted to single strands.

(4) *Sonication:* Sonication at 75 per cent full power for 1.5 and 5 min gave the same results. The relative amount and the sedimentation coefficient of I were essentially unchanged in neutral 1 *M* NaCl (Fig. 4). Components II and III formed a relatively broad band with decreased sedimentation coefficient, 12.2 S. At full power, sonication for 2 and 4 min resulted in the conversion of 50 and 80 per cent, respectively, of I into a slower material, 8.3 S. In previous studies with native DNA's, the resistance to shearing increases with decreasing sedimentation coefficient.[15] We have here the reverse effect, presumably because the extended length of the cyclic form is less than that of the linear form.

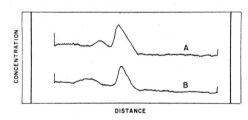

Fig. 4.—Effect of sonication of polyoma DNA. (*A*) untreated; (*B*) sonicated 1.5 min. Lamella: 15 μl, 25 μg/ml DNA. Bulk solution: 1.0 *M* NaCl, 0.01 *M* NaHPO$_4$, pH 8.0. Photographed 64 min after reaching full speed. 35,600 rpm, 20°.

(5) *Effects of heating at 100°C:* Component III is selectively melted out and becomes polydisperse in neutral sedimentation velocity experiments performed after heating solutions of polyoma DNA in 0.4 *M* NaCl 1–2 min at 100° followed by rapid cooling. The skewness disappears from the main band in buoyant density experiments at neutral pH. Component III forms a band of denatured DNA with a buoyant

density 1.721 gm cm^{-3}, 0.003 gm cm^{-3} less than the buoyant density of denatured polyoma DNA. In these experiments the relative amounts and the sedimentation velocities of I and II were unchanged. Longer heating leads to gradual conversion of I into a 16 S DNA.

When fresh polyoma DNA at a concentration of 1–2 μg/ml was heated at 100° 1–2 min in buffers of low ionic strength followed by rapid cooling, I renatured nearly completely, while an amount corresponding to II and III remained denatured as judged by buoyant density. This result indicates that intact cyclic coils renature readily. Upon further heating, increasing amounts of denatured DNA were formed. Heated materials analyzed in alkali contained decreasing amounts of 53 S DNA and increasing amounts of 15 S DNA. Depurination which occurs on heating and leads to chain scission in alkali[16–18] is in part responsible for the conversion of double-stranded cyclic coils to random coils.

Electron microscopy: The striking and important result is that cyclic molecules are observed (see *Appendix*).

Discussion.—The major fraction of polyoma DNA denatures in alkali and remains double-stranded. A smaller fraction denatures in alkali to form single strands with about half the molecular weight of the double-stranded form. Comparison of the sedimentation coefficients in alkali of these two denatured DNA's with each other and with the sedimentation coefficients in alkali of other DNA single strands of known molecular weight leads to the conclusion that the denatured double-stranded polyoma DNA has an abnormally low frictional coefficient. A simple model for the native DNA which can account for the properties mentioned is a cyclic helix with separately continuous strands (Fig. 3).

The requirement that the approximately 500 turns originally present in the helical structure be conserved as long as the strands remain continuous leads necessarily to a configuration for the denatured molecule which is more compact than the random-coil configuration. Any spatial rearrangement of the phosphodiester chains upon denaturation gives rise to supercoiling. This new type of DNA coil, which we refer to as the *double-stranded cyclic coil*, has an abnormally high sedimentation coefficient and a higher buoyant density in alkaline cesium chloride than the random coil.

The introduction of one discontinuity in one of the strands in the native or denatured double-stranded molecule results in the formation in alkali of single strands, one linear and one cyclic. The finding that the molecular weight is always first reduced by a factor of 2 upon degradation shows that the strands in polyoma DNA are separately continuous. None of the foregoing excludes the additional presence of a cross-link labile in alkali. Further evidence for a cyclic structure for component I is the relatively high resistance to shearing by sonication, and the presence in the electron micrographs of about 90 per cent cyclic molecules.

We now examine the evidence that II is an open or linear form of the cyclic helix I. The buoyant density of II is essentially the same as I. Upon thermal denaturation of II, a buoyant density shift characteristic for double-stranded DNA is observed. The alkaline buoyant density shift is also normal for a linear helix. We conclude therefore that II is a linear helix. I is converted into 16 S helical DNA by the action of concentrated CsCl or heating at 100° in 0.4 M NaCl.

The sedimentation coefficient of II is slightly lower than I, which is reasonable for the proposed relation between I and II. That II has the same molecular weight as I is indicated by the formation from II in alkali of 15 S single strands of one half the molecular weight of I. These arguments all suggest that II is an open or linear form of the cyclic helix I. Consistent with the above conclusion is the greater sensitivity of II to sonication, and the appearance in the electron micrographs of linear and cyclic molecules with the same mean length.

The origin of III, which was present in small amounts in all preparations, is not known. Its homogeneity with respect to sedimentation velocity and buoyant behavior rules out an adventitious contamination by host cell DNA. Component III was not found by Dulbecco and Vogt[3] in radioactively labeled polyoma DNA preparations, or by Crawford (cf. ref. 10) with boundary-sedimentation velocity experiments. Either the methods used by these investigators were not sufficiently sensitive or III represents a well-defined host cell DNA molecule which was synthesized prior to viral infection and thus remained unlabeled.

The thermal stability[1] of the infective titer of polyoma DNA upon heating at 100° can now be interpreted. The initial threefold increase in infective titer is the

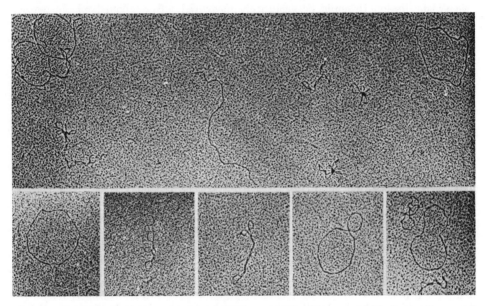

Fig. 5.—Electron micrographs of polyoma DNA, ×44,000. *Upper:* a field showing linear and cyclic forms. *Lower:* selected cyclic forms.

result of a more efficient attachment of the DNA to the mouse cells in the assay for infectivity.[1] The essentially unchanged titer during the next 10–20 min is attributed to a steady state between a thermal activation and an inactivation process. In the activation process the cyclic double-stranded molecule is opened to produce a molecule of higher infectivity. If a cleavage occurs at a phosphodiester bond, a pair of linear and cyclic single strands is formed. These separate unless held together by a cross link which we have not ruled out. If a cleavage at the site of a linker connecting four chain ends occurs, either single- or double-stranded linear molecules are formed. When all cyclic double-stranded molecules have been opened, the reservoir of potential infectivity is exhausted. Further heating leads to an exponential decrease in the infective titer.

The opening of the ring at high pH or as a result of heating may proceed by well-known reactions such as hydrolysis or depurination followed by hydrolysis; and these may occur at any site along the chains. The slow cleavage of the ring in concentrated neutral CsCl solutions is a surprising result which suggests that polyoma DNA contains a labile site in the chain not present in other biologically active DNA's which have been routinely studied in and recovered from equilibrium density gradient experiments in concentrated CsCl solutions.

APPENDIX BY WALTHER STOECKENIUS

THE ROCKEFELLER INSTITUTE

Three different configurations of polyoma viral DNA are visible in the electron micrographs (Fig. 5): coiled molecules without visible free ends, fully extended cyclic molecules, and linear molecules. If a coiled form contained not more than two twists, its length could be accurately measured. In a given typical area, 83 coiled, 8 extended, and 9 linear molecules were observed. The mean lengths of 66 cyclic and 68 linear molecules were 1.58 μ $\pm$ 0.16 S.D. and 1.53 μ $\pm$ 0.16 S.D., respectively. These lengths correspond to a molecular weight of 3.0×10^6, if we assume that the mass per unit length of DNA molecules in the protein film is the same as in the B-form in fibers of NaDNA.[19-21]

Electron microscopy thus confirms that polyoma DNA molecules exist in linear and cyclic forms of equal molecular weight. The proportion of I and II by electron microscopy corresponds to that found by sedimentation analysis, as described below.

A striking feature of the electron micrographs is the high proportion of tightly coiled molecules. In otherwise identical preparations of T-7 DNA, linear molecules and a very few tightly coiled molecules were found. The latter form is probably due mainly to intramolecular cross-linking by protein.[21] The absence of free ends in the tightly coiled polyoma DNA molecules indicates that they are cyclic; the high proportion of such forms suggests that cyclic molecules are more readily cross-linked by protein. Component III could not be identified in the electron micrographs.

A sample containing 2 μg/ml polyoma DNA in 0.006 M KCl, 0.01 M Tris, pH 7.5 was shipped in dry ice, thawed, and frozen three times during these studies, and the remainder returned in dry ice. Re-examination (by R. W. and J. V.) at neutral pH in 1.0 M NaCl and 3.5 M CsCl showed an unchanged composition, 85% I and about equal amounts of II and III. In alkaline NaCl only slow polydisperse strands were found.

Grids were prepared by the method of Kleinschmidt and Zahn,[22] except that diisopropyl-phosphoryl trypsin (Worthington Biochemical Corp., no. TRL-DIP104) was used instead of cytochrome C and the solution was delivered to the surface via a glass rod.[23] One ml of solution (0.24 μg DNA, 100 μg DIP-trypsin in 1.0 M NH$_4$ acetate, pH 8.0) was delivered onto 0.015 M NH$_4$ acetate, pH 8.0. The resulting surface film was compressed to 1.0 dyne/cm at about 450 cm^2. Randomly selected portions of the surface film were picked up on platinum grids and shadowed with uranium at an angle of 7° while the specimen was rotating.

Summary.—The results obtained in this study show that polyoma viral DNA contains linear and cyclic helical molecules. The double-stranded cyclic helix denatures in alkaline solutions to form a double-stranded cyclic coil, a new type of coiled molecule in which all of the turns originally present in the helix are conserved.

It is a pleasure to thank R. W. Watson for expert technical assistance, R. D. Bruner for helpful discussions during the preparation of the manuscript, R. Dulbecco for providing the polyoma virus, R. L. Sinsheimer and P. F. Davison for gifts of materials, and R. Dulbecco and M. Vogt for a copy of their unpublished manuscript. This work was supported in part by grants HE 03394 and GM 08812 from the U.S. Public Health Service.

* Contribution 3018 of Gates and Crellin Laboratories of Chemistry.

[1] Weil, R., these Proceedings, **49**, 480 (1963).

[2] Dulbecco, R., in *Monograph of the 17th Annual Symposium on Fundamental Cancer Research*, M. D. Anderson Hospital, Houston, Texas, Feb. 20, 1963 (Baltimore: Williams and Wilkins), in press.

[3] Dulbecco, R., and M. Vogt, these Proceedings, **50**, 236 (1963).

[4] Winocour, E., *Virology*, **19**, 158 (1963).

[5] Vinograd, J., R. Bruner, R. Kent, and J. Weigle, these Proceedings, **49**, 902 (1963).

[6] Vinograd, J., and J. Hearst, *Prog. in Chem. Org. Nat. Prods.*, **20**, 372 (1962).

[7] Vinograd, J., J. Morris, N. Davidson, and W. F. Dove, Jr., these Proceedings, **49**, 12 (1963).

[7a] These solutions were not free of carbon dioxide.

[8] Freifelder, D., and P. F. Davison, *Biophys. J.*, **3**, 49 (1963).

[9] Schildkraut, C. L., J. Marmur, and P. Doty, *J. Mol. Biol.*, **4**, 430 (1962).

[10] This value was obtained from the relative values of sedimentation coefficients of I and III in CsCl, $\rho = 1.35$, and the value for $s^{\circ}_{20,w}$ for I in 1.0 M NaCl. L. V. Crawford in *Virology*, **19**, 279 (1963), obtained 21 S and 14 S for two components in polyoma DNA.

[11] Similar results were obtained independently by Dulbecco and Vogt.[3]

[12] Sinsheimer, R. L., *J. Mol. Biol.*, **1**, 43 (1959).

[13] Davison, P. F., and D. Freifelder, *J. Mol. Biol.*, **5**, 643 (1962).

[14] Fiers, W., and R. L. Sinsheimer, *J. Mol. Biol.*, **5**, 424 (1962).

[15] Rosenkranz, H. S., and A. Bendich, *J. Am. Chem. Soc.*, **82**, 3198 (1960).

[16] Greer, S., and S. Zamenhof, *J. Mol. Biol.*, **4**, 123 (1962).

[17] Tamm, C., H. S. Shapiro, R. Lipshitz, and E. Chargaff, *J. Biol. Chem.* **203**, 673 (1953).

[18] Fiers, W., and R. L. Sinsheimer, *J. Mol. Biol.*, **5**, 420 (1962).

[19] Wilkins, M. H. F., G. Zubay, and H. R. Wilson, *J. Mol. Biol.* **1**, 179 (1959).

[20] Zubay, G., and M. H. F. Wilkins, *J. Mol. Biol.*, **4**, 444 (1962).

[21] Stoeckenius, W., *J. Biophys. Biochem. Cytol.*, **11**, 297 (1961).

[22] Kleinschmidt, A. K., and R. K. Zahn, *Z. Naturforsch.*, **14b**, 770 (1959).

[23] Trurnit, H. J., *J. Colloid Sci.*, **15**, 1 (1960).

ARTICLE **42**

From the *Proceedings of the National Academy of Sciences*, Vol. 53, No. 5, 1104–1111, 1965. Reproduced with permission of the authors.

THE TWISTED CIRCULAR FORM OF POLYOMA VIRAL DNA*

By J. Vinograd, J. Lebowitz, R. Radloff, R. Watson, and P. Laipis

GATES AND CRELLIN LABORATORIES OF CHEMISTRY AND
THE NORMAN W. CHURCH LABORATORY OF CHEMICAL BIOLOGY,
CALIFORNIA INSTITUTE OF TECHNOLOGY

Communicated by Norman Davidson, March 30, 1965

The major part of the DNA from polyoma virus has been shown to consist of circular base-paired duplex molecules without chain ends.[1-3] The intertwined circular form accounts for the ease of renaturation[4] of this DNA and the failure of the strands to separate in strand-separating solvents.[1-3]

In previous studies[1-3] a minor component, II, observed in variable amounts in sedimentation analyses of preparations of polyoma DNA at neutral pH, was regarded to be a linear form of the viral DNA. Both the major component I (20S) and II (16S) were infective.[1, 5] In our further investigations of the minor component the following results, which are reported below, have been obtained: (1) The minor component is a ring-shaped duplex molecule. (2) It is generated by introducing one single-chain scission in component I by the action of pancreatic DNAase or chemical-reducing agents. (3) The sedimentation coefficient of II is insensitive to several single-strand scissions. (4) The conversion products, when not excessively attacked, are infective.

The foregoing results raised a new problem. Why does the viral DNA, an intact duplex ring, sediment 20 per cent faster than a similar duplex ring containing one or more single-strand scissions? Experiments bearing on this problem, presented below, indicate the presence of a *twisted circular structure* in polyoma DNA I. A mechanism for the formation of this locked-in twisted structure is proposed.

Methods.—Isolation and purification of the virus and extraction of the DNA: Two methods[6, 7] for purification of the virus were used. The DNA was isolated by Weil's method[4] except that the phenol was freshly distilled under argon.

Ultracentrifugation: Sedimentation analyses were performed in a Spinco model E ultracentrifuge by band centrifugation.[8] Some of the results were recorded with the photoelectric scanning attachment.[9, 10] Sucrose density gradient experiments were performed at 4°, 30,000 rpm, and 9 hr. The 3% and 20% sucrose solutions contained SSC (0.15 M NaCl and 0.015 M Na citrate) and 0.05 M Tris chloride pH 8.0.

Enzymes: Pancreatic DNAase, 1 × crystallized, was obtained from Worthington Biochemicals Corp. *E. coli* endonuclease I, 1000 units/ml,[11] and *E. coli* phosphodiesterase, 2000 units/ml,[12] were gifts from Professor I. R. Lehman. BSA, 30% bovine albumin solution, sterile, was obtained from Armour Pharmaceutical Co. The endonuclease I, 0.12 units/μg DNA, converted 60% of I into linear molecules in 8 min at 20° in the incubation mixture described by Lehman.[11]

Sedimentation velocity-pH titration: Fifteen μl, 40 μg/ml DNA in SSC/10, flowed from the sample well of the type III[13] band-forming centerpiece onto an alkaline CsCl bulk-solution. This solution was prepared by titrating 10 ml (Harshaw Chemical Co.) optical grade CsCl, $\rho = 1.35$, with 1 M KOH in CsCl, $\rho = 1.35$, at 20° under argon, and was transferred to the cell assembly under argon. Usually four samples in a pH series were analyzed simultaneously. A Beckman research model pH meter, a general purpose probe glass electrode, and a calomel reference electrode modified with a ground glass junction[14] were used.

Plaque assay: Infectivity of polyoma DNA was measured as described by Weil.[4]

Electron microscopy: Specimens were prepared by the method of Kleinschmidt and Zahn.[15]

Results.—Preparation of polyoma DNA II: Polyoma II can be prepared from I by treatment with several mild chemical-reducing agents (Table 1). These reagents

1104

TABLE 1

ACTION OF REDUCING AGENTS ON POLYOMA DNA

Reagent	M^a	pH	Time (min)	Conversion[b] of I to II (%)
Hydroquinone[c]	0.0002	8.5	30	100
FeCl$_2$	0.001	8.6	30	90
Na$_2$SO$_3$ (1 × 10^{-3} M Cu$_2$SO$_4$, 0.1 M NH$_4$OH)[d]	0.01	10.8	60	45
Thiols[e]	0.01–0.02	3.8–4.2	60	80–90

[a] Final concentration of reducing agents which were diluted fivefold with SSC/10, 0.01 M Tris pH 8.5 containing 40 μg/ml DNA. The thiols were first dissolved in 0.4 M acetic acid before addition to the DNA solution.
[b] Sedimentation analysis at neutral pH.
[c] The reaction product was assayed for infectivity: 0.5 × 10³ pfu/μg DNA compared with 1.5 × 10³ obtained for untreated DNA in the same assay.
[d] Cu^{++} acts as a catalyst.[16] In the absence of Cu^{++} or NH$_4$OH 5% conversion was observed. In the absence of SO$_3^-$ no conversion occurred.
[e] Mercaptoethanol, cysteine, and glutathione. In experiments with acetic acid without thiols, 10% conversion occurred.

were suggested by the observation that rigorous exclusion of impurities from the phenol, used in the isolation of the DNA, diminished the amount of the minor component II in the final DNA preparation. A simple conversion of I to II without intermediates and without detectable degradation products was observed in sedimentation analyses at pH 8.0 (Table 1). Based on the earlier assignment[1, 2] of a linear form to component II, the reactions with reducing agents and the infective nature of the products indicated a specific duplex cleavage. Dulbecco and Vogt[1] reported a similar conversion of I to II with low concentrations of pancreatic DNAase. These authors postulated that a bond opposite the single-strand scission introduced by the enzyme hydrolyzed under the influence of ring strain. The foregoing puzzling results are clarified by the experiments below.

Structure of Polyoma II.—The products from the action of pancreatic DNAase and of the reducing agents (Table 1) were examined in the electron microscope. They were found to be in the circular form. Figure 1b is typical of electron micrographs of materials obtained by treatment of I with pancreatic DNAase or with the several reducing agents. The possibility that linear molecules were selectively excluded in the preparation of the grids was eliminated by the results of a reconstruction experiment. A synthetic mixture of 10 per cent II and 90 per cent linear polyoma III (cf. below) showed the expected proportions of linear molecules.

The circular form for II is compatible with the proposal that the reducing agents and pancreatic DNAase introduce single-strand scissions into polyoma DNA I. It may be calculated that the circular form of the molecule should be retained until on the average about 50 single-strand scissions per molecule have been introduced.[17] The material shown in Figure 1b contained, on the average, about three breaks per molecule as calculated from the Poisson distribution.

A still milder treatment with pancreatic DNAase should give rise to single-stranded rings and uniform single-stranded linear molecules in strand-separating solvents, such as alkaline NaCl or CsCl.[2, 18] An example of such a result with 0.6 breaks per molecule is given in Figure 2b. These two components, $s^\circ_{w,20}$ = 18.4S ± 0.4, 15.7S ± 0.3 for the alkaline Na DNA, have been identified as single-stranded rings (18S) and single-stranded linear molecules (16S), respectively, by the following variation of an experiment originally performed with φX DNA by Fiers and Sinsheimer.[19] An aliquot of the product of the pancreatic DNAase digestion

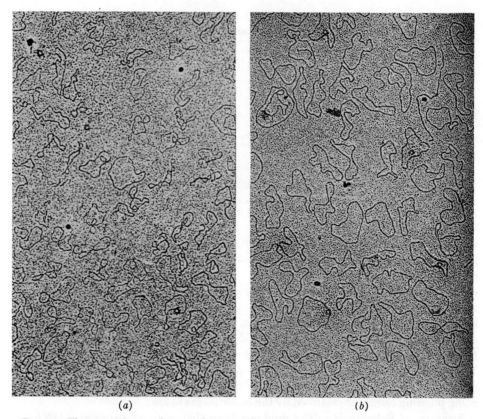

FIG. 1.—Electron micrographs of polyoma DNA × 21,000. The materials in (*a*) and (*b*) were prepared by treatment of polyoma I with pancreatic DNAase, as described under Fig. 3. (*a*) was withdrawn from the reaction mixture after 5% conversion of I to II; (*b*) after 95% conversion.

(40% conversion) was heat-denatured and treated with *E. coli* phosphodiesterase. This enzyme attacks single-stranded DNA with a free 3′OH group. It is seen that the amount of the 16S component was substantially diminished, while the 18S component was resistant to the exonuclease (Fig. 2*c*). Thus polyoma DNA II can contain a wholly intact circular strand. At this level of digestion the second strand in the molecule will contain only one or two breaks if the attack is statistical.

We now examine the possibility that one single-strand scission in the duplex is adequate to convert polyoma ring I to ring II. If only one-chain scission is necessary, the rate of conversion of I to II should be the same as the rate of conversion of the 53S component (intact, denatured, double-stranded, cyclic molecules) to the slower moving single-stranded molecules in alkali. If more than one break were necessary to convert I to II, a faster rate of conversion would be seen in alkali. Dulbecco and Vogt[1] have already reported that the two rates are alike. In view of the importance of the result, we have repeated this experiment with the analyses performed in the analytical ultracentrifuge.

In Figure 3*a* it is seen that the alkaline analyses and the neutral analyses give, within the experimental error, the same extent of conversion, a result which confirms the Dulbecco and Vogt finding. Therefore, the conversion of I to II occurs

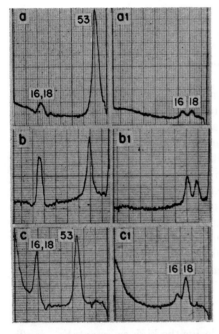

FIG. 2.—Sedimentation velocity patterns of polyoma DNA in alkaline CsCl. The *left* and *right* patterns are scans at about 30 min and 90 min after sedimentation begins. The field is directed toward the right. CsCl, ρ = 1.35 gm cm^{-3}, pH 12.5, 44,000 rpm. (*a, a1*) *Control*: Component I isolated in a sucrose gradient experiment treated identically as in (*b*) and (*c*) except for the absence of enzymes and BSA. Separate experiments showed the BSA to be free of DNAase activity under the conditions used. (*b, b1*) *Pancreatic DNAase treatment*: 106 µg/ml pure I in 0.048 *M* NaCl, 0.0075 *M* MgCl$_2$, 0.01 *M* Tris pH 8.0, 40 µg/ml BSA, and 2.7 × 10^{-4} µg/ml enzyme, 20 min at 20°. Reaction stopped by $^1/_{15}$ volume 1 *M* glycine buffer, pH 9.8. The leading band in (*a*), (*b*), and (*c*) is the 53S component. The resolved slower bands in (*a1*), (*b1*), and (*c1*) are the 16S and 18S components. (*c, c1*) *Effect of heat denaturation followed by E. coli phosphodiesterase treatment*: Product of (*b*) heated 5 min 100°, cooled rapidly. 70 µg/ml DNA in 0.03 *M* NaCl, 0.005 *M* MgCl$_2$, 0.007 *M* Tris, 0.067 *M* glycine pH 9.8, 0.90 mg/ml BSA, 71 units/ml *E. coli* phosphodiesterase, 90 min at 37°. Reaction stopped by $^1/_{10}$ volume 0.1 *M* EDTA.

whenever the first single-strand scission is introduced. The conversion appears to be first order. While the infectivity (Fig. 3*b*) declines at a slower rate than the conversion of I to II, the scatter in the data precludes any conclusions regarding the kinetics of inactivation. It is clear, however, that the first single-strand scission in this duplex DNA is not lethal.

More extensive treatment with pancreatic DNAase or the chemical-reducing agents so as to completely convert I to II (>4 average breaks per molecule) caused no detectable change in the sedimentation coefficient of II.

Preparation of Linear Polyoma DNA with E. coli Endonuclease I.—Polyoma I was partially converted into the linear form with *E. coli* endonuclease I, which is known to cleave duplex DNA.[11] The sedimentation velocity of the homogeneous linear molecules was 14.5S

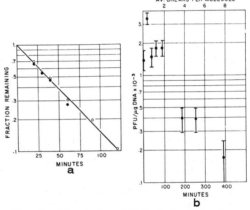

FIG. 3.—Chemical and biological effects of pancreatic DNAase treatment. (*a*) Analyses for single-stranded and double-stranded DNA. Extent of conversion was determined by band-sedimentation velocity experiments with photoelectric scanner. ⊙, (I)/(I + II) in neutral CsCl bulk solutions. ●, (53S)/(total) in alkaline CsCl pH 12.3. Areas under bands were corrected for radial dilution. Incubation mixture and conditions were the same as those given in legend to Fig. 2*b*, except for enzyme concentration, 2.0 × 10^{-4} µg/ml. 20-µl samples were withdrawn at the indicated times and added to 4 µl 0.1 *M* EDTA pH 8.5. The samples were frozen prior to analyses. (*b*) Infectivity of samples withdrawn from incubation in (*a*). The time for a unit average number of hits was obtained from (*a*) at 63% conversion. The error bars give the standard deviations from 16 replicate plates.

at pH 8.0, the same as previously reported[2] for the minor component III, and 16S in alkali. Polyoma II was not produced in detectable amounts in the above conversion of I to linear molecules. Electron micrographs confirmed the assignment of a linear form to the enzymatic product and also to the minor component III isolated by sucrose gradient sedimentation.

Structure of Component I.—The high sedimentation coefficient of I relative to II indicates that the viral component is either more compact or larger in mass than the circular conversion product. In the extreme case of no increase in friction, a 20 per cent reduction in mass is required to account for the change in *s*. An equal amount of mass would have to have been lost as a result of the action of pancreatic DNAase and the variety of reducing agents used. An excision of viral DNA would have been detected by Dulbecco and Vogt,[1] who were unable to find small fragments of labeled DNA after preparative band sedimentation of polyoma DNA treated with pancreatic DNAase. The identical buoyant densities of I and II[2] make it unlikely that a nonlabeled, non-DNA mass is removed.

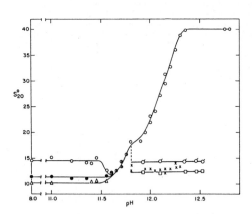

FIG. 4.—Sedimentation velocity-pH titration of the three components in polyoma DNA. I, ○; II, ●; III, △; ◑ only band present, cf. text; mixture of unresolved single strands in alkali, x; 29,500 rpm. Single linear, □, and single circular, ◌, strands in alkali, 44,770 rpm. Sedimentation coefficients at 20° in CsCl, $\rho = 1.35$ gm cm^{-3}, are not corrected for solvent viscosity, $\eta_r = 0.925$, or buoyancy effects. The values at pH 8.0 and 12.4 are the means of 12 and 7 determinations, respectively.

Three kinds of experiments suggest that a particular kind of compact structure—*a twisted circular form*—is responsible for the high sedimentation coefficient of polyoma DNA. (1) The electron micrographs of the grids prepared from polyoma I contained[1] twisted circles to a variable extent (Fig. 1*a*). Grids prepared from polyoma II contained extended circles and no twisted configurations (Fig. 1*b*). The DNA samples used to prepare the grids for Figures 1*a* and *b* were identical except for the time of incubation with the enzyme. The spreading forces acting on the macromolecules in the monolayer appear to remove loops and crossovers that are not locked in the structure of the DNA.

(2) A study of the sedimentation velocity in 3 *M* CsCl of a mixture of the three components of polyoma DNA as a function of pH from 8 to 12.5 revealed a complicated pH-melting curve for component I (Fig. 4). Component II behaved normally[18, 20] and moved faster as denaturation increased until strand separation occurred with an attendant sudden drop in sedimentation velocity at pH 11.8. Component I, like II, was at first insensitive to pH. At pH 11.5, however, the sedimentation coefficient first dropped, and then in the pH range 11.6 to 11.8 was the same as for polyoma II. Only one moving band was observed in this pH range. The sedimentation coefficient of I then increased to the very high value characteristic of the *double-stranded cyclic coil* previously reported.[2] Essentially the same results were obtained in 1.0 *M* KCl solutions. The dip in the sedimentation velocity-

pH curve was initially unexplainable.   If, however, polyoma DNA I contains left-handed tertiary turns, such a dip in the pH-melting profile would be required.   In the early stages of denaturation some of the duplex turns, which are known to be right-handed, unwind.   The unwinding of the duplex must be accompanied by a right-handed twisting of the remainder of the molecule.   If the tertiary turns were originally left-handed (Fig. 5), progressive unwinding would cause the molecule to pass through configuration I′ characterized by the absence of tertiary turns. The extended configuration I′ is similar to that in polyoma II (Fig. 5) and both I′ and II would have similar sedimentation velocities.   Further unwinding of duplex I′ is accompanied by continued right-hand twisting of the whole molecule until finally the double-stranded cyclic coil[2] configuration develops.

(3) The twisted circular structure provides a satisfactory explanation for the configurational change that occurs when one single-strand scission is introduced into the molecule.   Such a scission generates a site for the rotation of the helix in the complementary strand opposite the break.   The swivel relieves the topological restraint responsible for the twisted configuration.

Discussion.—A mechanism for the formation of the twisted circular structures suggested by the above analysis of the pH-melting curve.   According to this mechanism the last closure of chain ends occurs before all of the winding of the two DNA strands into the Watson-Crick structure is completed.   The closing leaves the duplex in the configuration I′ (Fig. 5) restrained from converting to I by an as yet unknown factor participating in the DNA synthesis.   Removal of the restraint then allows I′ to wind spontaneously into a complete Watson-Crick structure and form the twisted circular structure, I, with no change in winding number.

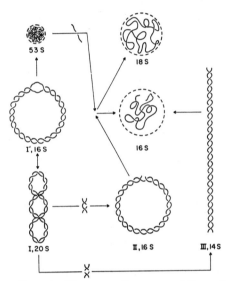

FIG. 5.—Diagrammatic representation of the several forms of polyoma DNA.   The duplex segments shown contain 12 turns, about one fortieth of the total number.   The twisted circular duplex shown contains one left-hand tertiary turn.   8% of the right-hand duplex turns in the model are unwound to form *I′*.   The dashed circles around the denatured forms indicate the relative hydrodynamic diameters.   The sedimentation coefficients were measured in neutral and alkaline NaCl solutions.

An alternative proposal is that the molecule in form II, which contains a swivel, is twisted by some organizer, e.g., the virus protein.   The last covalent backbone bond is then made while the DNA remains twisted under the constraint of the organizer.   This alternative is unlikely in view of Dulbecco's[21] finding that the polyoma DNA made before virus production begins has the sedimentation velocity of component I.

It is not possible at the present time to estimate reliably the number of tertiary turns.   A turn is defined as a 360° rotation of the helix.   The electron micrographs of I usually contain some molecules that are completely extended; these may have suffered a single break during grid preparation or may have been in state I′ due to denaturation induced by the spreading forces. This latter action would result in the unwinding of the twisted circles to form extended circles. The maximum number of crossovers that can be distinguished is 8, which corresponds to 4 turns or the unwinding of 40 base pairs. This limit may be low because it is difficult to count crossovers in tightly coiled forms. That the number of tertiary turns in the molecule is not large is suggested by the fact that the transition of I to I′ occurs before substantial melting of II takes place, as indicated by the s versus pH plot.

With the new assignment of structure to the three components of polyoma DNA, it is found that satisfactory agreement obtains between the observed sedimentation coefficient of the linear form, 14.5S, and the 15.3 ± 0.5S predicted by Studier's relation[18] for a molecular weight[2] of 3.0 ± 0.3 × 10⁶. The values, 18.4S and 15.7S for the alkaline single-stranded circular and linear forms similarly agree with the predicted values of 17.4S and 15.6S, respectively. The effect of ring closure of III to form component II is to increase S by 10 per cent. An effect of similar magnitude has been reported[22] for the cyclization of λ DNA.

The twisted circular structure observed here for polyoma DNA may be a common characteristic of covalently closed, circular duplex DNA. A part of the DNA from rabbit papilloma virus,[23] SV40 virus,[3] and the replicating form of φX DNA[24-26] have all been shown to be circular duplex molecules which do not strand-separate in alkali or after heating in formaldehyde. Two sedimentation velocity components differing by 20–30 per cent have been reported for the above DNA's.[27] Crawford and Black[3] observed sedimentation velocity-denaturation curves that are similar to our pH-melting curve upon heating SV40 DNA and polyoma DNA in formaldehyde solutions to various temperatures. No explanation was offered for this behavior, which we interpret as indicating the presence of a left-handed, twisted circular structure.[28]

Burton and Sinsheimer[29] have shown that the slow component II in RF-φX DNA dissociates in alkali to form linear and circular single-stranded molecules and have concluded that both of the undenatured forms of the DNA are circular. While this communication was in preparation, Jansz and Pouwels[30] reported that the pancreatic DNAase-induced conversion of I to II in RF-φX DNA represents a conversion between circular duplex molecules. No explanation for the change in the sedimentation coefficient was offered. In view of the results described here, it is likely that RF-φX DNA is in the twisted circular form. A common mechanism for the incorporation of the tertiary turns during replication is a strong possibility, and allows us to predict that the tertiary turns in the RF-φX DNA will be found to be left-handed.

Summary.—The results of this study show that circular duplex polyoma DNA may be converted to a less compact circular duplex by introducing a single-strand scission. The viral form contains tertiary turns which appear to have been locked in during replication.

It is a pleasure to thank I. R. Lehman for the generous gift of the *E. coli* enzymes, R. L. Sinsheimer, J. Petruska, and M. Fried for helpful discussions, R. Dulbecco for allowing us to quote his unpublished results, and T. Benjamin, L. Wenzel, and A. Drew for advice and assistance in the culture of the virus and the assay of the DNA. This work was supported in part by grants HE 03394 and CA 08014 from the U.S. Public Health Service.

* This work was reported in part at the 9th Annual Meeting of the Biophysical Society, February 24, 1965. Contribution 3227 of Gates and Crellin Laboratories of Chemistry.

[1] Dulbecco, R., and M. Vogt, these PROCEEDINGS, **50**, 236 (1963).

[2] Weil, R., and J. Vinograd, these PROCEEDINGS, **50**, 730 (1963).

[3] Crawford, L. V., and P. H. Black, *Virology*, **24**, 388 (1964).

[4] Weil, R., these PROCEEDINGS, **49**, 480 (1963).

[5] Crawford, L. V., R. Dulbecco, M. Fried, L. Montagnier, and M. Stoker, these PROCEEDINGS, **52**, 148 (1964).

[6] Winocour, E., *Virology*, **19**, 158 (1963).

[7] Murikami, W., *Science*, **142**, 56 (1963).

[8] Vinograd, J., R. Bruner, R. Kent, and J. Weigle, these PROCEEDINGS, **49**, 902 (1963).

[9] Beckman Instruments Co., Spinco Division, Palo Alto, Calif.

[10] Hanlon, S., K. Lamers, G. Lauterbach, R. Johnson, and H. K. Schachman, *Arch. Biochem. Biophys.*, **99**, 157 (1962).

[11] Lehman, I. R., G. G. Roussos, and E. A. Pratt, *J. Biol. Chem.*, **237**, 819 (1962).

[12] Lehman, I. R., *J. Biol. Chem.*, **235**, 1479 (1960).

[13] Vinograd, J., R. Radloff, and R. Bruner, *Biopolymers*, in press.

[14] Lebowitz, J., and M. Laskowski, Jr., *Biochemistry*, **1**, 1044 (1962).

[15] Kleinschmidt, A. K., and R. K. Zahn, *Z. Naturforsch.*, **14b**, 770 (1959).

[16] Swan, J. M., *Nature*, **180**, 643 (1957).

[17] Thomas, C. A., Jr., *J. Am. Chem. Soc.*, **78**, 1861 (1956). The estimate of 50 was obtained for the case of a 10% lowering of the weight average molecular weight.

[18] Studier, F. Wm., *J. Mol. Biol.*, **11**, 373 (1965).

[19] Fiers, W., and R. L. Sinsheimer, *J. Mol. Biol.*, **5**, 408 (1962).

[20] Davidson, P. F., and D. Freifelder, *J. Mol. Biol.*, **5**, 643 (1962).

[21] Dulbecco, R., private communication.

[22] Hershey, A. D., E. Burgi, and L. Ingraham, these PROCEEDINGS, **49**, 748 (1963).

[23] Crawford, L. V., *J. Mol. Biol.*, **8**, 489 (1964).

[24] Kleinschmidt, A., A. Burton, and R. L. Sinsheimer, *Science*, **142**, 961 (1963).

[25] Burton, A., and R. L. Sinsheimer, *Science*, **142**, 962 (1963).

[26] Burton, A., and R. L. Sinsheimer, *Abstracts*, 8th Annual Meeting of the Biophysical Society, 1964.

[27] R. Weil has informed us that the slow component II in rabbit papilloma DNA is circular as seen in the electron microscope.

[28] Human papilloma DNA appears to be similar in configuration to SV40 DNA and polyoma DNA (Crawford, L. V., manuscript submitted for publication).

[29] Burton, A., and R. L. Sinsheimer, private communication.

[30] Jansz, H. S., and P. H. Pouwels, *Biochem. Biophys. Res. Commun.*, **18**, 589 (1965).

ARTICLE 43

From the *Proceedings of the National Academy of Sciences*, Vol. 57, No. 5, 1514–1521, 1967. Reproduced with permission of the authors.

A DYE–BUOYANT-DENSITY METHOD FOR THE DETECTION AND ISOLATION OF CLOSED CIRCULAR DUPLEX DNA: THE CLOSED CIRCULAR DNA IN HELA CELLS*

By Roger Radloff, William Bauer, and Jerome Vinograd

Norman W. Church Laboratory for Chemical Biology,†
California Institute of Technology, Pasadena

Communicated by James Bonner, March 2, 1967

Covalently closed circular duplex DNA's are now known to be widespread among living organisms. This DNA structure, originally identified in polyoma viral DNA,[1,2] has been assigned to the mitochondrial DNA's in ox[3] and sheep heart,[4] in mouse and chicken liver,[3] and in unfertilized sea urchin egg.[5] The animal viral DNA's—polyoma, SV40,[6] rabbit[7] and human[8] papilloma—the intracellular forms of the bacterial viral DNA's—ϕX174,[9,10] lambda,[11,12] M13,[13] and P22[14]—and a bacterial plasmid DNA, the colicinogenic factor E_2,[15] have all been shown to exist as closed circular duplexes. Other mitochondrial DNA's[16,17] and a portion of the DNA from boar sperm[18] have been reported to be circular, but as yet have not been shown to be covalently closed.

The physicochemical properties of closed circular DNA differ in several respects from those of linear DNA or of circular DNA containing one or more single-strand scissions.[19] The resistance to denaturation,[2,20] the sedimentation velocity in neutral and alkaline solution, and the buoyant density in alkaline solution are all enhanced in the closed circular molecules. These three effects are a direct consequence of the topological requirement that the number of interstrand crossovers must remain constant in the closed molecule.

The principal methods currently used for the detection and the isolation of closed circular DNA are based on the first two general properties. In this communication we describe a method based on the buoyant behavior of closed circular DNA in the presence of intercalating dyes.

The binding of intercalative dyes has recently been shown to cause a partial unwinding of the duplex structure in closed circular DNA.[22–24] In such molecules any unwinding of the duplex causes a change in the number of superhelical turns, so that the total number of turns in the molecule remains constant. A small and critical amount of dye-binding reduces the number of superhelical turns to zero. Further dye-binding results in the formation of superhelices of the opposite sign or handedness. The creation of these new superhelices introduces mechanical stresses into the duplex and a more ordered conformation into the molecule. These effects increase the free energy of formation of the DNA-dye complex. The maximum amount of dye that can be bound by the closed molecule is therefore *smaller* than by the linear or nicked circular molecule. Correspondingly, since the buoyant density of the DNA-dye complex[23,25] is inversely related to the amount of dye bound, the buoyant density of the closed circular DNA-dye complex at saturation is *greater* than that of the linear or nicked circular DNA-dye complex.[23] Bauer and Vinograd have shown that the above effect results in a buoyant density difference of approximately 0.04 gm/ml in CsCl containing saturating amounts of ethidium bromide, an intercalating dye extensively studied by Waring[26] and Le Pecq.[27]

1514

The method has been tested with known mixtures of nicked and closed circular viral DNA's, and has been used to isolate closed circular DNA from the mitochondrial fraction of HeLa cells and from extracts of whole HeLa cells.

Materials and Methods.—Preparation of HeLa cell extracts: HeLa S3 cells were grown on Petri dishes in Eagle's medium containing 10% calf serum. H^3-thymidine, 18 c/mmole, was obtained from the New England Nuclear Corporation. Ten μc per ml of medium were added to each plate 20 hr before the cells were harvested. After washing with isotonic buffer and decanting, the cells were treated by the method described by Hirt[28] for the separation of polyoma DNA from nuclear DNA. Approximately 2 ml of 0.6% sodium dodecylsulphate (SDS), in 0.01 M ethylenediaminetetraacetate (EDTA), 0.01 M tris, pH 8, were added to the plates. After 30 min at room temperature, the viscous extracts were gently scraped from the plates with a rubber policeman and transferred with a widemouth pipet to a centrifuge tube. Either 5 M NaCl or 7 M CsCl was then added with gentle mixing to a final salt concentration of 1 M. The resulting solution was cooled to 4° and centrifuged for 15 min at 17,000 × g in a Servall preparative ultracentrifuge. The supernatant solution was dialyzed at 4° overnight against two changes of 0.01 M EDTA, 0.01 M tris, pH 8 buffer in order to remove H^3-thymidine. The mitochondria from HeLa cells were isolated by differential centrifugation of an homogenate followed by banding in a sucrose gradient.

Preparation of viral DNA: Polyoma viral DNA was prepared as described previously.[19] The intracellular lambda DNA was kindly supplied by John Kiger and E. T. Young, II, of the Biology Division.

Chemicals: The ethidium bromide was obtained as a gift from Boots Pure Drug Co., Ltd., Nottingham, England. Harshaw optical grade cesium was used. The SDS was obtained from Matheson Company. All other chemicals were of reagent grade.

Preparative ultracentrifugation: The experiments were performed in SW50 rotors in a Beckman Spinco model L preparative ultracentrifuge at 20°C. The CsCl solutions, in either cellulose nitrate or polyallomer tubes, were overlaid with light mineral oil. After centrifugation, the tubes were fractionated with a drop-collection unit obtained from Buchler Instruments. The drops were collected in small vials or on Whatman glass-fiber GF/A filters. The dried filters were immersed in 10 ml of toluene-PPO-POPOP, and the samples were counted in a Packard Tri-Carb scintillation counter.

Electron microscopy: Specimens for electron microscopy were prepared by the method of Kleinschmidt and Zahn[29] and were examined in a Phillips EM200 electron microscope. All electron micrographs were made at a magnification of ×5054. The magnification factor was checked with a grating replica. Cytochrome c was obtained from the California Biochemical Corporation.

Fluorescence: Prior to drop collection, the centrifuge tubes were examined in a darkened room with 365 mμ light from a Mineral Light Lamp or preferably from a "Transilluminator" supplied by Ultraviolet Products, Inc., San Gabriel, California. The tubes were photographed on Polaroid type-146L film through a "contrast filter" from Ultraviolet Products, Inc. The fluorescence measurements were performed with a double-monochromator apparatus constructed in this laboratory by W. Galley and N. Davidson. The instrument was calibrated with solutions of ethidium bromide (5×10^{-3} to 1.0 $\mu g/ml$) in calf thymus DNA, 20 $\mu g/ml$, 1 M CsCl. The intensity of fluorescence was measured at 590 mμ with an exciting wavelength of 548 mμ.[30]

Results.—Selection of initial dye concentration, cesium chloride concentration, and centrifugation variables: The conditions for the experiments in the swinging-bucket rotor in the model L ultracentrifuge were selected with the object of obtaining separations similar to those obtained at equilibrium within a reasonable period of time. At equilibrium, the separation between components is approximately constant at initial dye concentrations between 50 and 100 $\mu g/ml$.[23] At these dye levels the buoyant densities are between 1.57 and 1.62 gm/ml. Figure 1 presents the dye and density distributions in CsCl solutions centrifuged for 24 and 48 hours. The initial dye concentration in both cases was found at a distance from the meniscus corresponding to four-tenths the length of the liquid column. In

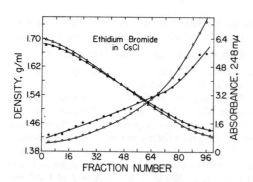

FIG. 1.—Density and dye distributions in CsCl density gradient columns, 3.00 ml 1.550 gm/ml CsCl, 100 µg/ml ethidium bromide, 24 (●, ▲) and 48 (○, △) hr at 43 krpm in a SW50 rotor at 20°.

this region of the cell, the time dependence of the dye concentration is minimal, and the constant density gradient is equal to approximately 010 gm/cm⁴.

Results with purified mixtures of nicked and closed circular DNA: Figure 2 presents the results obtained with 1.5 µg of tritiated polyoma DNA. Substantially complete resolution was obtained with this DNA which has a molecular weight of three million daltons. Since the fractional amount of closed circular material corresponded to the fraction obtained in analytical sedimentation velocity analyses,[21] it is concluded that single-strand scissions did not occur during the course of the experiments. An experiment in which the DNA was contained in a thin lamella at the top of the liquid column was performed to test the effect of the initial DNA distribution. The results were substantially the same as those obtained when the DNA was uniformly distributed in the CsCl solution.

The results obtained with a mixture of approximately 60 µg of covalently closed and nicked circular lambda DNA, mol wt = 3×10^7, are presented in Figure 3.

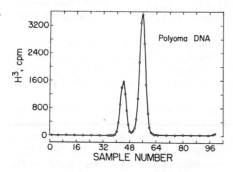

FIG. 2.—A mixture of purified polyoma DNA I and II 1.5 µg in buoyant CsCl, 3.00 ml 1.566 gm/ml CsCl, 100 µg/ml ethidium bromide, 24 hr at 43 krpm, 20°. The band maxima were separated by 12 fractions (four 7.5-µl drops per fraction). The buoyant densities of I and II are 1.588 and 1.553 gm/ml, respectively. The sample contains 30% I as indicated above compared with 32% by analytical band centrifugation.[21]

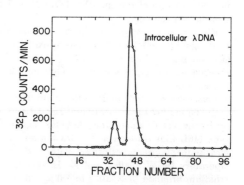

FIG. 3.—A mixture of purified intracellular lambda DNA I and II, 3.00 ml 1.55 gm/ml CsCl, 100 µg/ml ethidium bromide, 24 hr at 43 krpm, 20°. The centroids are separated by 9.5 fractions and 0.31 ml. Component I accounts for 16% of the total counts.

In this experiment, the nicked circular DNA formed a band of relatively high viscosity that may have distorted and broadened the light band during drop collection. An experiment to determine the effect of ethidium bromide on the protoplast assay for lambda DNA[32] was performed in collaboration with J. Kiger and E. T. Young, II. It was found that linear lambda had a normal titer when diluted by a factor of 1000 from a solution containing ethidium bromide and CsCl at the concentrations which occur at band center.

Detection of DNA by absorbance, fluorescence, and scintillation counting: The data in Figures 2 and 3 were obtained by scintillation counting of dried filter papers containing dye, cesium chloride, and labeled DNA. To examine the effect of the presence of the dye upon counting efficiency, a series of sixteen 50-μl samples of H^3-thymidine in a 1.55 gm/ml CsCl solution containing ethidium bromide in concentrations varying from 0 to 286 μg/ml were counted. The relative counting efficiency decreased linearly with a least-squares slope of 8.6×10^{-4} ml/μg. The dye gradient thus caused a difference of 1 per cent in relative counting efficiency between the two bands in Figure 2, while the dye depressed the relative counting efficiency by 8 per cent.

Red bands containing about 5 μg of DNA can be observed visually in the centrifuge tube. If an adequate amount of DNA is present, the fractionated gradient may be assayed spectrophotometrically at 260 mμ. At saturation, the increase in absorbance caused by dye-binding is about 40 per cent with linear DNA and about 20 per cent with closed circular DNA.

The photograph in Figure 4 shows the fluorescent emission from two DNA bands. The fluorescence from DNA in amounts as low as 0.5 μg per band are detectable visually. Since the separation between the two bands corresponds reliably to 0.30–0.36 ml, it was possible to use the fluorescent, less dense band as a marker to locate closed circular DNA present in amounts below the limit of detectability by spectroscopic or radioactive procedures. The method is thus especially suitable for examining nonradioactive preparations from tissues of higher animals.

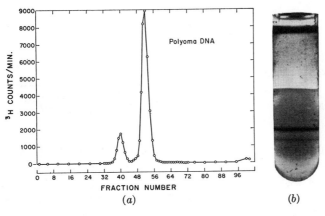

FIG. 4.—A mixture of purified polyoma DNA I and II, 3.00 ml, 1.558 gm/ml CsCl, 100 μg/ml ethidium bromide, 48 hr at 43 krpm, 20°. (*a*) The band maxima are separated by 12 fractions and 0.36 ml. The calculated buoyant densities are 1.592 and 1.556 gm/ml. (*b*) A photograph of the centrifuge tube prior to drop collection. The tube contains a total of 4 μg of DNA, and 0.64 μg of component I.

Removal of ethidium bromide from DNA solutions: It is often desirable to remove the dye quantitatively from a DNA sample. This was accomplished in a single passage of 1.0 ml of polyoma I DNA (40 μg/ml DNA, 100 μg/ml dye, 1 *M* CsCl) through a 0.8 $\times$ 4.5-cm column of analytical grade Dowex-50 resin. The fractions containing DNA were consolidated and were found to contain less than 1×10^{-2} μg dye in a fluorometric analysis.

Isolation of closed circular DNA from H^3-thymidine-labeled HeLa cells: Hela cell monolayers in Petri dishes containing 10^7 cells were treated with SDS by the procedure described by Hirt.[28] The results obtained when the extract was purposely sheared and when the shear was minimized (Figs. 5a and b) show that the relative amount of material in the light band was decreased when shear was minimized. The dense band in Figure 5a is clearly contaminated by material from the light band and reprocessing the dense band in a second CsCl-dye gradient would be necessary to obtain resolved bands.

Electron-microscope examination of the circular DNA in the dense band of HeLa cell extracts: Examination of electron micrographs of specimens prepared from fractions 33–36 in the dense band of HeLa cell extracts (Fig. 5c) showed less than 0.1

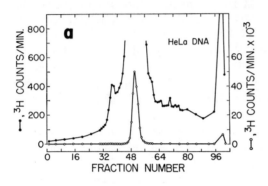

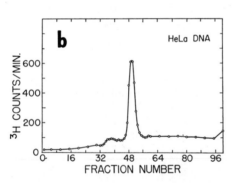

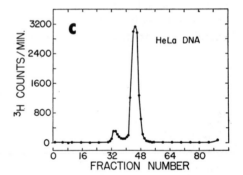

Fig. 5.—Buoyant-density profiles from extracts of whole HeLa cells, prepared as described in the text. Centrifuge conditions were the same as in Fig. 2. (a) DNA extract sheared during preparation. (b) DNA extract prepared with minimum shear. (c) Prepared with minimum shear for electron microscopy.

per cent of distinguishably linear DNA. The light band contained long linear DNA. The dense band contained an array of circles in three size groups: a homogeneous group of molecules with a mean length of 4.81 ± SE 0.24 microns; a heterogeneous population with lengths from 0.2 to 3.5 microns; and a paucidisperse set 2–4 times the length of the DNA in the homogeneous group. Figure 6 presents micrographs selected to illustrate the three size classes, and Figure 7 shows the frequency distribution among the first two groups. A survey of several hundred molecules on sparsely populated specimen grids revealed that the frequencies of the small and the large size classes were each about one-tenth the frequency of the homogeneous size class. In the large size class, the dimers were observed more frequently than the trimers or tetramers. The DNA from mitochondria isolated from HeLa cells consisted principally of molecules in the homogeneous size group. We conclude, therefore, that the homogeneous size class in Figure 7 is of mito chondrial origin.   Small circular DNA's were not observed in significant amounts in the preparation of DNA from the isolated mitochondria.

Discussion.—The method described in this paper has proved to be a simple and direct procedure for obtaining closed circular DNA from extracts of whole cells and cell fractions.   The method employs chemically mild conditions;   single-strand scissions are not introduced by interaction with the dye, nor is there any permanent rearrangement of the DNA structure.   All the steps of the separation

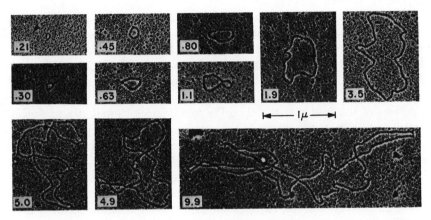

Fig. 6.—Electron micrographs of circular DNA from HeLa cells.   Fractions 33–36 from the dense band in Fig. 5c were pooled and used in the specimen preparations. The top photographs present selected molecules of the small size range.   The number in each insert gives the length in microns of the molecule.   The first two molecules in the second row are of mitochondrial size;   the third is twice the mitochondrial length.

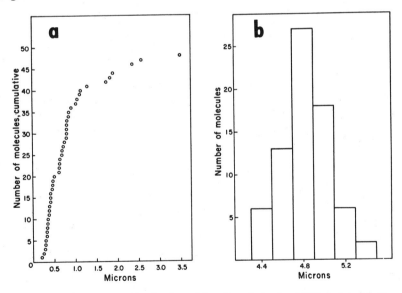

Fig. 7.—Frequency distribution of lengths of circular DNA isolated from the HeLa cell band referred to in the legend for Fig. 6.   (*a*) A cumulative frequency distribution of lengths of molecules in the submitochondrial size class. (*b*) A histogram of the distribution of lengths of molecules in the mitochondrial size class.

are carried out at room temperature, or below, and at neutral pH. In the experiments performed so far, the maximum amount of DNA introduced in a single tube was 125 μg of polyoma DNA.

It has become clear, in the course of this initial study of the closed circular DNA in malignant human cells, that a portion of the mitochondrial DNA in HeLa cells is in the form of closed duplexes. These circular molecules are similar in form, size, and homogeneity to the mitochondrial DNA's from invertebrate and vertebrate species previously investigated.[3-5, 16, 17]

The heterogeneous group of small molecules, 0.2–3.5 microns, are to be compared with the size range, 0.5–16.8 microns, reported by Hotta and Bassel[18] to be present in preparations of unfractionated boar sperm DNA. The conclusion that the small circles, seen in our electron micrographs, represent DNA molecules rests on three observations. (1) The small circles were found at the same level in a CsCl-dye gradient as were the closed circular mitochondrial DNA molecules. (2) The grain pattern of the shadowed metal on the small circles viewed at a magnification of 1×10^5 was indistinguishable from the pattern on mitochondrial DNA molecules in the same field. (3) The small circles were seen only very infrequently in DNA preparations from isolated HeLa cell mitochondria. It is thus unlikely that the small circles arise from an artifact in the electron-microscope procedure. A more definitive characterization of these small DNA molecules, which can code for only 200–3500 amino acid residues, requires the preparation of larger quantities of material. The frequency of occurrence of the small molecules in the SDS extracts we have studied does not necessarily represent the frequency of occurrence in the HeLa cell. It is emphasized here that the dye-buoyant method segregates only *closed* circular duplexes. Molecules that contain even one single-strand scission, for whatever reason, find their way to the less dense band and, in whole cell extracts, intermingle with the large excess of linear DNA.

The larger-size circles, which were also seen at a frequency of about 10 per cent relative to mitochondrial DNA in the dense band, are clearly multiples of the mitochondrial length. The mean lengths from measurements of 43 double-length circles and a smaller number of larger multiples were 9.56 ± 0.42, 14.1 ± 0.4, and 19.8 ± 1.1 microns. We have so far not found a fully extended large circle. Nor have we found, in the several large circles which contain infrequent crossovers, any one which could not have arisen from a pairing of two or more mitochondrial molecules. An investigation of the significance of these multiple-size circles in HeLa DNA is in progress. Nass[17] has reported measurements of multiple lengths of mitochondrial molecules liberated from mitochondria by osmotic shock during specimen preparation. She attributed the multiples to the superposition of DNA molecules which originated from single mitochondria.

Summary.—A buoyant-density method for the isolation and detection of closed circular DNA is described. The method is based on the reduced binding of the intercalating dye, ethidium bromide, by closed circular DNA. In an application of this method we have found that HeLa cells contain, in addition to closed circular mitochondrial DNA of mean length, 4.81 microns, a heterogeneous group of smaller DNA molecules which vary in size from 0.2 to 3.5 microns and a paucidisperse group of multiples of the mitochondrial length.

Note added in proof: That double-length closed mitochondrial DNA molecules do occur has been shown by D. Clayton in this laboratory with preparations from leucocytes obtained from the blood of a donor with chronic granulocytic leukemia. Circular DNA molecules free of crossovers and of twice the mitochondrial length were observed in electron micrographs.

It is a pleasure to thank J. Huberman for providing us with the labeled HeLa cells; J. Kiger and E. T. Young, II, for their gift of closed circular lambda DNA and for allowing us to quote their unpublished analysis; G. Attardi for advice and assistance in the preparation of HeLa mitochondria; L. Wenzel and J. Eden for their assistance in the culture of the polyoma virus; R. Watson for his several technical contributions; and R. Kent for assistance in the preparation of the manuscript.

* This work was supported in part by grants HE 03394 and CA 08014 from the U.S. Public Health Service and by fellowships from the U.S. Public Health Service and the National Science Foundation.

† Contribution 3490 from the Laboratories of Chemistry.

[1] Dulbecco, R., and M. Vogt, these Proceedings, **50**, 236 (1963).

[2] Weil, R., and J. Vinograd, these Proceedings, **50**, 730 (1963).

[3] Borst, P., and G. J. C. M. Ruttenburg, *Biochim. Biophys. Acta*, **114**, 645 (1965).

[4] Kroon, A. M., P. Borst, E. F. J. Van Bruggen, and G. J. C. M. Ruttenburg, these Proceedings, **56**, 1836 (1966).

[5] Pikó, L., A. Tyler, and J. Vinograd, *Biol. Bull.*, in press.

[6] Crawford, L. V., and P. H. Black, *Virology*, **24**, 388 (1964).

[7] Crawford, L. V., *J. Mol. Biol.*, **8**, 489 (1964).

[8] *Ibid.*, **13**, 362 (1965).

[9] Burton, A., and R. L. Sinsheimer, *J. Mol. Biol.*, **14**, 3276 (1965).

[10] Kleinschmidt, A. K., A. Burton, and R. L. Sinsheimer, *Science*, **142**, 1961 (1963).

[11] Young, E. T., II, and R. L. Sinsheimer, *J. Mol. Biol.*, **10**, 562 (1964).

[12] Bode, V. C., and A. D. Kaiser, *J. Mol. Biol.*, **14**, 399 (1965).

[13] Play, D. S., A. Preuss, and P. H. Hofschneider, *J. Mol. Biol.*, **21**, 485 (1966).

[14] Rhoades, M., and C. A. Thomas, private communication.

[15] Roth, T., and D. Helinski, private communication.

[16] David, I. B., these Proceedings, **56**, 269 (1966).

[17] Nass, M. M. K., these Proceedings, **56**, 1215 (1966).

[18] Hotta, Y., and A. Bassel, these Proceedings, **53**, 356 (1965).

[19] Vinograd, J., J. Lebowitz, R. Radloff, R. Watson, and P. Laipis, these Proceedings, **53**, 1104 (1965).

[20] Vinograd, J., and J. Lebowitz, *J. Gen. Physiol.*, **49**, 103 (1966).

[21] Vinograd, J., R. Bruner, R. Kent, and J. Weigle, these Proceedings, **49**, 902 (1963).

[22] M. Gellert, these Proceedings, **57**, 148 (1967).

[23] Bauer, W. R., and J. Vinograd, in preparation.

[24] Crawford, L. V., and M. J. Waring, *J. Mol. Biol.*, in press.

[25] Kersten, W., H. Kersten, and W. Szybalski, *Biochemistry*, **5**, 236 (1966).

[26] Waring, M. J., *Biochim. Biophys. Acta*, **114**, 234 (1966).

[27] Le Pecq, J., Dissertation, University of Paris (1965).

[28] Hirt, B., these Proceedings, **55**, 997 (1966).

[29] Kleinschmidt, A. K., and R. K. Zahn, *Z. Naturforsch.*, **14b**, 770 (1959).

[30] LePecq, J., and C. Paoletti, *Anal. Biochem.*, **17**, 100 (1966).

[31] Young, E. T., II, and R. L. Sinsheimer, private communication.

ARTICLE 44

From *Journal of Molecular Biology*, Vol. 71, 127–147, 1972. Reproduced with permission of the authors and publisher.

On the Structure of the Folded Chromosome† of *Escherichia coli*

A. Worcel and E. Burgi

Department of Biochemical Sciences
Princeton University, Princeton, N.J., U.S.A.

(*Received 8 June 1972*)

"Comme les autres sciences de la nature, la biologie a perdu aujourd'hui nombre de ses illusions. Elle ne cherche plus la vérité. Elle construit la sienne. . . . Il n'y a pas une organisation du vivant, mais une serie d'organisations emboîtées les unes dans les autres comme des poupées russes. Derrière chacune s'en cache une autre."

Francois Jacob
La logique du vivant

After careful lysis of exponentially growing *Escherichia coli* cells, the DNA is released in a highly-folded conformation (Stonington & Pettijohn, 1971), sedimenting as an heterogeneous population of structures with sedimentation coefficients ranging from 1300 s to 2200 s. Chromosomes arrested at initiation by amino-acid starvation sediment as an homogeneous 1300 s species. The gradual bimodal decrease in sedimentation rate seen after amino-acid starvation suggests that the faster sedimenting structures in exponentially growing cells represent the folded chromosomes at more advanced stages in their replication cycle.

The folded chromosomes are supercoiled, as evidenced by the biphasic response of their sedimentation rates to increasing concentrations of ethidium bromide. The 1300 s folded chromosome lined up at initiation, as well as the 1300 s to 2200 s replicating chromosomes, have about the same concentration of superhelices as previously observed for intracellular λ and SV40 DNA's, namely, about one negative superhelical turn per 400 base pairs. Nicking with DNase releases these superhelical twists and relaxes the folded chromosome to slower sedimenting forms. However, somewhere between 6 and 40 nicks per DNA strand are required before a fully "relaxed complex" is obtained; this complex sediments 40% slower than the original non-DNase-treated chromosome, but still five times faster than the corresponding unfolded DNA. It appears to have lost all of its superhelical character, since it behaves exactly like linear DNA upon ethidium bromide intercalation.

A preliminary model for the folded bacterial chromosome has the DNA looped around a *core*, probably an RNA species. The RNA–DNA interactions stabilize the folded structure and topologically divide the chromosome into a limited number of *loops*. There are somewhere between 12 and 80 loops per chromosome, each with the same superhelical concentration. A single-strand nick with DNase allows free rotation of the DNA chains within a loop, eliminating the superhelices of that particular loop without affecting the superhelical content of the rest of the chromosome.

† The terms folded chromosome, folded DNA, folded genome, DNA–RNA protein complex (or plain complex) are used interchangeably and they all refer to the compact, fast sedimenting, DNA-containing structures released from *Escherichia coli* after careful lysis of the cells by the Stonington & Pettijohn (1971) procedure.

The bacterial chromosome can then be seen in three distinct, well-defined conformations. (a) *Compact*, folded and supercoiled. (b) *Relaxed* and folded, with no superhelices but retaining its RNA core and DNA loops. This relaxed complex is observed after extensive nicking with DNase, or after eliminating the super-helices by ethidium bromide intercalation. (c) *Unfolded* chromosome à la Cairns, observed after breaking the RNA core or otherwise disrupting the stabilizing interactions.

1. Introduction

The DNA of *Escherichia coli* is a single molecule, a double-stranded closed ring with an over-all length of more than one millimeter (Cairns, 1963). Inside the cell this DNA is packaged into nuclear bodies of about 1 μm in diameter which occupy only a fraction of the cell volume (Ryter, 1968). No nuclear membrane is seen at the boundary of these chromosome-like structures, and the mechanism of folding the DNA into its compact *in vivo* conformation remains to be elucidated.

Although much information on genetic mechanisms has been obtained from careful and elegant studies on DNA fragments, i.e. the semi-conservative manner of DNA replication (Meselson & Stahl, 1958), the sequential and ordered pattern of chromosome replication (Lark, 1969), the events at the replicating fork (Okazaki *et al.*, 1968), and at the chromosome origin (Sueoka & Quinn, 1968), other important biological problems have been inaccessible to this approach. For instance, study of the termination of chromosome replication, and the segregation of the daughter chromosomes may require handling the whole intact bacterial chromosome in its "native" highly-folded conformation.

Recently, Stonington & Pettijohn devised a lysis procedure which releases the DNA from *E. coli* D10 (an RNase I⁻ strain) in a highly-folded conformation. The "folded genome" is complexed with RNA and protein, and is very sensitive to RNase. The DNA–RNA protein complex is 80% DNA by weight. The RNA in it, 10% by weight, is nascent messenger and ribosomal RNA; the protein, 10% by weight (less than 1% of the total cellular protein) is mostly RNA polymerase (α, β and β' subunits) (Stonington & Pettijohn, 1971). We have succeeded in extending the Stonington & Pettijohn lysis procedure, with slight modifications, to other *E. coli* strains. In this communication we shall report on some unusual properties of the folded chromosome of *E. coli*, which help to define its tertiary structure.

2. Materials and Methods

(a) *Bacterial strains and growth conditions*

E. coli K12 CR34 (Thy⁻, Thre⁻, Leu⁻) was grown in a shaking bath at 30°C in minimal medium M63 (Pardee, Jacob & Monod, 1959), containing 2 mg glucose/ml., 80 μg each of the required amino acids/ml., and 40 μg thymine/ml. In some experiments, Hfr Hayes, C600, D10, and various derivatives of CR34 carrying temperature-sensitive DNA mutations were used. Similar results were obtained with all strains tested when grown at 30°C. The properties of the folded DNA isolated from temperature-sensitive DNA mutants grown at the non-permissive temperature will be reported elsewhere.

(b) *Preparation of complex*

In most experiments, exponentially growing cells were labeled by adding 100 μCi of [³H]thymidine (NEN, 50 Ci/m-mole) and harvested by centrifugation at 2°C 30 min later. Lysis was carried out as described by Stonington & Pettijohn (1971) with slight modifications. The pellet (between 10^9 and 3×10^9 cells) was resuspended at 0°C in 0·2 ml. of solution A (0·01 M-Tris·HCl (pH 8·2), 0·01 M-sodium azide, 20% w/v sucrose (Mann

Enzyme grade) and 0·1 M-NaCl); 0·05 ml. of a freshly made solution B (0·12 M-Tris·HCl (pH 8·2), 0·05 M-trisodium-EDTA and 4 mg egg-white lysozyme/ml. (Worthington)) was then added. After careful mixing, the suspension was taken out of the ice, and the cells were lysed by the addition of 0·25 ml. of solution C (1% Brij-58, 0·4% deoxycholate (Sigma), 2·0 M-NaCl and 0·01 M-EDTA). In most experiments 0·1% diethylpyrocarbonate was added just before addition of solution C to protect against nucleases. The lysis mixture was held at room temperature until it cleared (between 10 to 30 min). Extreme care was taken in handling the suspension, and components were mixed by a slow rotatory motion. The lysate was spun in a Sorvall centrifuge at 4000 *g* in the cold for 5 min; the pellet of this low spin usually contained between 10 and 30% of the DNA of the lysate. If more than 3×10^9 cells were used per lysate, or if the lysis was carried out on ice, a larger proportion of the cellular DNA precipitated. The precipitation appears to be due to aggregation of the folded DNA. No difference could be detected between the DNA in the pellet and the DNA in the clear non-viscous supernatant: the same distribution was observed when (a) DNA was uniformly labeled; (b) DNA was pulse-labeled for a few seconds and the cells immediately lysed (fork-labeled chromosomes); (c) DNA was labeled after recovery from amino-acid starvation or after release of a block in DNA synthesis (initiation-labeled chromosomes).

(c) *Sedimentation through neutral sucrose gradients*

The supernatant of the Sorvall spin was layered on a 5 ml. 10 to 30% w/v sucrose gradient containing 0·01 M-Tris·HCl (pH 8·2), 1·0 M-NaCl, 1 mM-EDTA and 1 mM-β-mercaptoethanol. Centrifugation was carried out in an SW50 rotor of a Beckman ultracentrifuge for 30 min at 17,000 rev./min and 4°C. In most experiments, [^{14}C]leucine-labeled T4 phage was added to the lysate as an internal marker. A typical sucrose gradient profile is shown in Fig. 1(a). In this Figure, as in most others, total radioactivity is plotted. The broad peak sedimenting faster than the T4 phage marker represents the bulk of the trichloroacetic acid-precipitable counts of the lysate. The radioactivity at the top of the gradient in this run is due to free thymidine. In early experiments, fractions were precipitated with 5% trichloroacetic acid and counted after filtering through Millipore filters. After the lysis procedure was perfected, no significant trichloroacetic acid-precipitable counts remained at the top of the gradients (they represent unfolded or fragmented DNA). Therefore, the fractions are routinely collected directly into vials and counted in a Triton-containing Scintillation fluid (Patterson & Greene, 1965).

Fractions to be rerun were collected at 0°C, after puncturing the tubes with a 1-mm internal diameter needle, and diluted as needed with 0·2 M-Tris, pH 8·2. (NaCl at 1·0 M is required to preserve the folded conformation of the DNA in the lysate, but after purification the fast sedimenting forms remain relatively stable in low salt (Stonington & Pettijohn, 1971).) Reruns were carried out immediately after this preparative run. The folded DNA is not very stable; even in 1·0 M-NaCl, sedimentation coefficients are significantly lower after 24 hr storage in the cold, and after 4 to 5 days the DNA unfolds, probably because of breaks accumulated in the DNA molecule.

(d) *Ethidium bromide sedimentation*

Sedimentation velocity runs were carried out by layering the fractions containing the folded chromosomes on top of a standard 10 to 30% sucrose gradient containing various concentrations of ethidium bromide. Portions of less than 1 μg of DNA were used per each 5 ml. gradient, and therefore the concentration of free ethidium is nearly the same as the indicated total ethidium concentrations (Waring, 1965; Wang, 1969).

(e) *Alkaline sucrose gradients*

The 5 to 20% w/v alkaline sucrose gradients contained 0·3 N-NaOH, 1·0 M-NaCl and 1 mM-EDTA. The fractions containing the folded chromosomes were layered on top of the gradient, together with a sample of [^{14}C]thymidine-labeled T4 DNA (the DNA was extracted by heating T4 phage for 15 min at 65°C in 1% Sarcosyl–0·05 M-EDTA). The alkaline sucrose gradients were spun for 30 min at 40,000 rev./min and 4°C.

Pancreatic RNase 2700 units/mg protein, was obtained from Worthington. The enzyme was boiled (10 min at 100°C) to denature any contaminating traces of DNase. Pancreatic

9

DNase I, 2500 units/mg protein, RNase free, was obtained from Worthington. Ethidium bromide (2,7-diamino-10-ethyl-9-phenyl-phenanthridium bromide) was obtained from Calbiochem. U. Laemmli provided us with a stock of T4 and with procedures for labeling and purifying the phage.

3. Results

(a) *Folded DNA from exponentially growing cells*

After careful lysis of the cells as described, the *E. coli* DNA sediments very quickly through a sucrose gradient, giving a rather broad profile as shown in Figure 1(a).

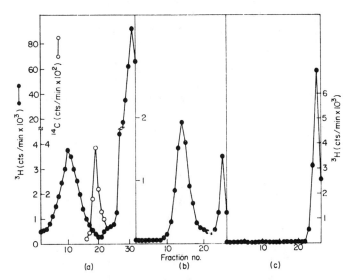

Fig. 1. Sucrose-gradient profiles of whole *E. coli* lysate (a) and reruns of the peak fraction before (b) and after shear (c). Experimental conditions are described in the text. Unless otherwise stated, all centrifugations were carried out at 17,000 rev./min for 30 min at 4°C. —O—O—, ¹⁴C-labeled T4 phage.

From the position of the internal phage T4 marker (1·025 s; Cummings, 1964) we estimate that the fast sedimenting structures peak at about 1700 s. This value is lower than the 3200 s reported by Stonington & Pettijohn (1971) for the folded genome of *E. coli* cells growing in enriched Casamino acids medium and at 37°C; it is possible that the differences in growth conditions are the reason for the discrepancy although we did not attempt to determine this point. It is known that fast growing cells in enriched media possess more replicating forks and more DNA per chromosome than cells growing in minimal media (Helmstetter & Cooper, 1968; Yoshikawa, O'Sullivan & Sueoka, 1964; Quinn & Sueoka, 1970).

The fast sedimenting, DNA-containing structures are very sensitive to mechanical shear. Figure 1(b) and (c) shows a rerun of the peak fraction in Figure 1(a), before (b) and after (c) being ejected through the tip of an Ependorf pipette; the ejected DNA sediments slowly like the commonly extracted unfolded *E. coli* DNA.

The heterogeneity in sedimentation velocities of the DNA containing structures is real and reproducible. Figure 2(a) shows a profile of a preparative run, with reruns of selected fractions from the preparative run shown in Figure 2(b) to (f). Note that fraction 9, the first fraction with counts significantly above background in the

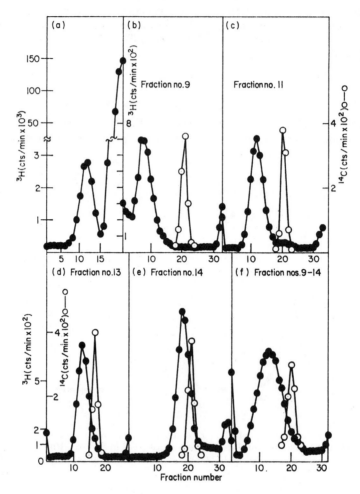

FIG. 2. Heterogeneity in sedimentation velocities of the folded DNA of exponentially growing cells. Fractions from a preparative run of an *E. coli* lysate (a) were rerun with internal ^{14}C-labeled T4 marker. Equal numbers of counts were applied to each gradient, except the last one where equal volumes of fractions 9 to 14 were mixed and centrifuged together.

preparative run, sediments fastest; and that fraction 14, the last fraction above background in the preparative run, sediments slowest, with the other fractions sedimenting in between. From the internal T4 marker, a sedimentation coefficient of 2200 s can be estimated for the DNA complexes in fraction 9, and one of 1300 s for those in fraction 14. A composite of equal samples of fractions 9 to 14, inclusive, regenerates the broad profile of Figure 1(a) (Fig. 2(f)).

Assuming that we are dealing with similar compact spherical structures, a twofold increase in mass should give a 67% increase in sedimentation velocity (Tanford, 1961). The 2200 s DNA particles, 70% faster than the 1300 s ones, could well represent the chromosomes nearing the end of their replication cycle and containing twice as much DNA as the 1300 s structures. According to this interpretation, chromosomes at the beginning of their replication cycle should belong to the 1300 s category.

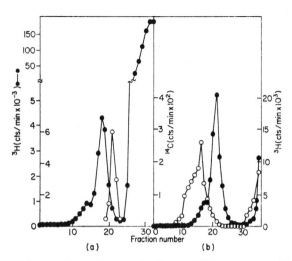

FIG. 3. Folded DNA from amino-acid-starved cells. (a) Cell lysate prepared 90 min after amino-acid starvation and run with ^{14}C-labeled T4 internal marker. (b) Double-label experiment, ^{14}C-labeled exponentially growing cells and ^{3}H-labeled amino-acid-starved cells (2 hr amino-acid starvation); the trichloroacetic acid-precipitable counts are shown. —●—●—, ^{3}H (cts/min × 10^3); —○—○—, ^{14}C(cts/min × 10^2).

(b) Folded DNA from amino-acid-starved cells

In bacteria, amino-acid starvation *lines up* the chromosomes at initiation, presumably because they lack some essential proteins required for initiation of DNA replication (Maaløe & Hanawalt, 1961; Cerda-Olmedo, Hanawalt & Guerola, 1968). Exponentially growing cells were therefore pulsed with [^{3}H]thymidine as before, and 30 minutes later the cells were filtered, washed, and resuspended in a fresh medium with glucose and thymine but without the required amino acids. Control experiments showed that total DNA synthesis continued at a reduced rate for about one hour and then stopped. Figure 3 shows the sucrose gradient profiles of lysates of the amino-acid-starved cells. The first frame shows the DNA after 90 minutes of amino-acid starvation. The profile is very sharp; most of the DNA is in an homogeneous peak with a sedimentation coefficient of 1300 s.

The profile of the lysate of amino-acid-starved cells is very different from the profile of the lysate of exponentially growing cells shown in Figure 1(a), and strikingly similar to the profile of the rerun fraction 14 in Figure 2. However, a small proportion of faster sedimenting forms are still present; they were always observed, even after three hours of amino-acid starvation where they accounted for less than 5% of the total DNA. We interpret them as belonging to chromosomes which could not complete their rounds of replication in the absence of protein synthesis. As suggested previously from other data, amino-acid starvation may not line up all of the chromosomes of exponentially growing cells (Caro & Berg, 1968).

In most of our experiments, T4 phage was used as an internal marker. It was thought desirable to detect the difference in sedimentation by some other means, particularly to rule out a conformational change in the folded DNA brought about by the different composition of the lysate of amino-acid-starved cells *vis-à-vis* the lysate of exponentially growing cells. Therefore, doubly-labeled experiments were designed, in which exponentially growing cells were labeled either with [¹⁴C]thymine or with [³H]-thymidine. After amino-acid starvation of the ³H-labeled bacteria, both cultures were mixed and processed jointly.

Figure 3(b) shows the results of such an experiment in which the ¹⁴C-labeled DNA belongs to exponentially growing cells, and the ³H-labeled DNA to cells starved of amino acids for two hours. As expected, the position of the sharp ³H-labeled DNA peak coincides with the position of the slowest sedimenting forms of the ¹⁴C-labeled DNA. In this picture trichloroacetic acid-precipitable counts are shown; the small amount of radioactivity at the top of the gradient represents the unfolded or fragmented DNA observed in our lysates.

(c) *Kinetics of the transition after amino-acid starvation*

The change observed after amino-acid starvation could be due to a conformational change in the folded DNA, or to a loss of RNA and protein from the complex brought about by the block in protein synthesis in this stringent strain. To rule out these possibilities, we examined the kinetics of the transition after amino-acid starvation. A gradual accumulation of the 1300 s species at the expense of the faster sedimenting forms would strongly support our original idea that the 1300 s species represent the chromosomes lined up at initiation, particularly if this accumulation parallels the slow kinetics of completion of chromosome replication under these conditions.

Figure 4 shows the sucrose gradient profiles of lysates prepared 15 minutes, 30 minutes, and 60 minutes after amino-acid starvation (only the first 25 fractions, out of a total of 32, are shown). After 15 minutes of amino-acid starvation, there is already a 1300 s species of folded DNA. The position of this peak does not change with time, it just increases. The profile after 15 minutes of amino-acid starvation is clearly bimodal; the slower sedimenting forms are 1300 s, while the faster ones have a maximum at about 1700 s. At later times the proportion of the 1300 s species

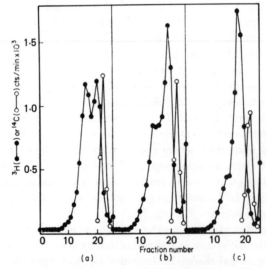

FIG. 4. Transition after amino-acid starvation. Sucrose gradient profiles of lysates at 15 (a), 30 (b), and 60 (c) min after amino-acid starvation. The top 7 fractions with the high levels of free [³H]thymidine have been left out. —O—O—, ¹⁴C-labeled T4 phage.

increases while the amount of faster sedimenting structures decreases, exactly as expected for a lining up of the chromosomes at initiation under these conditions.

The simplest explanation for these results is that the 1300 s folded DNA species is the folded chromosome at initiation, and that the 1300 s to 2200 s folded DNA structures of exponentially growing cells are the folded chromosomes at more advanced stages in their replication cycle.

(d) *Interaction of the folded chromosomes with ethidium bromide*

If the DNA-containing structures represent the intact highly-folded bacterial chromosome, we may probe into their structure by studying their interaction with intercalating dyes, such as ethidium bromide. Ethidium bromide intercalates between the base pairs of the DNA and causes a change in the pitch of the Watson–Crick double helix which in turn creates positive superhelical turns in a closed circular DNA molecule (Waring, 1965; Crawford & Waring, 1967). All of the closed circular double-stranded DNA molecules examined to date *in vitro* possess negative super-helices (deficiency in superhelical turns; Vinograd, Lebowitz, Radloff, Watson & Lapis, 1965; Vinograd & Lebowitz, 1966; Wang, 1969) and ethidium bromide inter-calation causes first a loss of the negative superhelical turns, and then as the amount of ethidium bromide intercalated increases, positive superhelices are created. These conformational changes are reflected in the sedimentation coefficient of the DNA, first by a decrease in S value (the lowest S value corresponding to the molecule with zero superhelical turns) followed by an increase in S value as the ethidium bromide concentration is increased and the DNA molecules become supertwisted in the opposite, positive sense. The strands of DNA fragments, or of nicked-circular DNA, are free to rotate and the only change in sedimentation velocity observed is a small continuous decrease in S value as the ethidium bromide intercalated gradually increases the frictional coefficient of the DNA. Sedimentation studies in the presence

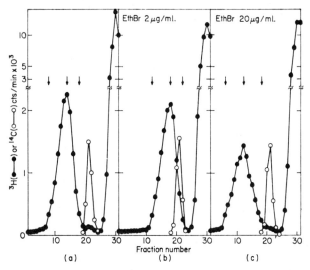

FIG. 5. Effect of ethidium bromide on the sedimentation velocity of the folded chromosomes. *E. coli* lysates were run in standard 10 to 30% sucrose gradients containing 0 (a), 2 (b), and 20 μg ethidium bromide/ml. (c). The arrows point to the fractions (first, peak, and last) whose sedimentation coefficients are plotted against ethidium bromide concentration in Fig. 6. —O—O—, ^{14}C-labeled T4 phage; EthBr, ethidium bromide.

of ethidium bromide should then tell us whether or not the DNA chains in the isolated folded *E. coli* genome are free to rotate.

We were surprised to find that, upon ethidium bromide intercalation, all of the *E. coli* folded DNA structures behave like intact closed-circular DNA molecules, i.e. as if there was not a single nick in the whole *E. coli* chromosome! Figure 5 shows the sucrose gradient profiles of samples of the same lysate of exponentially growing cells, run in the absence of ethidium bromide, with 2 μg ethidium bromide/ml., and with 20 μg ethidium bromide/ml. The whole broad DNA peak moves first back at the low ethidium bromide concentration, and then forward at the higher ethidium bromide concentration. In this experiment, care was taken to collect exactly the same number of drops; and since the internal phage T4 marker was always at fraction 21, it is clear that it is the *E. coli* folded DNA whose sedimentation rate changes and not the T4 phage.

As indicated by the three arrows in each profile, the S values of (a) the highest point in the peak of the DNA complexes of the exponentially growing cells; (b) the faster DNA complexes (roughly, the first point significantly above background); and (c) the slower DNA complexes (last point above background) were selected at each concentration of ethidium bromide. These S values were plotted against ethidium bromide concentration in Figure 6. Note that all of the curves show a dip at 2 μg

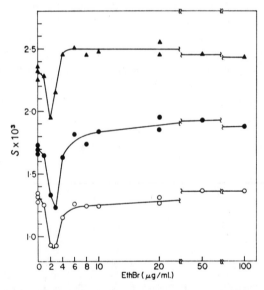

Fig. 6. Sedimentation coefficients as a function of ethidium bromide concentration. Each point represents the S value obtained from a run of the lysate of exponentially growing cells with ^{14}C-labeled T4 phage as internal marker. Portions of the same cell lysate were used for all the runs. The fractions chosen in each run for this plot are indicated in Fig. 5. EthBr, ethidium bromide.

ethidium bromide/ml., suggesting that the folded DNA of *E. coli* has superhelices. The curves are actually very similar to the ones obtained with other closed-circular DNA's isolated from *E. coli* (Wang, 1969), except that in our case, the S values measured are 50 to 200 times greater. The superhelical concentration of the *E. coli* chromosome (number of superhelical turns per base pair) estimated from the concentration of ethidium bromide which gives the lowest S value, thus turns out to be about the same as the superhelical concentration of these other closed-circular DNA molecules. Moreover, all *E. coli* chromosomes, at no matter what stage in their replication cycle, are seen to possess similar superhelical concentrations inasmuch as the three curves shown in Figure 6 have a minima at the same ethidium bromide concentration within experimental error.

The effect of ethidium bromide is even more evident with more homogeneous folded DNA complexes, purified by a preparative run of a lysate of exponentially growing cells (cf. Figs 8 and 9) or lined up at initiation by amino-acid starvation. These latter homogeneous 1300 s folded chromosomes are converted to slower sedimenting structures of about 800 s with 2 μg ethidium bromide/ml. At higher ethidium bromide concentrations the material sediments again faster (1400 s at 100 μg ethidium bromide/ml.), a homogeneous narrow peak being observed at all ethidium bromide concentrations tested (results not shown).

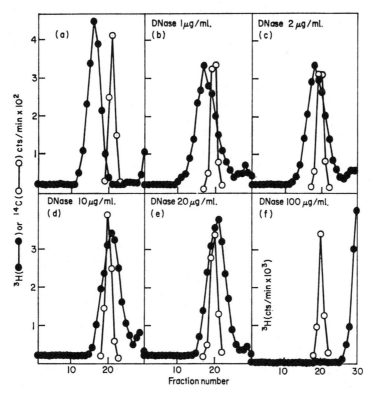

FIG. 7. Effect of DNase on the sedimentation velocity of the folded chromosomes. A fraction of about 1500 s was obtained from a preparative run similar to the one shown in Fig. 2(a). The 1500 s folded chromosomes were treated in an incubation mixture containing 20 μl. of the purified 1500 s folded chromosomes, 100 μl. of 0·2 M-Tris·HCl buffer (pH 8·2), 5 μl. of 0·1 M-MgSO$_4$ and DNase as indicated. Under these conditions (without Ca^{2+}) DNase I creates only single-strand cuts in DNA (P. Price, personal communication). After 10 min on ice, the reaction was terminated by the addition of 20 μl. of 0·1 M-trisodium-EDTA, and the mixture was layered on top of a sucrose gradient. Controls were treated in the same way, except that MgSO$_4$ and DNase were omitted. —O—O—, ^{14}C-labeled T4 phage.

(e) *Effect of DNase treatment on the folded chromosome*

If the changes in sedimentation velocity with ethidium bromide are due to changes in the superhelical content of the folded chromosome, it should be possible to fully relax the structure by single-strand nicking with DNase. By analogy with the smaller double-stranded closed-circular DNA molecules studied previously, we expected an all-or-none effect whereby the supercoil would relax to a slower sedimenting form after only one single-strand nick per chromosome. This relaxed molecule should react to ethidium bromide intercalation like linear DNA. As we shall now show, the actual experimental data fulfilled some of these expectations, but also revealed important differences between the folded *E. coli* chromosome and the more simple DNA models.

All of the experiments reported below were carried out with purified DNA complexes obtained from preparative runs of lysates of exponentially growing cells (as in Fig. 2). An average fraction of about 1500 s was chosen, and samples of the fraction were rerun, either with or without DNase treatment, and with or without ethidium bromide in the gradients. Figure 7 shows the results of a typical experiment. The first frame shows the profile of the isolated 1500 s native complex, and the other frames show the effect of incubation with increasing DNase concentrations for 10 minutes on ice. Slower sedimenting forms are produced, and with about 10 μg DNase/ml., a 900 s relaxed complex is obtained. The complexes treated with 1 and 2 μg DNase/ml. appear to be real intermediate forms between the 1500 s native structure and the 900 s relaxed one. Samples were also treated with less DNase: at 0·1 μg DNase/ml. no difference was seen between the treated complex and the non-treated one. At DNase concentrations higher than 10 μg/ml. (and up to at least 25 μg/ml.), the sedimentation coefficient does not decrease further. The 900 s relaxed complex therefore appears to be an unique structure, reflecting a particular conformation of the folded chromosome rather than an arbitrary intermediate state in the progressive unfolding of the folded chromosome. Figure 7 also shows the effect of 100 μg DNase/ml.; the DNA now behaves like unfolded or fragmented *E. coli* DNA and remains at the top of the gradient in our standard runs.

These DNase results are very reproducible. They suggest that more than one nick may be required to relax the chromosome, since relatively high DNase concentrations had to be used, and intermediate forms were observed at the lower DNase concentrations. Note that the DNase-relaxed complex retains considerable structural organization, as it still sediments much faster than the corresponding unfolded DNA.

(f) *Interaction of the DNase-relaxed complex with ethidium bromide*

The DNase-relaxed complex behaves like linear DNA in the presence of ethidium bromide. Figure 8 shows the results of an experiment in which a 1500 s complex, obtained from a preparative run of a lysate of exponentially growing cells, was treated with 10 μg DNase/ml. as described in section (e) above, and then centrifuged in the presence of ethidium bromide. The upper frames show the profile of the control, non-DNase treated complex, at 0, 2 and 6 μg ethidium bromide/ml. We can see the typical biphasic effect on the sedimentation velocity, a slowing down of the folded DNA as the number of supercoils diminishes at low ethidium bromide concentration, and then a speeding up due to the reversed supercoils generated at high ethidium bromide concentration. The actual sedimentation coefficients are 1500 s at 0 ethidium bromide, 850 s at 2 μg/ml., and 1650 at 6 μg ethidium bromide/ml. The lower frames

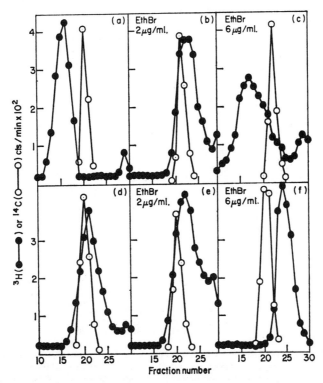

FIG. 8. Effect of ethidium bromide on the sedimentation velocity of the folded chromosomes before and after DNase treatment. A fraction containing 1500 s folded chromosomes was purified as explained in the legend to Fig. 7, and samples run on sucrose gradients containing 0, 2, and 6 μg ethidium bromide/ml., both before (upper panels) and after incubation with 10 μg DNase/ml. (lower panels). The DNase treatment was as indicated [in the legend to Fig. 7. The bottom 9 fractions have been left out in all the profiles; they had no radioactivity above background. —O—O—, ^{14}C-labeled T4 phage; EthBr, ethidium bromide.

show the DNase-relaxed complex under identical conditions. The first run, without ethidium bromide, shows that the DNase caused relaxation; the complex sediments with a rate only 60% of the rate of the same non-DNase treated chromosome (compare with frame directly above). At 2 μg ethidium bromide/ml., the profile is indistinguishable from the profile of the control at 2 μg ethidium bromide/ml., and at 6 μg ethidium bromide/ml. the relaxed particle sediments even slower than at 2 μg ethidium bromide/ml. Since the relaxed particle cannot "wind up" at the high ethidium bromide concentrations it behaves throughout like linear DNA. The sedimentation coefficients of the relaxed complex are 900 s at 0 ethidium bromide concentration, 800 s at 2 μg/ml., and 600 s at 6 μg ethidium bromide/ml.

Figure 9 plots the variation in these sedimentation velocities as a function of ethidium bromide concentration. Samples of the same 1500 s folded complex were used for all of the runs. The S value of the native folded chromosome at its minimum is indistinguishable from the S value of the DNase relaxed complex, which strongly suggests that the folded chromosome has a superhelical density of 0 at 2 μg ethidium bromide/ml. This is the equivalence region, since the number of turns removed by intercalation here equals the number of superhelical turns in the original DNA

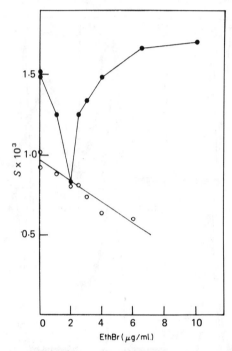

Fig. 9. Sedimentation coefficients of purified folded chromosomes as a function of ethidium bromide concentration. Each point represents an individual run of the ^{3}H-labeled folded chromosomes with ^{14}C-labeled T4 internal marker as shown in Fig. 8. Samples of the same 1500 s fraction obtained from a preparative run of a lysate of exponentially growing cells were used for all of the runs. —●—●—, Native supercoiled folded complex; —○—○—, DNase-treated relaxed complex; EthBr, ethidium bromide.

molecule (Crawford & Waring, 1967). The concentration of ethidium bromide at this minimum (2 μg/ml.) is the same as the ethidium concentration at the equivalence region of intracellular SV40 and λ closed-circular DNA molecules (Wang, 1969), indicating that the *E. coli* folded chromosome has the same superhelical concentration. The superhelical density σ (defined as the number of superhelical turns per 10 base pairs by Bauer & Vinograd (1968)) of closed-circular SV40 and λ DNA is -0.026 (Bauer & Vinograd, 1968; Wang, 1969). Like the closed-circular SV40 and λ DNA, the folded *E. coli* chromosome must also have, therefore, about one negative superhelical turn per 400 base pairs. The relative change in S values as the ethidium bromide intercalates is however much greater in the case of the *E. coli* folded chromosome, probably reflecting its much more complex tertiary structure.

(g) *Number of DNA nicks required to relax the supercoiled* E. coli *chromosome*

The relatively high levels of DNase required for full relaxation, and the existence of intermediate forms, suggest that more than one single-strand nick is required to obtain a fully relaxed complex from a native folded one. The actual number of single-strand breaks introduced by DNase under our conditions was therefore determined by measuring the length of the DNA single strands released from the treated complex by alkaline sucrose-gradient centrifugation.

Figure 10 shows alkaline sucrose profiles of the DNA from a 1500 s complex

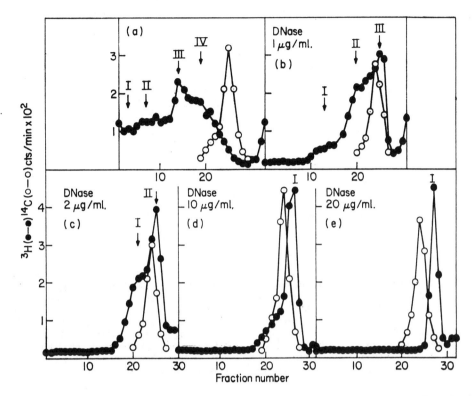

Fig. 10. Alkaline sucrose-gradient profile of the DNA fragments obtained from the folded chromosomes before and after DNase treatment. Denatured ^{14}C-labeled T4 DNA was used as internal marker. The 10 to 20% alkaline sucrose gradients were spun 30 min at 40,000 rev./min and 4°C.

purified and treated with DNase as previously described for Figure 7. [14C]Thymidine-labeled DNA extracted from phage T4 was run as an internal marker. The first frame shows the profile for the native, non-DNase treated, 1500 s complex. The broad profile indicates that DNA strands of very different lengths are present. Most strands are very long, none being smaller than the T4 marker (65×10^6 daltons). This profile changes drastically after DNase treatment of the complex; all fast sedimenting molecules disappear, and new slower sedimenting species appear. The fact that the profiles after DNase treatment are definitely non-Gaussian suggests either that some stretches of the folded DNA are more accessible to the DNase than other DNA segments buried inside the highly folded structure, or that, once a particular DNA section is unfolded, it becomes much more susceptible to further DNase attack.

Table 1 gives molecular-weight estimates for the single-strand DNA fragments observed in Figure 10. The molecular weights of the longest single strands observed appear to be somewhat higher than the molecular weight expected for a full-length single strand: taking Cairns' value of 25×10^6 for the molecular weight of *E. coli* DNA (Cairns, 1963), the maximum molecular weight of each single strand should be $12 \cdot 5 \times 10^6$. Possibly, the Studier (1965) relationship between molecular weight and sedimentation coefficients for alkaline denatured DNA which was used is not valid for such large DNA strands. However, the size of the fragments sedimenting slightly

TABLE 1

Molecular-weight estimates of the DNA single-strand fragments obtained from the
E. coli *folded chromosome before and after DNase treatment*

DNase (μg/ml.)	Type of DNA complex	Molecular weight-classes ($\times 10^8$) in Fig. 10		Breaks/strand of *E. coli* DNA
0	Compact (native)	I	17·7	Intact
		II	12·4	Intact
		III	5·63	1–2
		IV	2·64	4
1	Intermediate	I	6·90	1
		II	2·01	5
		III	0·44	∼30
2	Intermediate	I	1·80	6
		II	0·41	∼30
10	Relaxed	I	0·29	∼40
20	Relaxed	I	0·21	∼60

The values were obtained by assuming that the relationship between molecular weight (M) and sedimentation velocity (S) obeys the empirical Studier (1965) relationship for denatured DNA: $S1/S2 = (M1/M2)^{0.40}$ and that the mol. wt of the T4 single-strands marker is 65×10^6 (Mosig, 1963). Roman numerals refer to the fractions in the alkaline-sucrose gradients shown in Figure 10. The numbers of breaks per strand were calculated by taking a value of 12.5×10^8 as the molecular weight of the intact *E. coli* single-strand DNA (Cairns, 1963). We assume that all the DNA single strands observed in Fig. 10 are linear molecules. It is, however, possible that the abnormally fast sedimenting forms (classes I and II of Fig. 10(a)) represent closed-circular DNA molecules.

slower than T4 single strands are of primary interest, and these calculations should be quite reliable (Studier, 1965).

Table 1 also lists the number of breaks introduced by DNase into an average DNA strand as a function of the DNase concentration used. In the native folded complex, most of the chromosomes have less than 3 nicks. With increasing concentrations of DNase, many breaks are found per strand, as expected. The folded chromosome treated with 10 μg DNase/ml. is converted to a fully relaxed complex (Fig. 7), so that the 40 breaks per DNA strand observed under this condition is probably an upper estimate for the number of breaks needed to relax the chromosome. A lower estimate of 6 breaks per DNA strand can be derived from the 2 μg DNase/ml. experiment, which does not suffice to relax the complex fully. In this case, the actual number of breaks per DNA strand cannot be estimated precisely because of the non-Gaussian profiles in the alkaline sucrose gradient (Fig. 10(c)). The 6 breaks per DNA strand is a conservative estimate taken from the larger molecular weight class in Figure 10(c) which accounts for almost 50% of the total DNA. The smallest molecules may represent DNA stretches very susceptible to DNase attack as discussed previously. Therefore, we estimate that somewhere between 6 and 40 nicks per DNA strand are needed to relax the folded *E. coli* chromosome.

(h) *Nature of the forces which create and stabilize the folded chromosome*

In agreement with the original observations of Stonington & Pettijohn (1971), we have found that the complex remains folded after treatment with 1 mg pronase/ml. (both DNase-relaxed and native complex) but unfolds with mechanical shear, 1% sodium dodecyl sulfate, 1% Sarcosyl, and 5 μg pancreatic ribonuclease/ml. (all incubations carried out on ice for 15 min). The ribonuclease effect is most intriguing. Contrary to the DNase action, the RNase unfolds the complex in an all-or-none manner: at low RNase concentrations only a fraction of the chromosomes unfold, while the rest remain in their native folded conformation.

Figure 11 shows the sucrose-gradient profiles of folded chromosomes (obtained from a preparative run of a lysate of amino-acid-starved cells) before (Fig. 11(a)) and after (Fig. 11(b) and (c)) RNase treatment. It is clear that a fraction of the chromosomes

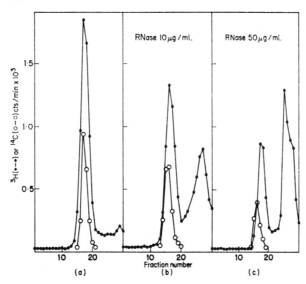

FIG. 11. Effect of RNase on the sedimentation velocity of the folded chromosomes. The complexes were obtained from a preparative run of a lysate of amino-acid-starved cells as described in the legend to Fig. 3(a). 20 μl. of the peak fraction of the preparative run were incubated with 100 μl. of 0·2 M-Tris·HCl buffer (pH 8·2), 20 μl. 0·1 M-EDTA, and 0, 5, or 25 μl. of a 0·3 mg/ml. solution of pancreatic RNase. After 15 min on ice, the mixture was layered on top of a sucrose gradient. The low sedimentation coefficient of the control run (1000 s instead of the usual 1300 s of *lined-up* chromosomes of amino-acid-starved cells) suggests that the chromosomes may have had more nicks than usual, resulting in a slightly relaxed structure. —○—○, ^{14}C-labeled T4 phage.

is completely unfolded by RNase, while the rest remain folded (although the sedimentation velocity of the folded species after RNase treatment may be slightly lower).

Similar RNase effects have been obtained with folded chromosomes isolated from exponentially growing cells; in the case of the replicating chromosomes, lower RNase concentrations (1 μg/ml. or less) are sufficient to cause unfolding (L. Carlson & A. Worcel, unpublished experiments).

The unfolding by RNase suggests an all-or-none effect of the enzyme. It appears therefore that an RNA species, and perhaps a single RNA molecule, stabilizes the folded chromosome.

4. Discussion

A model for the folded chromosome of Escherichia coli

The data presented can be readily interpreted if we assume that the chromosome is folded into a limited number of large *loops*, each of the same superhelical concentration. The interactions which stabilize the loops and prevent unfolding could also

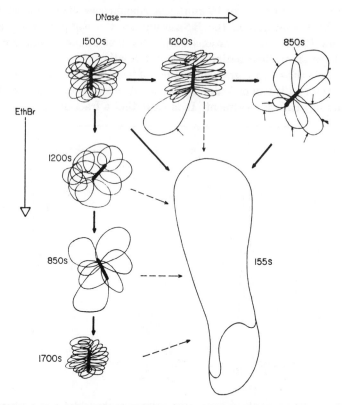

FIG. 12. Schematic representation of the folded *E. coli* chromosome. The chromosome is shown as made up of seven loops (the actual number is greater, see text) and the DNA of each loop is coiled into a "slinky" (this coiling will result in superhelices when the DNA is extended, see Fig. 13). Single-strand nicks by DNase are indicated by the small arrows. The black rod holding the loops together (both in their supercoiled and relaxed conformation) represents the RNA core. Heavy arrows indicate observed transitions, the broken ones represent transitions which probably also take place.

The chromosome of *E. coli* is seen in its three main configurations:

(1) A *compact* folded supercoil, either in its native negative sense (DNA chains underwound), or supercoiled in the opposite, positive sense (overwound) at high levels of ethidium bromide intercalation.

(2) A *relaxed* folded complex, without superhelices, arrived at after an intermediate level of ethidium bromide intercalation or after extensive nicking with DNase. The relaxed complex is still folded into loops.

(3) An *unfolded* chromosome *à la Cairns*, which has lost all its organized tertiary structure. EthBr, ethidium bromide.

prevent propagation of the rotational events from one loop to the others. A single-strand nick would allow the free rotation of the DNA strands and unwind the supercoils within a particular loop, without affecting the superhelical concentration in the rest of the chromosome. Such a structure is shown schematically in Figure 12. In the upper left-hand corner of this Figure, a typical 1500 s replicating chromosome is represented coiled into its compact conformation. Each nick with DNase releases the rotational constraints and *abruptly* eliminates the superhelices in a *localized* stretch of DNA or loop. As drawn, the chromosome has seven such loops, although the actual number is greater (see below). Only when all of the loops are nicked is the complex fully relaxed (upper right-hand corner); this structure behaves like linear DNA upon ethidium bromide intercalation.

In contrast, ethidium intercalation in the native (non-DNase treated) complex eliminates the negative superhelices *gradually throughout* the complex. At the concentration of ethidium bromide where the superhelical content is zero, the ethidium-relaxed complex has the same sedimentation velocity as the DNase relaxed one. At higher ethidium concentrations, the native complex winds up again (in the opposite, positive sense), and the sedimentation velocity increases as the superhelical content of the complex increases.

The loops, both in the supercoiled and in the relaxed conformation, are held together by interaction with a *core* (designated by a black rod in Fig. 12) which is, in all likelihood, an RNA molecule. Any disruption of the RNA core will result in the unfolding of the DNA. Possibly, DNA–RNA interactions give a rigidity to the folded chromosome which manifests itself as a series of independent loops. In other words, the rotation of the DNA chains may not propagate through such DNA–RNA interactions. It is interesting to note that Lark has recently postulated the existence of an RNA core based on *in vivo* studies with inhibitors of RNA synthesis (Lark, 1972). According to him, the replication apparatus, not the DNA, would be assembled around the hypothetical RNA core.

The sedimentation coefficient value of 155 s assigned to the Cairns-type unfolded chromosome has been calculated from the empirical Burgi–Hershey relationship (1963): $S_1/S_2 = (M_1/M_2)^{0.35}$ taking an S value for T4 DNA of 63, a molecular weight for T4 DNA of 1.3×10^8 (Mosig, 1963) and a molecular weight for the *E. coli* DNA of 2.5×10^9 (Cairns, 1963). The actual values of the sedimentation velocity of intact bacterial DNA obtained experimentally agree well with that estimate (Kavenoff, 1971; Stonington & Pettijohn, 1971). The fully relaxed complexes thus sediment more than five times faster than the unfolded DNA, and, therefore, must still have a compact structural organization, as diagrammed in Figure 12. We shall now attempt to define our model in more quantitative terms with respect to both the number of loops per chromosome, and the tertiary structure of the DNA within each loop.

(i) *Number of loops per chromosome*

This number can be estimated directly from the data of Table 1. As each chromosome has two DNA strands, and as a single break in one of the two strands suffices to relax a loop, the number of loops per chromosome should be equal to twice the number of breaks per strand needed to relax the folded complex. Therefore, a chromosome must have somewhere between 12 and 80 loops. It is not possible to give a more precise estimate because some loops will have more than one nick by the time the whole folded chromosome is relaxed. This is due not only to a random probability factor which could be calculated from a Poisson distribution, but also to the apparent non-random nicking by DNase as evidenced by the non-Gaussian profiles of the alkaline sucrose gradients. For the sake of simplicity in the discussion which follows, we shall assume that there are 50 loops per chromosome (a round average number between 12 and 80).

As the chromosome appears to be made up of independent loops, it is now clear why we could detect the superhelices in the first place: the alkaline sucrose-gradient profiles indicate that most of the DNA single strands obtained from the native folded chromosome have about two breaks per strand (four single-strand nicks per chromosome), and, therefore, only four loops out of the total of fifty would have lost their superhelices. In other words, well over 90% of the superhelical content of the chromosome would still be retained under those conditions. In the absence of the postulated loops, the rotation of the DNA strands around the nicks would result in the loss of all the superhelices.

The broadening of the sedimentation profiles of the native folded chromosome at high ethidium bromide concentration (Figs 5(c) and 8(c)) can be explained by the fact that the nicked loops will behave like fragmented DNA, and their sedimentation velocity will tend to decrease, while the sedimentation velocity of the rest of the chromosome will tend to increase with increasing ethidium concentrations. Therefore, different numbers of nicks per chromosome will create an heterogeneous profile, chromosomes with more nicks will sediment slower than chromosomes with fewer nicks. This means that our values for the sedimentation velocity at high ethidium concentrations are minimum estimates, and the actual increase in sedimentation velocity due to the reverse supercoiling after ethidium intercalation may even be greater.

The postulated existence of the chromosome loops is the only way in which we can explain all the experimental evidence presented here.

As isolated, there is a total of about 10,000 negative superhelical turns per *E. coli* chromosome (4×10^6, number of base pairs per chromosome, divided by 400, number of base pairs per negative superhelical turn). If each chromosome has 50 loops, each loop will have, on the average, 200 negatives superhelices.

(ii) *Tertiary structure within each loop*

As we only have about 50 loops per chromosome, each loop is still, on the average, about 20 μm long, and we are still a long way from our 1-μm diameter nuclear body. Therefore, we would like to suggest a possible model for the folding within each loop, based on the known superhelical concentration of the DNA.

It is a reasonable and generally accepted assumption that the DNA inside the cell has no superhelices as such and that they are generated during isolation (Vinograd & Lebowitz, 1966; Wang, 1969). Two likely explanations have been forwarded with regard to their origin in isolated DNA: (a) they result from a difference in pitch of the Watson–Crick double helix inside and outside the cell (Wang, Baumgarten & Olivera, 1967; Wang, 1969); (b) they reflect the presence of some organized tertiary structure *in vivo* which is destroyed during chromosome isolation (Vinograd & Lebowitz, 1966; Kayser, 1966). So far, it has not been possible to distinguish experimentally between the two.

We would like to elaborate on the second possibility and suggest a model of folding which can account for the superhelices. If the DNA is coiled in a broad helix (see Fig. 13(a)), each 360° turn of this coil will generate one superhelix when the closed-circular molecule, or our DNA loop, is extended (Fig. 13(b)). We would thus predict, from the known superhelical concentration of the DNA, that each 360° turn of such a coil will be 400 nucleotide pairs long. Such a conformation of the "worm-like" DNA helix is clearly possible, without melting or breaking the double helix; estimates of the DNA persistence length in solution (which may be seen as the length of helix needed to create a 90° bend) are around 450 Å (Triebel, Reinert & Strassburger, 1972), or 132 nucleotide pairs. To make a circle (four 90° bends), 528 nucleotide pairs would

10

FIG. 13. Schematic representation of the postulated helical coiling of the DNA. (a) Compact *in vivo* conformation; (b) supercoiled form observed *in vitro* by extending the DNA.
Note that this Fig. could have been drawn in 4 different ways (both the helices shown in (a) and (b) could have been arbitrarily drawn as right-handed or left-handed). However, this is the only correct way to draw this Fig.: as the Watson–Crick helix is right-handed, the tertiary helical coiling of the DNA has to be left-handed (as shown in (a)) in order to cause a topological deficiency in total helical turns (negative superhelices). After extending the DNA, the left-handed helical coiling will generate right-handed superhelices (as shown in (b)).

be required, a figure not very different from the 400 nucleotide pairs in our model (Fig. 13(a)). According to this model of folding, each one of our 50 chromosome loops contains, on the average, 200 turns of the DNA coil (with 400 nucleotide pairs per turn).

All of the closed-circular intracellular DNA's examined so far have about the same superhelical concentration (Wang, 1969) and, therefore, they may all have the same basic tertiary structure inside the cell (as shown in Fig. 13(a)).

From the ratio of the volume of the DNA circumference to the volume of the enclosed hole, one can calculate that such packaging could generate a structure which, if all the DNA rings were closely packed upon each other, would have one-fifth of the theoretical maximal compactness of the DNA. The compactness of the folded complex is actually much less than this: assuming that the complex sediments as a sphere, the equation $r = M (1 - \nu\rho)/N S 6\pi\eta$ (Tanford, 1961), gives a radius of 0·8 μm for the folded chromosomes arrested at initiation ($M = 2·5 \times 10^9$ daltons, $S = 1300$; $\nu = 0·53$). The concentration of DNA in such a particle is then about 2 mg/cm³, or less than 1/500th of the density of the DNA in the head of T2 phage (Stonington & Pettijohn, 1971; Zarybnicky, 1969; Tikchonenko, 1969). Therefore, the observed compactness of the complex is easily compatible with our postulated DNA coiling.

The radius of the bacterial nuclear body is about 0·5 μm (Ryter, 1968). The slightly larger radius observed *in vitro* may be due to the extensive loss of proteins, normally associated with the DNA, which would be expected after lysis in 1·0 M-NaCl (Alberts, Amodio, Jenkins, Gutman & Ferris, 1968). Those proteins could conceivably maintain the chromosome in a somewhat more compact configuration *in vivo*.

An important question about the postulated coiling is why the direction of the coiling is unique (as shown in Fig. 13(a)). As DNA always has negative superhelices when examined *in vitro*, the coiling *in vivo* would have to take place only in one direction, the one which causes a topological deficiency in helical turns. This could be explained if it is much easier to underwind than to overwind the double helix. Alternatively, it may be argued that this direction is imposed by proteins which interact with the DNA inside the cell, and that once the DNA starts rolling in one direction, it will continue to do so.

The model for the folding presented here is very general and preliminary. Actually, if we consider the replicating chromosome instead of the resting chromosome of stationary cells, a model with three nucleation points should be considered, one for the unreplicated segment of the chromosome and the other two for the partially synthesized daughter chromosomes; the older complex can then be visualized as gradually decreasing in size as the two new complexes grow. A necessary event before initiation of chromosome replication would then be the synthesis of the two new RNA cores, around which the newly replicated DNA could fold (see also Lark, 1972).

One last comment about the folded chromosome of *Escherichia coli* has to do with the protein recently purified by Wang from *E. coli* which eliminates the superhelices of λ and other small circular twisted DNA's (Wang, 1971). As the enzyme is present in uninfected *E. coli* cells, the folded supercoiled chromosome described here should be, in all likelihood, its natural substrate. This prediction should be easy to test.

We thank B. Alberts for a critical reading of the manuscript. This research has been supported by a grant from the National Science Foundation and a grant-in-aid from Princeton University.

REFERENCES

Alberts, B., Amodio, F., Jenkins, M., Gutman, E. & Ferris, F. (1968). *Cold Spr. Harb. Symp. Quant. Biol.* **33**, 289.

Bauer, W. R. & Vinograd, J. (1968). *J. Mol. Biol.* **33**, 141.

Burgi, E. & Hershey, A. D. (1963). *Biophys. J.* **3**, 309.

Cairns, J. (1963). *J. Mol. Biol.* **6**, 208.

Caro, L. G. & Berg, C. M. (1968). *Cold Spr. Harb. Symp. Quant. Biol.* **33**, 559.

Cerda-Olmedo, E., Hanawalt, P. C. & Guerola, N. (1968). *J. Mol. Biol.* **33**, 705.

Crawford, L. V. & Waring, J. J. (1967). *J. Mol. Biol.* **25**, 23.

Cummings, D. (1964). *Virology*, **23**, 408.

Helmstetter, C. E. & Cooper, S. (1968). *J. Mol. Biol.* **31**, 507.

Kavenoff, R. (1971). Ph.D. Thesis, Albert Einstein College of Medicine.

Kayser, A. D. (1966). *J. Gen. Physiol.* **49**, 171.

Lark, K. G. (1969). *Ann. Rev. Biochem.* **38**, 569.

Lark, K. G. (1972). *J. Mol. Biol.* **64**, 47.

Maaløe, O. & Hanawalt, P. (1961). *J. Mol. Biol.* **3**, 144.

Meselson, M. & Stahl, F. W. (1958). *Proc. Nat. Acad. Sci., Wash.* **44**, 671.

Mosig, G. (1963). *Cold Spr. Harb. Symp. Quant. Biol.* **28**, 35.

Okazaki, R., Okazaki, T., Sakobe, K., Sugimoto, K., Kainuma, R., Sugina, A. & Iwatsuki, N. (1968). *Cold Spr. Harb. Symp. Quant. Biol.* **33**, 129.

Pardee, A. B., Jacob, F. & Monod, L. (1959). *J. Mol. Biol.* **1**, 165.

Patterson, M. S. & Greene, R. C. (1965). *Analyt. Chem.* **37**, 854.

Quinn, W. G. & Sueoka, N. (1970). *Proc. Nat. Acad. Sci., Wash.* **67**, 717.

Ryter, A. (1968). *Bact. Rev.* **32**, 39.

Stonington, G. O. & Pettijohn, D. E. (1971). *Proc. Nat. Acad. Sci., Wash.* **68**, 6.

Studier, F. W. (1965). *J. Mol. Biol.* **11**, 373.

Sueoka, N. & Quinn, W. G. (1968). *Cold Spr. Harb. Symp. Quant. Biol.* **33**, 695.

Tanford, C. (1961). *Physical Chemistry of Macromolecules.* New York: Wiley.

Tikchonenko, T. (1969). In *Advances in Virus Research*, ed. by K. Smith, M. Lauffer & F. Bang, vol. 15, p. 201. New York: Academic Press.

Triebel, H., Reinert, K. E. & Strassburger, J. (1972). *Biopolymers*, **10**, 2619.

Vinograd, J. & Lebowitz, J. (1966). *J. Gen. Physiol.* **49**, 103.

Vinograd, J., Lebowitz, J., Radloff, R., Watson, R. & Lapis, P. (1965). *Proc. Nat. Acad. Sci., Wash.* **53**, 1104.

Wang, J. C. (1969). *J. Mol. Biol.* **43**, 263.

Wang, J. C. (1971). *J. Mol. Biol.* **55**, 523.

Wang, J. C., Baumgarten, D. & Olivera, B. M. (1967). *Proc. Nat. Acad. Sci., Wash.* **58**, 1852.

Waring, M. J. (1965). *J. Mol. Biol.* **13**, 269.

Yoshikawa, H., O'Sullivan, A. & Sueoka, N. (1964). *Proc. Nat. Acad. Sci., Wash.* **66**, 365.

Zarybnicky, V. (1969). *J. Theoret. Biol.* **22**, 33.

Questions

ARTICLES 36 and 37   Fiers and Sinsheimer

1. State the argument from which it is concluded that a single-strand break in ϕX174 DNA does not change the molecular weight.
2. The discontinuity in the ϕX174 DNA circle has now been explained by the observation that if the DNA is boiled in very low ionic strength there seems to be no discontinuity. Explain.
3. Is linear ϕX174 DNA infective? Explain.

ARTICLE 38   Kleinschmidt, Burton, and Sinsheimer
ARTICLE 39   Freifelder, Kleinschmidt, and Sinsheimer

1. Why does the addition of formaldehyde cause the ϕX174 DNA to disentangle? Could substances such as formamide and dimethylsulfoxide be just as effective? Why would boiling the DNA not have this effect?
●2. Would you expect the double-stranded DNA described in Article 38 and the single-stranded DNA described in Article 39 to have the same length? Would any of the conditions used to prepare the DNA for electron microscopy affect the length? Explain.

ARTICLE 40   Dulbecco and Vogt

1. What are the F and S forms? How do you know that neither is a linear molecule?
2. Why should F elute from a MAK column at a lower salt concentration than S?

ARTICLE 41   Weil and Vinograd

1. What are the structures of forms I, II, and III? Why did Weil and Vinograd identify form II incorrectly?
●2. Explain the conversion of I to II in concentrated CsCl.

ARTICLE 42   Vinograd et al.

●1. How should the shear sensitivities of forms I, II, and III be related? Should the shear sensitivity of form II and a variant of form I containing no supercoiling be the same?
●2. If radioactive formaldehyde is added to normal DNA, the binding of the formaldehyde can be detected by the appearance of acid-precipitable radioactivity (because DNA is acid-insoluble). By sampling at various times after the addition of formaldehyde, the rate of binding can be measured. Propose an explanation for the fact that the rate of binding is greater with form I than with form II. As an aid, note that the binding to boiled forms II or III is extremely rapid.
●3. An endonuclease known as S1 makes a strand break only with single but not with double-stranded DNA. However, it can cleave one of the forms of polyoma DNA, usually making only a single break. Which form is cleaved and why? (Note that this and the preceding question are related.)
4. Explain why the curve of Figure 3a is linear.

1. Why is the density of DNA lower when ethidium bromide is present than when it is absent? What would you expect the relation to be between the density in ethidium fluoride and that in ethidium iodide?

2. Why does a supercoil have a higher density than a linear molecule when ethidium bromide is present?

3. Would you expect the density to decrease, when ethidium bromide is present, with decreasing GC content?

4. In CsCl containing ethidium bromide, a supercoiled DNA molecule has a density of 1.592 g/ml, and a linear DNA molecule a density of 1.556 g/ml. What would be the density of a supercoil linked, as in a chain, with an open circle of (1) identical mass or (2) twice the mass? (Linked circles do exist and are called *catenanes*.)

•5. Could a density shift occur as a result of the binding of a nonintercalating substance to a phosphate group or the deoxyribose? Under what circumstances could it be used to separate supercoiled from linear DNA?

1. Summarize the argument that the *E. coli* chromosome consists of supercoiled units that are in some way isolated from one another.

2. In Fig. 9, in the curve for the DNase-treated relaxed complex, why does the sedimentation coefficient decrease continuously?

•3. Suppose you had introduced two nicks into the complex and then measured the sedimentation coefficient as a function of the concentration of ethidium bromide. Would a curve such as that with the solid circles of Fig. 9 result? How would the sedimentation coefficient at 0 and 10 μg/ml ethidium bromide differ from that shown in the figure? Would there be a minimum-value inflection of the sedimentation coefficient and, if so, at what concentration?

4. How does the *E. coli* DNA supercoil differ from the polyoma DNA supercoil?

5. Would there be any supercoiling after treatment with RNase? Is there any evidence for this?

6. Do you think that the molecules studied by Cairns in Article 6 had nicks? Explain your answer.

References

DeWitt, S., and E. A. Adelberg. 1962. "Transduction of the Attached Sex Factor of *Escherichia coli*." *J. Bact.* **83**, 673–678. Evidence for linkage between proximal and distal parts of an Hfr chromosome.

Stonington, O. G., and D. E. Pettijohn. 1971. "The Folded Genome of *Escherichia coli*: Isolation of a Protein-DNA-RNA Complex." *Proc. Nat. Acad. Sci.* **68**, 6–12.

Szybalski, W., R. L. Erikson, G. A. Gentry, L. G. Gafford, and C. C. Randall. 1963. "Unusual Properties of Fowlpox Virus DNA." *Virology* **19**, 586–589.

Weil, R. 1963. "The Denaturation and the Renaturation of the DNA of Polyoma Virus." *Proc. Nat. Acad. Sci.* **49**, 480–487.

Additional Readings

Clayton, D. A., and J. Vinograd. "Circular Dimer and Catenated Forms of Mitochondrial DNA in Human Leukaemic Lymphocytes."*Nature* **216**, 652–657. An early report of twisted dimer circles and of linked circles.

Clewell, D. B., and D. R. Helinski. 1969. "Supercoiled Circular DNA-Protein Complex in *Escherichia coli*: Purification and Induced Conversion to an Open Circular DNA Form," *Proc. Nat. Acad. Sci.* **62**, 1159–1166. The discovery of a unique protein bound to certain supercoiled plasmids.

Clowes, R. C. 1972. "Molecular Structure of Bacterial Plasmids." *Bact. Rev.* **36**, 361–405.

Hickson, F. T., T. F. Roth, and D. R. Helinski. 1967. "Circular DNA Forms of a Bacterial Sex Factor." *Proc. Nat. Acad. Sci.* **58**, 1731–1738.

Jacob, F., and E. L. Wollman. 1961. *Sexuality and the Genetics of Bacteria*. Academic Press. A classic showing the development of the concept of a circular genetic map of *E. coli*.

EPILOGUE

What Do We Know?

The development of our ideas on DNA structure has been presented throughout this book. It is curious that some of these ideas were accepted only after rigorous proof whereas others were too attractive to be ignored. Nils Barricelli once had posted on the door to his office the words, "To say *I believe* is good religion but bad science." Perhaps this motto is true, but it is the belief, the hunch, or the disbelief that is frequently, if not usually, the driving force behind good experiments.

We now know that the Watson-Crick structure is correct and that, except for the few examples of naturally single-stranded DNA, the structure is a double-strand helix. Four-stranded structures do not occur. The principal deviation from the classic model is that an unbroken DNA is frequently circular and supercoiled. In fact, it is probably true that almost all DNA molecules (we know of one exception so far) are supercoiled at some stage of their life cycle.

Chemical structure can vary also in that, in some organisms, one or more of the standard bases is replaced by another. The replacement however always has the same hydrogen bonding properties. Also all DNA molecules studied thus far contain methylated bases, and the methylation occurs after the standard base is incorporated into the DNA. These methylated bases apparently protect specific sequences from attack by nucleases whose function may be to destroy foreign DNA.

The base composition of naturally occurring DNA molecules ranges from 25% to 75% adenine-thymine (AT) in prokaryotes. As one goes up the evolutionary scale, the range becomes smaller until with mammals, values less than 45% or greater than 53% AT are rare. However, some higher organisms, such as the crab, contain a small fraction of DNA that is nearly 100% AT; the significance of this high AT content is not yet known.

It is believed, though not yet definitively

proved, that in cells containing several chromosomes, each chromosome contains a single DNA molecule. This means that the size of natural DNA ranges from a molecular weight of 10^6 for the smallest bacterial plasmids to 10^{11} for some animal chromosomes. A few species of DNA with molecular weight as low as 25,000 were reported in the early 1960's, but have not been studied carefully; their significance is obscure. In general the amount of DNA per cell increases with complexity, viruses having the least, then bacteria, then the yeasts, and so on to the vertebrates. The rule is not altogether applicable though, since many lower organisms, including the frog, have more DNA per cell than man.

It is probably accurate to say that little remains to be understood about the structure of free DNA although some very interesting aspects, not discussed in this book, are still being explored. How DNA is arranged into a bacterial cell or in a eukaryotic chromosome is another question. The precise manner of folding of a 50-micron piece of phage DNA into a phage whose head is 0.1 micron in diameter is far from clear. The work of Pettijohn and the subsequent findings of Worcel and Burgi indicate that inside a bacterium the DNA is not naked but seems to be complexed with RNA and probably protein as well. Other work has indicated that DNA is also bound somehow to the cell membrane. Recent work has shown that in eukaryotes the DNA is in the form of chromatin, a complex of DNA and a class of proteins called histones. Furthermore, when chromatin condenses to form a chromosome, RNA and other proteins may be contained in the chromosome structure. This is a major topic of current research.

The mechanism of DNA replication and the structure of a replicating molecule remain poorly understood. The list of complexities of replicating DNA lengthens continually—Okazaki fragments, bidirectional replication, rolling circles, three polymerases and at least twelve other genes, initiators, terminators, switches, swivels, and so on. As more research is performed, our understanding of these topics will surely grow.

Structural Formulas of the Components of DNA and RNA

The three-dimensional structure of DNA is shown in Fig. A-1, and the chemical structure of one strand of the double helix in Fig. A-2. Note that the single strand consists of an alternating sugar-phosphate chain (called the backbone) to which the organic bases are attached. The sugar-base unit is called a *nucleoside;* when the phosphate is attached, it is a *nucleotide.* A molecule is called a de-

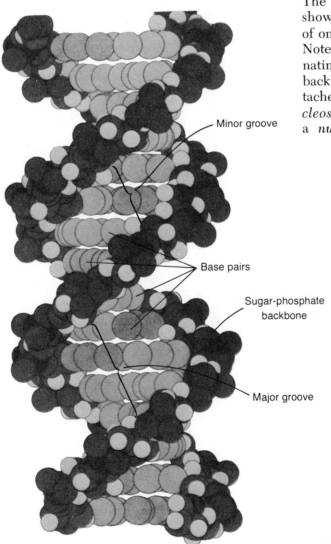

Figure A-1
A three-dimensional space-filling model of the DNA double helix [From *DNA Synthesis* by A. Kornberg. W. H. Freeman and Company. Copyright © 1974.]

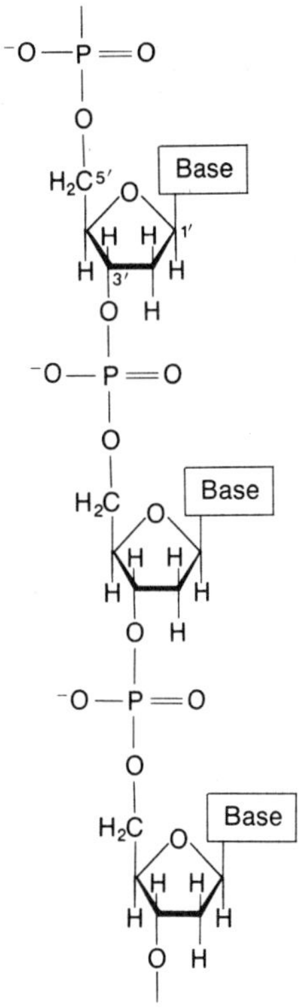

Figure A-2
The chemical structure of part of a DNA chain.

485

oxyribonucleotide or deoxynucleotide if the sugar is deoxyribose and ribonucleotide when it is ribose. The basic repeating unit of the DNA strand is the deoxynucleotide. The chemical structures of the bases and of a nucleotide are shown in Fig. A-3.

RNA differs from DNA in three ways: (1) it is a single-stranded polynucleotide; (2) each nucleotide contains the sugar, ribose, instead of deoxyribose; (3) RNA contains the base uracil rather than thymine.

The names (with their common abbreviations in parentheses) that are used to describe the components of DNA and RNA are listed in Table A-1. Note that the prefix "r" refers to the ribo- form, and "d" to the deoxy- form.

The base pairs, adenine-thymine and guanine-cytosine are those found in naturally occurring DNA. They are shown in Fig. A-4.

TABLE A-1
Nucleic acid nomenclature

Base	Nucleoside	Nucleotide[1]
Purines (Pu)		
Adenine (A)	Adenosine (rA)	Adenylic acid, or adenosine monophosphate (AMP)
	Deoxyadenosine (dA)	Deoxyadenylic acid, or deoxyadenosine monophosphate (dAMP)
Guanine (G)	Guanosine[2] (rG)	Guanylic acid, or guanosine monophosphate (GMP)
	Deoxyguanosine (dG)	Deoxyguanylic acid, or deoxyguanosine monophosphate (dGMP)
Pyrimidines (Py)		
Cytosine (C)	Cytidine (rC)	Cytidylic acid, or cytidine monophosphate (CMP)
	Deoxycytidine (dC)	Deoxycytidylic acid, or deoxycytidine monophosphate (dCMP)
Thymine (T)	Thymidine[3] (dT)	Thymidylic acid, or thymidine monophosphate (TMP)
Uracil (U)	Uridine[4] (rU)	Uridylic acid, or uridine monophosphate (UMP)

[1]Note that each nucleotide has two names for the same substance.
[2]Guanosine should not be confused with guanidine, which is not a nucleic acid base.
[3]Thymidine is the deoxy- form. The ribo- form, ribosylthymine, is not generally found in nucleic acids.
[4]Uridine is the ribo- form. Deoxyuridine is not usually found as such, although it is present in the biosynthetic pathway of thymidylic acid—i.e., deoxyuridylic acid is methylated to yield thymidylic acid.

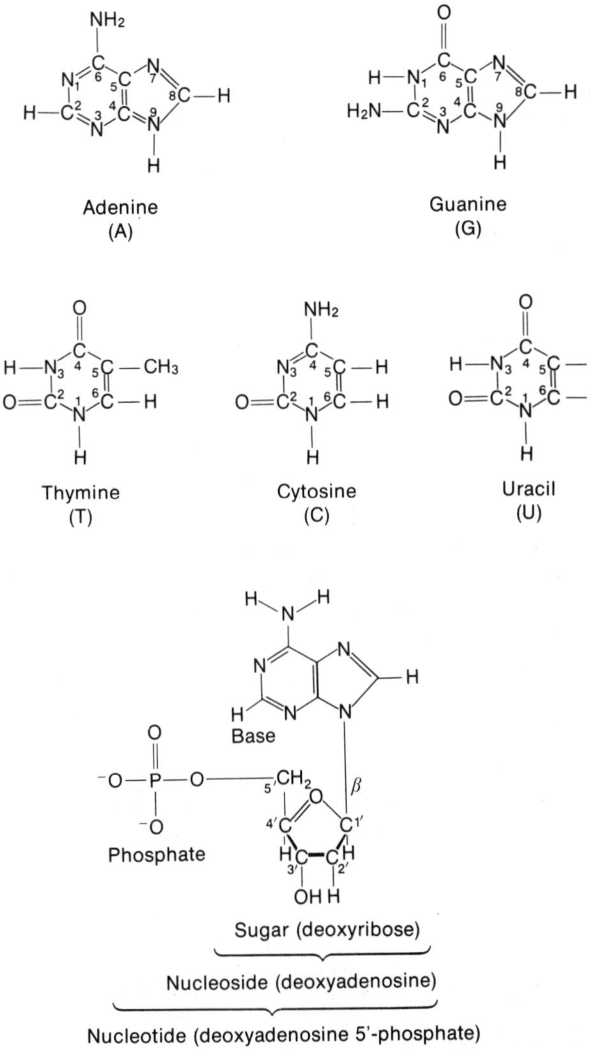

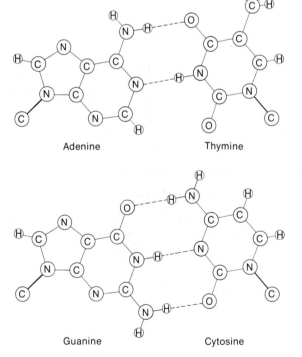

Adenine
(A)

Guanine
(G)

Thymine
(T)

Cytosine
(C)

Uracil
(U)

Base

Phosphate

Sugar (deoxyribose)

Nucleoside (deoxyadenosine)

Nucleotide (deoxyadenosine 5'-phosphate)

Figure A-3
Structure of the bases of DNA and RNA and a nucleotide.

Adenine   Thymine

Guanine   Cytosine

Figure A-4
Structure of the base pairs of DNA. Hydrogen bonds are shown as dotted lines. One carbon atom of the deoxyribose residues is also shown.

Life Cycle of *E. coli* phages T2 and T4*

An electron micrograph of *E. coli* phage T4 is shown in Fig. B-1. The phage particle consists of two units—the head, which contains the DNA; and the tail, which is responsible for delivering the DNA to the host bacterium. The tail is quite a complex structure, consisting of a hollow core, a surrounding sheath, and a base plate to which are attached six tail fibers that

*Phages T2 and T4 are very closely related. Except for a few minor points their life cycles are identical.

absorb to the bacterial cell wall. The phage, which can reproduce only in a living *E. coli* bacterium, has a life cycle that can be simply divided into six stages: (1) adsorption of the phage to the bacterium, (2) injection of DNA into the host bacterium, (3) destruction of bacterial DNA and synthesis of phage DNA, (4) synthesis of phage proteins, (5) packaging of the DNA in heads and assembly of the phage particle, and (6) release of the phage particles by lysis of the bacterium. This cycle is shown schematically in Fig. B-2. A unique

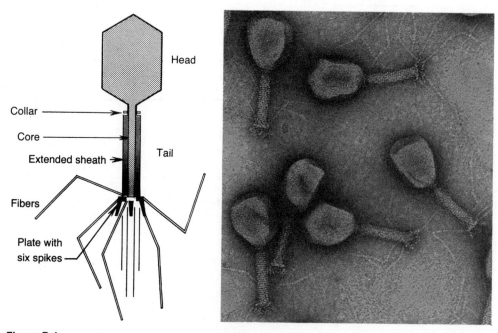

Figure B-1
An electron micrograph and interpretive drawing of a T4 phage. [From *Molecular Biology of Bacterial Viruses* by G. S. Stent. W. H. Freeman and Company. Copyright © 1963.]

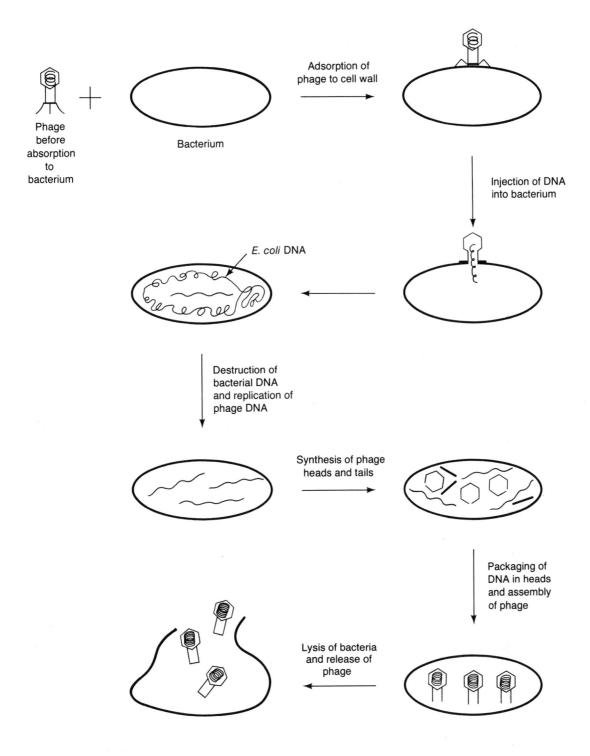

Phage
before
absorption
to
bacterium

Bacterium

Adsorption of
phage to cell wall

Injection of DNA
into bacterium

E. coli DNA

Destruction of
bacterial DNA
and replication of
phage DNA

Synthesis of phage
heads and tails

Packaging of
DNA in heads
and assembly
of phage

Lysis of bacteria
and release of
phage

Figure B-2
Outline of the sequence of events in the replication of *E. coli* phage T2 or T4. Note that the phage and bacteria are not drawn to scale, the phage being roughly 100 times too large.

feature of the replication and packaging systems is shown in Fig. B-3. Note that the order of genes (i.e., ABC⋯XYZ) is terminally redundant. By a mechanism probably involving both DNA replication and recombination, a multi-unit DNA molecule is constructed. This molecule, termed a *concatemer,* is cut into lengths of DNA exactly equal to the content of one head. Since the genomic length (i.e., genes ABC⋯XYZ) is less than one headful, the packaged DNA is both terminally redundant and cyclically permuted.

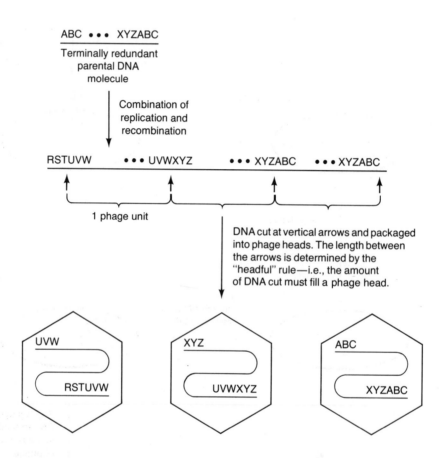

Progeny phage particles contain a cyclically permuted, terminally redundant set of DNA molecules.

Figure B-3
The mode of DNA replication that results in a population of T4 phages containing cyclically permuted, terminally redundant DNA.

The Sedimentation Coefficient *s*

The sedimentation coefficient *s* is a useful parameter for characterizing a macromolecule. It is measured with an ultracentrifuge.

If a particle having molecular weight M and suspended in a solvent of density ρ is in a centrifugal force field generated by a spinning rotor with angular velocity ω, and is at a distance r from the axis of rotation, the velocity v of the particle is described by the equation

$$v = \frac{\omega^2 r\, M\, (1 - \bar{v}\rho)}{f}$$

where $\bar{v}$ is the partial specific volume of the particle (a quantity related to the density of the particle) and f is the frictional coefficient (a parameter that is related to the shape of the particle and that is a measure of the friction experienced as the particles move through the solvent).

Because the velocity of the particle is proportional to the magnitude of the centrifugal force, i.e., $\omega^2 r$, sedimentation properties are usually described by the sedimentation coefficient, i.e., the velocity per unit field, or

$$s = \frac{v}{\omega^2 r} = \frac{M\, (1 - \bar{v}\rho)}{f}$$

For molecules of a given class, e.g., nucleic acids, $\bar{v}$ is unaffected by M and only weakly affected by chemical composition. However, f is a strong function of M, so that s does not vary linearly with M. The usual relation between s and M is found to be of the type

$$s = kM^a \text{ or } s = c + kM^a$$

where a, c, and k are constants which apply to the particular class of molecules.

Determination of s is simple and usually rapid whereas measurement of M is almost always tedious.* Therefore, it is common to use s to describe the molecular weight of biological macromolecules and particles.

The unit of s is the *svedberg* (named after The Svedberg who designed and built the first ultracentrifuge in 1923). The usual notation is the following: if a particle has an s value of 30 svedbergs, it is described as a 30s particle.

---

*A technique has been developed for the rapid measurement of M of DNA if M is reasonably small. But this method—gel electrophoresis—has been in use for only a few years.

Chromatography of Nucleic Acids on Methylated Serum Albumin-Kieselguhr (MAK) Columns

In the early days of nucleic acid research the fractionation of DNA into size classes was a necessary part of many experiments, especially since, as isolated, DNA was rarely homogeneous in its molecular weight M. Various materials used in chromatographic separation of other substances were tried but proved to be unsatisfactory. Joseph Mandel and Alfred Hershey discovered that if bovine serum albumin (a protein obtained from cow blood) is methylated, the negative charges of the protein are eliminated and the resulting positively charged protein has the ability to bind DNA molecules (which are negatively charged) by ionic bonds—i.e., bonds utilizing solely the attraction of positive and negative charges. They then found that methylated albumin can be tightly adsorbed to diatomaceous earth (kieselguhr), and thus made available a new chromatographic system (MAK) capable of fractionating DNA according to M. The basis of fractionating with MAK is the following. DNA is a negatively charged polymer having a constant charge density—i.e., one negative charge per nucleotide. Therefore, the *net* charge is proportional to M. When DNA is in a NaCl solution, the positively charged sodium (Na^+) ions bind to the negatively charged phosphates in the DNA backbone; the fraction of the negative charges neutralized by the Na^+ ion increases with the NaCl concentration. Thus, for a given NaCl concentration the *number* of remaining negative charges per DNA molecule (i.e., the net charge) depends upon M, so that the larger DNA molecules are bound more tightly to MAK than smaller molecules. This means that if a population of DNA molecules is bound to MAK when the NaCl concentration is low and the NaCl concentration is slowly and continuously increased, smaller molecules will become released from the MAK before the larger molecules will.

The operation of the MAK system is as follows. MAK is suspended in a NaCl solution usually between 0.1 and 0.4 M. This suspension is placed in a vertical tube (a chromatographic column) containing a small orifice at the bottom, which can be opened and closed at will. Most of the liquid is allowed to drain out, leaving the MAK as a fairly well packed layer called the *bed of the column*. The DNA solution to be fractionated is layered on top of the MAK bed and the orifice is opened. As the DNA solution passes through the MAK bed, DNA molecules bind tightly to the methylated albumin. After all of the DNA molecules are bound, the column is washed free of any unbound molecules and then the column is *eluted* by applying a NaCl solution, which is prepared so that the concentration increases with the volume of solution added (Fig. E-1 of Appendix E shows how a concentration gradient can be prepared). The liquid that drains from the base of the column contains DNA molecules, whose molecular weight increases with salt concentration. In the usual operation of the column, the eluted liquid is collected in a series of samples of equal volume by means of an instrument called a fraction collector. The optical density at 260 mμ (OD$_{260}$) is determined for each sample with a spectrophotometer. The 260-mμ wavelength is always used in DNA experiments because the DNA concentration can be calculated on the basis of the fact that a solution with OD$_{260}$ = 1 contains 50 μg DNA per milliliter. One of the first DNA samples fractionated by this method was a mixture of whole and half molecules of T2 DNA; this separation or fractionation is illustrated in Fig. D-1.

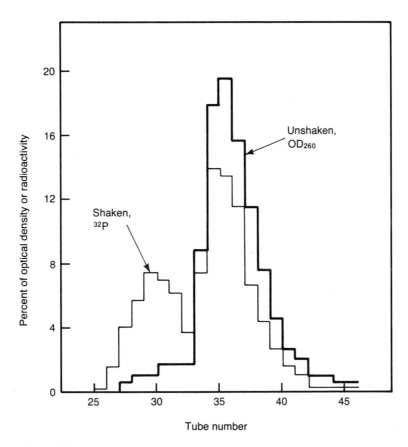

Figure D-1

Chromatography on MAK of a mixture of whole (unshaken) and whole and half (shaken) molecules of T2 DNA. The tube number refers to the aliquots of liquid recovered from the column. The unshaken molecules (solid curve) are detected by their optical density (OD$_{260}$) and the shaken molecules (broken curve) by their radioactivity ^{32}P. The shaken sample contains both whole and half molecules. [Reprinted from J. Mandel and A. D. Hershey, *Analyt. Biochem.* **1** (1960), 66–77.]

Zonal Centrifugation in a Preformed Density Gradient

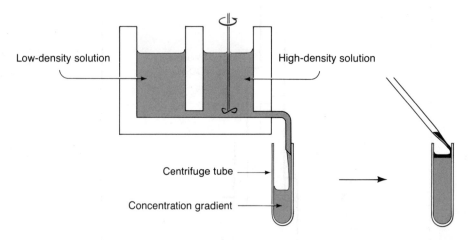

a. Formation of gradient

b. The sample is layered on top of the gradient

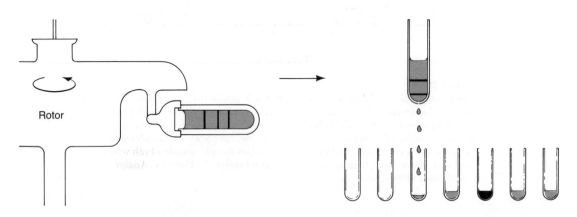

c. The tube is placed in a swinging bucket rotor and centrifuged. The components of the sample separate according to their *s* values.

d. A hole is made in the bottom of the tube with a needle, and the drops are collected in a series of tubes.

Figure E-1
Operations in zonal centrifugation. [From *Physical Biochemistry* by D. Freifelder. W. H. Freeman and Company. Copyright © 1976.]

The method of zonal centrifugation in a preformed density gradient is described in a number of the papers reprinted in this book. Since the density gradient is most frequently prepared with sucrose solutions, the term *sucrose gradient centrifugation* frequently appears in the scientific literature. The phrase *density gradient centrifugation* (also laboratory jargon) must not be confused with *zonal centrifugation*—the first one refers to equilibrium centrifugation, usually in solutions of cesium salts, and the second to a continuously moving system in which sedimentation stops only when the molecules reach the bottom of the centrifuge tube, or when the centrifuge is stopped after a measured time.

The methods used in zonal centrifugation are shown in Fig. E-1. A small volume of a solution containing the molecules to be characterized is layered on top of a preformed concentration gradient contained in a centrifuge tube. The solution being layered always has a density less than that at the top of the gradient (otherwise it would sink into the gradient). The tube is then centrifuged and the molecules in the starting layer sediment through the gradient. If the molecules have the same sedimentation coefficient, they will sediment within a narrow zone as shown in Fig. E-2. If there are molecules of differing sedimentation coefficients, they will separate from one another as sedimentation proceeds. The different components will then be resolved into a series of zones or bands, which then sediment only through solvent and independently of one another. After centrifugation is complete, the contents of the tube are fractionated, most commonly by drop collection from the tube bottom. If the dripping rate is slow enough that no turbulence occurs, each drop represents a single layer in the tube. The fractions then consist of one or more drops each. A great variety of techniques can be used to assay the materials and thereby to determine the concentration distribution.

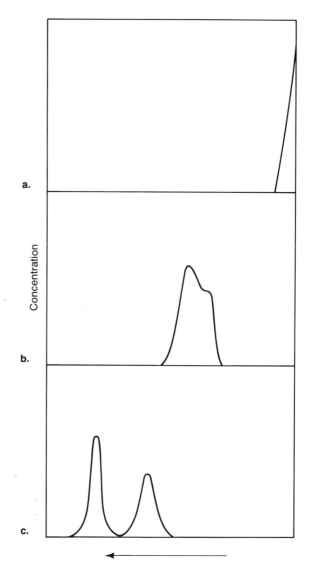

Figure E-2
Separation of two components by zonal centrifugation in a sucrose density gradient. Sedimentation is from right to left: (a) before centrifugation; (b and c) at successively longer periods of centrifugation. [From *Physical Biochemistry* by D. Freifelder. W. H. Freeman and Company. Copyright © 1976.]

INDEX